VARIATIONAL ANALYSIS AND SET OPTIMIZATION

Developments and Applications in Decision Making

Editors:

Akhtar A. Khan
Center for Applied and Computational Mathematics
Rochester Institute of Technology
Rochester, NY, USA

Elisabeth Köbis
Institute of Mathematics
Martin-Luther-University Halle-Wittenberg
Halle, Germany

Christiane Tammer
Institute of Mathematics
Martin-Luther-University Halle-Wittenberg
Halle, Germany

CRC Press
Taylor & Francis Group
Boca Raton London New York

CRC Press is an imprint of the
Taylor & Francis Group, an **informa** business

A SCIENCE PUBLISHERS BOOK

CRC Press
Taylor & Francis Group
6000 Broken Sound Parkway NW, Suite 300
Boca Raton, FL 33487-2742

First issued in paperback 2021

© 2019 by Taylor & Francis Group, LLC
CRC Press is an imprint of Taylor & Francis Group, an Informa business

No claim to original U.S. Government works

Version Date: 20190304

ISBN-13: 978-0-367-77972-6 (pbk)
ISBN-13: 978-1-138-03726-7 (hbk)

This book contains information obtained from authentic and highly regarded sources. Reasonable efforts have been made to publish reliable data and information, but the author and publisher cannot assume responsibility for the validity of all materials or the consequences of their use. The authors and publishers have attempted to trace the copyright holders of all material reproduced in this publication and apologize to copyright holders if permission to publish in this form has not been obtained. If any copyright material has not been acknowledged please write and let us know so we may rectify in any future reprint.

Except as permitted under U.S. Copyright Law, no part of this book may be reprinted, reproduced, transmitted, or utilized in any form by any electronic, mechanical, or other means, now known or hereafter invented, including photocopying, microfilming, and recording, or in any information storage or retrieval system, without written permission from the publishers.

For permission to photocopy or use material electronically from this work, please access www.copyright.com (http://www.copyright.com/) or contact the Copyright Clearance Center, Inc. (CCC), 222 Rosewood Drive, Danvers, MA 01923, 978-750-8400. CCC is a not-for-profit organization that provides licenses and registration for a variety of users. For organizations that have been granted a photocopy license by the CCC, a separate system of payment has been arranged.

Trademark Notice: Product or corporate names may be trademarks or registered trademarks, and are used only for identification and explanation without intent to infringe.

Library of Congress Cataloging-in-Publication Data

Names: Khan, Akhtar A., author. | Kèobis, Elisabeth, author. | Tammer, Christiane, author.
Title: Variational analysis and set optimization : developments and applications in decision making / Akhtar A. Khan (Center for Applied and Computational Mathematics, School of Mathematical Sciences, Rochester Institute of Technology, Rochester, NY, USA), Elisabeth Kèobis (Institute of Mathematics, Martin-Luther-University, Halle-Wittenberg, Faculty of Natural Sciences II, Halle, Germany), Christiane Tammer (Institute of Mathematics, Martin-Luther-University Halle-Wittenberg, Halle, Germany).
Description: Boca Raton, FL : CRC Press, 2019. | "A science publishers book." | Includes bibliographical references and index.
Identifiers: LCCN 2019007546 | ISBN 9781138037267 (hardback)
Subjects: LCSH: Calculus of variations. | Mathematical analysis. | Variational inequalities (Mathematics) | Mathematical optimization.
Classification: LCC QA315 .K43 2019 | DDC 515/.64--dc23
LC record available at https://lccn.loc.gov/2019007546

Visit the Taylor & Francis Web site at
http://www.taylorandfrancis.com

and the CRC Press Web site at
http://www.crcpress.com

Dedicated to our dear friend Rosalind Elster

Preface

During the last few decades, variational analysis and set optimization has witnessed explosive growth in research activities that have resulted in a wealth of powerful theoretical tools, efficient and reliable numerical techniques, and novel and important applications. The primary object of this volume is to present, in a unified framework, a wide variety of topics in variational analysis and set optimization with particular emphasis on its application in decision sciences. We firmly believe that this volume will provide an insight into some of the most active research directions in these dynamic fields.

This volume contains twelve chapters. It begins with a contribution by B. Mordukhovich and A. Soubeyran who explore applications of variational analysis in local aspects of behavioral science modeling by developing an effective variational rationality approach. G. Colajanni and P. Daniele present a network-based mathematical model that formulates a multi-period portfolio selection problem as a Markowitz mean-variance optimization problem with intermediaries and the addition of transaction costs and taxes on the capital gain. A. Soubeyran, J. C. Souza, and J. X. C. Neto explore Nash games using tools from the variational analysis. An engaging treatment of scalarization techniques for sets relations is presented by K. Ike, Y. Ogata, T. Tanaka, and H. Yu. The focus of P. Weidner is on extended real-valued functions with uniform sublevel sets which constitute an essential tool for scalarization in nonlinear functional analysis, vector optimization, and mathematical economics. R. Kipka presents a new proof of the Pontryagin Maximum Principle for problems with state constraints by employing a technique of exact penalization. V. A. Tuan and T. T. Le generalize the well-known result on the Lipschitz continuity of convex scalar functions. M. Durea, E. Florea, and R. Strugariu provide a survey and some improvements to the theory of vector optimization with variable ordering structures. C. Günther deals with penalization approaches in multi-objective optimization. G. Eichfelder and T. Gerlach study special set-valued maps and show that some of them are simple enough that they can equivalently be expressed as a vector optimization problem. A. R. Doagooei presents the basic notions of abstract convex analysis and studies three classes of abstract convex functions, namely increasing and positively homogeneous (IPH) functions, increasing and co-radiant (ICR) functions and increasing and plus-homogeneous (topical) functions. B. Jadamba, A. A. Khan,

M. Sama, and Chr. Tammer present an overview of some of the recent developments in the regularization methods for optimal control problems for partial differential equations with pointwise state constraints.

Given the wide range of the topics covered, we believe that this volume will be of great interest to researchers in applied mathematics, decision sciences, economics, engineering, variational analysis, optimization, and other related fields.

Finally, we express our most profound gratitude to all the researchers who contributed to this volume. We are immensely thankful to the referees of the enclosed contributions.

December, 2018

Akhtar A. Khan
Elisabeth Köbis
Christiane Tammer

Contents

Preface v

1. Variational Analysis and Variational Rationality in Behavioral Sciences: Stationary Traps 1
Boris S. Mordukhovich and *Antoine Soubeyran*

   1.1 Introduction 1
   1.2 Variational Rationality in Behavioral Sciences 3
       1.2.1 Desirability and Feasibility Aspects of Human Behavior 3
       1.2.2 Worthwhile (Desirable and Feasible) Moves 4
   1.3 Evaluation Aspects of Variational Rationality 5
       1.3.1 Optimistic and Pessimistic Evaluations 5
       1.3.2 Optimistic Evaluations of Reference-Dependent Payoffs 9
   1.4 Exact Stationary Traps in Behavioral Dynamics 11
   1.5 Evaluations of Approximate Stationary Traps 13
   1.6 Geometric Evaluations and Extremal Principle 17
   1.7 Summary of Major Finding and Future Research 20
References 22

2. A Financial Model for a Multi-Period Portfolio Optimization Problem with a Variational Formulation 25
Gabriella Colajanni and *Patrizia Daniele*

   2.1 Introduction 25
   2.2 The Financial Model 28
   2.3 Variational Inequality Formulation and Existence Results 35
   2.4 Numerical Examples 39
   2.5 Conclusions 41
References 42

3.	**How Variational Rational Agents Would Play Nash: A Generalized Proximal Alternating Linearized Method**			44
	Antoine Soubeyran, João Carlos Souza and *João Xavier Cruz Neto*			
	3.1	Introduction		44
	3.2	Potential Games: How to Play Nash?		45
		3.2.1	Static Potential Games	45
			3.2.1.1 Non Cooperative Normal Form Games	46
			3.2.1.2 Examples	46
			3.2.1.3 Potential Games	48
		3.2.2	Dynamic Potential Games	49
			3.2.2.1 Alternating Moves and Delays	49
			3.2.2.2 The "Learning How to Play Nash" Problem	49
	3.3	Variational Analysis: How to Optimize a Potential Function?		50
		3.3.1	Minimizing a Function of Several Variables: Gauss-Seidel Algorithm	50
		3.3.2	The Problem of Minimizing a Sum of Two Functions Without a Coupling Term	51
		3.3.3	Minimizing a Sum of Functions With a Coupling Term (Potential Function)	52
			3.3.3.1 Mathematical Perspective Proximal Regularization of a Gauss-Seidel Algorithm	52
			3.3.3.2 Game Perspective: How to Play Nash in Alternation	52
			3.3.3.3 Cross Fertilization Between Game and Mathematical Perspectives	53
	3.4	Variational Rationality: How Human Dynamics Work?		54
		3.4.1	Stay/stability and Change Dynamics	54
		3.4.2	Worthwhile Changes	54
			3.4.2.1 One Agent	55
			3.4.2.2 Two Interrelated Agents	56
		3.4.3	Worthwhile Transitions	57
		3.4.4	Ends as Variational Traps	58
			3.4.4.1 One Agent	58
			3.4.4.2 Two Interrelated Agents	59
	3.5	Computing How to Play Nash for Potential Games		59
		3.5.1	Linearization of a Potential Game with Costs to Move as Quasi Distances	59
	References			68
4.	**Sublinear-like Scalarization Scheme for Sets and its Applications to Set-valued Inequalities**			72
	Koichiro Ike, Yuto Ogata, Tamaki Tanaka and *Hui Yu*			
	4.1	Introduction		72
	4.2	Set Relations and Scalarizing Functions for Sets		74
	4.3	Inherited Properties of Scalarizing Functions		83
	4.4	Applications to Set-valued Inequality and Fuzzy Theory		85

		4.4.1 Set-valued Fan-Takahashi Minimax Inequality	85
		4.4.2 Set-valued Gordan-type Alternative Theorems	87
		4.4.3 Application to Fuzzy Theory	88

References 90

5. Functions with Uniform Sublevel Sets, Epigraphs and Continuity 92
Petra Weidner

- 5.1 Introduction 92
- 5.2 Preliminaries 93
- 5.3 Directional Closedness of Sets 94
- 5.4 Definition of Functions with Uniform Sublevel Sets 97
- 5.5 Translative Functions 98
- 5.6 Nontranslative Functions with Uniform Sublevel Sets 103
- 5.7 Extension of Arbitrary Functionals to Translative Functions 106

References 110

6. Optimality and Viability Conditions for State-Constrained Optimal Control Problems 112
Robert Kipka

- 6.1 Introduction 112
  - 6.1.1 Statement of Problem and Contributions 113
  - 6.1.2 Standing Hypotheses 115
- 6.2 Background 115
  - 6.2.1 Elements of Nonsmooth Analysis 115
  - 6.2.2 Relaxed Controls 117
- 6.3 Strict Normality and the Decrease Condition 119
  - 6.3.1 Overview of the Approach Taken 119
  - 6.3.2 The Decrease Condition 120
- 6.4 Metric Regularity, Viability, and the Maximum Principle 123
- 6.5 Closing Remarks 125

References 126

7. Lipschitz Properties of Cone-convex Set-valued Functions 129
Vu Anh Tuan and *Thanh Tam Le*

- 7.1 Introduction 129
- 7.2 Preliminaries 130
- 7.3 Concepts on Convexity and Lipschitzianity of Set-valued Functions 134
  - 7.3.1 Set Relations and Set Differences 134
  - 7.3.2 Cone-convex Set-valued Functions 138
  - 7.3.3 Lipschitz Properties of Set-valued Functions 141
- 7.4 Lipschitz Properties of Cone-convex Set-valued Functions 145
  - 7.4.1 (C, e)-Lipschitzianity 145
  - 7.4.2 C-Lipschitzianity 147
  - 7.4.3 G-Lipschitzianity 152
- 7.5 Conclusions 154

References 155

8. Efficiencies and Optimality Conditions in Vector Optimization with Variable Ordering Structures — 158

Marius Durea, Elena-Andreea Florea and Radu Strugariu

- 8.1 Introduction — 158
- 8.2 Preliminaries — 160
- 8.3 Efficiency Concepts — 167
- 8.4 Sufficient Conditions for Mixed Openness — 180
- 8.5 Necessary Optimality Conditions — 192
- 8.6 Bibliographic Notes, Comments, and Conclusions — 203
- References — 206

9. Vectorial Penalization in Multi-objective Optimization — 210

Christian Günther

- 9.1 Introduction — 210
- 9.2 Preliminaries in Generalized Convex Multi-objective Optimization — 212
- 9.3 Pareto Efficiency With Respect to Different Constraint Sets — 215
- 9.4 A Vectorial Penalization Approach in Multi-objective Optimization — 218
  - 9.4.1 Method by Vectorial Penalization — 219
  - 9.4.2 Main Relationships — 222
- 9.5 Penalization in Multi-objective Optimization with Functional Inequality Constraints — 228
  - 9.5.1 The Case of a Not Necessarily Convex Feasible Set — 229
  - 9.5.2 The Case of a Convex Feasible Set But Without Convex Representation — 234
- 9.6 Conclusions — 237
- References — 238

10. On Classes of Set Optimization Problems which are Reducible to Vector Optimization Problems and its Impact on Numerical Test Instances — 241

Gabriele Eichfelder and Tobias Gerlach

- 10.1 Introduction — 241
- 10.2 Basics of Vector and Set Optimization — 243
- 10.3 Set Optimization Problems Being Reducible to Vector Optimization Problems — 246
  - 10.3.1 Set-valued Maps Based on a Fixed Set — 246
  - 10.3.2 Box-valued Maps — 251
  - 10.3.3 Ball-valued Maps — 254
- 10.4 Implication on Set-valued Test Instances — 258
- References — 264

11. Abstract Convexity and Solvability Theorems — 266

Ali Reza Doagooei

- 11.1 Introduction — 266
- 11.2 Abstract Convex Functions — 268
- 11.3 Solvability Theorems for Real-valued Systems of Inequalities — 272
  - 11.3.1 Polar Functions of IPH and ICR Functions — 273
  - 11.3.2 Solvability Theorems for IPH and ICR Functions — 274
  - 11.3.3 Solvability Theorem for Topical Functions — 277
- 11.4 Vector-valued Abstract Convex Functions and Solvability Theorems — 278
  - 11.4.1 Vector-valued IPH Functions and Solvability Theorems — 279
  - 11.4.2 Vector-valued ICR Functions and Solvability Theorems — 281
  - 11.4.3 Vector-valued Topical Functions and Solvability Theorems — 283
- 11.5 Applications in Optimization — 285
  - 11.5.1 IPH and ICR Maximization Problems — 285
  - 11.5.2 A New Approach to Solve Linear Programming Problems with Nonnegative Multipliers — 289

References — 294

12. Regularization Methods for Scalar and Vector Control Problems — 296

Baasansuren Jadamba, Akhtar A. Khan, Miguel Sama and *Christiane Tammer*

- 12.1 Introduction — 296
- 12.2 Lavrentiev Regularization — 299
- 12.3 Conical Regularization — 302
- 12.4 Half-space Regularization — 304
- 12.5 Integral Constraint Regularization — 306
- 12.6 A Constructible Dilating Regularization — 309
- 12.7 Regularization of Vector Optimization Problems — 312
- 12.8 Concluding Remarks and Future Research — 314
  - 12.8.1 Conical Regularization for Variational Inequalities — 314
  - 12.8.2 Applications to Supply Chain Networks — 315
  - 12.8.3 Nonlinear Scalarization for Vector Optimal Control Problems — 315
  - 12.8.4 Nash Equilibrium Leading to Variational Inequalities — 315

References — 316

Index — 323

Chapter 1

Variational Analysis and Variational Rationality in Behavioral Sciences: Stationary Traps

Boris S. Mordukhovich
Department of Mathematics, Wayne State University, Detroit, MI 48202, USA.
`boris@math.wayne.edu`

Antoine Soubeyran
Aix-Marseille School of Economics, Aix-Marseille University, CNRS and EHESS, Marseille 13002, France. `antoine.soubeyran@gmail.com`

1.1 Introduction

Recent years have witnessed broad applications of advanced tools of variational analysis, generalized differentiation, and multiobjective (vector and set-valued) optimization to real-life models, particularly those related to economics and finance; see, e.g., [1, 3, 8, 10, 11, 12] and the references therein. Lately [4, 5], certain variational principles and techniques have been developed and applied to models of behavioral sciences that mainly concern human behavior. The latter applications are based on the *variational rationality* approach to behavioral sciences initiated by Soubeyran in [17, 18, 19, 20]. Major concepts of the variational rationality approach include the notions of (stationary and variational) *traps*, which describe underlying positions of individual or group behavior related to making worthwhile decisions on changing or staying at the current position; see Section 2 for more details. Mathematically, these notions correspond to points of equilibria, optima, aspirations, etc., and thus call to employ and develop powerful machinery of variational analysis and optimization theory for their comprehensive study and applications.

Papers [4, 5] mostly dealt with the study of *global variational traps* in connection with dynamical aspects of the variational rationality approach and its applications to goal systems in psychology [4] and to capability theory of wellbeing in behavioral sciences [5]. Appropriated tools of variational analysis developed in [4, 5] for the study and applications of the aforementioned global dynamical issues were related to set-valued extensions of the Ekeland variational principle in the case of set-valued mappings defined on quasimetric spaces and taken values in vector spaces with variable ordering structures; see also [1, 8, 10] for multiobjective optimization problems of this type.

The main goal of this paper is largely different from those in [4, 5]. Here, we primarily focus on the study and variational analysis descriptions of *local stationary traps* in the variational rationality framework. It will be shown that the well-understood subdifferential notions for convex and nonconvex extended-real-valued functions, as well as generalized normals to locally closed sets, are instrumental in describing and characterizing various types of such traps. Furthermore, the fundamental *variational* and *extremal principles* of variational analysis provide deep insights into efficient descriptions and evaluations of stationary traps and their approximate counterparts.

The rest of the paper is organized as follows: In Section 2, we review relevant aspects of the variational rationality approach to behavioral sciences and formulate the major problems of our interest in this paper. Section 3 is devoted to global and local *evaluation aspects* of variational rationality. Here, we introduce and investigate new notions of *linear evaluation* (both *optimistic* and *pessimistic*) of advantages and disadvantages to changing payoff and utility functions in various model structures. Then, we discuss relationships between optimistic (resp. pessimistic) evaluations with *majorization-minimization* (resp. *minorization-maximization*) algorithms of optimization. The main results here establish *complete subdifferential descriptions* of the *rates of change* for linear optimistic evaluations that are global for convex functions and (generally) local for nonconvex ones.

Section 4 conducts the study of *exact stationary traps* in behavioral dynamics. We discuss the meanings of such traps (which relate to a worthwhile choice between change and stay at the current position; see [19, 20] for more details) and consider both global and local ones while paying the main attention to local aspects. The usage of appropriate subgradients of variational analysis and linear optimistic evaluations allow us to derive efficient *certificates* for such stationary traps in general settings. Based on them, we reveal and discuss remarkable special cases of such traps in behavioral models.

Next, Section 5 deals with *approximate stationary traps*, which naturally appear in the framework of variational rationality and allow us to employ to their study powerful variational principles of variational analysis. Proceeding in this way, we derive efficient descriptions of approximate stationary traps and establish verifiable conditions for the existence of such traps, satisfying certain optimality and subdifferential properties.

Section 6 is devoted to *geometric* aspects of stationary trap evaluations by using generalized normals to nonconvex sets and their variational descriptions, as well as by employing the fundamental *extremal principle* of variational analysis. Besides applications to behavior science modeling via the variational rationality approach, we provide *behav-*

ioral interpretations of generalized normals and the basic ε-extremal principle around stationary trap positions. Furthermore, the important notion of *tilt perturbations* is included in this section in order to evaluate stationary traps in behavioral dynamics associated with linear utility functions and the corresponding proximal payoffs.

The concluding Section 7 summarises our major finding in this paper. Then, we discuss some open problems and directions of future research in this exciting area of interrelationships between variational analysis and variational rationality in behavioral sciences.

1.2 Variational Rationality in Behavioral Sciences

The variational rationality approach to human behavior addresses a complex changing world under a long list of human limitations, including limited cognitive resources, limited perception, unknown environment, etc.; see [17, 18] and the references therein for more details. All of this leads to the imperfect evaluation of payoffs. It has been well realized in behavioral sciences (see, e.g., [9, 22]) that agents evaluate their payoffs relative to moving reference points while depending on agents' experience and current environment. This definitely complicates the evaluation process and calls for possible simplifications.

In this paper, we mostly concentrate on *local linear optimistic evaluations* (see below for the exact definitions) of nonlinear payoff and utility functions in the framework of the variational rationality approach to behavioral sciences. This will be done by using the basic tools and results of variational analysis and generalized differentiation.

First, we briefly go over relevant aspects of variational rationality in the framework of *linear normed spaces*; this setting is assumed in what follows, unless otherwise stated.

The main emphasis of the variational rationality approach to human behavior is the concept of a *worthwhile change*. This approach considers a change to be worthwhile if the motivation to change rather than to stay is higher in comparison to the resistance to change rather than to stay at the current position. To formalize/modelize this concept and related ones, we first clarify the terminology and present the required definitions.

1.2.1 Desirability and Feasibility Aspects of Human Behavior

Starting with **desirability aspects** of human behavior, we denote by $g\colon X \to \overline{\mathbb{R}} := (-\infty, \infty)$ the so-called "to be increased" payoffs that reflect profit, satisfaction, utility, valence, revenue functions and the like. "To be decreased" payoffs are denoted by $\varphi(x) = -g(x)$ while reflecting unsatisfied needs, disadvantages, losses, etc. Then:

- *Advantage to change* (rather than to stay) payoff from x to y is defined by

$$A(y/x) := g(y) - g(x) = \varphi(x) - \varphi(y) \geq 0. \tag{1.1}$$

- *Disadvantage to change* payoff from x to y means accordingly that $A(y/x) \leq 0$ for the quantity $A(x/y)$ given in (1.1).
- *Motivation to change* from x to y is defined by $M(y/x) := U[A(y/x)]$, where $U\colon A \to \mathbb{R}_+$ is a given nonnegative function of A reflecting the attractiveness of change. In this paper, we confine ourselves for simplicity to the linear dependence $U(A) = A$.

Proceeding further with **feasibility aspects** of human behavior, define the linear *cost of changing* from x to y by

$$C(x,y) := \eta \|y - x\|, \text{ where the quantity } \eta = \eta(x) > 0 \tag{1.2}$$

denotes the cost of changing per unit of the distance $d(x,y) = \|y - x\|$. Since $\eta(y) \neq \eta(x)$ in many practical situations (e.g., while moving up and down the hill), the cost of changing in (1.2) is generally *asymmetric*, i.e., $C(y,x) \neq C(x,y)$. We suppose in this paper that $C(x,x) = 0$ for all $x \in X$, although it might be that $C(x,x) > 0$ in more general settings.

- *Inconvenience to change* from x to y (rather than to stay at x) is defined by

$$I(y/x) := C(x,y) - C(x,x) = C(x,y) \in \mathbb{R}_+. \tag{1.3}$$

- *Resistance to change* from x to y (rather than to stay at x) is described by $R(y/x) := D[I(y/x)]$ via (1.3) and the disutility function $D: I \to \mathbb{R}_+$. For simplicity, we consider the linear case, where $D(I) = I$ for all $I \geq 0$.

1.2.2 Worthwhile (Desirable and Feasible) Moves

Balancing desirability and feasibility aspects. The variational rationality approach considers different payoffs in order to balance desirability and feasibility aspects of human dynamics. Among the most important payoffs are the following:

- *Proximal payoff* functions are defined by

$$P_\xi(y/x) := g(y) - \xi C(x,y) \text{ and } Q_\xi(y/x) := \varphi(y) + \xi C(x,y) \tag{1.4}$$

for "to be increased" and for "to be decreased" payoffs g and φ, respectively, where the cost of changing $C(x,y)$ is taken from (1.2). The given *weight factor* $\xi = \xi(x) \geq 0$ in (1.4) reflects the status quo at x, as well as some other aspects of the current situation.

- *Worthwhile to change gain* is defined by

$$A_\xi(y/x) := A(y/x) - \xi I(y/x) \geq 0 \tag{1.5}$$

while representing the difference between advantages (1.1) and inconveniences (1.3) to change rather than to stay. Based on the definitions, (1.5) can be represented as the difference of proximal payoffs (1.4) by

$$A_\xi(y/x) = \big(g(y) - g(x)\big) - \xi \big(C(x,y) - C(x,x)\big) = P_\xi(y/x) - P_\xi(x/x). \tag{1.6}$$

- *Not worthwhile to change loss* is defined similarly to (1.5) via the *loss function* $L(y/x) := -A(y/x)$ by

$$L_\xi(y/x) := L(y/x) + \xi I(y/x) \geq 0, \tag{1.7}$$

i.e., it represents the sum of losses and weighted inconveniences to change rather than to stay. It follows from the definitions that

$$L_\xi(y/x) = \big(\varphi(y) - \varphi(x)\big) + \xi \big(C(x,y) - C(x,x)\big)$$
$$= \big(\varphi(y) + \xi C(x,y)\big) - \big(g(x) + \xi C(x,x)\big) = Q_\xi(y/x) - Q_\xi(x/x),$$

which verifies the relationship $L_\xi(y/x) = -A_\xi(y/x)$ between functions (1.5) and (1.12).

Worthwhile changes. Based on the above discussions, we say that it is *worthwhile to change* from x to y rather than to stay at the current position x if advantages to change rather than to stay are "high enough" relative to inconveniences to change rather than to stay. Mathematically, this is expressed via the quantities

$$A(y/x) \geq \xi I(y/x), \text{ or equivalently as } A_\xi(y/x) \geq 0, \tag{1.8}$$

with an appropriate weight factor $\xi = \xi(x) > 0$ in (1.8).

Note that the proximal payoff functions (1.4) may be treated as particular cases of the so-called *reference-dependent payoffs* $\Gamma(\cdot/r) \colon x \in X \mapsto \Gamma(x/r)$, $r \in \mathbb{R}$, which depend on a chosen reference point r even if the original payoffs $g(\cdot)$ and $\varphi(\cdot)$ do not depend on r. Observe also that the variational rationality approach also addresses the situations where the original payoff function depends on references points; see [17, 18] for more details. The latter case covers reference-dependent utilities that were investigated, in particular, in psychology and economics; see, e.g., [9, 22] and the bibliographies therein.

1.3 Evaluation Aspects of Variational Rationality

Due to the extreme complications and numerous uncertain factors of human behavior, reliable *evaluation procedures* play a crucial role in behavioral science modeling. The variational rationality approach, married to the constructions and results of variational analysis and generalized differentiation, offers efficient tools to provide such evaluations.

We start with the relevant evaluation notions from both behavioral and mathematical viewpoints. Roughly speaking, an agent is said to be *bounded rational* if he/she merely tries to improve ("improve enough") his/her "to be increased" payoff or tries to decrease ("decrease enough") his/her "to be decreased" payoff. Optimization refers to *perfect rationality*. A bounded rational agent has limited capabilities to evaluate advantages and disadvantages to change if the agent knows his/her payoff $g(x)$ or $\varphi(x) = -g(x)$ only at the current position while not knowing the values $g(y)$ or $\varphi(y) = -g(y)$ at $y \neq x$.

1.3.1 Optimistic and Pessimistic Evaluations

(Optimistic evaluations for "to be decreased" and "to be increased" payoffs) When confronted to applications in Behavioral Sciences, we must make the distinction between two cases: A "to be decreased" payoff $\varphi(.)$, like an unsatisfaction and a cost function, and a "to be increased" payoff $g(.) = -\varphi(.)$, like a utility, valence, value, profit and revenue function. Mathematicians usually use "to be decreased" payoffs, while Economists and Psychologists consider "to be increased" payoffs. Because the audience of this book is more mathematically oriented, moving from Variational Analysis to applications, after having defined both cases and given an example 1.3.2, relative to "to be increased" payoffs in Economics, we will focus the attention on "to be decreased" payoffs to make mathematicians more comfortable with applications.

Consider a realistic case of human behavior, where the payoff function $\varphi(.)$ or $g(\cdot) = -\varphi(.)$ is known at the current position bx, but unknown (at least precisely) elsewhere for $y = x$. First we introduce the notion of local optimistic evaluations and their linear version.

Definition 1.3.1 *(a) Given a "to be decreased" function $\varphi(.) : X \longmapsto \overline{R}$ and a current position $x \in X$, we say that the function $l^\varphi(./x) : X \longmapsto \overline{R}$ provides a local optimistic evaluation of $\varphi(.)$ around x if $l^\varphi(x/x) = \varphi(x)$ and there is a neighborhood $V(x)$ of x such that $\varphi(y) \geq l^\varphi(y/x)$ for all $y \in V(x)$. That is, what, ex ante, an agent expects to lose, $l^\varphi(y/x)$, is lower than what, ex post, he will really lose, $\varphi(y)$.*

(b) Given a "to be increased" function $g(.) : X \longmapsto \overline{R}$ and a current position $x \in X$, we say that the function $l^g(./x) : X \longmapsto \overline{R}$ provides a local optimistic evaluation of $g(.)$ around x if $l^g(x/x) = g(x)$ and there is a neighborhood $V(x)$ of x such that $g(y) \leq l^g(y/x)$ for all $y \in V(x)$. That is, what, ex ante, an agent expects to gain, $l^g(y/x)$, is higher than what, ex post, he will really gain, $g(y)$.

The linear case. If there is $x^* \in X^*$ such that $l^\varphi_{x^*}(y/x) = \varphi(x) + <x^*, y-x>$ or $l^g_{x^*}(y/x) = g(x) + <x^*, y-x>$, the optimistic evaluation is linear. In this linear setting, $x^* \in \partial \varphi(x)$ if $\varphi(y) \geq l^\varphi_*(y/x)$ for all $y \in V(x)$ defines the classic local subgradient of $\varphi(.)$ at x.

Definition 1.3.1 bis: Optimistic evaluation of a loss and an advantage.

Let $E^\varphi(y/x) = l^\varphi(y/x) - l^\varphi(x/x)$ and $E^g(y/x) = l^g(y/x) - l^g(x/x)$ be the evaluations of the loss to change $L(y/x) = \varphi(y) - \varphi(x) = -A(y/x)$ and the advantage to change $A(y/x) = g(y) - g(x) = \varphi(x) - \varphi(y)$ payoffs (see paragraph 1.2.1).

Then, i) an optimistic evaluation of a loss to change is $L(y/x) \geq E^\varphi(y/x) = E(y/x)$, that is, $L(y/x) = \varphi(y) - \varphi(x) \leq l^\varphi(y/x) - l^\varphi(x/x) =: E^\varphi(y/x) = E(y/x)$, for all $y \in V(x)$,(1.3.1.a)

and ii) , an optimistic evaluation of a loss and an advantage to change is $A(y/x) \leq E^g(y/x)$, that is, $A(y/x) = g(y) - g(x) \leq l^g(y/x) - l^g(x/x) =: E^g(y/x)$, for all $y \in V(x)$, (1.3.1.b)

The optimistic evaluation $E(y/x)$, either $E^\varphi(y/x)$ in (i.a) or $E^g(y/x)$ in (i.b) is linear if there is $x^* \in X^*$ such that $E(y/x) = <x^*, y-x>$ for all $y \in V(x)$, where x^* provides a rate of change in the linear optimistic evaluation.

Global evaluations. This is the case if $V(x) = X$.

To illustrate the notions in Definition 1.3.1, we consider the following example that addresses a realistic situation in behavioral economics.

Example 1.3.2 (linear optimistic evaluation in economics). Let $g: \mathbb{R}_+ \to \mathbb{R}$ be a utility function, which depends on consuming a quantity $x \geq 0$ of goods and is defined by

$$g(x) := x - \frac{1}{2}x^2 \text{ for all } x \in \mathbb{R}_+. \tag{1.9}$$

We can see that the utility function (1.9) attends its maximum at $x = 1$ while increasing on the interval $[0, 1]$ and decreasing on $[0, \infty)$.

• Take $x = 1/2$ as our reference point and suppose that the agent knows the value of the utility function (1.9) and its derivative therein:

$$g(1/2) = 1/2 \text{ and } \nabla g(1/2) = 1 - 1/2 = 1/2.$$

Then the local linear evaluation $E_{x^*}(y/x)$ of the advantage to change $A(y/x)$ around the reference position $x = 1/2$ is calculated by

$$E_{x^*}(y/x) = x^*(y-x) = \nabla g(x)(y-x) = (1/2)(y-x). \tag{1.10}$$

If, in this case, we have $x \leq y \leq 1$, then the computation shows that $0 \leq A(y/x) \leq E_{x^*}(y/x)$. This gives us an optimistic linear evaluation of the realized gain before moving.

If $x \leq y \leq 1$ with $x = 1/2$ as before, then we have the realized loss $-A := -A(y/x) = g(x) - g(y) \geq 0$ and compute the expected local loss $-E := -E(y/x) = (1/2)(x-y) \geq 0$ from (1.10). Since $-A \geq -E$, the linear evaluation is optimistic in this case as well.

• Consider now, another initial position $x = 3/2$ of the agent with the same utility function g in (1.9). In this setting, we have

$$g(3/2) = 3/8 \text{ and } \nabla g(x) = -1/2.$$

Proceeding as above, we examine the following two cases:

Let $1 \leq y \leq x$. Then $A = g(y) - g(x) \geq 0$ is the realized gain and $E = (-1/2)(y-x) \geq 0$ is the expected gain. Since $A \leq E$, we have a linear local optimistic evaluation.

Finally, let $x = 3/2 \leq y$. Then, we have the realized loss $-A = g(x) - g(y) \geq 0$ and the expected loss $-E = -(-1/2)(y-x) \geq 0$ in this case. Since $-A \geq -E$, it again gives us a linear local optimistic evaluation.

Note that similar justifications can be done for the inequality $L(y/x) \geq E(y/x)$, which describes linear optimistic evaluations of "to be decreased" payoffs.

Summarizing, we see that in all the cases under consideration, the realized gains are *lower* than the expected gains, while the realized losses are *higher* than the expected ones. This is the characteristic feature of *optimistic* evaluations.

Next we introduce the notion of *pessimistic* evaluations where, in contrast to optimistic ones, realized gains are *higher* than expected gains and realized losses are *lower* than expected ones around the reference point.

Definition 1.3.3 (pessimistic evaluations). *Given a "to be decreased" payoff function $\varphi : X \longmapsto R$ and a current position $x \in X$, we say that the function $l(./x) = l^\varphi(./x) : X \longmapsto R$ provides a local pessimistic evaluation of $\varphi(.)$ around x if the following two conditions hold: $l(x/x) = \varphi(x)$ and there is a neighborhood $V(x)$ of x such that $\varphi(y) \leq l(y/x)$ for all $y \in V(x)$. That is, ex ante, expected losses $l(y/x)$ are higher than, ex post, realized loss $\varphi(y)$.*

This results in $L(y/x) \leq E(y/x)$ for such y. The notions of linear and global evaluations are formulated similarly to cases (ii) and (iii) of Definition 1.3.1. The case of "to be increased" payoffs work similarly and results in $A(y/x) \geq E(y/x)$.

Let us now discuss some connections of the above evaluation concepts and their variational interpretations via relevant aspects of optimization and subdifferential theory.

• **MM procedures in optimization.** Procedures of this kind and the corresponding algorithms have been well recognized in optimization theory and applications; see, e.g., [6] and the references therein. Comparing the general scheme of such algorithms with the above Definitions 1.3.1 and 1.3.3 of optimistic and pessimistic evaluations shows that *majorization-minimization* procedures to minimize a cost function $g: X \to \bar{R}$ agree with *optimistic* evaluation of $g(y) - g(x)$, while *minorization-maximization* procedures to maximize $g(\cdot)$ agree with *pessimistic* evaluations of the difference $g(y) - g(x)$.

These observations and the corresponding results of MM optimization allow us to deduce, in particular, that optimistic (resp. pessimistic) evaluations of convex (resp. concave) functions can always be chosen to be *global* and *quadratic*.

- **Subgradient evaluations of payoffs.** Here, we begin a discussion on the usage of subgradients of convex and variational analysis in order to provide constructive linear evaluations of payoff functions. Let us consider, for definiteness, optimistic evaluations, while observing that we can proceed symmetrically with pessimistic ones for symmetric classes of functions (e.g., for concave vs. convex). Following the notation and terminology of variational analysis and optimization [11, 16], the reference point is denoted by $\bar{x}$ and the moving one by x. Given $\vartheta: X \to \bar{\mathbb{R}}$ finite at $\bar{x}$ (i.e., with $\bar{x} \in \text{dom}\,\vartheta$ from the domain of $\bar{x}$), a *classical subgradient* $x^* \in \partial \vartheta(\bar{x})$ of ϑ at $\bar{x}$ is defined by

$$\langle x^*, x - \bar{x}\rangle \leq \vartheta(x) - \vartheta(\bar{x}) \quad \text{for all } x \in V(\bar{x}), \tag{1.11}$$

where $V(\bar{x})$ is some neighborhood of $\bar{x}$, and where $\partial \vartheta(\bar{x})$ stands for the (local) *subdifferential* of ϑ at $\bar{x}$ as the collection of all its subgradients at this point. If $V(\bar{x}) = X$ in (1.11), we get back to the classical definition of the subdifferential of convex analysis, which is nonempty under mild assumptions on a convex function ϑ; e.g., when $\bar{x}$ is an interior point of $\text{dom}\,\vartheta$ (or belongs to the relative interior of $\text{dom}\,\vartheta$ in finite dimensions).

Proposition 1.3.4 (subgradient linear optimistic evaluations of "to be decreased" payoffs). *Let $\varphi: X \to \bar{\mathbb{R}}$ be a "to be decreased" payoff function, let $\bar{x} \in \text{dom}\,\varphi$ be a reference current position, and let*

$$L(x/\bar{x}) := \varphi(x) - \varphi(\bar{x}) \quad \text{with } x \in V(\bar{x}) \tag{1.12}$$

be the loss function when moving from $\bar{x}$ to the new position x in a neighborhood $V(\bar{x})$ of $\bar{x}$. Then, $x^* \in X^*$ provides a rate of change in the local linear optimistic evaluation of the payoff function φ around $\bar{x}$ if and only if $x^* \in \partial \varphi(\bar{x})$ in (1.11). Furthermore, this gives us a complete subdifferential description of the global linear optimistic evaluation of φ around any $\bar{x} \in \text{dom}\,\varphi$ when $V(\bar{x}) = X$ in (1.11) as in the case of convex functions $\varphi(\cdot)$.

Proof. It follows from the definitions of subgradients (1.11) and loss function (1.12) that

$$L(x/\bar{x}) \geq \langle x^*, x - \bar{x}\rangle \quad \text{for all } x \in V(\bar{x}) \tag{1.13}$$

with the same neighborhood $V(\bar{x})$ in (1.11) and (1.12). Invoking definition of the linear quantity $E(x/\bar{x})$ tells us that

$$L(x/\bar{x}) \geq E(x/\bar{x}) \quad \text{for all } x \in V(\bar{x}),$$

which means by Definition 1.3.1(iv) that we get a local linear optimistic evaluation of the "to be decreased" payoff $\varphi(\cdot)$ with the rate of change x^*. It clearly follows from the above that this evaluation is global if $V(\bar{x}) = X$. □

In the next subsection, we obtain linear optimistic evaluations of *proximal payoffs* associated with general *nonconvex* functions via appropriate subgradient extensions.

1.3.2 Optimistic Evaluations of Reference-Dependent Payoffs

Here, we address evaluation problems for *reference-dependent payoffs* $\Gamma(\cdot/\bar{x}) \colon x \in X \mapsto \Gamma(x/\bar{x}) \in \overline{\mathbb{R}}$ discussed in Subsection 1.2.2. They include, in particular, the proximal payoff functions (1.4) of our special interest in this paper. Similarly to the setting of Definition 1.3.1, we say that $x^* \in X^*$ is the *rate of change* of the local linear optimistic evaluation for the reference-dependent loss function $L_\Gamma(x/\bar{x}) := \Gamma(x/\bar{x}) - \Gamma(\bar{x},\bar{x})$ if there is a neighborhood $V(\bar{x})$ of the reference point $\bar{x}$ such that

$$L_\Gamma(x/\bar{x}) \geq E_{x^*}(x/\bar{x}) := \langle x^*, x - \bar{x}\rangle \text{ for all } x \in V(\bar{x}). \tag{1.14}$$

It is clear that x^* in (1.14) can be interpreted as a (local) classical/convex subgradient (1.11) of the reference-dependent payoff function $\Gamma(\cdot/\bar{x})$ at $\bar{x}$. Consider now, the special case of the reference-dependent *proximal payoff function*

$$\Gamma(x/\bar{x}) = Q_\xi(x/\bar{x}) := \varphi(x) + \xi \|x - \bar{x}\|, \quad \xi \geq 0, \tag{1.15}$$

defined for the "to be decreased" original payoff $\varphi(\cdot)$ in accordance with (1.4) by taking into account the expression for the changing cost $C(\bar{x},x)$ in (1.2). Then, the local linear optimistic evaluation (1.14) is written as

$$L_\xi(x/\bar{x}) := Q_\xi(x/\bar{x}) - Q_\xi(\bar{x}/\bar{x}) \geq E_{x^*}(x/\bar{x}) := \langle x^*, x - \bar{x}\rangle \text{ for all } x \in V(\bar{x}), \tag{1.16}$$

where $L_\xi(x/\bar{x})$ can be interpreted as a "not worthwhile to change loss function". We show next, based on (1.16) and the structure of the proximal payoff $Q_\xi(x/\bar{x})$ in (1.15), that the rate of change x^* in the linear optimistic evaluation (1.14) can be completely characterized via ε-subgradients of the *arbitrary original payoff* $\varphi(\cdot)$ in contrast to the convex-like subgradients of the proximal payoff function $Q_\xi(x/\bar{x})$ as in (1.16).

Given an arbitrary function $\vartheta \colon X \to \overline{\mathbb{R}}$ finite at $\bar{x}$ and following [11, Definition 1.83], we say that x^* is an ε-*subgradient* of ϑ at $\bar{x}$ for some $\varepsilon \geq 0$ if it belongs to the (analytic) ε-*subdifferential* of ϑ at this point defined by

$$\hat{\partial}_\varepsilon \vartheta(\bar{x}) := \left\{ x^* \in X^* \;\middle|\; \liminf_{x \to \bar{x}} \frac{\vartheta(x) - \vartheta(\bar{x}) - \langle x^*, x - \bar{x}\rangle}{\|x - \bar{x}\|} \geq -\varepsilon \right\}. \tag{1.17}$$

The next theorem establishes a *two-sided relationship* between local linear optimistic evaluations of "to be decreased" payoff functions in the variational rationality theory for behavioral sciences and ε-subgradients of variational analysis. From one side, it reveals a *behavioral sense* of ε-subgradients for general functions, while from the other side, it provides an efficient ε-*subdifferential mechanism* to calculate rates of change in local linear optimistic evaluations of "to be decreased" payoffs. Note that the second assertion of the theorem allows us to keep a similar description with the replacement of the original payoff by its differentiable approximation with the same rate of change at the reference point. This makes the result more convenient for implementations and applications.

Theorem 1.3.5 (ε-**subgradient description of rates of change in linear optimistic evaluations of "to be decreased" payoffs**). *Let $\varphi \colon X \to \overline{\mathbb{R}}$ be a "to be decreased" payoff in behavioral dynamics, and let $\varepsilon \geq 0$.*

(i) We have that $x^* \in \hat{\partial}_\varepsilon \varphi(\bar{x})$ if and only if for every weight factor $\xi > \varepsilon$ in the proximal payoff (1.15) there is a neighborhood $V(\bar{x})$ of $\bar{x}$ such that the local linear optimistic

evaluation (1.16) holds around $\bar{x}$ with the rate of change x^* for the not worthwhile to change loss function $L_\xi(x/\bar{x})$, represented as

$$L_\xi(x/\bar{x}) = \big(\varphi(x) - \varphi(\bar{x})\big) + \xi\|x - \bar{x}\|. \tag{1.18}$$

(ii) *If $\varepsilon = 0$, then in addition to the behavioral description of $x^* \in \hat{\partial}_0 \varphi(\bar{x}) := \hat{\partial}\varphi(\bar{x})$ in (i) we claim the existence of a "more decreased" function $s\colon V(\bar{x}) \to \mathbb{R}$, which is Fréchet differentiable at $\bar{x}$ with $\nabla s(\bar{x}) = x^*$, $s(\bar{x}) = \varphi(\bar{x})$, and $s(x) \le \varphi(x)$ whenever $x \in V(\bar{x})$.*

Proof. To justify (i), we first fix $\varepsilon \ge 0$ and pick an arbitrary ε-subgradient $x^* \in \hat{\partial}_\varepsilon \varphi(\bar{x})$ for the "to be decreased" payoff $\varphi(\cdot)$ under consideration. Given $\xi > \varepsilon$, denote $\nu := \xi - \varepsilon > 0$ and employ the *variational description* of ε-subgradients from [11, Proposition 1.84], which says that for every $\nu > 0$ there is a neighborhood $V(\bar{x})$ of $\bar{x}$ such that the function

$$\psi(x) := \varphi(x) - \varphi(\bar{x}) + (\varepsilon + \nu)\|x - \bar{x}\| - \langle x^*, x - \bar{x}\rangle$$

attains it minimum on $V(\bar{x})$ at $\bar{x}$. Thus, $\psi(x) \ge \psi(\bar{x}) = 0$ whenever $x \in V(\bar{x})$. This yields

$$\varphi(x) - \varphi(\bar{x}) + \xi\|x - \bar{x}\| \ge \langle x^*, x - \bar{x}\rangle \quad \text{for all } x \in V(\bar{x}) \tag{1.19}$$

by taking into account the choice of ν. Recalling the definition of the proximal payoff $Q_\xi(x/\bar{x})$ in (1.15) with $Q_\xi(\bar{x}/\bar{x}) = \varphi(\bar{x})$, we deduce from (1.19) and (1.16) that x^* is a rate of change in the local linear optimistic evaluation of $L_\xi(x/\bar{x})$ from (1.18) around $\bar{x}$.

To verify the converse implication in (i), we simply reverse the argumentations above while observing that the variational description of ε-subgradients in [11, Proposition 8.4] provides a complete characterization of these elements.

It remains to justify assertion (ii) of the theorem. To proceed, we employ a *smooth variational description* of *regular subgradients* $x^* \in \hat{\partial}\varphi(\bar{x})$ (known also as Fréchet and viscosity ones) from [11, Theorem 1.88](i), which ensures the existence of a real-valued function $s\colon V(\bar{x}) \to \mathbb{R}$ satisfying the listed properties. The aforementioned result is stated in Banach spaces X while its proof holds without the completeness requirement on the normed space X under consideration. $\square$

Remark 1.3.6 (further results and discussions on linear subgradient evaluations).
(i) The properties of the supporting function $s(\cdot)$ in Theorem 1.3.5 can be significantly improved if the space X is Banach and satisfies additional "smoothness" requirements; see [11, Theorem 1.88(ii,iii)] for more details.

(ii) It has been well recognized in variational analysis that ε-subgradients ($\varepsilon \ge 0$) from (1.17) used in linear optimistic evaluations of Theorem 1.3.5 are not robust and do not satisfy major calculus rules for subdifferentiation of sums, compositions, etc., unless ϑ is convex and belong to some other restrictive classes of functions. This may complicate applications of subgradient linear evaluations established in Theorem 1.3.5. The situation is dramatically improved for the limiting construction

$$\partial \vartheta(\bar{x}) := \Big\{ x^* \in X^* \;\Big|\; \exists \text{ seqs. } \varepsilon_k \downarrow 0,\; x_k \xrightarrow{\vartheta} \bar{x},\; \text{and } x_k^* \xrightarrow{w^*} x^* \\ \text{such that } x_k^* \in \hat{\partial}_{\varepsilon_k}\vartheta(x_k) \text{ for all } k = 1,2,\ldots \Big\} \tag{1.20}$$

known as the Mordukhovich *basic/limiting subdifferential* of φ at $\bar{x} \in \text{dom } \vartheta$. The symbol $x \xrightarrow{\vartheta} \bar{x}$ in (1.20) indicates that $x \to \bar{x}$ with $\vartheta(x) \to \vartheta(\bar{x})$ while w^* stands for the weak* topology of X^*. The limiting subdifferential construction (1.20) can be viewed as a *robust regularization* of ε-subdifferentials (1.17), and the validity of *full calculus* for it in finite-dimensions and broad infinite-dimensional settings (particularly in the class of *Asplund spaces*, which includes every reflexive Banach space, etc.) is due to *variational/extremal principles* of variational analysis; see [11] for a comprehensive study. Having this in mind, we may treat robust subdifferential evaluations of behavioral payoffs expressed in terms of (1.20) as *asymptotic* versions of those obtained in Theorem 1.3.5 and in what follows.

1.4 Exact Stationary Traps in Behavioral Dynamics

One of the most important questions in the framework of variational rationality can be formulated as follows: *When is it worthwhile to change the current position rather than to stay at it?* This issue is closely related to appropriate notions of *traps*. Among the main aims of this paper is to study some versions of *stationary traps* and their efficient descriptions by using adequate tools of variational analysis and generalized differentiation. We start our study by considering *exact* versions of stationary traps in behavioral models and then proceed with their *approximate* counterparts in the next section. To begin with, we formally introduce the notions under consideration.

Definition 1.4.1 (stationary traps). *Let $\bar{x} \in X$ be a reference point, let $\xi \geq 0$ be a weight factor, and let $Q_\xi(x/\bar{x}) := \varphi(x) + \xi \|x - \bar{x}\|$ be the reference-dependent proximal payoff built upon a given "to be decreased" original payoff $\varphi: X \to \bar{\mathbb{R}}$. We say that:*

(i) *$\bar{x}$ is a* LOCAL STATIONARY TRAP *in behavior dynamics with the weight factor ξ if there is a neighborhood $V(\bar{x})$ such that*

$$L_\xi(x/\bar{x}) := Q_\xi(x/\bar{x}) - Q_\xi(\bar{x}/\bar{x}) \geq 0 \text{ for all } x \in V(\bar{x}) \tag{1.21}$$

via the not worthwhile to change loss function $L_\xi(x/\bar{x})$ that can be directly defined by (1.18). Equivalently, condition (1.21) can be expressed in the form

$$A_\xi(x/\bar{x}) := A(x/\bar{x}) - \xi I(x/\bar{x}) \leq 0 \text{ for all } x \in V(\bar{x}), \tag{1.22}$$

where $A(x/\bar{x})$ and $I(x/\bar{x})$ are the advantage to change and the inconvenience to change from $\bar{x}$ to x defined in terms of the "to be increased" payoff function $g(x) = -\varphi(x)$ in (1.1) and (1.3), respectively, via the cost of changing (1.2) with $\eta = 1$ for simplicity.

(ii) *$\bar{x}$ is a* STRICT LOCAL STATIONARY TRAP *in behavioral dynamics with the weight factor ξ if there exists a neighborhood $V(\bar{x})$ such that a counterpart of (1.21) holds with the replacement of "$\geq$" by "$>$" for all $x \in V(\bar{x}) \setminus \{\bar{x}\}$.*

(iii) *If $V(\bar{x}) = X$, then the conditions in (i) and (ii) define $\bar{x}$ as a* GLOBAL STATIONARY TRAP *and its* STRICT GLOBAL *version, respectively.*

Roughly speaking, stationary traps are positions, which are *not worthwhile to quit*. For definiteness, we confine ourselves in what follows to considering only stationary traps, while observing that their strict counterparts can be studied similarly.

The subdifferential optimistic evaluations of reference-dependent proximal payoffs obtained in Theorem 1.3.5 allow us to derive efficient *certificates* (sufficient conditions) for stationary traps in the general framework of normed spaces X. Again for definiteness, the results below are presented only in terms of "to be decreased" payoff functions.

Proposition 1.4.2 (ε-subdifferential certificates for stationary traps from optimistic evaluations). *Given a "to be decreased" payoff $\varphi \colon X \to \overline{\mathbb{R}}$ and a point $\bar{x} \in \operatorname{dom}\varphi$, fix any $\varepsilon \geq 0$ and assume that there exist an ε-subgradient $x^* \in \hat{\partial}_\varepsilon \varphi(\bar{x})$ and a neighborhood $V(\bar{x})$ of $\bar{x}$ such that we have*

$$E_{x^*}(x/\bar{x}) := \langle x^*, x - \bar{x}\rangle \geq 0 \text{ for all } x \in V(\bar{x}). \tag{1.23}$$

Then $\bar{x}$ is a local stationary trap in behavioral dynamics with any weight factor $\xi > \varepsilon$. Furthermore, this trap is global if $V(\bar{x}) = X$ in (1.23).

Proof. It follows from Theorem 1.3.5(i) that for any $\varepsilon \geq 0$, any ε-subgradient $x^* \in \hat{\partial}_\varepsilon \varphi(\bar{x})$, and any weight $\xi > \varepsilon$ there is a neighborhood $V(\bar{x})$ such that

$$L_\xi(x/\bar{x}) := Q_\xi(x/\bar{x}) - Q_\xi(\bar{x}/\bar{x}) \geq \langle x^*, x - \bar{x}\rangle \text{ for any } x \in V(\bar{x}) \tag{1.24}$$

Combining (1.24) with assumption (1.23) and using the stationary trap Definition 1.4.1(i,iii) justifies the claimed conclusions. $\square$

Since we clearly have by definition (1.17) that

$$\hat{\partial}_{\varepsilon_1}\varphi(\bar{x}) \subset \hat{\partial}_{\varepsilon_2}\varphi(\bar{x}) \text{ whenever } 0 \leq \varepsilon_1 \leq \varepsilon_2,$$

for each $\bar{x}$ and x^* there is the *minimal subdifferential factor* $\varepsilon_{\min} = \varepsilon_{\min}(\bar{x}, x^*) = \min\{\varepsilon\}$ over all ε such that (1.23) holds. Note that we may have $\varepsilon_{\min} > 0$ for some $\bar{x}$ and x^*.

It follows from the subgradient estimate (1.23) that the analysis of stationary traps via Proposition 1.4.2 depends on the two major parameters:

(a) an ε-subgradient x^* that can be chosen from the subdifferential set $\hat{\partial}_\varepsilon \varphi(\bar{x})$ together with an appropriate number $\varepsilon \geq 0$;

(b) a weight factor $\xi > \varepsilon$ that is generally unknown, or at least not known exactly.

Let us discuss some situations that emerge from Proposition 1.4.2 when $\xi > \varepsilon$ and also directly from Definition 1.4.1 when the latter condition fails.

1. Flat linear optimistic evaluations. This is the case when there exists $\varepsilon \geq 0$ such $0 \in \hat{\partial}_\varepsilon \varphi(\bar{x})$. Choosing $x^* = 0$, we get the flat evaluation

$$E_{x^*}(x/\bar{x}) = \langle x^*, x - \bar{x}\rangle = 0 \text{ for all } x \in V(\bar{x}),$$

which ensures that $\bar{x}$ is a local/global stationary trap by Proposition 1.4.2.

• If $\xi = 0$ in this setting, Proposition 1.4.2 does not apply, although $E_{x^*}(x/\bar{x}) = 0$. However, the very definitions of local (global) traps in (1.21) amount to saying that

$$\varphi(x) - \varphi(\bar{x}) \geq 0 \text{ for all } x \in V(\bar{x}),$$

which means that $\bar{x}$ is a local (global) *minimizer* of the *original payoff* $\varphi(\cdot)$. Thus, we have in this case that local (global) stationary traps in behavioral dynamics correspond to local (global) minima of "to be decreased" payoff functions.

• If $\xi > 0$ in this setting, then stationary traps mean by (1.21) that

$$Q_\xi(x/\bar{x}) - Q_\xi(\bar{x}/\bar{x}) \geq 0 \text{ for all } x \in V(\bar{x}),$$

i.e., local (global) stationary traps reduce in this to local (global) *minimizers* of the *proximal payoff* $Q(x/\bar{x})$ instead of the original one independently of $\xi > 0$. If $\xi > \varepsilon$, it can be also detected via the linear optimistic evaluation of Proposition 1.4.2.

2. Subdifferential evaluations of variational analysis. This is the setting where $x^* \neq 0$ for a given ε-subgradient from the set $\hat{\partial}_\varepsilon \varphi(\bar{x})$ as $\varepsilon \geq 0$.

• If $\xi = 0$ (no inertia and costs of changing in the moving process), Proposition 1.4.2 does not apply, while the application of (1.21) is exactly the same as in the case of $x^* = 0$.

• If $\xi > 0$, then there are inertia and costs of changing in the moving process, while the application of (1.21) is not different from the case of $x^* = 0$ and does not actually provide verifiable information for the stationary trap determination. This is due to the fact that the very definition (1.21) with $\xi > 0$ is merely conceptional, not constructive, and that the weight factor ξ is generally *unknown*. In contrast to it, the subgradient linear optimistic evaluation (1.23) obtained in Proposition 1.4.2 ensures that the position $\bar{x}$ is a local or global trap *for any weight factor* $\xi > \varepsilon$ without determining this factor a priori. This is a *serious advantage* of the subgradient trap evaluation from variational analysis.

• The relationship $\xi > \varepsilon$ between the weight factor ξ (a parameter of *behavioral dynamics*) and the subdifferential factor ε (a parameter of *variational analysis*) plays a crucial role in the obtained evaluation of stationary traps. Let us present a striking interpretation of this relationship from the viewpoint of *variational stationarity*.

It is reasonable to identify the subdifferential factor ε with the *unit cost of changing* $\eta = \eta(\bar{x})$ in $C(\bar{x},x)$ from $\bar{x}$ to x in (1.2). This means that the value $\varepsilon = \eta$ reflects the *resistance aspect* of behavioral dynamics near $\bar{x}$. The lower number ε, the less resistance of the agent to change locally around the reference point is. On the other hand, the size of ξ determines how much it is *worthwhile to move* from $\bar{x}$ to x. Thus, the required condition $\xi > \varepsilon$ shows that the advantage to change should be *high enough* in comparison to the resistance to change at the reference position of the agent.

1.5 Evaluations of Approximate Stationary Traps

In this subsection, we consider more flexible *approximate* versions of stationary traps from Definition 1.4.1 (omitting their strict counterparts) and show that these notions admit verifiable subdifferential evaluations similar to those obtained above for exact traps, as well as new variational descriptions derived by using fundamental *variational principles*.

Definition 1.5.1 (approximate stationary traps). *Staying in the framework of Definition 1.4.1, take any* $\gamma > 0$. *It is said that* $\bar{x}$ *is a* γ-APPROXIMATE LOCAL STATIONARY TRAP *in behavioral dynamics with the weight factor* $\xi \geq 0$ *if there is a neighborhood* $V(\bar{x})$ *of* $\bar{x}$ *on which we have the inequality*

$$L_\xi(x/\bar{x}) := Q_\xi(x/\bar{x}) - Q_\xi(\bar{x}/\bar{x}) \geq -\gamma \text{ whenever } x \in V(\bar{x}). \tag{1.25}$$

The γ-*approximate stationary trap* $\bar{x}$ *is* GLOBAL *if* $V(\bar{x}) = X$ *in* (1.25).

Similarly to the case of stationary traps we derive efficient certificates of approximate stationary traps from optimistic subdifferential evaluations of proximal payoffs.

Proposition 1.5.2 (ε-subdifferential certificates for approximate stationary traps from optimistic evaluations). *Given a "to be decreased" payoff $\varphi \colon X \to \bar{\mathbb{R}}$, a point $\bar{x} \in \mathrm{dom}\,\varphi$ and a rate $\gamma > 0$, fix any $\varepsilon \geq 0$ and assume that there exist an ε-subgradient $x^* \in \hat{\partial}_\varepsilon \varphi(\bar{x})$ and a neighborhood $V(\bar{x})$ of $\bar{x}$ such that we have*

$$E_{x^*}(x/\bar{x}) := \langle x^*, x - \bar{x} \rangle \geq -\gamma \text{ for all } x \in V(\bar{x}). \tag{1.26}$$

Then, $\bar{x}$ is a γ-approximate local stationary trap with any weight factor $\xi > \varepsilon$. Furthermore, the γ-approximate stationary trap is global if $V(\bar{x}) = X$ in (1.26).

Proof. As in the proof of Proposition 1.4.2, we deduce from Theorem 1.3.5(i) that estimate (1.24) holds for any $\varepsilon \geq 0$, any ε-subgradient $x^* \in \hat{\partial}_\varepsilon \varphi(\bar{x})$, and any weight $\xi > \varepsilon$ with some neighborhood $V(\bar{x})$, which can be identified with the one in (1.26) without loss of generality. Then, substituting the assumed estimate (1.26) into (1.24) shows that $\bar{x}$ is a γ-approximate stationary trap (local or global) by definition (1.25). □

Based on the certificate obtained in Proposition 1.5.2 when $\xi > \varepsilon$ and on definition (1.26) otherwise, we discuss some remarkable features of approximate stationary traps.

Observe first that *flat evaluations* of approximate stationary traps is about the same as for their exact counterparts discussed above. Indeed, in this case, we simply replace exact minimizers of the original payoff $\varphi(\cdot)$ by its γ-*approximate minimizers*

$$\varphi(\bar{x}) \leq \varphi(x) + \gamma \text{ for all } x \in V(\bar{x})$$

if $\xi = 0$, and correspondingly for the proximal payoff $Q_\xi(x/\bar{x})$ if $\xi > 0$.

If $\xi > \varepsilon$, we can apply the *subdifferential evaluations* from Proposition 1.5.2 with the estimate $E_{x^*}(x/\bar{x})$ from (1.26), we always have

$$\left| E_{x^*}(x/\bar{x}) \right| = \left| \langle x^*, x - \bar{x} \rangle \right| \leq \|x^*\| \cdot \|x - \bar{x}\|, \tag{1.27}$$

which allows us to single out the case of *flat enough evaluations* for approximate stationary traps. It is the case when the value of $\|x^*\|$ is *sufficiently small* for some ε-subgradient $x^* \in \hat{\partial}_\varepsilon \varphi(\bar{x})$. In this case, the estimating value $|E_{x^*}(x/\bar{x})|$ can be made small enough even for a large set $V(\bar{x})$ in (1.26). This verifies by Proposition 1.5.2 that there exists $\gamma > 0$ such that $\bar{x}$ is a γ-approximate stationary trap (1.25)

Note that, since the expected gain or loss $E_{x^*}(x/\bar{x})$ is small when $\|x^*\|$ is small, the agent has no incentive to change from $\bar{x}$ to x even in the case of gain. Changing a bit (a marginal change) leads to *disappointment* in both cases of loss and gain. Thus, an approximate local stationary trap is a position where it is *not worthwhile to change*.

Next, we proceed with applying a different machinery of variational analysis to study approximate traps by involving powerful *variational principles* instead of subdifferential estimates as above. The new machinery essentially relies on the fact that $\gamma \neq 0$ in Definition 1.5.1 of γ-approximate traps, i.e., it does not apply to the study of exact traps. Furthermore, in contrast to the results of Propositions 1.4.2 and 1.5.2, the statements below reveal conditions that are satisfied at the approximate trap point $\bar{x}$, i.e., *necessary conditions* for this concept. Recall that Proposition 1.5.2 offers *sufficient conditions* of a different type that ensure the validity of the approximate trap property (1.25).

To proceed in the new direction, we first employ an appropriate version of the fundamental *Ekeland variational principle* in order to establish the *existence* of approximate stationary traps with certain *optimality* properties to minimize *perturbed* proximal payoffs.

Theorem 1.5.3 (existence of approximate stationary traps minimizing perturbed proximal payoffs). *Pick any $\gamma > 0$ and let $\bar{x} \in \mathrm{dom}\, \varphi$ be a local γ-approximate stationary trap from Definition 1.5.1 with the "to be decreased" payoff $\varphi \colon X \to \bar{\mathbb{R}}$ and the neighborhood $V(\bar{x})$ in (1.25). Assume that the space X is Banach, and that the function $\varphi(\cdot)$ is lower semicontinuous and bounded from below around $\bar{x}$. Then for every $\xi \geq 0$ and every $\lambda > 0$ there exists $x_\gamma = x_\gamma(\xi, \lambda)$ such that:*
(a) *x_γ is a local γ-approximate stationary trap;*
(b) *$\|x_\gamma - \bar{x}\| \leq \lambda$;*
(c) *x_γ is an exact minimizer of the perturbed proximal payoff, meaning that*

$$Q_\xi(x_\gamma/\bar{x}) \leq Q_\xi(x/\bar{x}) + \frac{\gamma}{\lambda}\|x - x_\gamma\| \text{ for all } x \in V(\bar{x}). \tag{1.28}$$

Proof. Assume without loss of generality that $V(\bar{x})$ is closed and define the function

$$\theta(x) := L_\xi(x/\bar{x}) \text{ for all } x \in V(\bar{x}), \tag{1.29}$$

which is lower semicontinuous and bounded from below on the complete metric space $V(\bar{x})$. It follows from (1.18) and the γ-approximate stationary trap definition (1.25) that

$$\theta(\bar{x}) = 0 \text{ and } \theta(\bar{x}) \leq \inf_{V(\bar{x})} \theta(x) + \gamma \text{ for all } x \in V(\bar{x}).$$

Thus, we are in a position to apply the Ekeland variational principle (see, e.g., [11, Theorem 2.26]) to θ on $V(\bar{x})$ with the initial data $x_0 := \bar{x}$, $\varepsilon := \gamma$, and $\lambda := \lambda$ taken from the formulation of the theorem. This ensures the existence of a perturbed point $x_\gamma \in V(\bar{x})$ such that $\|x_\gamma - \bar{x}\| \leq \lambda$, that the minimality condition

$$\theta(x_\gamma) \leq \theta(x) + \frac{\gamma}{\lambda}\|x - x_\gamma\| \text{ for all } x \in V(\bar{x}) \tag{1.30}$$

is satisfied, and that $\theta(x_\gamma) \leq \theta(\bar{x})$. The latter condition immediately implies that x_γ is also a local γ-approximate stationary trap in (1.25), while (1.30) readily yields the minimality condition (1.28). This therefore completes the proof of the theorem. □

Condition (c) (see 1.30), can be rewritten as
(c) $g(x) - g(x_\gamma) = \varphi(x_\gamma) - \varphi(x) \leq \delta C(x_\gamma, x)$ for all $x \in V(\bar{x})$, where $\delta = \xi + (\gamma/\lambda) > 0$.
This equivalent formulation of condition (c) gives a striking result. It shows that, starting with an arbitrary local approximate trap x, it is possible to reach in one arbitrary small step $\|x_\gamma - x\| \leq \lambda$, an exact stationary trap x_γ such that $A(x/x_\gamma) \leq \delta C(x_\gamma, x)$ for all $x \in V(\bar{x})$, where $A(x/x_\gamma) = g(x) - g(x_\gamma)$ and $C(x, y) = \|y - x\|$.

Proof: After some manipulations, (1.30) is equivalent to
$$g(x) - g(x_\gamma) \leq \xi\left[C(\bar{x},x) - C(\bar{x},x_\gamma)\right] + (\gamma/\lambda)C(x_\gamma,x) \leq \xi C(x_\gamma,x) + (\gamma/\lambda)C(x_\gamma,x),$$
that is,
$$g(x) - g(x_\gamma) \leq [\xi + (\gamma/\lambda)]C(x_\gamma,x) \text{ for all } x \in V(\bar{x}).$$
Then, $A(x/x_\gamma) \leq \delta C(x_\gamma,x)$ for all $x \in V(\bar{x})$.

Note that the perturbation term in (1.28) can be made *regulated* by an arbitrary choice of $\lambda > 0$ under the fixed rank $\gamma > 0$ of both approximate stationary traps $\bar{x}$ and x_γ. Note furthermore that the perturbation structure in (1.28) relates to the *changing cost* $C(x_\gamma,x)$ from (1.2) at the minimizing γ-approximate stationary trap x_γ.

The following variational result establishes the existence of a local γ-approximate stationary trap in the modified framework of Theorem 1.5.3 with a regulated *rate of change* at this trap, which can be made sufficiently small if needed. The proof is based on applying the *lower subdifferential variational principle* [11] that is actually equivalent by [11, Theorem 2.28] to the (geometric) *approximate extremal principle* considered in the next section. Furthermore, the aforementioned result from [11] states that a large subclass of Banach spaces, known as Asplund spaces, is the most adequate setting for the validity of these fundamental principles of variational analysis.

Recall that a Banach space X is *Asplund* if each of its separable subspaces has a separable dual. This class is sufficiently broad while including, in particular, each reflexive spaces, Banach spaces admitting equivalent norms that are Fréchet differentiable at nonzero points, those with separable topological duals, etc. Besides the aforementioned lower subdifferential variational principle, the following theorem exploits the subdifferential *sum rules* that are valid for regular subgradients (1.17) as $\varepsilon = 0$ under the Fréchet differentiability of one summand and for our basic/limiting subgradients (1.20) in Asplund spaces without any differentiability assumptions. The latter sum rule fails for the regular subdifferential even in simple nonsmooth settings.

Theorem 1.5.4 (subgradient rates of change at local approximate stationary traps). *Staying in the framework of Theorem 1.5.3, assume in addition that the space X is Asplund. Then for every $\xi \geq 0$ and every $\lambda > 0$ there exists $x_\gamma = x_\gamma(\xi,\lambda)$ such that conditions* (a) *and* (b) *therein hold, while condition* (c) *is replaced by the following:*

$$\text{there exists } x_\gamma^* \in \hat{\partial}Q_\xi(x_\gamma/\bar{x}) \text{ with } \|x_\gamma^*\| \leq \frac{\gamma}{\lambda}. \tag{1.31}$$

Furthermore, we have the following inclusion for the rate of change in (1.31):

$$x^* \in \partial\varphi(x_\gamma) + \begin{cases} \{v^* \in X^* \mid \|v^*\| \leq 1\} & \text{if } x_\gamma = \bar{x}, \\ \{v^* \in X^* \mid \|v^*\| = 1, \langle v^*, x_\gamma - \bar{x}\rangle = \|x_\gamma - \bar{x}\|\} & \text{if } x_\gamma \neq \bar{x}. \end{cases} \tag{1.32}$$

Finally, the inclusion for the rate x^* reads as

$$x^* \in \nabla\varphi(x_\gamma) + \begin{cases} \{v^* \in X^* \mid \|v^*\| \leq 1\} & \text{if } x_\gamma = \bar{x}, \\ \{v^* \in X^* \mid \|v^*\| = 1, \langle v^*, x_\gamma - \bar{x}\rangle = \|x_\gamma - \bar{x}\|\} & \text{if } x_\gamma \neq \bar{x} \end{cases} \tag{1.33}$$

provided that the original payoff $\varphi(\cdot)$ is Fréchet differentiable at $\bar{x}$.

Proof. To proceed, we consider the lower semicontinuous and bounded from below function $\theta(\cdot)$ from (1.29) defined on an Asplund space and apply to it the lower subdifferential variational principle from [11, Theorem 2.28] with the parameters $x_0 := \bar{x}$, $\varepsilon := \gamma$, and $\lambda := \lambda$. This gives us a perturbed point $x_\gamma \in V(\bar{x})$ satisfying $\|x_\gamma - \bar{x}\| \leq \lambda$ and $\theta(x_\gamma) \leq \theta(\bar{x})$ as well as a regular subgradient $x_\gamma^* \in \hat{\partial}\theta(x_\gamma)$ with $\|x_\gamma^*\| \leq \gamma/\lambda$. The latter inclusion clearly gives us (1.31) due to (1.29) and the form of $L_\xi(x/\bar{x})$ in (1.25).

To justify further the subdifferential inclusion (1.31) with our basic subdifferential (1.20) therein, we first observe from (1.20) that

$$\hat{\partial} Q_\xi(x_\gamma/\bar{x}) \subset \partial Q_\xi(x_\gamma/\bar{x}).$$

Then we apply the basic subdifferential sum rule from [11, Theorem 2.33] to the proximal payoff function defined as the sum

$$Q_\xi(x/\bar{x}) = \varphi(x) + \xi \|x - \bar{x}\|, \quad \xi \geq 0$$

at $x = x_\gamma$ with taking into account that the function $\xi \|x - \bar{x}\|$ is Lipschitz continuous around $\bar{x}$ while the original payoff $\varphi(\cdot)$ is assumed to be lower semicontinuous around this point. Using the sum rule and the well-known formula for subdifferentiation of the norm function in convex analysis, we deduce from (1.31) the subdifferential inclusion (1.32).

It remains to verify the final rate of change inclusion (1.33) when $\varphi(\cdot)$ is Fréchet differentiable at x_γ. It is in fact a direct consequence of the equality sum rule from [11, Proposition 1.107(i)] applied to $\hat{\partial} Q_\xi(x_\gamma/\bar{x})$ in (1.31) due to the summation form of $Q_\xi(\cdot/\bar{x})$. This completes the proof of the theorem. $\square$

Note that the last inclusion (1.33) is generally independent of the subdifferential one (1.32) since $\partial \varphi(x_\gamma)$ may not reduce to the gradient $\nabla \varphi(x_\gamma)$ when $\varphi(\cdot)$ is merely Fréchet differentiable at this point. A simple example is provided by the function $\varphi(x) := x^2 \sin(1/x)$ if $x \neq 0$ and $\varphi(0) := 0$ with $x_\gamma = 0$. For this function we have

$$\nabla \varphi(0) = 0 \quad \text{while} \quad \partial \varphi(0) = [-1, 1].$$

For the validity of $\partial \varphi(x_\gamma) = \{\nabla \varphi(x_\gamma)\}$ we require the *strict differentiablity* of $\varphi(\cdot)$ at the point in question; this holds, in particular, when $\varphi(\cdot)$ continuously differentiable around this point; see [11, Definition 1.13 and Corollary 1.82].

Observe finally that in the case where $x_\gamma \neq \bar{x}$ and the norm on X is Fréchet differentiable at nonzero points (we can reduce to this setting any Banach space with a Fréchet differentiable renorm; in particular, any reflexive space [11]), the second summand in (1.32) and (1.33) is a *singleton*. Thus, in this case formula (1.33) holds as equality, and we get a *precise computation* of the rate of change x^* at the γ-approximate stationary trap x_γ.

1.6 Geometric Evaluations and Extremal Principle

In this section, we continue the investigation of stationary traps in behavioral dynamics, while invoking for these purposes *geometric* constructions and results of variational analysis. Given a set $\Omega \subset X$, a point $\bar{x} \in \Omega$, and a number $\varepsilon \geq 0$, we consider the collection of

ε-normals to Ω at $\bar{x}$ defined by

$$\hat{N}_\varepsilon(\bar{x};\Omega) := \left\{ x^* \in X^* \,\Big|\, \limsup_{x \xrightarrow{\Omega} \bar{x}} \frac{\langle x^*, x - \bar{x}\rangle}{\|x - \bar{x}\|} \le \varepsilon \right\}, \tag{1.34}$$

where the symbol $x \xrightarrow{\Omega} \bar{x}$ signifies that $x \to \bar{x}$ with $x \in \Omega$. Let $u_{x^*}(x) := \langle x^*, x\rangle$ be a linear *utility function* with the rate of change x^*, and let $\xi \ge 0$. Define the *proximal payoff* of type (1.4) for to "to be increased" utility function $u_{x^*}(x)$ by

$$P_{\xi,x^*}(x/\bar{x}) := u_{x^*}(x) - \xi \|x - \bar{x}\|. \tag{1.35}$$

The following result gives us a generalized normal description of local stationary traps in behavioral dynamics with linear utility functions. We say that $\bar{x}$ is a local stationary trap *relative to* some set Ω if x belongs to Ω in the relationships of Definition 1.4.1(i). Our study below is based on the "to be increased" payoff form (1.22) of stationary traps.

Proposition 1.6.1 (ε-normal description of local stationary traps). *Let $\bar{x} \in \Omega$, and let $\varepsilon \ge 0$. Then $x^* \in \hat{N}_\varepsilon(\bar{x};\Omega)$ if and only if $\bar{x}$ is a local stationary trap relative to Ω with the linear utility function $u_{x^*}(x) = \langle x^*, x\rangle$ and any weight factor $\xi > \varepsilon$.*

Proof. We employ the *variational description* of ε-normals from [11, Proposition 1.28] saying that $x^* \in \hat{N}_\varepsilon(\bar{x};\Omega)$ if and only if for any $\nu > 0$ the function

$$\psi(x) := \langle x^*, x - \bar{x}\rangle - (\varepsilon + \nu)\|x - \bar{x}\|$$

attains its local maximum $\psi(\bar{x}) = 0$ at $\bar{x}$ over Ω. This means that there is a neighborhood $V(\bar{x})$ of $\bar{x}$ relative to Ω such that the proximal payoff (1.35) satisfies the condition

$$P_{\xi,x^*}(x/\bar{x}) := u_{x^*}(x) - \xi\|x - \bar{x}\| \le P_{\xi,x^*}(\bar{x}/\bar{x}) = \langle x^*, \bar{x}\rangle \text{ for all } x \in V(\bar{x}) \cap \Omega \tag{1.36}$$

with $\xi := \varepsilon + \nu > \varepsilon$. The latter tells us by definition (1.22) that $\bar{x}$ is a local stationary trap relative to Ω in behavioral dynamics with the linear utility function $u_{x^*}(x)$ whenever $\xi > \varepsilon$ for the weight factor ξ in the proximal payoff (1.35). $\square$

The next theorem makes a connection between local stationary traps and the *extremal principle* for closed set systems that is yet another fundamental result of variational analysis with numerous applications; see, e.g., [11, 12] and the references therein. We primarily use here the (approximate) ε-extremal principle from [11, Definition 2.5(i)], which holds in any Asplund space by [11, Theorem 2.20].

First we recall the notion of local extremal points of extremal systems of sets in normed spaces that is taken from [11, Definition 2.1].

Definition 1.6.2 (set extremality). *We say that $\bar{x}$ is a LOCALLY EXTREMAL POINT of the system of finitely many sets $\Omega_1, \ldots, \Omega_n$ in the normed space X if:*

(a) $\bar{x}$ is a common point of the sets Ω_i for all $i = 1, \ldots, n$;

(b) there exist a neighborhood V of $\bar{x}$ and sequences $\{a_{ik}\} \subset X$ for $i = 1, \ldots, n$ such that $a_{ik} \to 0$ as $k \to \infty$ whenever $i = 1, \ldots, n$ and that

$$\bigcap_{i=1}^{n} \left(\Omega_i - a_{ik}\right) \cap V = \emptyset \text{ for all large natural numbers } k.$$

In this case we say that $\{\Omega_1, \ldots, \Omega_n, \bar{x}\}$ is an EXTREMAL SYSTEM in X.

It has been well recognized in variational analysis and its applications that the concept of set extremality encompasses various notions of optimal solutions in problems of scalar, vector, and set-valued optimization, equilibria, game theory, systems control, models of welfare economics, etc. On the other hand, local extremal points naturally appear in developing generalized differential calculus, geometric aspects of functional analysis, and related disciplines. We refer the reader to both volumes of [11] and to the new book [12] for the extensive theoretical material on these issues and numerous applications.

Now we are ready to establish the aforementioned relationship between set extremality and stationary traps via the ε-extremal principle in Asplund spaces.

Theorem 1.6.3 (local extremal points and stationary traps via the ε-extremal principle). *Let* $\{\Omega_1,\ldots,\Omega_n,\bar{x}\}$ *be an extremal system in an Asplund space X, where the sets* Ω_i *are locally closed around* $\bar{x}$. *Then for any* $\varepsilon > 0$ *there exist points*

$$x_i \in \Omega_i \text{ with } \|x_i - \bar{x}\| \leq \varepsilon \text{ as } i = 1,\ldots,n \qquad (1.37)$$

and dual elements $x_i^* \in X^*$, *which are rates of change for the linear utility functions*

$$u_{x_i^*}(x) := \langle x_i^*, x \rangle, \quad i = 1,\ldots,n, \qquad (1.38)$$

while satisfying the conditions

$$x_1^* + \ldots + x_n^* = 0 \text{ and } \|x_1^*\| + \ldots + \|x_n^*\| = 1, \qquad (1.39)$$

such that each x_i *is a local stationary trap relative to* Ω_i *with respect to the linear utility function* $u_{x_i^*}(x)$ *in* (1.38) *for any weight factor* $\xi > \varepsilon$.

Proof. Since $\bar{x}$ is a locally extremal point of the system $\{\Omega_1,\ldots,\Omega_n\}$ of locally closed sets Ω_i in the Asplund space X, we can employ the ε-extremal principle from [11, Theorem 2.20] and find x_i satisfying (1.37) and x_i^* satisfying (1.39) such that

$$x_i^* \in \hat{N}_\varepsilon(x_i;\Omega_i) \text{ for all } i = 1,\ldots,n. \qquad (1.40)$$

Then, we use the stationary trap description from Proposition 1.6.1 of ε-normals x_i^* to Ω_i at points x_i that are sufficiently close to the given local extremal point $\bar{x}$. □

Remark 1.6.4 (further discussions on geometric and analytic evaluations of stationary traps). It is worth mentioning the following interpretations and extensions of the above evaluations of local stationary traps:

(i) Taking into account the definitions of the advantage of change $A(\bar{x}/x)$ from $\bar{x}$ to x in (1.1) and the corresponding cost of changing $C(\bar{x},x)$ in (1.2), we can interpret the local stationary trap description of ε-normals $x^* \in \hat{N}_\varepsilon(\bar{x};\Omega)$ from Proposition 1.6.1 as

$$A(\bar{x}/x) \leq \xi C(\bar{x},x) \text{ for all } v \in V(\bar{x}) \cap \Omega \qquad (1.41)$$

with the corresponding weight factor ξ exceeding ε, the linear utility/evaluation function $u_{x^*}(x)$ from (1.38) with the rate of change x^*, and some neighborhood $V(\bar{x})$ of $\bar{x}$ relative to Ω that depends on the factors above. Then (1.41) says that it is *not worthwhile to move* from the local stationary trap position $\bar{x}$ to a point $x \in \Omega$ nearby.

(ii) Given a payoff function $\varphi \colon X \to \bar{\mathbb{R}}$, consider its *tilt perturbation*

$$\varphi_v(x) := \varphi(x) - \langle v, x \rangle \text{ with some } v \in X^*. \tag{1.42}$$

The importance of tilt perturbations and the related notion of *tilt stability*, introduced by Poliquin and Rockafellar [15], have been well recognized in variational analysis and optimization; see, e.g., [7, 13, 14] with the references therein for more recent publications. In behavioral economics, tilt perturbations were proposed and developed by Thaler [21] in the framework of *acquisition utility*. Then, the evaluation results obtained above in terms of ε-subgradients (1.17) can be interpreted in the way that, instead of minimizing the original payoff φ, we actually minimize its *tilt-perturbed* counterpart (1.42) shifted by the *cost of changing/resistance* term $\xi \|x - \bar{x}\|$. This reflects behavioral aspects well understood in psychology, where agents balance between *desirability* issues (minimizing $\varphi_v(\cdot)$ in our case) and *feasibility* ones, i.e., minimizing their costs of changing $C(\bar{x}, x)$.

(iii) In various behavioral situations, described by models in finite-dimensional or Hilbert spaces $(X^* = X)$, it is reasonable to replace the term $\|x - \bar{x}\|$ in the costs of changing by its *quadratic* modification of the resistance to change $R = D[I] = C(\bar{x}, x)^2$. Such settings can be investigated similarly to the above developments with replacing the collections of ε-subgradients (1.17) by the *proximal subdifferential*

$$\partial_P \vartheta(\bar{x}) := \left\{ v \in X \mid \vartheta(x) \geq \vartheta(\bar{x}) + \langle v, x - \bar{x} \rangle - (\rho/2)\|x - \bar{x}\|^2 \text{ for all } x \text{ near } \bar{x} \right\}.$$

This construction is useful for the design and investigation of *proximal-type algorithms* in optimization and related areas with applications to behavioral sciences; see, e.g., [2].

1.7 Summary of Major Finding and Future Research

This paper reveals two-sided relationships between some basic notions and results of variational analysis with variational rationality in behavioral sciences. On one hand, we apply well-recognized constructions and principles of variational analysis and generalized differentiation to the study of stationary traps and related aspects of human dynamics. On the other hand, our new results provide valuable behavioral interpretations of general notions of variational analysis that is done for the first time in the literature.

Among the *major finding* in this paper we underline the following:

• Introducing the notions of optimistic and pessimistic evaluations of payoff functions and derive efficient linear optimistic evaluation of the original payoffs and proximal payoffs via subgradients of convex analysis for original payoffs and ε-subgradients of variational analysis for reference-dependent proximal payoff functions.

• Establishing a relationship between subgradient and weight factors that ensures simultaneously a linear evaluation of rates of change for proximal payoffs and a behavioral interpretation of ε-subgradients for general extended-real-valued functions.

• Deriving certificates (sufficient conditions) for exact stationary traps in behavioral dynamics expressed in terms of ε-subgradients of payoff functions.

• Similar subgradient certificates are obtained for approximate stationary traps.

• The obtained results allow us to determine and classify various special types of exact and approximate stationary traps in behavioral dynamics.

• By using powerful variational principles of Ekeland and lower subdifferential types, we establish the existence of approximate traps that satisfy certain optimality and subdifferential properties. This sheds light on the very nature of such traps under perturbations and on rates of change in behavioral dynamics at and around such positions.

• Besides the aforementioned analytic evaluations of exact and approximate stationary traps by using subgradients, we develop their geometric evaluations in terms of generalized normals to closed sets. These developments allow us also to describe generalized normals of variational analysis via stationary traps in behavioral dynamics with linear utility functions and high enough weight factors in costs of changing.

• Finally, we relate local stationary traps in behavioral processes described via linear utility functions with local extremal points of set systems and the fundamental extremal principle of variational analysis that plays a crucial role in both theory and applications.

The established two-sided relationships between variational analysis and variational rationality in behavioral sciences open the gate for further developments in this direction, which we plan to pursue in our *future research*. Besides the questions mentioned in the remarks above, they include while are not limited to:

• Considering *variational traps* in behavioral dynamics, which indicate positions that are worthwhile to approach and reach by a successions of worthwhile moves, but not worthwhile to quit. Helpful insights into the study of variational traps are given by the constructive dynamic *proofs* of the extremal principle in both finite and infinite dimensions; see [11, Theorems 2.8 and 2.10].

• Implementing the established relationships and proposed variational ideas into developing *numerical algorithms* of proximal, subgradient, and majorization-minimization type with applications to behavioral science modeling; compare, e.g., [2, 6].

• Developments and applications of the extremal principle and related tools of variational analysis to problems with *variable cone preferences*, which naturally arise in behavioral sciences via the variational rationality approach; see [4, 5] and also compare it with the books [1, 8, 10] treated general multiobjective optimization problems of this type.

Acknowledgments. Research of the first author was partly supported by the National Science Foundation under grants DMS-1512846 and DMS-1808978, and by the Air Force Office of Scientific Research under grant #15RT0462.

References

[1] Q. H. Ansari, E. Köbis and J.-C. Yao. *Vector Variational Inequalities and Vector Optimization. Theory and Applications.* Springer, Berlin, 2018.

[2] H. Attouch and A. Soubeyran. Local search proximal algorithms as decision dynamics with costs to move. *Set-Valued Var. Anal.*, 19: 157–177, 2011.

[3] T. Q. Bao and B. S. Mordukhovich. Set-valued optimization in welfare economics. *Adv. Math. Econ.*, 13: 114–153, 2010.

[4] T. Q. Bao, B. S. Mordukhovich and A. Soubeyran. Variational analysis in psychological modeling. *J. Optim. Theory Appl.*, 164: 290–315, 2015.

[5] T. Q. Bao, B. S. Mordukhovich and A. Soubeyran. 2015. Fixed points and variational principles with applications to capability theory of wellbeing via variational rationality. *Set-Valued Var. Anal.*, 23: 375–398, 2015.

[6] J. Bolte and E. Pauwels. Majoration-minimization procedures and convergence of SQP methods for semi-algebraic and tame programs. *Math. Oper. Res.*, 41: 442–465, 2016.

[7] D. Drusvyatskiy and A. S. Lewis. Tilt stability, uniform quadratic growth, and strong metric regularity of the subdifferential. *SIAM J. Optim.*, 23: 256–267, 2013.

[8] G. Eichfelder. *Variable Ordering Structures in Vector Optimization.* Springer, Berlin, 2014.

[9] D. Kahneman and A. Tversky. Prospect theory: An analysis of decision under risk. *Economet.*, 47: 263–291, 1979.

[10] A. A. Khan, C. Tammer and C. Zălinescu. *Set-Valued Optimization. An Introduction with Applications.* Springer, Berlin. 2015.

[11] B. S. Mordukhovich. *Variational Analysis and Generalized Differentiation, I: Basic Theory; II: Applications.* Springer, Berlin, 2006.

[12] B. S. Mordukhovich. *Variational Analysis and Applications*. Springer, Cham, Switzerland, 2018.

[13] B. S. Mordukhovich and T. T. A. Nghia. Second-order characterizations of tilt stability with applications to nonlinear programming. *Math. Program.*, 149: 83–104, 2015.

[14] B. S. Mordukhovich and R. T. Rockafellar. Second-order subdifferential calculus with applications to tilt stability in optimization. *SIAM J. Optim.*, 22: 953–986, 2012.

[15] R. A. Poliquin and R. T. Rockafellar. Tilt stability of a local minimum. *SIAM J. Optim.*, 8: 287–299, 1998.

[16] R. T. Rockafellar and R. J-B. Wets. *Variational Analysis*. Springer, Berlin, 1998.

[17] A. Soubeyran. *Variational Rationality: A Theory of Individual Stability and Change, Worthwhile and Ambidextry Behaviors*. Preprint, Aix-Marseille University, 2009.

[18] A. Soubeyran. *Variational Rationality and the Unsatisfied Man: Routines and the Course Pursuit between Aspirations, Capabilities and Beliefs*. Preprint, Aix-Marseille University, 2010.

[19] A. Soubeyran. *Variational rationality. Part 1. A theory of the unsatisfied men, making worthwhile moves to satisfy his needs*. Preprint. AMSE. Aix-Marseille University, 2019a.

[20] A. Soubeyran. *Variational rationality. Part 2. A unified theory of goal setting, intentions and goal striving*. Preprint. AMSE. Aix-Marseille University, 2019b.

[21] R. H. Thaler. Mental accounting and consumer choice. *Marketing Sci.*, 27: 15–25, 2018.

[22] A. Tversky and D. Kahneman. Loss aversion in riskless choice: A reference-dependent model. *Quarterly J. Econ.*, 106: 1039–1061, 1991.

Annex: Optimistic evaluations of proximal payoffs and proximal subgradients

Here, we address evaluation problems for reference-dependent payoffs $\Gamma(./x): y \in X \longmapsto \Gamma(y/x) \in R$ discussed in 1.2.2 and 1.3.2. They include, in particular, the proximal payoff functions (1.4) of our special interest in this paper. Results 1 and 2 given below must be linked with the proof of theorem 1.5.3. Here, we emphasize the link between subgradients and linear optimistic evaluations (local of global) of reference dependent payoffs. In theorem 1.5.3, we emphasize the other side, the link between subgradients of reference dependent proximal payoffs (proximal subgradients) and stationary traps.

Optimistic local evaluation of a reference dependent payoff

Extending the definition of an optimistic local evaluation $l^\varphi(./x)$ of the "to be decreased" payoff $\varphi(.)$ at $x \in X$, such that $l^\varphi(x/x) = \varphi(x)$ and $\varphi(y) \geq l^\varphi(x/x)$ for all $y \in V(x)$, we have,

Definition. $l^\Gamma(./x): y \in X \longmapsto l^\Gamma(y/x) \in R$ is an optimistic local evaluation $l^\Gamma(./x)$ of the "to be decreased" reference dependent payoff $\Gamma(./x)$ at $x \in X$, if $l^\Gamma(x/x) = \Gamma(x/x)$ and $\Gamma(y/x) \geq l^\Gamma(x/x)$ for all $y \in V(x)$.

Definition. $l^\Gamma_{x^*}(./x) = \Gamma(x/x) + <x^*, .-x>$ is a linear optimistic local evaluation of $\Gamma(./x)$ at $x \in X$ if it exists $V(x) \subset X$ such that $\Gamma(y/x) \geq l^\Gamma_{x^*}(y/x) = \Gamma(x/x) + <x^*, y-x>$ for all $y \in V(x)$. In this case $l^\Gamma_{x^*}(x/x) = \Gamma(x/x)$.

Definition. $x^* \in \partial \Gamma(x/x)$ is a local subgradient of a reference dependent payoff $\Gamma(./x)$ if it exists $V(x) \subset X$ such that

$\Gamma(y/x) \geq l^\Gamma_{x^*}(y/x) = \Gamma(x/x) + <x^*, y-x>$ for all $y \in V(x)$. Then,

RESULT 1. $x^* \in \partial \Gamma(x/x)$ is a local subgradient at x of the reference dependent payoff $\Gamma(./x)$ if it exists a neighborhood of x, $V(x) \subset X$ such that $l^\Gamma_{x^*}(./x) = \Gamma(x/x) + <x^*, .-x>$ is a linear optimistic local evaluation of $\Gamma(./x)$ at $x \in X$.

Linear optimistic local evaluation of a "to be decreased" proximal payoff

Let $\Gamma(y/x) = Q_\xi(./x) = \varphi(.) + \xi \|.-x\|$ be a reference dependent "to be decreased" proximal payoff. Then, a linear optimistic local evaluation $l^{Q_\xi}_{x^*}(y/x)$ of this proximal payoff $Q_\xi(./x)$ at $x \in X$ is such that

$Q_\xi(y/x) \geq l^{Q_\xi}_{x^*}(y/x) = Q_\xi(x/x) + <x^*, y-x>$ for all $y \in V(x)$. Then,

RESULT 2. $x^* \in \partial Q_\xi(x/x)$ is a local subgradient at x of the "to be decreased" proximal payoff $Q_\xi(./x)$ (that is a proximal subgradient) if it exists a neighborhood of x, $V(x) \subset X$ such that $l^{Q_\xi}_{x^*}(./x) = Q_\xi(x/x) + <x^*, .-x>$ is a linear optimistic local evaluation of $Q_\xi(./x)$ at x, i.e., such that

$Q_\xi(y/x) \geq l^{Q_\xi}_{x^*}(y/x) = Q_\xi(x/x) + <x^*, y-x>$ for all $y \in V(x)$.

This last inequality is equivalent to the important final inequality

$\psi(y) = \varphi(y) - \varphi(x) - <x^*, y-x> + \xi \|y-x\| \geq \psi(x) = 0$ for all $y \in V(x)$.

This final inequality is identical, for a given $\xi > 0$, to the very important inequality given in (Mordukhovich, 2006, Proposition 1.84) in order to characterize an epsilon subgradient $x^* \in \partial_\varepsilon \varphi(x)$. See the proof of theorem 1.3.5 where the condition $\xi > \varepsilon$ is added.

Similar optimistic evaluations hold for "to be increased" proximal payoff $P_\xi(./x)$.

Chapter 2

A Financial Model for a Multi-Period Portfolio Optimization Problem with a Variational Formulation

Gabriella Colajanni
Department of Mathematics and Computer Science, University of Catania, Viale A. Doria, 6 - 95125 Catania, Italy. `colajanni@dmi.unict.it`

Patrizia Daniele
Department of Mathematics and Computer Science, University of Catania, Viale A. Doria, 6 - 95125 Catania, Italy. `daniele@dmi.unict.it`

2.1 Introduction

In financial literature, a portfolio is considered as a set of financial assets or investments which are owned by an individual (an investor) or a financial institution and consist of various financial instruments, such as shares of a company (often referred as equities), government bonds, and so on.

Given a financial portfolio, it is possible to obtain different combinations of expected returns and risks depending on the choices related to the placement of their investments.

Therefore, it is important to find the combination that allows us to obtain the best possible strategy (i.e. the best performance for a given level of risk). To this end, the principle of Dominance is introduced.

Let us assume that we have two portfolios A and B, and denote by $\mathbb{E}[u_A]$ and $\mathbb{E}[u_B]$ their expected yields, respectively, and by r_A^2 and r_B^2 their risks. Portfolio A is said to be efficient and dominant on B ($A \succ B$) if it satisfies the following properties:

- $\mathbb{E}[u_A] \geq \mathbb{E}[u_B]$;
- $r_A^2 \leq r_B^2$;

where at least one of the two inequalities must be strictly satisfied. If both properties are satisfied as equalities, then the two portfolios are equivalent.

The foundation of portfolio optimization and asset allocation problem is always attributed to the Markowitz's Modern Portfolio Theory (MPT) (see [16] and [17]) based on the mean-variance analysis.

The underlying principle behind Markowitz's theory is that, in order to build an efficient portfolio, a combination of securities should be identified in order to maximize the performance and minimize the total risk by choosing as few correlated securities as possible. The fundamental assumptions of the portfolio theory are as follows:

- investors intend to maximize their ultimate wealth and are at risk;
- the investment period is unique (for Markowitz model, time is not a significant variable);
- transaction costs and taxes are zero and the assets are perfectly divisible;
- expected value and standard deviation are the only parameters that guide the choice;
- the market is perfectly competitive.

Ever since then, in numerous research papers, modifications, extensions, and alternatives to MPT have been introduced in order to simplify and reduce the limitations of Markowitz's model. In [29] and [7], the authors present the inclusion of a risk-free asset in the traditional Markowitz formulation and the optimal risk portfolio can be obtained without any knowledge of the investor's preferences (this is also known as Separation theorem), whereas, in [26] the author, taking into account the risk-free asset and the mean-variance analysis, develops the Capital Asset Pricing Model (CAPM), studied also by Lintner in [14] and Mossin in [21], in which he shows that not all the risk of an asset is rewarded by the market in the form of a higher return, but only the part which can not be reduced by diversification. He also describes how expected portfolios can be calculated by summing the pure (risk-free) rate of interest and the multiplication between the price of risk reduction for efficient portfolios and the portfolio standard deviation, known as Capital Market Line (see [27]).

CAPM limits are:

- the investment horizon is one-period;
- you can negotiate any amount of securities (which is almost unrealistic);

- absence of taxes and transaction costs;
- all investors analyze securities in the same way with the same probability estimates;
- regular performance distribution.

It is precisely the existence of this risk-free title in the CAPM that is the main and most significant difference with the Markowitz portfolio selection model because the utility curves are eliminated and, thus, the strong subjective component in the efficient portfolio selection; indeed, all individuals invest in the same portfolio of tangency, while the weights inside it to the various titles, and in particular to the risk-free title, change.

One of the newest models is the one developed by Black and Litterman for calculating optimal portfolio weights (see [3] and [4]). The innovative aspect of the Black-Litterman model lies in the fact that, thanks to the Bayes theorem, it is able to put together two types of information different from each other, namely the market equilibrium and the investor's views on the future trend of the market. The obtained results are then used by the classic media-variance optimization approach in order to calculate the mean, variance and, consequently, the excellent portfolio composition.

Other extensions of the Markowitz model are studied, in which the variance has been replaced by the Value-at-Risk (with threshold) (see [1]) or with the Conditional Value-at-Risk (CVaR) (see [24]). In an Optimization Portfolio Problem, the multiperiod theory must be taken into account and becomes crucial. A formulation neglecting this feature can easily become misleading. Therefore, in this paper, Markovitz's portfolio theory is reviewed for investors with long-term horizons.

In 1969, Samuelson (see [25]) and Merton (see [19]), taking inspiration from Mossin's work (see [21]), formulated and solved a many-period generalization, corresponding to lifetime planning of consumption and investment decisions. Samuelson and Merton were, therefore, the first authors to study the problem with long-term horizons, but in their case the investment horizon was irrelevant and the choice of portfolio was considered short-sighted because investors ignored what was going to happen the next period and continued to choose the same portfolio, as opposed to what is studied in this paper, in which we consider the predictable and variable returns (or profits) over time.

In [28], multiperiod mean-variance models are analyzed and the final goal lies in constructing an approximate downside risk minimization through appropriate constraints. In financial markets, buying and selling securities entail brokerage fees and sometimes lump sum taxes are imposed on the investors.

In [2], the authors used Mixed Integer Programming (MIP) methods to construct portfolios, reducing the number of different stocks and assuming that it is desirable to have a portfolio with a small number of transactions (considered as the processes of rebalancing the portfolio). Mao (see [15]), Jacob (see [10]) and Levy (see [13]) have examined the fixed transaction costs problem by placing restrictions on the number of securities in the optimal portfolio. In 2000, Kellerer, Mansini, and Speranza (see [12]) introduced some Mixed Integer Linear Programming (MILP) problems with the presence of transaction costs and studied the problem of portfolio selection with fixed and proportional costs and possibly with minimum transaction lots, but they only allow linear objective function and linear and integer constraints on transaction amounts, so nonlinear constraints cannot be managed. In 2013, Greco, Matarazzo, and Slowinski (see [9]), considering the quantities as evaluation criteria of the portfolios, solved a multiobjective optimization problem by using a Multiple Criteria Decision Aiding method.

In some models, investors can negotiate any amount of securities, but this hypothesis is unrealistic, as each investor has a maximum budget limit available to invest. In this article, however, we impose that the resources used are not greater than the available ones, making the model more realistic. The objective of this paper is to formulate the multi-period portfolio selection problem as a Markowitz mean-variance optimization problem with the addition of transaction costs and taxes (on the capital gain). Moreover, by means of the proposed Integer Nonlinear Programming (INLP) Problem, it is possible to establish when it is convenient to buy and to sell financial securities, while maximizing the profits and minimizing the risk.

The paper is organized as follows: In Section 2.2, we present the financial model consisting of financial securities, issuers, investors, and intermediaries. We derive the optimization problem of each investor based on the maximization of his expected gain and the minimization of his risk portfolio. In Section 2.3, we characterize the optimality conditions for all investors simultaneously by means of a variational inequality, introducing the Lagrange multipliers associated with the constraints and report existence and uniqueness results. In Section 2.4, we apply the model to some numerical examples consisting of a financial network with two issuers, two financial securities and an investor. Section 2.5 summarizes the obtained results and contains ideas for future research.

2.2 The Financial Model

We consider a financial network consisting of: n financial securities, the typical one being denoted by i; S issuers of financial securities, such as companies, banks, countries, etc., the typical one being denoted by s; K investors (security purchasers), the typical one being denoted by k; B financial intermediaries, the typical one being denoted by b. In addition, we consider a partition of the financial securities by means of the sets $\mathcal{A}_1, \ldots, \mathcal{A}_s, \ldots, \mathcal{A}_S$, where $\mathcal{A}_s$ represents the set of financial securities made available by issuer s. A representation of the financial network is given in Figure 2.1.

We can remark that in the network the financial intermediaries are denoted by parallel edges, since they are not decision makers. We analyze the model in a discrete time horizon: $1, \ldots, j, \ldots, t$.

Every investor k aims to determine which securities he has to buy and sell, which financial intermediary he has to choose and at what time it is more convenient to buy and sell a security in order to maximize his own profit and minimize his own risk.

For every security i, there is a purchase cost, $C_{i,j}$, which varies over time; moreover, it is necessary to pay a commission to the chosen financial intermediary (often the banks), which consists of a percentage of the purchase cost, $\gamma_k^b \cdot C_{i,j}$, and a flat fee C_k^b.

During the ownership time of the security, it is possible (not necessary) to obtain funds (such as dividends in the case of shares, interests in the case of bonds) $D_{i,j}$ or pay money (for example in the case of an increase in the corporate capital) $P_{i,j}$. Obviously, in the event that one does not get or does not have to pay anything until the expiration or sale of the security, these quantities vanish.

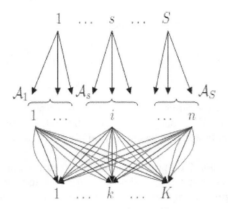

Figure 2.1: Financial Network.

Each investor has the opportunity to sell his own securities and, in this case, he will receive the sum $R_{i,j}$, but he will have to pay a charge to the chosen financial intermediary $\beta_k^b \cdot R_{i,j} + F_k^b$ (similar to purchase) and a taxation on the capital gain or a percentage on the gain obtained from the title. In case of loss, no taxation will be carried out, whereas, on the contrary, you generally have the compensation, but such a situation is not examined in this paper. So, we have:

$$\alpha_i^k \left(\frac{|\mathbb{E}[u_{i,j}]| + \mathbb{E}[u_{i,j}]}{2} \right),$$

where $|\mathbb{E}[u_{i,j}]|$ denotes the absolute value of the expected gain, which coincides with the capital gain.

We note that the tax treatment of the capital gain in Italy varies according to the subject k, making the gain (individual or individual company or company) and the type of financial security i.

In this work, we will refer to the declarative regime and, moreover, we will assume that, for each security, the financial intermediary of the sale coincides with that of the purchase.

Therefore, we introduce the following binary variables

$$x_{i,j}^k = \begin{cases} 1 & \text{if security i is purchased by k at time j} \\ 0 & \text{otherwise} \end{cases}$$

$$y_{i,j}^k = \begin{cases} 1 & \text{if security i is sold by k at time j} \\ 0 & \text{otherwise} \end{cases}$$

$$z_i^b = \begin{cases} 1 & \text{if security i is purchased and sold by b} \\ 0 & \text{otherwise} \end{cases},$$

defined for all $i = 1, \ldots, n$, $j = 1, \ldots, t$, $k = 1, \ldots, K$, and $b = 1, \ldots, B$.

Since at the initial time the values $P_{i,j}$, $D_{i,j}$ and $R_{i,j}$ of the subsequent times are unknown, we will use their expected values: $\mathbb{E}[P_{i,j}]$, $\mathbb{E}[D_{i,j}]$ and $\mathbb{E}[R_{i,j}]$.

Thus, the capital gain of a security i which has been purchased or sold, $\mathbb{E}[u_{i,j}]$, if positive, will be given by the difference between the selling price $\mathbb{E}[R_{i,j}]$ and the purchasing price $C_{i,\bar{j}}$ plus all the dividends (interests) and minus all the paid fees (if any)

$\mathbb{E}[D_{i,j}] - \mathbb{E}[P_{i,j}]$ while holding the title, that is:

$$\mathbb{E}[u_{i,j}] = \mathbb{E}[R_{i,j}] - C_{i,\bar{j}} + \sum_{\bar{j}=\bar{j}+1}^{j} \left(\mathbb{E}[D_{i,\bar{j}}] - \mathbb{E}[P_{i,\bar{j}}]\right),$$

where $\bar{j}$ and j indicate the purchase and selling time respectively, with $1 \leq \bar{j} < j \leq t$.

Some financial securities have a length which we denote by τ_i. So $S = \bar{j} + \tau_i$, which means the time given by the purchasing period plus the length of the title, represents its expiration time.

In order to determine the optimal portfolio of securities, it is necessary to establish the time interval such that $t > \tau_i \quad \forall i = 1, \ldots, n$. Therefore, if a title does not have a pre-established length, we impose $\tau_i = t - 1$; in such a way, the time of their fictitious expiration coincides with t or will be greater than t.

If a financial security i has not been sold before it expires, the investor who owns this security, when it expires, will receive an amount equal to its expected nominal value $\mathbb{E}[N_{i,\bar{j}+\tau_i}]$ and will have to pay the tax on capital gain, $\mathbb{E}[g_{i,\bar{j}+\tau_i}]$, if positive, as in the case of sale:

$$\alpha_i^k \left(\frac{|\mathbb{E}[g_{i,\bar{j}+\tau_i}]| + \mathbb{E}[g_{i,\bar{j}+\tau_i}]}{2} \right),$$

where, in this case, $\mathbb{E}[g_{i,\bar{j}+\tau_i}] = \mathbb{E}[N_{i,\bar{j}+\tau_i}] - C_{i,\bar{j}} + \sum_{\bar{j}=\bar{j}+1}^{\bar{j}+\tau_i} \left(\mathbb{E}[D_{i,\bar{j}}] - \mathbb{E}[P_{i,\bar{j}}]\right).$

If the expiration time of the unsold security exceeds t, or in the case of non-expiration securities, the investor at time t will own the security whose expected value is $\mathbb{E}[N_{i,t}]$.

Every investor k aims to determine the decision variables $x_{i,j}^k, y_{i,j}^k, z_i^b \in \{0,1\} \quad \forall i = 1, \ldots, n, \quad \forall j = 1, \ldots, t, \quad \forall b = 1, \ldots, B$, which means deciding, at every time, which securities are convenient to buy and to sell and through which financial intermediary, in order to maximize the profit of each security, which is obtained by taking into account:

- the purchase cost and the commission to be given to the chosen financial intermediary (given by a percentage on the purchasing cost plus a fixed fee)

$$-C_{i,\bar{j}} - \sum_{b=1}^{B} z_i^b \cdot (\gamma_k^b C_{i,\bar{j}} + C_k^b),$$

if security i was purchased at time $\bar{j}$;

- if security i was purchased at time $\bar{j}$, the investor has the possibility to sell it at time j (with $\bar{j}+1 \leq j \leq \min\{\bar{j}+\tau_i, t\}$, where τ_i is the length of the title); if investor k sells his security, he will receive the selling price, but he will have to pay the tax on the capital gain and the commission to the chosen financial intermediary:

$$\mathbb{E}[R_{i,j}] - \alpha_i^k \left(\frac{|\mathbb{E}[u_{i,j}]| + \mathbb{E}[u_{i,j}]}{2} \right) - \sum_{b=1}^{B} z_i^b \cdot (\beta_k^b \mathbb{E}[R_{i,j}] + F_k^b);$$

- during the period of ownership of his security, investor k may receive dividends (or interests) and pay some amounts of money:

$$\sum_{j=\bar{j}+1}^{\min\{\bar{j}+\tau_i,t\}} \left(\mathbb{E}[-P_{i,j}+D_{i,j}] - y_{i,j}^k \sum_{j=j+1}^{\min\{\bar{j}+\tau_i,t\}} (\mathbb{E}[-P_{i,\hat{j}}+D_{i,\hat{j}}]) \right);$$

- if financial security i has not been sold and
    - if the security expires before the final term t, then investor k, at time $\bar{j}+\tau_i$ receives the nominal value of the security and pays the tax α_i^k in the event that there is a positive capital gain

$$\sum_{\bar{j}=1}^{t-\tau_i-1} x_{i,\bar{j}}^k \left[(1 - \sum_{j=\bar{j}+1}^{\bar{j}+\tau_i} y_{i,j}^k) \left(\mathbb{E}[N_{i,\bar{j}+\tau_i}] - \alpha_i^k \left(\frac{|\mathbb{E}[g_{i,\bar{j}+\tau_i}]| + \mathbb{E}[g_{i,\bar{j}+\tau_i}]}{2} \right) \right) \right];$$

- if the expiration of this security is such that $S_i = \bar{j}+\tau_i \geq t$ or the title does not expire (we set $\tau_i = t-1 \Rightarrow \bar{j}+\tau_i = \bar{j}+t-1 \geq t$), then investor k, at time t, holds a security that has a certain nominal value

$$\sum_{\bar{j}=t-\tau_i}^{t-1} x_{i,\bar{j}}^k \left[(1 - \sum_{j=\bar{j}+1}^{t} y_{i,j}^k) \mathbb{E}[N_{i,t}] \right].$$

Therefore, we are dealing with maximization of the expected gain of the portfolio:

$$\mathbb{E}[e_p^k] = \sum_{i=1}^{n} \sum_{\bar{j}=1}^{t-1} x_{i,\bar{j}}^k \mathbb{E}[e_{i,\bar{j}}^k] = \sum_{i=1}^{n} x_i^k \mathbb{E}[e_i^k],$$

namely:

$$\max \mathbb{E}[e_p^k] = \max \sum_{i=1}^{n} \sum_{\bar{j}=1}^{t-1} x_{i,\bar{j}}^k \mathbb{E}[e_{i,\bar{j}}^k]$$

$$= \max \sum_{i=1}^{n} \Bigg\{ \sum_{\bar{j}=1}^{t-1} x_{i,\bar{j}}^k \Bigg[-C_{i,\bar{j}} - \sum_{b=1}^{B} z_i^b \cdot (\gamma_k^b C_{i,\bar{j}} + C_k^b)$$

$$+ \sum_{j=\bar{j}+1}^{\min\{\bar{j}+\tau_i,t\}} \left(\mathbb{E}[-P_{i,j}+D_{i,j}] + y_{i,j}^k \left(\mathbb{E}[R_{i,j}] - \alpha_i^k \left(\frac{|\mathbb{E}[u_{i,j}]| + \mathbb{E}[u_{i,j}]}{2} \right) \right. \right.$$

$$\left. \left. - \sum_{b=1}^{B} z_i^b \cdot (\beta_k^b \mathbb{E}[R_{i,j}] + F_k^b) - \sum_{\hat{j}=j+1}^{\min\{\bar{j}+\tau_i,t\}} (\mathbb{E}[-P_{i,\hat{j}}+D_{i,\hat{j}}]) \right) \right) \Bigg]$$

$$+ \sum_{\bar{j}=1}^{t-\tau_i-1} x_{i,\bar{j}}^k \left[(1 - \sum_{j=\bar{j}+1}^{\bar{j}+\tau_i} y_{i,j}^k) \left(\mathbb{E}[N_{i,\bar{j}+\tau_i}] - \alpha_i^k \left(\frac{|\mathbb{E}[g_{i,\bar{j}+\tau_i}]| + \mathbb{E}[g_{i,\bar{j}+\tau_i}]}{2} \right) \right) \right]$$

$$+ \sum_{\bar{j}=t-\tau_i}^{t-1} x_{i,\bar{j}}^k \left[(1 - \sum_{j=\bar{j}+1}^{t} y_{i,j}^k) \mathbb{E}[N_{i,t}] \right] \Bigg\}.$$

Another objective of investor k is to minimize his risk portfolio. In [23], Nagurney and Ke also assumed that the decision-makers seek not only to increase their net revenues but also to minimize risk, with the risk being considered as the possibility of suffering losses compared to the expected profit. It can be measured through the use of statistical indices, such as the variance or standard deviation of the asset's earnings distribution.

Given the aleatory gain on security i, e_i^k, the risk of the security, as a variance, is

$$(\sigma_i^k)^2 = \frac{\sum_{m=1}^{M}(e_m^k - \mathbb{E}[e_i^k])^2}{M-1}.$$

The risk on the portfolio is given by:

$$(\sigma_p^k)^2 = \sum_{i=1}^{n}(x_i^k)^2(\sigma_i^k)^2 + 2\sum_{i=1}^{n-1}\sum_{h>i}^{n}x_i^k x_h^k \sigma_{ih}^k,$$

where σ_{ih}^k is the covariance between securities i and h.

As it is well known, covariance lies in $]-\infty, +\infty[$, hence, it is often more useful to take into account correlation $\rho_{ih}^k = \dfrac{\sigma_{ih}^k}{\sigma_i^k \sigma_h^k}$ since it lies in $[-1,1]$ and it measures the correlation or discrepancy between the gains of the securities i and h.

As a consequence, the minimization of the portfolio risk can be expressed as:

$$\min(\sigma_p^k)^2 = \min\left[\sum_{i=1}^{n}(x_i^k)^2(\sigma_i^k)^2 + 2\sum_{i=1}^{n-1}\sum_{h>i}^{n}x_i^k x_h^k \rho_{ih}^k \sigma_i^k \sigma_h^k\right].$$

The main objective of investor k is to maximize his profit and, at the same time, to minimize his portfolio risk. Hence, we introduce the aversion degree or risk inclination, η_k, which depends on subjective evaluations of the single investor k and on the influences of the external environment that surrounds it (see [11]), and add the term

$$-\eta_k(\sigma_p^k)^2$$

to the objective function to be maximized, obtaining:

$$\max \sum_{i=1}^{n} \left\{ \sum_{\bar{j}=1}^{t-1} x_{i,\bar{j}}^{k} \left[-C_{i,\bar{j}} - \sum_{b=1}^{B} z_{i}^{b} \cdot (\gamma_{k}^{b} C_{i,\bar{j}} + C_{k}^{b}) \right. \right.$$

$$+ \sum_{j=\bar{j}+1}^{\min\{\bar{j}+\tau_{i},t\}} \left(\mathbb{E}[-P_{i,j} + D_{i,j}] + y_{i,j}^{k} \left(\mathbb{E}[R_{i,j}] - \alpha_{i}^{k} \left(\frac{|\mathbb{E}[u_{i,j}]| + \mathbb{E}[u_{i,j}]}{2} \right) \right) \right.$$

$$\left. - \sum_{b=1}^{B} z_{i}^{b} \cdot (\beta_{k}^{b} \mathbb{E}[R_{i,j}] + F_{k}^{b}) - \sum_{\hat{j}=j+1}^{\min\{\bar{j}+\tau_{i},t\}} \left(\mathbb{E}[-P_{i,\hat{j}} + D_{i,\hat{j}}] \right) \right) \right]$$

$$+ \sum_{\bar{j}=1}^{t-\tau_{i}-1} x_{i,\bar{j}}^{k} \left[\left(1 - \sum_{j=\bar{j}+1}^{\bar{j}+\tau_{i}} y_{i,j}^{k} \right) \left(\mathbb{E}[N_{i,\bar{j}+\tau_{i}}] - \alpha_{i}^{k} \left(\frac{|\mathbb{E}[g_{i,\bar{j}+\tau_{i}}]| + \mathbb{E}[g_{i,\bar{j}+\tau_{i}}]}{2} \right) \right) \right]$$

$$\left. + \sum_{\bar{j}=t-\tau_{i}}^{t-1} x_{i,\bar{j}}^{k} \left[(1 - \sum_{j=\bar{j}+1}^{t} y_{i,j}^{k}) \mathbb{E}[N_{i,t}] \right] \right\} - \eta_{k}(\sigma_{p}^{k})^{2}.$$

The problem formulation is as follows:

$$\max \mathbb{E}[e_{p}^{k}] - \eta_{k}(\sigma_{p}^{k})^{2} \qquad (2.1)$$

subject to the constraints

$$\sum_{k=1}^{n} \sum_{j=1}^{t-1} x_{i,j}^{k} \leq 1 \quad \forall i = 1, \ldots, n \qquad (2.2)$$

$$y_{i,j}^{k} \leq \sum_{\bar{j}=j-\tau_{i}+1}^{j-1} x_{i,\bar{j}}^{k}, \quad \forall i = 1, \ldots, n, \quad \forall j = 2, \ldots, t \qquad (2.3)$$

$$y_{i,j}^{k} \leq \frac{\sum_{\bar{j}=2}^{j-1}(1 - y_{i,\bar{j}}^{k})}{j-2} \quad \forall i = 1, \ldots, n, \quad \forall j = 3, \ldots, t \qquad (2.4)$$

$$(\sigma_{p_{k}})^{2} \leq \bar{R}_{k} \qquad (2.5)$$

$$\sum_{i=1}^{n} \sum_{j=1}^{t-1} x_{i,j}^{k} C_{i,j} \leq \bar{B}_{k} \qquad (2.6)$$

$$\sum_{b=1}^{B} z_{i}^{b} = \sum_{j=1}^{t-1} x_{i,j}^{k} \quad \forall i = 1, \ldots, n \qquad (2.7)$$

$$\sum_{\substack{\max\{j<\bar{j}:\\D_{i,\bar{j}}>0\}}}^{\bar{j}} \sum_{z \in \mathcal{A}_{s}} \sum_{k=1}^{K} (x_{z,j}^{k} - y_{z,j}^{k}) \geq 1 \quad \forall s \in \mathcal{S}, D_{i,\bar{j}} > 0 \qquad (2.8)$$

$$x_{i,j}^{k}, y_{i,j}^{k}, z_{i}^{b} \in \{0,1\}, \forall i = 1, \ldots, n, \quad \forall j = 1, \ldots, t, \quad \forall b = 1, \ldots, B. \qquad (2.9)$$

It is interesting to note that:

(2.2) means that it is possible to buy the same security only once and it can be purchased by a single investor (but there are numerous coincident securities);

(2.3) means that it is possible to sell a security only if it has been purchased previously and has not yet expired;

(2.4) means that you can sell a stock only if it has not yet been sold;

(2.5) means that there is a risk limit, $\bar{R}_k$, which represents the maximum risk limit that the investor is willing to accept;

(2.6) means that there is a budget limit, $\bar{B}_k$, which represents the maximum available budget for an investor;

(2.7) means that for each security, only one financial intermediary can be chosen for purchasing and selling activities;

(2.8) means that each issuer must sell at least one security during the dividend distribution periods, where the dividend $D_{i,\bar{j}}$ at time $\bar{j}$ of security $i \in \mathcal{A}_s$ is given by:

$$D_{i,\bar{j}} = \frac{U_{\bar{j}}^s - R_{\bar{j}}^s}{\sum_{\max\{j<\bar{j}:\, D_{i,j}>0\}}^{\bar{j}} \sum_{z \in \mathcal{A}_s} \sum_{k=1}^{K} (x_{z,j}^k - y_{z,j}^k)}.$$

In some particular cases, additional constraints could be included in the model, for example:

- $\sum_{j=1}^{t} x_{i,j}^k = 1$, if security i must be purchased;

- $\sum_{j=1}^{t} (x_{i,j}^k + x_{h,j}^k + x_{w,j}^k) \leq 1$, if only one security among i, h and w can be purchased;

- $\sum_{j=1}^{t} (x_{i,j}^k + x_{h,j}^k + x_{w,j}^k) \geq 1$, if only one security among i, h and w must be purchased;

- $\sum_{j=1}^{t} x_{i,j}^k \leq \sum_{j=1}^{t} x_{h,j}^k$, if security i can be purchased only when security h has also been purchased;

- $x_{i,j}^k \leq \sum_{\bar{j}=1}^{j} x_{h,\bar{j}}^k$, $\forall j = 1,\ldots,t$, if security i can be purchased only when h has already been purchased;

- $x_{i,j}^k \leq \prod_{\bar{j}=1}^{j} (1 - x_{h,\bar{j}}^k)$, $\forall j = 1,\ldots,t$, if security i can be purchased only when security h has not yet been purchased;

- $\sum_{j=1}^{t} x_{i,j}^k \leq \frac{1}{2} \sum_{j=1}^{t} \left(x_{h,j}^k + x_{w,j}^k \right)$ if security i can be purchased only if h and w are purchased too.

2.3 Variational Inequality Formulation and Existence Results

The relaxation of problem (2.1)–(2.9) can be rewritten replacing the binary constraint on the variables with constraints (2.17) and (2.18):

$$\max \mathbb{E}[e_p^k] - \eta_k (\sigma_p^k)^2$$

subject to

$$\sum_{k=1}^{n} \sum_{j=1}^{t-1} x_{i,j}^k \leq 1 \quad \forall i = 1,\ldots,n \tag{2.10}$$

$$y_{i,j}^k \leq \sum_{\bar{j}=j-\tau_i+1}^{j-1} x_{i,\bar{j}}^k \quad \forall i = 1,\ldots,n, \quad \forall j = 2,\ldots,t \tag{2.11}$$

$$y_{i,j}^k \leq \frac{\sum_{\bar{j}=2}^{j-1}(1-y_{i,\bar{j}}^k)}{j-2} \quad \forall i = 1,\ldots,n, \quad \forall j = 3,\ldots,t \tag{2.12}$$

$$(\sigma_{p_k})^2 \leq \bar{R}_k \tag{2.13}$$

$$\sum_{i=1}^{n} \sum_{j=1}^{t-1} x_{i,j}^k C_{i,j} \leq \bar{B}_k \tag{2.14}$$

$$\sum_{b=1}^{B} z_i^b = \sum_{j=1}^{t-1} x_{i,j}^k \quad \forall i = 1,\ldots,n \tag{2.15}$$

$$\sum_{\substack{\max\{j<\bar{j}:\\D_{i,\bar{j}}>0\}}}^{\bar{j}} \sum_{z\in\mathcal{A}_s} \sum_{k=1}^{K} (x_{z,j}^k - y_{z,j}^k) \geq 1 \quad \forall s \in \mathcal{S}, D_{i,\bar{j}} > 0 \tag{2.16}$$

$$\sum_{i=1}^{n} \sum_{j=1}^{t-1} x_{i,j}^k (1 - x_{i,j}^k) + \sum_{i=1}^{n} \sum_{j=2}^{t} y_{i,j}^k (1 - y_{i,j}^k) + \sum_{i=1}^{n} \sum_{b=1}^{B} z_i^b (1 - z_i^b) \geq 0 \tag{2.17}$$

$$x_{i,j}^k, y_{i,j}^k, z_i^b \in [0,1] \; \forall i = 1,\ldots,n, \quad \forall j = 1,\ldots,t, \quad \forall b = 1,\ldots,B. \tag{2.18}$$

Now, we group the variables $x_{i,\bar{j}}^k$, $i=1,\ldots,n$, $\bar{j}=1,\ldots,t-1$, $k=1,\ldots,K$ into the vector $\mathbf{x} \in [0,1]^{n(t-1)K}$, the variables $y_{i,j}^k$, $i=1,\ldots,n$, $j=2,\ldots,t$, $k=1,\ldots,K$ into the vector $\mathbf{y} \in [0,1]^{n(t-1)K}$ and the variables z_i^b, $i=1,\ldots,n$, $b=1,\ldots,B$ into the vector $\mathbf{z} \in [0,1]^{nB}$.

This allows us to apply the Lagrange theory, which has proved to be very productive for the study of the behavior of important aspects of the problem and has been applied in many fields, such as the organ transplant model, the cybersecurity investment supply chain game theory model, the elastic-plastic torsion problem, the bilevel problem, and so on (see [5], [6], [8], [18] for some applications).

In our financial model, by applying the classical Lagrange theory, we find that the optimality conditions (8.40)–(2.18) for all investors can be characterized simultaneously by means of the following variational inequality:

Find $(\mathbf{x}^*, \mathbf{y}^*, \mathbf{z}^*, \lambda_1^*, \lambda_2^*, \lambda_3^*, \lambda_4^*, \lambda_5^*, \lambda_6^*, \lambda_7^*) \in \mathbb{K} \times \mathbb{R}_+^{K(n+n(t-1)+n(t-2)+3)} \times \mathbb{R}^{Kn}$ such that

$$\sum_{k=1}^{K}\sum_{i=1}^{n}\sum_{\bar{j}=1}^{t-1}\left[\eta_k\frac{\partial(\sigma_p^k(x_{i,\bar{j}}^{k*}))^2}{\partial x_{i,\bar{j}}^k} - \frac{\partial\mathbb{E}[e_p^k](x_{i,\bar{j}}^{k*},y_{i,\bar{j}}^{k*},z_i^{b*})}{\partial x_{i,\bar{j}}^k} + \lambda_{1i}^{k*} - \sum_{j=2}^{\min\{\bar{j}+\tau_i-1,t\}}\lambda_{2ij}^{k*}\right.$$
$$\left. + \lambda_4^{k*}[2x_{i,\bar{j}}^{k*}(\sigma_{i,\bar{j}}^k)^2 + 2\sum_{h\neq i}x_{h,\bar{j}}^{k*}\rho_{ih\bar{j}}\sigma_{i\bar{j}}\sigma_{h\bar{j}}] + \lambda_5^{k*}C_{i,\bar{j}} - \lambda_{6i}^{k*} - \lambda_7^*(1-2x_{1,\bar{j}}^{k*})\right]$$
$$\times (x_{i,\bar{j}}^k - x_{i,\bar{j}}^{k*})$$
$$+\sum_{k=1}^{K}\sum_{i=1}^{n}\sum_{j=2}^{t}\left[-\frac{\partial\mathbb{E}[e_p^k](x_{i,j}^{k*},y_{i,j}^{k*},z_i^{b*})}{\partial y_{i,j}^k} + \lambda_{2ij}^{k*} + \lambda_{3ij}^{k*} + \sum_{\bar{j}=j+1}^{t}\frac{\lambda_{3i\bar{j}}^{k*}}{\bar{j}-2} - \lambda_7^*(1-2y_{i,j}^{k*})\right]$$
$$\times (y_{i,j}^k - y_{i,j}^{k*})$$
$$+\sum_{k=1}^{K}\sum_{i=1}^{n}\sum_{b=1}^{B}\left[-\frac{\partial\mathbb{E}[e_p^k](x_{i,j}^{k*},y_{i,j}^{k*},z_i^{b*})}{\partial z_i^b} + \lambda_{6i}^{k*} - \lambda_7^*(1-2z_i^{b*})\right]$$
$$\times (z_i^b - z_i^{b*})$$
$$-\sum_{k=1}^{K}\sum_{i=1}^{n}\left[\sum_{j=1}^{t-1}x_{i,j}^{k*}-1\right]\times(\lambda_{1i}^k - \lambda_{1i}^{k*}) - \sum_{k=1}^{K}\sum_{i=1}^{n}\sum_{j=2}^{t}\left[y_{i,j}^{k*} - \sum_{\bar{j}=j-\tau_i+1}^{j-1}x_{i,\bar{j}}^{k*}\right]$$
$$\times (\lambda_{2ij}^k - \lambda_{2ij}^{k*})$$
$$-\sum_{k=1}^{K}\sum_{i=1}^{n}\sum_{j=2}^{t}\left[y_{i,j}^{k*} - \frac{\sum_{\bar{j}=2}^{j-1}(1-y_{i,\bar{j}}^{k*})}{j-2}\right]\times(\lambda_{3ij}^k - \lambda_{3ij}^{k*}) - \sum_{k=1}^{K}\left[(\sigma_p^k)^2 - \bar{R}_k\right]$$
$$\times (\lambda_4^k - \lambda_4^{k*})$$
$$-\sum_{k=1}^{K}\left[\sum_{i=1}^{n}\sum_{j=1}^{t-1}x_{i,j}^{k*}C_{i,j} - \bar{B}_k\right]\times(\lambda_5^k - \lambda_5^{k*})$$
$$-\sum_{k=1}^{K}\sum_{i=1}^{n}\left[\sum_{b=1}^{B}z_i^{b*} - \sum_{j=1}^{t-1}x_{i,j}^{k*}\right]\times(\lambda_{6i}^k - \lambda_{6i}^{k*})$$
$$+\sum_{k=1}^{K}\left[\sum_{i=1}^{n}\sum_{j=1}^{t-1}x_{i,j}^{k*}(1-x_{i,j}^{k*}) + \sum_{i=1}^{n}\sum_{j=2}^{t}y_{i,j}^{k*}(1-y_{i,j}^{k*}) + \sum_{i=1}^{n}\sum_{b=1}^{B}z_i^{b*}(1-z_i^{b*})\right]$$
$$\times (\lambda_7^k - \lambda_7^{k*}) \geq 0, \tag{2.19}$$

for all $(\mathbf{x}, \mathbf{y}, \mathbf{z}, \lambda_1, \lambda_2, \lambda_3, \lambda_4, \lambda_5, \lambda_6, \lambda_7) \in \mathbb{K} \times \mathbb{R}_+^{K(n+n(t-1)+n(t-2)+3)} \times \mathbb{R}^{Kn}$, where

$$\mathbb{K} = \left\{ (\mathbf{x}, \mathbf{y}, \mathbf{z}) \in [0,1]^{2n(t-1)K+nB} \text{ such that:} \right.$$

$$\left. \sum_{\substack{\max\{j<\bar{j}: \\ D_{i,\bar{j}}>0\}}}^{\bar{j}} \sum_{z \in \mathcal{A}_s} \sum_{k=1}^{K} (x_{z,j}^k - y_{z,j}^k) \geq 1 \quad \forall s \in \mathcal{S}, D_{i,\bar{j}} > 0 \right\},$$

and $\lambda_1, \lambda_2, \lambda_3, \lambda_4, \lambda_5, \lambda_6$ and λ_7 are the Lagrange multipliers associated with constraints (8.40), (8.44), (8.14), (8.15), (2.14), (2.15) and (2.17), respectively. In equilibrium, the solution $(\mathbf{x}^*, \mathbf{y}^*, \mathbf{z}^*)$ represents the optimal securities purchased by investors, the optimal purchasing and selling times, and the optimal financial intermediary chosen by the investors for every security. Then, using some suitable conditions obtained by studying the Lagrange multipliers, we are able to find the optimal variables with integer values.

Now we assume that all the involved functions (such as the purchase cost function, the commission cost function, the tax function and the selling price function) are continuously differentiable and convex. Variational inequality (2.19) can be put in a standard form (see [22]) as follows:

$$\text{Find } X^* \in \mathcal{K} \text{ such that: } \langle F(X^*), X - X^* \rangle \geq 0, \quad \forall X \in \mathcal{K}, \tag{2.20}$$

where

$$\mathcal{K} = \mathbb{K},$$
$$X = (\mathbf{x}, \mathbf{y}, \mathbf{z}, \lambda_1, \lambda_2, \lambda_3, \lambda_4, \lambda_5, \lambda_6, \lambda_7),$$
$$F(X) = (F_1(X), F_2(X), F_3(X), F_4(X), F_5(X), F_6(X), F_7(X), F_8(X), F_9(X), F_{10}(X))$$

and

$$F_1(X) = \eta_k \frac{\partial (\sigma_p^k(x_{i,j}^{k*}))^2}{\partial x_{i,\bar{j}}^k} - \frac{\partial \mathbb{E}[e_p^k](x_{i,\bar{j}}^{k*}, y_{i,j}^{k*}, z_i^{b*})}{\partial x_{i,\bar{j}}^k} + \lambda_{1i}^{k*} - \sum_{j=2}^{\min\{\bar{j}+\tau_i-1,t\}} \lambda_{2ij}^{k*}$$
$$+ \lambda_4^{k*}\left[2x_{i,\bar{j}}^{k*}(\sigma_{i,\bar{j}}^k)^2 + 2\sum_{h\neq i} x_{h,\bar{j}}^{k*}\rho_{ihj}\sigma_{ij}\sigma_{hj}\right] + \lambda_5^{k*}C_{i,\bar{j}} - \lambda_{6i}^{k*} - \lambda_7^*(1 - 2x_{1,\bar{j}}^{k*})$$

$$F_2(X) = -\frac{\partial \mathbb{E}[e_p^k](x_{i,j}^{k*}, y_{i,j}^{k*}, z_i^{b*})}{\partial y_{i,j}^k} + \lambda_{2ij}^{k*} + \lambda_{3ij}^{k*} + \sum_{\bar{j}=j+1}^{t} \frac{\lambda_{3i\bar{j}}^{k*}}{\bar{j}-2} - \lambda_7^*(1 - 2y_{i,j}^{k*})$$

$$F_3(X) = -\frac{\partial \mathbb{E}[e_p^k](x_{i,j}^{k*}, y_{i,j}^{k*}, z_i^{b*})}{\partial z_i^b} + \lambda_{6i}^{k*} - \lambda_7^*(1 - 2z_i^{b*})$$

$$F_4(X) = -\sum_{j=1}^{t-1} x_{i,j}^{k*} + 1$$

$$F_5(X) = -y_{i,j}^{k*} + \sum_{\bar{j}=j-\tau_i+1}^{j-1} x_{i,\bar{j}}^{k*}$$

$$F_6(X) = -y_{i,j}^{k*} + \frac{\sum_{\bar{j}=2}^{j-1}(1 - y_{i,\bar{j}}^{k*})}{j-2}$$

$$F_7(X) = -(\sigma_p^k)^2 + \bar{R}_k$$

$$F_8(X) = -\sum_{i=1}^{n}\sum_{j=1}^{t-1} x_{i,j}^{k*}C_{i,j} + \bar{B}_k$$

$$F_9(X) = -\sum_{b=1}^{B} z_i^{b*} + \sum_{j=1}^{t-1} x_{i,j}^{k*}$$

$$F_{10}(X) = \sum_{i=1}^{n}\sum_{j=1}^{t-1} x_{i,j}^{k*}(1 - x_{i,j}^{k*}) + \sum_{i=1}^{n}\sum_{j=2}^{t} y_{i,j}^{k*}(1 - y_{i,j}^{k*}) + \sum_{i=1}^{n}\sum_{b=1}^{B} z_i^{b*}(1 - z_i^{b*})$$

Since the feasible set $\mathcal{K}$ is bounded, closed and convex, we can obtain the existence of a solution to (2.19) based on the assumption of the continuity of F. Hence, we have the following theorem:

Theorem 2.3.1 (Existence) *If $\mathcal{K}$ is a compact and convex set and $F(X)$ is continuous on $\mathcal{K}$, then variational inequality (2.20) admits at least a solution X^*.*

In addition, we are able to provide a uniqueness result.

Theorem 2.3.2 (Uniqueness) *Under the assumptions of Theorem 2.3.1, if the function $F(X)$ is strictly monotone on $\mathcal{K}$, that is*

$$\langle (F(X_1) - F(X_2))^T, X_1 - X_2 \rangle > 0 \quad \forall X_1, X_2 \in \mathcal{K}, \ X_1 \neq X_2,$$

then variational inequality (2.20) and, hence, variational inequality (2.19), admits a unique solution.

2.4 Numerical Examples

In this section, we apply the model to some numerical examples that consist of a financial network with two issuers, two financial securities and an investor, as depicted in Figure 2.2.

We consider also two financial intermediaries and we analyze the model in the following time horizon: $1,\ldots,5$.

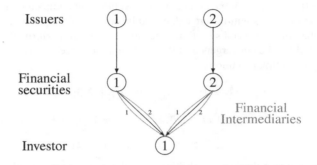

Figure 2.2: Network Topology for the Numerical Example

Since we want to report all the results for transparency purposes, we select the size of problems as reported. The numerical data are inspired by realistic values and are constructed for easy interpretation purposes.

To solve the examples we used Matlab on a laptop with an Intel Core2 Duo processor and 4 GB RAM.

We assume the following data are given:

C_{ij}	$j=1$	$j=2$	$j=3$	$j=4$
$i=1$	5	6	6	7
$i=2$	7	7	8	9

$\mathbb{E}[-P_{ij}]$	$j=2$	$j=3$	$j=4$	$j=5$
$i=1$	0	0	-2	0
$i=2$	0	0	0	0

$\mathbb{E}[U_{ij}-R_{ij}]$	$j=2$	$j=3$	$j=4$	$j=5$
$i=1$	0	3	0	15
$i=2$	0	0	0	0

$\mathbb{E}[R_{ij}]$	$j=2$	$j=3$	$j=4$	$j=5$
$i=1$	7	7	8	1
$i=2$	20	20	20	1

We also assume that $\tau_1 = 2$ and that the second financial security does not expire, so we require $\tau_2 = 4$. Further, we assume that the nominal value of each security at maturity or at time $j = 5$ coincides with its current value (cost) $\mathbb{E}[N_{i,\bar{j}+\tau_i}] = C_{i,\bar{j}+\tau_i}$, $\mathbb{E}[N_{i,5}] = C_{i,4}$, that the maximum budget and risk values are $\bar{B} = 25$ and $\bar{R} = 15$, respectively, that the percentages of taxation are $\alpha_1 = 15\%$ and $\alpha_2 = 10\%$, that commission costs are given by $\beta^1 = \gamma^1 = 5\%$, $\beta^2 = \gamma^2 = 15\%$, $C^1 = F^1 = 0.5$ and $C^2 = F^2 = 2$, that $\eta = 0.2$ is the risk aversion index, $(\sigma_{1j}) = (2,2,2,2,2)$, $(\sigma_{2j}) = (1,1,1,1,1)$ the variances of the titles and $\rho_{12j} = 0 \, \forall j = 1, \ldots, 5$, meaning that the two titles are completely unrelated.

The optimal solutions are calculated by solving the associated variational inequality described in the previous section. The calculations are performed using the Matlab program and for the convergence of the method a tolerance $\varepsilon = 10^{-4}$ was fixed. We get the following optimal solutions:

$$x_{14}^* = x_{21}^* = 1; \quad x_{1j}^* = 0 \, \forall j = 1,2,3; \quad x_{2j}^* = 0 \, \forall j = 2,3,4;$$

$$y_{1j}^* = 0 \, \forall j = 2,3,4,5; \quad y_{22}^* = 1, \, y_{2j}^* = 0 \, \forall j = 3,4,5;$$

$$z_1^{1*} = z_2^{1*} = 1, \quad z_1^{2*} = z_2^{2*} = 0.$$

These optimal solutions clearly show that the most convenient choice for the investor is to buy security 1 at time 4 and security 2 at time 1, to sell security 2 at time 2 but never sell security 1.

For both securities, it is better to choose the financial intermediary 1, reaching 22.5 as the total gain.

Now we consider a second example where the investor has a greater degree of risk aversion than the previous one and the variance of the securities is greater. We suppose, in this case, that $\eta = 0.9$ and $(\sigma_{1j}) = (4,4,3,3,2)$, $(\sigma_{2j}) = (2,2,1,1,1)$.

Then, we get the following optimal solutions:

$$x_{1j}^* = 0 \, \forall j = 1,2,3,4; \quad y_{1j}^* = 0 \, \forall j = 2,3,4,5;$$

$$z_1^{b*} = 0 \, \forall b = 1,2;$$

$$x_{21}^* = 1; \quad x_{2j}^* = 0 \, \forall j = 2,3,4;$$

$$y_{23}^* = 1, \quad y_{2j}^* = 0 \, \forall j = 2,4,5;$$

$$z_2^{1*} = 1, \quad z_2^{2*} = 0.$$

Therefore, in this case, it is never convenient for the investor to buy security 1, but to buy security 2 at time 1 and sell it at time 3, through the financial intermediary 1, thus obtaining a profit of 5.75.

A third example refers to the case when the degree of risk aversion of the investor is $\eta = 0.2$, as in the first example, but the maximum risk is smaller, that is 4. In this case, we get the following optimal solutions:

$$x_{1j}^* = 0 \, \forall j = 1,2,3,4; \quad y_{1j}^* = 0 \, \forall j = 2,3,4,5;$$

$$z_1^{b*} = 0 \, \forall b = 1,2;$$

$$x_{21}^* = 1; \quad x_{2j}^* = 0 \, \forall j = 2,3,4;$$

$$y_{23}^* = 1, \quad y_{2j}^* = 0 \,\forall j = 2,4,5;$$
$$z_2^{1*} = 1, \quad z_2^{2*} = 0.$$

We remark that such solutions are the same as the ones of the second example, but the total profit is now 8.55.

It is worth observing that not all the total budget is used.

2.5 Conclusions

In this paper, we focused our attention on an important problem which is studied by many researchers, namely the Portfolio Optimization problem. Specifically, Markovitz's portfolio theory is reviewed for investors with long-term horizons. We presented a financial model, taking into account that, in financial markets, buying and selling securities entail brokerage fees and sometimes lump sum taxes are imposed on the investors. Furthermore, in this paper, we imposed that the used resources are not greater than the available ones, making the model more realistic. Therefore, the objective of this paper was to formulate the multi-period portfolio selection problem as a Markowitz mean-variance optimization problem with the addition of transaction costs and taxes (on the capital gain). The presented financial model determines which securities every investor has to buy and sell, which financial intermediary he has to choose and at what time it is more convenient to buy and sell a security in order to maximize his own profit and minimize his own risk. For every security, we assumed that there is a purchase cost and that it is also necessary to pay a commission to the chosen financial intermediary (often the banks), which consists of a percentage of the purchase cost, and a flat fee. We also assumed that during the ownership time of the security, it is possible to obtain funds (such as dividends in the case of shares, interests in the case of bonds) or pay money (for example in the case of an increase in the corporate capital). Furthermore, each investor has the opportunity to sell his own securities and, in this case, he will receive a sum, but he will have to pay a charge to the chosen financial intermediary and a taxation on the capital gain or a percentage on the gain obtained from the title.

In this paper we also took into account that some financial securities have a length or a deadline. Therefore, inspired by reality, we supposed that if the security expires before the final term, then the investor receives the nominal value of the security and pays the tax in the event that there is a positive capital gain. We obtained the optimality conditions of the investors which have been characterized by a variational inequality and we studied some numerical examples.

In a future work, we intend to continue the study of this topic and, in particular, we will deal with short selling or financial transactions that consist in the sale of non-owned financial instruments with subsequent repurchase. In addition, we could examine the case of transfer of titles and we could also analyze the behavior of investors in the presence of the secondary market. The results in this paper add to the growing literature of operations research techniques for portfolio optimization modeling and analysis.

Acknowledgements. The research of the authors was partially supported by the research project "Modelli Matematici nell'Insegnamento-Apprendimento della Matematica" DMI, University of Catania. This support is gratefully acknowledged.

References

[1] S. Benati and R. Rizzi. A mixed-integer linear programming formulation of the optimal mean/Value-at-Risk portfolio problem. *European Journal of Operational Research,* v. 176: 423–434, 2007.

[2] D. Bertsimas, C. Darnell and R. Soucy. Portfolio construction through mixed-integer programming at Grantham, Mayo, Van Otterloo and company. *Interfaces,* 29(1), 49–66, 1999.

[3] F. Black and R. Litterman. Asset Allocation: combining investor views with market Equilibrium. *The Journal of Fixed Income,* 7–18, 1991.

[4] F. Black and R. Litterman. Portfolio optimization. *Financial Analysts Journal,* 48(5): 28–43, 1992.

[5] V. Caruso and P. Daniele. A network model for minimizing the total organ transplant costs. *European Journal of Operational Research,* 266: 652–662, 2018.

[6] G. Colajanni, P. Daniele, S. Giuffrè and A. Nagurney. Cybersecurity investments with nonlinear budget constraints and conservation laws: Variational equilibrium, marginal expected utilities, and Lagrange multipliers. *Intl. Trans. in Op. Res.,* 25: 1443–1464, 2018.

[7] D. H. Donald and J. Tobin. *Risk Aversion and Portfolio Choice.* John Wiley and Sons, Inc., 1967.

[8] S. Giuffrè, A. Maugeri and D. Puglisi. Lagrange multipliers in elastic-plastic torsion problem for nonlinear monotone operators. *J. Differential Equations,* 259: 817–837, 2015.

[9] S. Greco, B. Matarazzo and R. Slowinski. Beyond Markowitz with multiple criteria decision aiding. *Journal of Business Economics,* 83: 29–60, 2013.

[10] N. L. Jacob. A limited-diversification portfolio selection model for the small investor. *Journal of Finance,* 29(3): 847–856, 1974.

[11] D. Kahneman and A. Tversky. Prospect Theory: An analysis of decision under risk. *Econometrics,* 47(2): 263–291, 1979.

[12] H. Kellerer, R. Mansini and M. G. Speranza. Selecting portfolios with fixed costs and minimum transaction lots. *Annals of Operations Research,* 99: 287–304, 2000.

[13] H. Levy. Equilibrium in an imperfect market: A constraint on the number of securities in the portfolio. *American Economic Review,* 68(4): 643–658, 1978.

[14] J. Lintner. The valuation of risk assets and the selection of risky investments in stock portfolios and capital budgets. *Review of Economics and Statistics,* 47(1): 13–37, 1965.

[15] J. C. T. Mao. Essentials of portfolio diversification strategy. *Journal of Finance,* 25(5): 1109–1121, 1970.

[16] H. M. Markowitz. Portfolio selection. *The Journal of Finance,* 7: 77–91, 1952.

[17] H. M. Markowitz. *Portfolio Selection: Efficient Diversification of Investments.* John Wiley & Sons, Inc., New York, 1959.

[18] A. Maugeri and L. Scrimali. A new approach to solve convex infinite-dimensional bilevel problems: Application to the pollution emission price problem. *Journal of Optimization Theory and Applications,* 169: 370–387, 2016.

[19] R. C. Merton. Lifetime portfolio selection under uncertainty: The continuous-time case. *The Review of Economics and Statistics,* 51(3): 247–257, 1969.

[20] J. Mossin. Equilibrium in a capital asset market. *Econometrica,* 34: 768–783, 1966.

[21] J. Mossin. Optimal multiperiod portfolio policies. *The Journal of Business,* 41(2): 215–229, 1968.

[22] A. Nagurney. *Network economics: A variational inequality approach*, 2nd ed. (revised), Boston, Massachusetts: Kluwer Academic Publishers, 1999.

[23] A. Nagurney and K. Ke. Financial networks with intermediation: risk management with variable weights. *European Journal of Operational Research,* 172(1): 40–63, 2006.

[24] D. Roman, K. Darby-Dowman and G. Mitra. Mean-risk models using two risk measures: A multi-objective approach. *Quantitative Finance,* 7(4): 443–458, 2007.

[25] P. A. Samuelson. Lifetime portfolio selection by dynamic stochastic programming. *The Review of Economics and Statistics,* 51(3)(Aug.): 239–246.

[26] W. Sharpe. Capital Asset Prices: A Theory of Market Equilibrium Under Conditions of Risk. *Journal of Finance,* 1964.

[27] W. F. Sharpe. Imputing expected security returns from portfolio composition. *The Journal of Financial and Quantitative Analysis,* 9(3): 463–472, 1974.

[28] M. C. Steinbach. Markowitz revisited: Mean-variance models in financial portfolio analysis. *Society for Industrial and Applied Mathematics,* 43(1): 31–85, 2001.

[29] J. Tobin. Liquidity preference as behavior towards risk. *Review of Economic Studies,* XXV(2): 65–86, 1958.

Chapter 3

How Variational Rational Agents Would Play Nash: A Generalized Proximal Alternating Linearized Method

Antoine Soubeyran
Aix-Marseille School of Economics, Aix-Marseille University, CNRS and EHESS, Marseille 13002, France. `antoine.soubeyran@gmail.com`

João Carlos Souza
Federal University of Piauí, Brazil. `joaocos.mat@ufpi.edu.br`

João Xavier Cruz Neto
Federal University of Piauí, Brazil. `jxavier@ufpi.edu.br`

3.1 Introduction

In this chapter of the book, we want to show how variational rational agents (point 1, in Behavioral Sciences) would play Nash in alternation (point 2, in Game theory), giving rise to a generalized proximal alternating linearized method (point 3, in Variational analysis). Then, this paper clearly shows how Game theory, Variational analysis and the Variational rationality approach of human dynamics cross fertilize each other, that is, when behavioral applications (point 1) give new ideas to generalize existing theories (points 2 and 3) and when theories give new ideas to better understand applications and generalize their formulations.

The recent Variational rationality approach of human dynamics (Soubeyran [48, 49, 50, 51]) (VR) proposed a general model for human stay and change dynamics. This model first defines a worthwhile proximal dynamic as a worthwhile change or stay within the current period. A change being worthwhile when motivation to change rather than to stay is higher enough with respect to resistance to change rather than to stay. The definition of these terms requires a lot of work, given the huge number of theories, both of motivation and resistance to change in Psychology and, more generally, in Behavioral sciences. Then, it defines a worthwhile distal dynamic as a worthwhile transition, that is, a succession of worthwhile changes or stay. Finally, it defines ends (traps or desires) of worthwhile transitions. Traps are positions worthwhile to approach and reach from an initial position (aspiration points, dynamical aspect), but not worthwhile to leave (equilibria, static aspect). Desired ends (desires) refer to ends not worthwhile to leave when resistance to change is absent. The (VR) problem is to know when a worthwhile distal dynamic ends in a trap, or in a desired end.

Potential games in Game theory (G) refer to normal form games between interrelated agents whose common interests are larger than their rival interests. Formally, there exists a best reply potential function, such that each player makes "as if" he would maximize this potential function, given the actions chosen previously by the other players.

Proximal alternating algorithms in Variational Analysis (VA) consider Gauss-Seidel alternating algorithms where, at each step, the objective is optimized, acting on one variable, and leaving the other variables fixed at their previous levels. The regularization of such algorithms consists of adding, each step, a specific perturbation term relative to this step, in order to make the objective function more friendly (regular, smooth, concave, convex, ...).

The second section defines the most important category of non cooperative normal form games, that is potential games, then, it considers how to play Nash processes, and gives examples. Section 3 considers proximal alternating algorithms. Section 4 gives a short and simplified version of the variational rationality approach of human dynamics. The last section proposes a generalized proximal alternating linearized method where players, at each period, make worthwhile moves. In this way, perturbation terms, which represent disutilities of costs of moving, become asymmetric. They refer to squares of quasi distances instead of squares of an euclidian distance. The conclusion follows.

3.2 Potential Games: How to Play Nash?

Games can be static or dynamic.

3.2.1 Static Potential Games

Static games include non cooperative normal form games, cooperative games, games with complete and incomplete information. We will consider only non cooperative normal form games.

3.2.1.1 Non Cooperative Normal Form Games

Let us consider static games where, i) players choose their actions simultaneously (without knowledge of the others' choices) (static), ii) players receive payoffs that depend on their own actions and on actions of others (games), iii) each player's payoff is common knowledge among the players.

Normal form games. For simplification, consider two players $j \in \{1,2\} = J$. Let $x = (x^j, x^{\smallsetminus j}) \in X = X^j \times X^{\smallsetminus j}$ be the profile of their feasible actions. Let $g^j(\cdot) : x \in X \longmapsto g^j(x) = g^j(x^j, x^{\smallsetminus j}) \in \mathbb{R}$, $j \in \{1,2\}$ be the payoff function of player j. Then, the normal form of the game is $g^j(x) = g^j(x^j, x^{\smallsetminus j}), j = 1, 2$. A very important case refers to aggregative normal form games where the payoff of each player j depends on their own action x^j and of an aggregate of all actions, for example $x^j + x^{\smallsetminus j}$, when $X^j, X^{\smallsetminus j} \subset \mathbb{R}$. In this case $g^j(x) = g^j(x^j, x^j + x^{\smallsetminus j}), j = 1, 2$.

Non cooperative Nash games. The profile of actions $x_* = (x_*^j, x_*^{\smallsetminus j}) \in X$ is a Nash equilibrium of this game if no player will want to change, given that his rival will stay there. That is, x_* satisfies the inequalities $g^j(x_*^j, x_*^{\smallsetminus j}) \geq g^j(y^j, x_*^{\smallsetminus j})$ for all $y^j \in X^j$. This means that, within the same period, each player j solves the program $\max \left\{ g^j(y^j, x_*^{\smallsetminus j}), y^j \in X \right\}, j = 1, 2$. Let $B^j(x_*^{\smallsetminus j}) \subset X^j$ be the subset of optimal solutions of program j, that is, the best response correspondance of player j at $x_*^{\smallsetminus j}$. Then, a Nash equilibrium is a fixed point of the best response system $x_*^j \in B^j(x_*^{\smallsetminus j})$, $j = 1, 2$.

Games with strategic substitutes (complements). They are such that each best response correspondence includes a non increasing (non decreasing) selection (Dubey et al. [27]). Games can present both strategic substitutes and strategic complements. Let us see a list of very famous examples with one or several equilibria.

3.2.1.2 Examples

Example 1. Cournot duopoly. Consider two firms 1 and 2 competing as Cournot duopolists. That is, they produce quantities x^1 and x^2 of the same good, $x^1, x^2 \in X^1 = X^2 = X = [0, a] \subset \mathbb{R}_+$. If they propose these quantities for sale on the market, consumers will buy the overall quantity $x^1 + x^2$ at the common price $p(x^1, x^2) = a - (x^1 + x^2)$, where $a > 0$ is the maximum price of one unit of the good. Costs of production $\Gamma^j(x^j) = c^j x^j$ $j = 1, 2$, are linear, unit costs of production being $c^j > 0, j = 1, 2$. Thus, profit functions are $g^j(x^j, x^{\smallsetminus j}) = p(x^j, x^{\smallsetminus j}) x^j - c^j x^j$, $j = 1, 2$ where $\smallsetminus j = 2$ if $j = 1$ and the reverse. Suppose there are two periods and consider the second period. Given the quantities $x = (x^1, x^2)$ they have produced and sold in the initial period, each firm j chooses to produce and offer (sell) to the market a quantity y^j which maximizes his profit, making as if the rival will produce and offer to the market the same quantity $y^{\smallsetminus j} = x^{\smallsetminus j}$ as before. Then, in the second period, each firm j solves the program $\max \left\{ g^j(y^j, x^{\smallsetminus j}), y^j \in X \right\}, j = 1, 2$, that is, each firm j constructs a best response $x^j \in B^j(x^{\smallsetminus j})$ to the given anticipated production level $\widehat{x}^{\smallsetminus j} = x^{\smallsetminus j}$ of firm $\smallsetminus j$. In this example, best reponses $x^j = B^j(x^{\smallsetminus j}) = (1/2) \left[(a - c^j) - x^{\smallsetminus j} \right], j = 1, 2$ are decreasing in the quantity produced and sold by the rival. There is a unique Cournot-Nash equilibrium $x_* = (x_*^j, x_*^{\smallsetminus j})$, such that best responses intersect once, $x_*^j = B^j(x_*^{\smallsetminus j}), j = 1, 2$. When unit costs of production $c^j = c^j(x^{\smallsetminus j})$ depend on the production level of the rival, these externalities make best responses non monotone, giving the opportunity of several Cournot-Nash equilibria. See Monaco, Sabarwal [39].

Example 2. Bertrand duopoly. Consider two firms j and $\searrow j$ which produce and sell $q^j = D^j \in \mathbb{R}_+, j = 1, 2$ units of a differentiated good. Demand of good j, $D^j = D^j(p^j, p^{\searrow j}) = \alpha^j - \beta^j p^j + \gamma p^{\searrow j}$ decreases with the price of good j and increases with the price of the other good, $\searrow j$, with $\alpha^j, \beta^j, \gamma > 0$. Let $\Gamma^j(q^j) = c^j q^j, c^j > 0$, be the production cost of firm j, its unit cost being $c^j > 0$. Margins are $m^j = p^j - c^j \geq 0$. Then, the profit of firm j is $g^j(m^j, m^{\searrow j}) = p^j q^j - \Gamma^j(q^j) = m^j \left[r^j - \beta^j m^j + \gamma m^{\searrow j} \right]$, where $r^j = \alpha^j - \beta^j c^j + \gamma c^{\searrow j} > 0$. Firm's j best response $p^j = B^j(p^{\searrow j}) = (1/2\beta^j) \left[r^j + \gamma m^{\searrow j} \right]$ is increasing in the price $p^{\searrow j}$ of the rival firm. There is a unique Bertrand-Nash equilibrium.

Example 3. Partnership games. Two players of a team join their efforts $(x^j, x^{\searrow j}) \in X^j \times X^{\searrow j} = \mathbb{R}_+ \times \mathbb{R}_+$ to produce $f(x^j, x^{\searrow j}) = x^j + x^{\searrow j} + x^j x^{\searrow j}$ units of a good sold at the given unit price $p = 1$. They share the joint revenue $pf(x^j, x^{\searrow j})$, the share of each member j being $0 < s^j < 1, s^j + s^{\searrow j} = 1$. Costs of efforts are $\Gamma^j(x^j) = (1/2)(x^j)^2$. Then, the payoffs of each member of the team are $g^j(x^j, x^{\searrow j}) = s^j p f(x^j, x^{\searrow j}) - (1/2)(x^j)^2, j = 1, 2$. Best responses $B^j(x^{\searrow j}) = s^j \left[1 + x^{\searrow j} \right], j = 1, 2$ are increasing. There is a unique Partnership-Nash equilibrium.

Example 4. Coordination games. In this case, two players $j = 1, 2$, must work together in order to achieve a commonly desired outcome, but neither player will benefit from his efforts $x^j, x^{\searrow j} \in \mathbb{R}_+$ if his partner does not do his part. For example, consider two members $j = 1, 2$ of a team. Each of them produces one unit of a specific input $j = 1, 2$ that they assemble to produce one unit of a final good. The quality $q(x^j, x^{\searrow j}) = \min\{q^j, q^{\searrow j}\}$ of the final good is the minimum of the quality $q^j = x^j$, $q^{\searrow j} = x^{\searrow j} \in \mathbb{R}_+$ of the two inputs. The revenue of the team $r(x^j, x^{\searrow j}) = vq(x^j, x^{\searrow j}), v > 0$, increases with the quality of the final good. Costs of efforts are $\Gamma^j(x^j) = (c/2)x^j$, $0 < c < V$. Let $g^j(x^j, x^{\searrow j}) = (1/2)r(x^j, x^{\searrow j}) - (c/2)x^j = (1/2) \left[v \min\{x^j, x^{\searrow j}\} - c x^j \right]$, be the payoff of player $j = 1, 2$, who gets one half of the revenue. Then,

- if $0 < x^j \leq x^{\searrow j}, g^j(x^j, x^{\searrow j}) = (1/2)(v - c)x^j > 0$ gives

$$x^j = B^j(x^{\searrow j}) = x^{\searrow j},$$

- if $x^j \geq x^{\searrow j} > 0, g^j(x^j, x^{\searrow j}) = (1/2)(-c)x^j < 0$ gives

$$x^j = B^j(x^{\searrow j}) = x^{\searrow j}.$$

This shows that this game has an infinity of Nash equilibria $x_*^j = x_*^{\searrow j} = x_* > 0$. For different exemples of coordination games see Goeree, Holt [30].

Example 5. The rent seeking contest. Consider two agents j and $\searrow j$ who fight in order to capture a resource $V > 0$ (Tullock [54]). If they acquire amount of arms $x^j, x^{\searrow j} \in \mathbb{R}_+$, rival's j probability of winning is $p^j(x^j, x^{\searrow j}) = x^j/(x^j + x^{\searrow j})$. Let $\Gamma^j(x^j) = cq^j, c^j > 0$, be his investment cost in arms. Then, the payoff of rival j is $g^j(x^j, x^{\searrow j}) = V p^j(x^j, x^{\searrow j}) - cq^j, j = 1, 2$. Best response of rival j is $x^j = B^j(x^{\searrow j}) = \gamma \sqrt{x^{\searrow j}} - x^{\searrow j}$, which first increases on $[0, V/4c]$ and then, decreases on $[V/4c, V/c]$. There is a unique Tullock-Nash equilibrium $x_* = (x_*^j, x_*^{\searrow j}) = (V/4c, V/4c)$.

Example 6. Private provision of a public good: A family game. Let $j = 1, 2$ be a man and a woman who share their available time $T < +\infty$ between the time $0 \leq x^j \leq T$ spent with their children to improve their education and the residual time $T - x^j \geq 0$ spent for their hobbies. The level of education given to children $L = L(x^j, x^{\searrow j}) = x^j x^{\searrow j}$

increases with the time spent by both parents, is zero if one parent does not spend time with children, and the more one parent spends time with children, the more the education level increases if the other parent spends some added time. Utility $U(L) = L$ to improve the education level of children increases with this level. Utility to spend time in hobbies is $H(T-x^j) = T-x^j$. The weighted utility of each parent is $g^j(x^j,x^{\smallsetminus j}) = \alpha^j H(T-x^j) + \beta^j L(x^j,x^{\smallsetminus j})$, with $\alpha^j, \beta^j > 0$. Then, $g^j(x^j,x^{\smallsetminus j}) = \alpha^j(T-x^j) + \beta^j x^j x^{\smallsetminus j}$. Taking partial derivatives shows that, i) if $\partial g^j/\partial x^j = -\alpha^j + \beta^j x^{\smallsetminus j} > 0$, then $x^j = T$, ii) if $\partial g^j/\partial x^j = -\alpha^j + \beta^j x^{\smallsetminus j} < 0$, then $x^j = 0$ and, iii) if $\partial g^j/\partial x^j = -\alpha^j + \beta^j x^{\smallsetminus j} = 0$, then, $x^j \in [0,T]$. Let $0 < \gamma^j = \alpha^j/\beta^j < 1$. This implies that

$$x^j = B^j(x^{\smallsetminus j}) = \left\{ \begin{array}{l} 0 \text{ if } x^{\smallsetminus j} < \gamma^j \\ \in [0,T] \text{ if } x^{\smallsetminus j} = \gamma^j \\ T \text{ if } x^{\smallsetminus j} > \gamma^j \end{array} \right\}, j = 1,2.$$

Best responses are non decreasing. They cross three times. There are three Nash equilibria: $x_* = (x_*^j, x_*^{\smallsetminus j}) \in \{(0,0),(1,1),(\gamma^j,\gamma^{\smallsetminus j})\}$. Either parents spend no time with their children, spend full time, or share their time between education and hobbies. The two first cases can lead to divorce! This game is an original version of the well known private provision of public good game, including the tragedy of the common (Falkinger et al. [28]).

Examples 8. Supermodular games. Examples of games with strategic complements are supermodular games. In this case, when one player takes a higher action, the others want to do the same (payoffs functions have increased differences). Then, best replies are not decreasing. For references see [20, 36, 38, 53, 55].

3.2.1.3 Potential Games

In games of common interests, it is "as if" players maximize the same objective ([2, 40, 47, 57]). Potential games are noncooperative games for which there exist auxiliary functions, called best response potentials, such that the maximizers of these potentials are also Nash equilibria of the corresponding game ([2, 47]). Best responses potentials include exact, weighted, ordinal, and generalized ordinal potential games. For a survey, explicit definitions and examples of potential functions in each of these cases, see [2, 31].

A canonical example of a (normal form) best response potential game that we will consider in this paper is the following: The payoff of each player j, $g^j(x^j,x^{\smallsetminus j}) = \alpha^j u^j(x^j) + \beta^j S(x^j,x^{\smallsetminus j})$, $\alpha^j, \beta^j > 0$, have a weighted individual part $u^j(x^j)$ which depends on his own action $x^j \in X^j$, and a weighted joint payoff $S(x^j,x^{\smallsetminus j})$ which depends on his own actions and actions of others. The potential function is $\Psi(x^j,x^{\smallsetminus j}) = \gamma^j u^j(x^j) + \gamma^{-j} u^{\smallsetminus j}(x^{\smallsetminus j}) + S(x^j,x^{\smallsetminus j})$, where $\gamma^j = \alpha^j/\beta^j > 0$. This is a best reponse potential game because it is "as if" each player j maximizes the same objective: $\max\{g^j(y^j,x^{\smallsetminus j}), y^j \in X^j\} = \max\{\Psi(y^j,x^{\smallsetminus j}), y^j \in X^j\}$. This nice property comes from the separability assumption of each payoff between a joint and an individual payoff. Interactions between players depend of the coupling term $S(x^j,x^{\smallsetminus j})$. In the most general class of pseudo potential games, the best reply correspondence of each player is included in the best reply correspondence of a unique function, a pseudo potential. Then, it is also "as if" each player maximizes a common objective. For references see [35, 57]. For approximate solutions of potential games see [11].

3.2.2 Dynamic Potential Games

Dynamic games refer to extensive form games, open loop and close loop games, Markov games, and so on. We limit our attention to alternating games.

3.2.2.1 Alternating Moves and Delays

In this case, players $j = 1, 2$ play in alternation. Each period includes two stages, where,
 - in stage 1: Player j moves and player $\smallsetminus j$ stays. The change is $(x^j, x^{\smallsetminus j}) \curvearrowright (y^j, x^{\smallsetminus j})$, player j giving his best response $y^j \in \arg\max \left\{ g^j(z^j, x^{\smallsetminus j}), z^j \in X^j \right\}$.
 - in stage 2: Player j stays and player $\smallsetminus j$ moves. The change is $(y^j, x^{\smallsetminus j}) \curvearrowright (y^j, y^{\smallsetminus j})$, player $\smallsetminus j$ giving his best response $y^{\smallsetminus j} \in \arg\max \left\{ g^{\smallsetminus j}(y^j, z^{\smallsetminus j}), z^{\smallsetminus j} \in X^{\smallsetminus j} \right\}$. In this paper, we will consider the "argmin formulation" to better fit with a mathematically oriented audience.

If the normal form game is a potential game, with a potential function $\Psi(\cdot, \cdot)$, then, it is "as if" there is one player who maximizes this potential function, using a Gauss-Seidel alternating algorithm.

Stackelberg games. In this simple case, there is only one period, where the leader j plays first $x_*^j = \arg\max \left\{ g^j(z^j, B^{\smallsetminus j}(z^j)), z^j \in X^j \right\}$ and the follower $\smallsetminus j$ gives his best response $x^{\smallsetminus j} = B^{\smallsetminus j}(x_*^j) = (1/2) \left[(a - c^{\smallsetminus j}) - x_*^j \right]$ if we use the Cournot normal form game.

3.2.2.2 The "Learning How to Play Nash" Problem

Four central questions. In non cooperative game theory, a central question is to know how players learn to play Nash equilibria (Chen and Gazzale [20]). Examples of learning dynamics are Bayesian learning, fictitious play, Cournot best reply, adaptive learning, evolutionary dynamics and reinforcement learning. If we consider normal form games,
 - players can play simultaneously each period (as in fictitious play, Monderer and Shapley [40]). They first form beliefs over what each other player will do this period. These beliefs are usually the Cesaro mean of the actions played in the past by each other player. Then, they give a best reply to these beliefs. In this case, they follow a forward dynamic. Beliefs are updated each period, as a mean of the previous actions of each other player and their previous beliefs.
 - players can also play sequentially, moving in alternation, one player at each period, as seen previously. In this case, they follow a backward dynamic if the deviating player has observed (knows) what all other players have done in the past. Each period, the deviating player chooses a best reply, taking as given the actions played just before by the non deviating players.

In this dynamic context, the "learning to play Nash" problem poses four central questions: i) How do these learning dynamics converge to the set of Nash equilibria (positive convergence) or converge to a Nash equilibrium? ii) Does the process converge in finite time? iii) What is the speed of convergence and do plays converge gradually or abruptly? iv) How can constraints on the spaces of actions be included for each player?

Literature. In the literature, answers to the three first questions have been given in two main cases: Supermodular games and potential games,

A) Supermodular games. The following hypothesis gives convergence towards a Nash equilibrium; see Levin [36]. a) strategy spaces X^j are subsets of a $X = \mathbb{R}^m$, each of them being a complete sublattice, b) the payoff function $g^j(x^j, x^{\searrow j})$ of each player $j \in J$ is supermodular in $x^j \in X^j$ with increasing differences in (x^j, x^{-j}). Results concerning convergence in finite time are given only for supermodular games with a finite number of actions for each player (finite games). To our knowledge, there exists no result about convergence in finite time and about the speed of convergence for infinite supermodular games.

B) Potential games. In this context, the most general convergence results seem to be the following Jensen and Oyama [35]. Consider a best reply potential game with single valued continuous best reply functions and compact strategy sets. Then, any admissible sequential best reply path converges to a set of pure strategy Nash equilibria. This result concerns the so called "positive convergence to a set", i.e. sequences for which every convergent subsequence converges to a point of this set, in this setting, the set of Nash equilibria. Admissible sequences require that whenever card J successive periods have passed, all $h \in H$ players have moved. Results concerning finite time and speed of convergence are very rare. One recent exception is the interesting but restrictive case of fast convergence to nearly optimal solutions in potential games with ≥ 2 (see Awerbuch et al. [11]). These authors consider approximate solutions to a Nash equilibrium (with rate of deviations lower than one for the payoff of each player, instead of a negative rate of deviation). But their assumptions are quite specific: A γ bounded jump condition, a β nice hypothesis. Convergence occurs in exponential time. For zero sum games see Hofbauer and Sorin [34].

Notice that our paper provides the more general result about the speed of convergence.

3.3 Variational Analysis: How to Optimize a Potential Function?

The importance of potential games in Behavioral Sciences pushes us to pose the question: How do we optimize a potential function? A lot of answers have been given in Variational analysis and Optimization theory.

3.3.1 Minimizing a Function of Several Variables: Gauss-Seidel Algorithm

Consider the minimization of a function of two variables $\Psi(\cdot, \cdot) : (u, v) \in U \times V \longmapsto \Psi(u, v) \in \mathbb{R}$. The Gauss-Seidel method, also known as block coordinate descent method, solves in alternation the problems

$$u_{n+1} \in \arg\min\{\Psi(u, v_n) : u \in U\} \text{ and } v_{n+1} \in \arg\min\{\Psi(u_{n+1}, v) : v \in V\}.$$

Convergence results for the Gauss-Seidel method can be found in [10, 17, 44, 45, 52, 60]. An important assumption necessary to prove convergence is that the minimum in each step is uniquely attained, see e.g., [60]. Otherwise, as shown in [45], the method may cycle indefinitely without converging. However very few general results ensure that the sequence $\{(u_n, v_n)\}$ converges to a global minimizer, even for strictly convex functions.

3.3.2 The Problem of Minimizing a Sum of Two Functions Without a Coupling Term

This problem is at the heart of mathematical optimization research, starting with the pioneering and early works [8, 17]. Interest for solving optimization problems involving sum of convex functions comes from applications in approximation theory (see [56]), image reconstruction (see [21, 26]), statistics (see [25, 33]), partial differential equations and optimal control (see [37, 58]), signal/image processing (see [21]), machine learning (see [43]), etc.

Example: convex feasibility problems. A well-known convex optimization splitting algorithm in signal processing is POCS (Projection onto convex sets); see Combettes and Pesquet [24]. This algorithm helps to recover/synthesize a signal, simultaneously satisfying several convex constraints. Such a problem can be formalized as min $\{f^1(x) + \cdots + f^m(x), x \in \mathbb{R}^m\}$, by letting each function f^j be the indicator function of a nonempty closed convex set C^j modeling a constraint. As shown by Acker and Prestel [1], this problem reduces to the classical convex feasibility problem, find $x \in \cap_{j=1}^m C^j$. Consider the projection $P_C(x)$ of $x \in X = \mathbb{R}^m$ onto a nonempty closed convex set $C \subset X$. It is the solution to the problem: $P_C(x) = \min \left\{ \delta_C(z) + (1/2) \|z - x\|^2, z \in X \right\}$, where $\delta_C(\cdot)$ is the indicator function of C. For simplification, take $m = 2$. Then, the POCS algorithm [56] successively activates each set $C^j, j = 1, 2$, individually by means of its projection operator $P_{C^j}(\cdot)$. Then, for any x_0 in X, the sequence $\{x_n\}_{n \in \mathbb{N}}$ obtained by alternatively projecting on C_1 and on C_2, namely $u_n = (P_{C_1} P_{C_2})^n u_0$, strongly converges to the projection of u_0 onto $C_1 \cap C_2$.

Acker and Prestel [1] generalized this Von Neumann algorithm by considering the proximal operator

$$P_f(x) = \min \left\{ f(z) + (1/2) \|z - x\|^2, z \in X \right\}.$$

This leads to the proximal alternating algorithm for $m = 2$,

$$u_{n+1} = \arg\min \left\{ f^1(u) + (1/2) \|u - v_n\|^2, u \in X \right\}$$

and

$$v_{n+1} = \arg\min \left\{ f^2(v) + (1/2) \|v - u_{n+1}\|^2, v \in X \right\},$$

where $f^j, j = 1, 2$ are two closed convex proper functions on the Hilbert space X.

Then, the sequence $\{u_n, v_n\}_{n \in \mathbb{N}}$ weakly converges to a solution of the joint minimization problem on $X \times X$,

$$\min \left\{ f^1(u) + f^2(v) + (1/2) \|u - v\|^2, u, v \in X \right\},$$

if we assume that the set of minimum points is non empty.

Several variants exist. See [13] for proximal gradient algorithms (forward, backward splitting algorithms, ...) with applications to signal recovery.

3.3.3 Minimizing a Sum of Functions With a Coupling Term (Potential Function)

3.3.3.1 Mathematical Perspective Proximal Regularization of a Gauss-Seidel Algorithm

At the mathematical level, the present paper focuses the attention on the minimization of a sum of functions $\Psi(x^1,...,x^j,...,x^m) = \sum_{j=1}^m f^j(x^j) + S(x)$, where $x = (x^1,...,x^j,...,x^m) \in X$, with a coupling term $S(\cdot) : x \in X \longmapsto S(x) \in \mathbb{R}$ and individual terms $f^j(\cdot) : x^j \in X^j \longmapsto f^j(x^j) \in \mathbb{R}$, $j = 1,...,m$. For simplification, we consider the case $m = 2$, $x = (u,v) \in X$, and $\Psi(u,v) = h(u) + k(v) + S(u,v)$, with $h(u) = f^1(u)$ and $k(v) = f^2(v)$.

Auslender [9] seems to be the first to solve this problem using a proximal regularization of the Gauss-Seidel alternating minimization algorithm, that is, if $m = 2$ for simplification,

$$u_{n+1} \in \arg\min\left\{\Psi(u,v_n) + (\lambda_n/2)\|u - u_n\|^2, u \in U\right\}$$

and

$$v^{n+1} \in \arg\min\left\{\Psi(u_{n+1},v) + (\mu_n/2)\|v - v_n\|^2, u \in U\right\},$$

where $\lambda_n, \mu_n > 0$ are positive real numbers.

Auslender [9] considered the convex case, where all these functions are proper, convex and closed. See also [15, 16] for partial proximal minimization algorithms for Convex programming, and [12, 32, 59] among others.

3.3.3.2 Game Perspective: How to Play Nash in Alternation

Later, Attouch et al. [4] seem to have been the first to remark that a function $\Psi(x) = \sum_{j=1}^m f^j(x^j) + S(x)$ is a best response potential for players $j = 1,2,...,m$ playing a Nash non cooperative game. They introduced a new class of alternating minimization algorithms with a coupling term $S(x)$ and costs to move. The introduction, each period n, of non autonomous disutilities of costs to move $\Gamma_n^j(x^j,y^j) \in \mathbb{R}_+$ from $x^j \in X^j$ to $y^j \in X^j$ for each player j comes from the recent Variational approach of human dynamics (Soubeyran [48, 49]). See, in section 4, a short presentation of the (VR) Variational rationality approach of human dynamics which justifies the introduction of costs of moving, where motivation to change (to be defined precisely) balances more or less, each move, resistance to change (to be defined). More explicitly, they consider two players $j = 1,2$, their respective actions $x^j = u \in U = X^j$ and $x^{\smallsetminus j} = v \in V = X^{\smallsetminus j}$, their payoffs $g^j(x^j, x^{\smallsetminus j}) = h(u) + S(u,v)$ and $g^{\smallsetminus j}(x^j, x^{\smallsetminus j}) = k(u) + S(u,v)$ and the best response potential function $\Psi(u,v) = h(u) + k(v) + S(u,v), u \in U, v \in V$ and, each period $n \in \mathbb{N}$, disutilities of costs to move $\Gamma_n^j(x^j,y^j) = (\lambda_n/2)C_U(u_n,u)^p$, and $\Gamma_n^{\smallsetminus j}(x^{\smallsetminus j},y^{\smallsetminus j}) = (\mu_n/2)C_V(v_n,v)^p \in \mathbb{R}_+$, $j = 1,2$, for $p = 1$ or $p = 2$. In this case, players move in alternation, giving their best response to the action done previously by the other player, that is, they solve in alternation, a potential game with costs to move (inertial game),

$$u_{n+1} \in \arg\min\{h(u) + S(u,v_n) + (\lambda_n/2)C_U(u_n,u)^p, u \in U\}$$

and

$$v^{n+1} \in \arg\min\{k(v) + S(u_{n+1},v) + (\mu_n/2)C_V(v_n,v)^p, v \in V\},$$

that is,
$$u_{n+1} \in \arg\min\{\Psi(u,v_n) + (\lambda_n/2)C_U(u_n,u)^p, u \in U\}$$
and
$$v^{n+1} \in \arg\min\{\Psi(u_{n+1},v) + (\mu_n/2)C_V(v_n,v)^p, v \in V\}.$$

The variational rationality approach (Soubeyran [48, 49]) of human behaviors suggested two different lists of hypothesis:

i) if resistance to change is high, $p = 1$ (see section 4), the list of hypotheses is weak: Spaces of action are complete metric spaces, payoff functions $h(\cdot)$ and $k(\cdot)$ are proper, bounded below, semicontinuous functions, the coupling term $S(\cdot,\cdot)$ is proper, lower semicontinuous, and costs of moving are continuous, and bounded from below by distance functions.

ii) if resistance to change is low, $p = 2$, the list of hypotheses is much more restrictive: Spaces of actions are Hilbert spaces, payoff functions $h(\cdot)$ and $k(\cdot)$ are proper, bounded below, semicontinuous functions and convex, the coupling term $S(\cdot,\cdot)$ and costs of moving are quadratic.

3.3.3.3 Cross Fertilization Between Game and Mathematical Perspectives

This initial paper (Attouch et al. [4]) starts a cross fertilization process between game theory (potential games) and variational analysis (prox-regularization of Gauss-Seidel algorithms), using the (VR) approach of human dynamics (Soubeyran [48, 49]). More precisely, it motivates several other papers, with two different perspectives, considering two players with different spaces of actions $(u,v) \in U \times V$, different payoffs $h(u)$ and $k(v)$, different coupling terms $S(u,v)$, and different costs of moving $C(u,v) \in \mathbb{R}_+$, in the particular case where $C(u,u) = 0$.

A) a behavioral game perspective. Attouch et al. [4] motivates the following three papers where,
- in Attouch et al. [5], U and V are Hilbert spaces (finite or not), $h(\cdot)$ and $k(\cdot)$ are proper, lower semicontinuous, closed convex functions, the coupling term $S(u,v)$ is a non negative quadratic form, hence convex, and costs of moving are $\Gamma_U(u,u') = \|u'-u\|_U^2$, $\Gamma_V(v,v') = \|v'-v\|_V^2$.
- in Flores Bazan et al. [29], actions spaces are complete metric spaces, players have variable (reference-dependent) preferences which balance, each period, their motivation and resistance to change, defined in the context of the variational rationality approach (Soubeyran [48, 49]), motivation to change functions are bounded above and superlinear in the action of each player, and resistance to change functions are higher than a distance function in the action of each player.
- in Cruz Neto et al. [23], U and V are finite dimensional complete Hadamard manifolds, $h(\cdot)$ and $k(\cdot)$ are proper, lower semicontinuous, $S(\cdot,v_0)$ is proper, $S(\cdot,\cdot)$ is bounded below, lower semicontinuous and KL, ∇S is Lipschitz continuous on bounded subsets of $M \times N$, and costs of moving are the square of not too asymmetric quasi distances on the manifolds U and V, with $\lambda_n, \mu_n > 0$ belonging to an interval of R_+.

B) a mathematical perspective. Attouch et al. [4] motivates two papers relative to the proximal regularization of a Gauss-Seidel algorithm and forward-backward splitting methods,

- in Attouch et al. [6], $U = V = X = \mathbb{R}^m$, $h(\cdot)$ and $k(\cdot)$ are proper, lower semi-continuous, the coupling term $S(\cdot,\cdot)$ is bounded below and C^1, $S(.,v_0)$ is proper, ∇S is Lipschitz continuous on bounded subsets, costs of moving are $\Gamma_n^U(u,u') = (\lambda_n/2)\|u'-u\|_X^2$, $\Gamma_n^V(v,v') = (\mu_n/2)\|v'-v\|_X^2$, with $\lambda_n, \mu_n > 0$ belonging to an interval of R_+, and $\Psi(\cdot,\cdot)$ have the Kurdyka-Łojasiewicz property (KL).

- in Attouch et al. [7], the authors considered forward-backward splitting methods as specific cases of regularized Gauss-Seidel algorithms in the KL context.

3.4 Variational Rationality: How Human Dynamics Work?

3.4.1 Stay/stability and Change Dynamics

The (VR) variational rationality approach (Soubeyran [48, 49, 50, 51]) modelizes and unifies a lot of different models of stay and change dynamics which appeared in Behavioral Sciences (Economics, Management Sciences, Psychology, Sociology, Political Sciences, Decision theory, Game theory, Artificial Intelligence, etc.). Stays refer to static execution/exploitation phases, that is, temporary repetitions of the same action, like temporary habits, routines, rules and norms, etc. Changes represent dynamic exploration phases, where agents, via search, exploration, learning and training, build new capabilities, conserve others, and stop using the last ones, forming and breaking habits and routines, changing doings (actions), havings and beings, etc. This dynamic approach considers entities (an agent, an organization or several interacting agents) which are, at the beginning of the story, in an undesirable initial position, and are unable to reach a final desired position immediately (in one period). The goal of this approach is to examine the transition problem: How such entities can find, build and use an acceptable and feasible transition which is able to overcome a lot of intermediate obstacles, difficulties and resistance to change, with not too many intermediate sacrifices and high enough intermediate satisfactions in order to sustain motivation to change and persevere until reaching the final desired position. This (VR) approach admits a lot of variants, based on the same short list of general principles and concepts. The five main concepts refer to changes and stays, worthwhile changes and stays, worthwhile transitions, variational traps, worthwhile to approach and reach but not worthwhile to leave, and desires. A stay and change dynamic refers to a succession of periods, where $n+1$ is the current period and n is the previous period, where $x = x_n \in X$ can be a previous bundle of activities (doings), havings or beings and $y = x_{n+1} \in X$ can be a current bundle of activities (doings), havings or beings. Within the current period, $x \curvearrowright y, y \neq x$ represents a single change from $x = x_n \in X$ to $y = x_{n+1} \in X$. A single stay at x is $x \curvearrowright y, y = x$.

3.4.2 Worthwhile Changes

The (VR) approach starts with a general definition of a worthwhile change: A change is worthwhile if motivation to change rather than to stay is "high enough" with respect to resistance to change rather than to stay. This definition allows a lot of variants, as many variants as we can find for the definitions of motivation (more than one hundred theories/aspects of motivations exist in Psychology), the definition of resistance (which includes a lot of different aspects) and what it means to be "high enough" (see Soubeyran [48, 49, 50, 51]). Let us give a very simple formulation of a worthwhile change, first for an agent, then, for two players in the context of a normal form game.

3.4.2.1 One Agent

Consider an agent who changes from having done a bundle of activities $[0,x]^{def} = x \subset \mathbb{R}_+$ in the previous period to carrying out a bundle of activities $[0,y]^{def} = y \subset \mathbb{R}_+$ in the current period, to improve his utility from $g(x) \in \mathbb{R}$ to $g(y) \in \mathbb{R}$. Activities are ranked in a given order, from zero (first activity to be done) to the last one, x or y. Then, changing from x to y, $y > x$ ($y < x$) means that, in the current period, the agent starts doing (stops doing) $y - x > 0$ ($x - y > 0$) activities. For simplification, each activity takes the same amount of time. Then, given a change $x = x_n \curvearrowright y = x_{n+1}$, the variational rationality approach defines the following concepts:

1) motivation to change rather than to stay $M(x,y) = U[A(x,y)] \in \mathbb{R}_+$ is the utility $U[A]$ of advantages to change $A = A(x,y) \geq 0$.

2) resistance to change rather than to stay $R(x,y) = D[I(x,y)] \in \mathbb{R}_+$ is the disutility $D[I]$ of inconveniences to change $I = I(x,y) \geq 0$.

We suppose that $U[\cdot] : A \in \mathbb{R}_+ \longmapsto U[A] \in \mathbb{R}_+$ and $D[\cdot] : I \in \mathbb{R}_+ \longmapsto D[I] \in \mathbb{R}_+$ are strictly increasing and zero at zero.

3) advantages to change rather than to stay refer, if they exist, to an improvement in utility $A(x,y) = g(y) - g(x) \geq 0$.

The mapping $g(\cdot) : y \in X \longmapsto g(y) \in R$ represents a utility (level of satisfaction) that the agent wants to increase. If $g^* = \sup\{g(z), z \in X\} < +\infty$ is the highest utility level the agent can hope to reach, then, $f(y) = g^* - g(y) \geq 0$ refers to the unsatisfaction level of this agent that he wants to decrease. Then, $A(x,y) = g(y) - g(x) = f(x) - f(y)$.

4) inconveniences to change rather than to stay are, if non negative, $I(x,y) = C(x,y) - C(x,x) \geq 0$.

The mapping $C(\cdot,\cdot) : (x,y) \in X.X \longmapsto C(x,y) \in \mathbb{R}_+$ modelizes a cost of moving from having done the bundle of activities $x =^{def} [0,x]$ in the previous period to doing the bundle of activities $y =^{def} [0,y]$ in the currend period. Costs of starting $y - x > 0$ new activities are $C(x,y) = \rho_+(y-x)$, where $\rho_+ > 0$ is the cost of being able to do and doing a new activity. Costs of stopping $x - y > 0$ previous activities are $C(x,y) = \rho_\searrow(x-y)$, where $\rho_\searrow > 0$ is the cost of stopping one of the previous activities. Costs of continuing the $y = x$ previous activities are $C(x,x) = \rho_= x$, where $\rho_= \geq 0$ is the cost of being able to do and repeat a previous activity. For simplification (this will require too long a discussion), we will suppose that $\rho_= = 0$. Then, $C(x,x) = 0$, and inconveniences to change are

$$I(x,y) = C(x,y) = \left\{ \begin{array}{l} \rho_+(y-x), \text{ if } y \geq x \\ \rho_\searrow(x-y), \text{ if } y \leq x \end{array} \right\} \geq 0.$$

Notice that the function $q(\cdot) : (x,y) \in X \times X \longmapsto I(x,y) = q(x,y) \in \mathbb{R}_+$ is a quasi distance on $X = \mathbb{R}_+$. More precisely, a quasi distance is a mapping $q : X \times X \to \mathbb{R}_+$ satisfying:

1. For all $x, y \in X$, $q(x,y) = q(y,x) = 0 \Leftrightarrow x = y$;
2. For all $x, y, z \in X$, $q(x,z) \leq q(x,y) + q(y,z)$.

Therefore, metric spaces are quasi metric spaces satisfying the symmetric property $q(x,y) = q(y,x)$. Quasi distances are not necessarily convex or differentiable. More examples of quasi distances can be found in Moreno et al. [42] and references therein.

A worthwhile change from x to y is such that motivation to change rather to stay $M(x,y)$ is higher enough with respect to resistance to change rather than to stay, that is $M(x,y) \geq \xi R(x,y)$. What is "high enough" is defined by the size of $\xi > 0$. Note that, in this simple formulation of the (VR) approach, to stay is worthwhile because $A(x,x) = I(x,x) = 0$ for all $x \in X$.

The present paper proposes a simple linear quadratic formulation where $M = U[A] = A$, $R = D[I] = I^2$. See Soubeyran [48, 49, 50, 51] for more general cases. In this context, a change $x \curvearrowright y$ is worthwhile if advantages to change are high enough with respect to resistances to change, i.e., if $C(x,x) = 0$,
$$A(x,y) = g(y) - g(x) \geq \xi C(x,y)^2, \text{ or } A(x,y) = f(x) - f(y) \geq \xi C(x,y)^2.$$
These worthwhile to change conditions are equivalent to $g(y) - \xi C(x,y)^2 \geq g(x)$ or $f(y) + \xi C(x,y)^2 \leq f(x)$.

We are now in a good position, taking advantage of this specific example where $U[A] = A$, and $D[I] = I^2$, to show how a worthwhile change can be seen as a proximal step of an inexact proximal algorithm.

Let $P_\xi(y/x) = g(y) - \xi C(x,y)^2$ be the proximal satisfaction of the agent that he wants to increase, which is the difference between his current satisfaction $g(y)$ and his weighted resistance to change $R(x,y) = C(x,y)^2$.

Let $Q_\xi(y/x) = f(y) + \xi C(x,y)^2$ be the proximal unsatisfaction of the agent that he wants to decrease, which is the sum of his current unsatisfaction $f(y)$ and his weighted resistance to change $R(x,y)$.

Then, in a linear quadratic setting, a change is worthwhile if, moving from x to y, the proximal satisfaction increases, $P_\xi(y/x) \geq P_\xi(x/x)$, and the proximal unsatisfaction decreases, $Q_\xi(y/x) \leq Q_\xi(x/x)$.

In this behavioral context, the term proximal satisfaction (unsatisfaction) refers to a "short term" or per period satisfaction (unsatisfaction), net of resistance to change rather than to stay. We see that this behavioral terminology coincides with the mathematical one, relative to a step of a proximal algorithm, where the regularization term refers to resistance to change.

3.4.2.2 Two Interrelated Agents

Consider two players j and $\smallsetminus j$ and expected Nash moves, that is, unilateral expected deviations from the previous position $x = (x^j, x^{\smallsetminus j})$. In this case, each player j considers, ex ante, a move $(x^j, x^{\smallsetminus j}) \curvearrowright (y^j, x^{\smallsetminus j})$, while, ex post, players moves are $x = (x^j, x^{\smallsetminus j}) \curvearrowright (y^j, y^{\smallsetminus j}) = y$. For a normal form game with two players, externalities (reciprocal influences) can be located both in the satisfaction/utility function $g^j(y^j, x^{\smallsetminus j})$ and in costs of moving $C^j(x,y)$. For simplification, we will suppose no influences of player $\smallsetminus j$ on the costs of moving of player j, that is, $C^j(x,y) = C^j(x^j, y^j)$. This is the case when, on the example, unit costs of moving $\rho_+^j(y)$ and $\rho_-^j(y)$ are independent of y. Then, ex ante, a worthwhile Nash move is such that

$$g^j(y^j, x^{\smallsetminus j}) - g^j(x^j, x^{\smallsetminus j}) = f^j(x^j, x^{\smallsetminus j}) - f^j(y^j, x^{\smallsetminus j}) \geq \xi^j \left[C^j(x^j, y^j) \right]^2,$$

with $\xi^j > 0$, for $j = 1, 2$.

Proximal satisfaction and unsatisfaction of player j are

$$P^j_{\xi^j}(y/x) = g^j(y) - \xi^j R^j(x, y) = g^j(y) - \xi^j C^j(x^j, y^j)^2$$

and

$$Q^j_{\xi^j}(y/x) = f^j(y) + \xi^j R^j(x, y) = f^j(y) + \xi^j C^j(x^j, y^j)^2,$$

respectively.

3.4.3 Worthwhile Transitions

One agent. A transition is a succession of moves $x_0 \curvearrowright x_1 \curvearrowright \ldots \curvearrowright x_n \curvearrowright x_{n+1} \curvearrowright$, where, each period, a move can be a change $x_{n+1} \neq x_n$ or a stay $x_{n+1} = x_n$. A worthwhile transition is such that each successive change is worthwhile, i.e.,

$$g(x_{n+1}) - g(x_n) \geq \xi_{n+1} C(x_n, x_{n+1})^2, \quad n \in \mathbb{N},$$

that is,

$$g(x_{n+1}) - \xi_{n+1} C(x_n, x_{n+1})^2 \geq g(x_n), \quad n \in \mathbb{N},$$

or,

$$f(x_n) - f(x_{n+1}) \geq \xi_{n+1} C(x_n, x_{n+1})^2, \quad n \in \mathbb{N},$$

that is,

$$f(x_{n+1}) + \xi_{n+1} C(x_n, x_{n+1})^2 \leq f(x_n), \quad n \in \mathbb{N}.$$

This clearly shows that exact and inexact proximal algorithms are specific examples of worthwhile transitions where $\xi_{k+1} = 1/2\lambda_k > 0$. A worthwhile change is exact if $x_{n+1} \in \arg\min \{ Q_{\xi_{n+1}}(y/x_n), y \in X \}, n \in \mathbb{N}$. Then, $Q_{\xi_{n+1}}(x_{n+1}/x_n) \leq Q_{\xi_{n+1}}(y/x_n)$, for all $y \in X$. Taking $y = x_n$ gives $f(x_{n+1}) + \xi_{n+1} C(x_n, x_{n+1})^2 \leq f(x_n)$. An inexact worthwhile change is any worthwhile change "close enough" to an exact worthwhile change, where the term "close enough" can have several different interpretations, depending of chosen reference points and frames.

The (VR) approach defines, starting from a position $x \in X$, a set of worthwhile changes, including a stay, as $W_\xi(x) = \{ y \in X, M(x, y) \geq \xi R(x, y) \} \subset X$, with $\xi > 0$. In our simple example, $x \in W_\xi(x)$ for all $x \in X$ and

$$W_\xi(x) = \left\{ y \in X, \; g(y) - g(x) = f(x) - f(y) \geq \xi C(x, y)^2 \right\}.$$

Then, a worthwhile transition is such that $x_{n+1} \in W_{\xi_{n+1}}(x_n), n \in \mathbb{N}$.

Two interrelated agents. Nash worthwhile transitions are such that

$$g^j(x^j_{n+1}, x^{\smallsetminus j}_n) - g^j(x^j_n, x^{\smallsetminus j}_n) = f^j(x^j_n, x^{\smallsetminus j}_n) - f^j(x^j_{n+1}, x^{\smallsetminus j}_n) \geq \xi^j_{n+1} C^j(x^j_n, x^j_{n+1})^2,$$

$j = 1, 2, n \in \mathbb{N}$, that is,

$$g^j(x_{n+1}^j, x_n^{\searcher j}) - \xi_{n+1}^j C^j(x_n^j, x_{n+1}^j)^2 \geq g^j(x_n^j, x_n^{\searcher j}),$$

or

$$f^j(x_{n+1}^j, x_n^{\searcher j}) + \xi_{n+1}^j C^j(x_n^j, x_{n+1}^j)^2 \leq f^j(x_n^j, x_n^{\searcher j}),$$

$j = 1, 2, n \in \mathbb{N}$.

Then, in our simple example where, in particular, $R^j(x,y) = R^j(x^j, y^j) = C^j(x^j, x^j)^2$, the (VR) approach defines, for each player $j = 1, 2$, starting from a profile of activities $x = (x^j, x^{\searcher j}) \in X$, an individual set of worthwhile Nash changes

$$W_{\xi^j}^j(x) = \left\{ (y^j, x^{\searcher j}) \in X, M^j\left[(x^j, x^{\searcher j}), (y^j, x^{\searcher j})\right] \geq \xi^j R^j(x^j, y^j) \right\} \subset X.$$

Then, a Nash worthwhile transition is such that,

$$x_{n+1}^j \in W_{\xi_{n+1}^j}^j(x_n) = \left\{ y^j \in X^j, g^j(y^j, x_n^{\searcher j}) - \xi_{n+1}^j C^j(x_n^j, y^j)^2 \geq g^j(x_n^j, x_n^{\searcher j}) \right\},$$

or

$$x_{n+1}^j \in W_{\xi_{n+1}^j}^j(x_n) = \left\{ y^j \in X^j, f^j(y^j, x_n^{\searcher j}) + \xi_{n+1}^j C^j(x_n^j, y^j)^2 \leq f^j(x_n^j, x_n^{\searcher j}) \right\},$$

$n \in \mathbb{N}$, $j = 1, 2$, where $x_n = (x_n^j, x_n^{\searcher j}) \in X$.

3.4.4 Ends as Variational Traps

Let us consider a worthwhile transition.

3.4.4.1 One Agent

In this case, a variational trap $x_* \in X$ of a worthwhile transition $x_{n+1} \in W_{\xi_{n+1}}(x_n)$, $n \in \mathbb{N}$ is both,

i) an aspiration point $x_* \in W_{\xi_{n+1}}(x_n), n \in \mathbb{N}$, worthwhile to reach from any position of the transition,

ii) a stationary trap $W_{\xi_*}(x_*) = \{x_*\}$, where it is not worthwhile to move to any other position $y \neq x_*$, given that the worthwhile to change ratio tends to a limit, $\xi_{n+1} \to \xi_* > 0$, $n \to +\infty$, and finally,

iii) worthwhile to approach, i.e., which converges to the aspiration point.

More explicitly, in our simple example, with one agent, x_* is a variational trap if,

i) $A(x_n, x_*) = g(x_*) - g(x_n) = f(x_n) - f(x_*) \geq \xi_{n+1} R(x_n, x_*) = \xi_{n+1} C(x_n, x_*)^2$, $n \in \mathbb{N}$,

ii) $A(x_*, y) = g(y) - g(x_*) = f(x_*) - f(y) < \xi_* R(x_*, y) = \xi_* C(x_*, y)^2$, $n \in \mathbb{N}$, for $y \neq x_*$,

iii) it is a limit point of the worthwhile transition, i.e., $x_n \to x_*$, $n \to +\infty$.

A stationary variational trap does not require to be an aspiration point.

3.4.4.2 Two Interrelated Agents

In this game situation, a variational trap $x_* = (x_*^j, x_*^{\searrow j}) \in X^j \in X^{\searrow j}$ of a Nash worthwhile transition

$$x_{n+1}^j \in W_{\xi_{n+1}^j}^j(x_n), \quad j = 1,2, \, n \in \mathbb{N},$$

is both,

i) an aspiration point $x_*^j \in W_{\xi_{n+1}^j}^j(x_n)$, $n \in \mathbb{N}$, for $j = 1, 2$.

ii) a stationary trap $W_{\xi_*^j}^j(x^*) = \{x^*\}$, for $j = 1, 2$, given that the worthwhile ratios ξ_{n+1}^j tend to a limit, $\xi_{n+1}^j \to \xi_*^j > 0, n \to +\infty, j = 1, 2$.

iii) a limit point of the worthwhile transition, i.e., $x_n \to x_*, n \to +\infty$.

3.5 Computing How to Play Nash for Potential Games

What we do in this last section is the following:

On the behavioral side, we consider two main points for agents playing a potential game in alternation. When agents move (dynamical aspects),

i) inertia matters. That is, agents take care of asymmetric costs to move;

ii) each period, agents do not know well their reference dependent payoffs, which change when other agents move. Hence, they must construct, each period, simple models of their current payoffs. The simpler model being a linear evaluation of each payoff in the current position (profile of actions).

On the mathematical side, we generalize the Palm algorithm (Bolte et al. [18]), replacing regularization terms (squares of a the Euclidian distance) by costs of moving (square of quasi distances). This modification is a necessity to modelize agents playing a potential game in alternation, because Palm algorithm makes "as if" players have symmetric costs to move. This is unrealistic. What is very realistic is that it is "as if" players linearize their common payoff $S(\cdot, \cdot)$, making their paper a great paper.

Then, convergence to a Nash equilibrium becomes both practical (from a computational point of view) and realistic if we consider the square of asymmetric costs to move as regularization terms instead of the squares of the Euclidian norm.

3.5.1 Linearization of a Potential Game with Costs to Move as Quasi Distances

Consider two agents who play in alternation the potential game with costs to move defined in Section 3,

$$u_{n+1} \in \arg\min \left\{ \Psi(u, v_n) + \frac{\lambda_n}{2} C_U(u_n, u)^2, u \in U = \mathbb{R}^l \right\}$$

$$v^{n+1} \in \arg\min \left\{ \Psi(u_{n+1}, v) + \frac{\mu_n}{2} C_V(v_n, v)^2, v \in V = \mathbb{R}^m \right\},$$

where $\Psi(u,v) = h(u) + S(u,v) + k(v)$, $x = (u,v) \in X = U \times V = \mathbb{R}^l \times \mathbb{R}^m$ and $c_n, d_n > 0$ are positive real numbers. In section 3, we show how payoffs of player j and $-j$ are $h(u) + S(u,v)$ and $k(v) + S(u,v)$. This allows them to make as if they maximize $\Psi(u,v) = h(u) + S(u,v) + k(v)$, player j choosing u and player $-j$ choosing v, in alternation.

Here, following the (VR) approach, we suppose that costs to move $C(u,u') = q_U(u,u')$ and $C(v,v') = q_V(v,v')$ refer to quasi distances in $U = \mathbb{R}^l$ and $V = \mathbb{R}^m$. In this case, costs to move are asymmetric, as required.

At the mathematical level, this game generalizes the Gauss-Seidel alternating algorithm,

$$u_{n+1} \in \arg\min \left\{ \Psi(u, v_n) + \frac{c_n}{2} \|u - u_n\|^2, u \in U \right\}$$

and

$$v_{n+1} \in \arg\min \left\{ \Psi(u_{n+1}, v) + \frac{d_n}{2} \|v - v_n\|^2, u \in U \right\},$$

taking squares of costs to move instead of squares of the Euclidian norm as regularization terms.

When regularization terms are squares of the Euclidian norm, costs to move $C_U(u_n, u) = \|u - u_n\|$ and $C_V(v_n, v) = \|v - v_n\|$ are symmetric. Then, they cannot refer to costs to move which are asymmetric. Hence, the interpretation of a regularized Gauss-Seidel alternating algorithm, in terms of a game with costs to move, is not valid.

Moreover, as the (VR) approach does, we suppose that agents are bounded rational. They do not know well their payoff functions. For simplification, we suppose that they know well their individual payoffs $h(\cdot)$ and $k(\cdot)$, but do not know well their common payoff $S(\cdot, \cdot)$. Hence, they will build a simple model of this common payoff $S(\cdot, \cdot)$, by taking a partial linearization of it at each point $x = (u, v)$.

In the sequel, for convenience, we use the notation $z^n = (u^n, v^n)$, for all $n \geq 0$. In order to obtain convergence results to our generalized proximal alternating linearized method, we consider the following set of assumptions:

i) The functions $h : \mathbb{R}^l \to \mathbb{R}$ and $k : \mathbb{R}^m \to \mathbb{R}$ are lower semicontinuous functions and $S : \mathbb{R}^l \times \mathbb{R}^m \to \mathbb{R}$ is a C^1 function, i.e., S is continuously differentiable.

ii) The lower boundedness of the functions under consideration, that is, $\inf_{\mathbb{R}^l \times \mathbb{R}^m} \Psi > -\infty$, $\inf_{\mathbb{R}^l} h > -\infty$ and $\inf_{\mathbb{R}^m} k > -\infty$.

iii) ∇S is Lipschitz continuous on bounded subsets of $\mathbb{R}^l \times \mathbb{R}^m$, i.e., for each bounded subsets $B_1 \times B_2$ of $\mathbb{R}^l \times \mathbb{R}^m$ there exists $M > 0$ such that

$$\|(\nabla_u S(x_1, y_1) - \nabla_u S(x_2, y_2), \nabla_v S(x_1, y_1) - \nabla_v S(x_2, y_2))\| \leq M \|(x_1 - x_2, y_1 - y_2)\|$$

$\forall (x_i, y_i) \in B_1 \times B_2, i = 1, 2$, where $\nabla_u S$ (resp. $\nabla_v S$) stands to the gradient of the mapping of $S(\cdot, \cdot)$ with respect to the first (resp. second) variable, the second (resp. first) one being fixed.

iv) For any fixed v, the function $u \mapsto S(u,v)$ is $C^{1,1}_{L_u(v)}$ which means that the mapping $\nabla_u S(\cdot, v)$ is $L_u(v)$-Lipschitz, that is,

$$\|\nabla_u S(x, v) - \nabla_u S(y, v)\| \leq L_u(v) \|x - y\|, \quad \forall x, y \in \mathbb{R}^l.$$

Likewise, for any fixed u the function $v \mapsto S(u,v)$ is assumed to be $C^{1,1}_{L_v(u)}$.

v) For $i = 1,2$ there exist constants $\kappa_i^-, \kappa_i^+ > 0$ such that

$$\inf_{v \in \mathbb{R}^m} L_u(v) \geq \kappa_1^- \quad \text{and} \quad \inf_{u \in \mathbb{R}^n} L_v(u) \geq \kappa_2^-$$

$$\sup_{v \in \mathbb{R}^m} L_u(v) \leq \kappa_1^+ \quad \text{and} \quad \sup_{u \in \mathbb{R}^n} L_v(u) \leq \kappa_2^+.$$

We consider the partial linearization of $S(\cdot,\cdot)$ as follows:

$$u_{n+1} \in \arg\min_{u \in \mathbb{R}^l} \left\{ h(u) + \langle u - u_n, \nabla_u S(u_n, v_n) \rangle + \frac{\lambda_n}{2} C_U(u_n, u)^2 \right\}, \tag{3.1}$$

where $\lambda_n = \gamma_1 \frac{L_u(v_n)}{\alpha_l^2}$, for some $\gamma_1 > 1$, and

$$v_{n+1} \in \arg\min_{v \in \mathbb{R}^m} \left\{ k(v) + \langle v - v_n, \nabla_v S(u_{n+1}, v_n) \rangle + \frac{\mu_n}{2} C_V(v_n, v)^2 \right\}, \tag{3.2}$$

where $\mu_n = \gamma_2 \frac{L_v(u_{n+1})}{\alpha_m^2}$, for some $\gamma_2 > 1$.

Our next hypothesis is on quasi distances. Given any $p \in \mathbb{N}$, we denote for convenience by q_p a quasi distance defined on $\mathbb{R}^p \times \mathbb{R}^p$. We assume that there exist real numbers $\alpha_p > 0$ and $\beta_p > 0$ such that

$$\alpha_p \|x - y\| \leq q_p(x,y) \leq \beta_p \|x - y\|, \quad \forall x, y \in \mathbb{R}^p. \tag{3.3}$$

This means that costs of moving are not too asymmetric.

Lemma 3.5.1 *Let $g : \mathbb{R}^p \to \mathbb{R}$ be a continuously differentiable function with gradient ∇g assumed L_g-Lipschitz continuous. Then,*

$$g(w) \leq g(v) + \langle w - v, \nabla g(v) \rangle + \frac{L_g}{2\alpha_p^2} q_p(w,v)^2, \quad \forall w, v \in \mathbb{R}^p.$$

Proof. Let t be a scalar parameter and let $f(t) = g(v + tu)$. Then,

$$g(v+u) - g(v) = f(1) - f(0) = \int_0^1 \frac{df(t)}{dt} dt = \int_0^1 \langle u, \nabla g(v+tu) \rangle dt$$

$$\leq \int_0^1 \langle u, \nabla g(v) \rangle dt + \left| \int_0^1 \langle u, \nabla g(v+tu) - \nabla g(v) \rangle dt \right|$$

$$\leq \langle u, \nabla g(v) \rangle + \|u\| \int_0^1 \|\nabla g(v+tu) - \nabla g(v)\| dt$$

$$\leq \langle u, \nabla g(v) \rangle + \|u\|^2 L_g \int_0^1 t \, dt = \langle u, \nabla g(v) \rangle + \frac{L_g}{2} \|u\|^2, \tag{3.4}$$

for any $u, v \in \mathbb{R}^p$. Taking $v + u = w$ in (3.4), we obtain, for any $w, v \in \mathbb{R}^p$,

$$\begin{aligned} g(w) &\leq g(v) + \langle w - v, \nabla g(v) \rangle + \frac{L_g}{2} \|w - v\|^2 \\ &\leq g(v) + \langle w - v, \nabla g(v) \rangle + \frac{L_g}{2\alpha_p^2} q_p(w,v)^2 \end{aligned}$$

where the second inequality comes from assumption (3.3). □

Lemma 3.5.2 *Let $g : \mathbb{R}^p \to \mathbb{R}$ be a continuously differentiable function with ∇g assumed L_g-Lipschitz continuous and let $\sigma : \mathbb{R}^p \to \mathbb{R}$ be a lower semicontinuous function with $\inf_{x \in \mathbb{R}^p} \sigma(x) > -\infty$. Fix $t > \frac{L_g}{\alpha_p^2}$, where α_p is given by (3.3). Given any $u \in \mathbb{R}^p$ there exists $u^+ \in \mathbb{R}^p$ such that*

$$u^+ \in \arg\min_{x \in \mathbb{R}^p} \left\{ \sigma(x) + \langle x - u, \nabla g(u) \rangle + \frac{t}{2} q_p(x, u)^2 \right\}, \quad (3.5)$$

and, hence, one has

$$g(u^+) + \sigma(u^+) \leq g(u) + \sigma(u) - \frac{1}{2}\left(t - \frac{L_g}{\alpha_p^2}\right) q_p(u^+, u)^2.$$

Proof. It follows from (3.5) that

$$\sigma(u^+) + \langle u^+ - u, \nabla g(u) \rangle + \frac{t}{2} q_p(u^+, u)^2 \leq \sigma(x) + \langle x - u, \nabla g(u) \rangle + \frac{t}{2} q_p(x, u)^2, \quad \forall x \in \mathbb{R}^p,$$

and, in particular, taking $x = u$ we have

$$\sigma(u^+) + \langle u^+ - u, \nabla g(u) \rangle + \frac{t}{2} q_p(u^+, u)^2 \leq \sigma(u).$$

Thus, applying (3.5) with $w = u^+$ and $v = u$ in the above inequality, we obtain the desired result. □

Proposition 3.5.3 *Let $\{z_n\} := \{(u_n, v_n)\}$ be a sequence generated by GPALM. The following assertions hold.*

i) *The sequence $\{\Psi(z_n)\}$ is nonincreasing and satisfies*

$$\theta_1 q_l(u_n, u_{n+1})^2 + \theta_2 q_m(v_n, v_{n+1})^2 \leq \Psi(z_n) - \Psi(z_{n+1}), \quad \forall n \geq 0,$$

for some $\theta_1, \theta_2 > 0$.

ii)

$$\sum_{n=0}^{\infty} \left[\theta_1 q_l(u_n, u_{n+1})^2 + \theta_2 q_m(v_n, v_{n+1})^2 \right] < \infty,$$

in particular, $\lim_{n \to \infty} q_n(u_n, u_{n+1}) = 0$ and $\lim_{n \to \infty} q_m(v_n, v_{n+1}) = 0$.

Proof. Under the assumptions made, functions $h(\cdot)$ and $k(\cdot)$ are lower semicontinuous and functions $u \mapsto S(u,v)$ (v is fixed) and $v \mapsto S(u,v)$ (u is fixed) are differentiable and have Lipschitz gradient with modulus $L_u(v)$ and $L_v(u)$, respectively. Thus, using the iterative steps (3.1) and (3.2), applying Lemma 3.5.2 twice, first with $\sigma(\cdot) = h(\cdot)$,

$g(\cdot) = S(\cdot, v_n)$ and $t = \lambda_n > (L_u(v_n)/\alpha_l^2)$, and secondly with $\sigma(\cdot) = k(\cdot)$, $g(\cdot) = S(u_{n+1}, \cdot)$ and $t = \mu_n > (L_v(u_{n+1})/\alpha_m^2)$, we obtain successively

$$S(u_{n+1}, v_n) + h(u_{n+1}) \leq S(u_n, v_n) + h(u_n) - \frac{1}{2}\left(\lambda_n - \frac{L_u(v_n)}{\alpha_l^2}\right) q_l(u_n, u_{n+1})^2$$

$$= S(u_n, v_n) + h(u_n) - \frac{L_u(v_n)}{2\alpha_l^2}(\gamma_1 - 1)q_l(u_n, u_{n+1})^2,$$

and

$$S(u_{n+1}, v_{n+1}) + k(v_{n+1}) \leq S(u_{n+1}, v_n) + k(v_n)$$
$$- \frac{1}{2}\left(\mu_n - \frac{L_v(u_{n+1})}{\alpha_m^2}\right) q_m(v_n, v_{n+1})^2$$
$$= S(u_{n+1}, v_n) + k(v_n)$$
$$- \frac{L_v(u_{n+1})}{2\alpha_m^2}(\gamma_2 - 1)q_m(v_n, v_{n+1})^2.$$

Adding last two inequalities, we obtain for all $n \geq 0$,

$$\Psi(u_n, v_n) - \Psi(u_{n+1}, v_{n+1}) \geq \frac{L_u(v_n)}{2\alpha_l^2}(\gamma_1 - 1)q_l(u_n, u_{n+1})^2$$
$$+ \frac{L_v(u_{n+1})}{2\alpha_m^2}(\gamma_2 - 1)q_m(v_n, v_{n+1})^2$$
$$\geq \frac{\kappa_1^-}{2\alpha_l^2}(\gamma_1 - 1)q_l(u_n, u_{n+1})^2$$
$$+ \frac{\kappa_2^-}{2\alpha_m^2}(\gamma_2 - 1)q_m(v_n, v_{n+1})^2 \geq 0.$$

Therefore, the first assertion is proved just taking

$$\theta_1 = \frac{\kappa_1^-}{2\alpha_l^2}(\gamma_1 - 1) \text{ and } \theta_2 = \frac{\kappa_2^-}{2\alpha_m^2}(\gamma_2 - 1).$$

Now, summing up the above inequality from $n = 0$ to k, we get

$$\sum_{n=0}^{k}\left[\theta_1 q_l(u_n, u_{n+1})^2 + \theta_2 q_m(v_n, v_{n+1})^2\right] \leq \Psi(z_0) - \Psi(z_{k+1}).$$

Letting k goes to ∞ and taking into account the lower boundedness of $\Psi(\cdot)$, we obtain

$$\sum_{n=0}^{\infty}\left[\theta_1 q_l(u_n, u_{n+1})^2 + \theta_2 q_m(v_n, v_{n+1})^2\right] < \infty,$$

and hence,

$$\sum_{n=0}^{\infty} \theta_1 q_l(u_n, u_{n+1})^2 < \infty \text{ and } \sum_{n=0}^{\infty} \theta_2 q_m(v_n, v_{n+1})^2 < \infty.$$

In particular, we have $\lim_{n\to\infty} q_l(u_n, u_{n+1}) = 0$ and $\lim_{n\to\infty} q_m(v_n, v_{n+1}) = 0$, and the proof is concluded. □

Next, we recall some concepts and properties of subdifferential of a function, which will be useful throughout this section; for more details see [41, 46]. Let $f : \mathbb{R}^n \to \mathbb{R}$ be a lower semicontinuous function.

1. The Fréchet subdifferential of f at x, denoted by $\hat{\partial} f(x)$, is defined as follows

$$\hat{\partial} f(x) = \begin{cases} \{v \in \mathbb{R}^n : \liminf_{y \to x, y \neq x} \frac{f(y) - f(x) - \langle v, y-x \rangle}{\|x-y\|} \geq 0\}, & \text{if } x \in \text{dom}(f); \\ \emptyset, & \text{if } x \notin \text{dom}(f). \end{cases}$$

2. The limiting-subdifferential of f at x, denoted by $\partial f(x)$, is defined as follows

$$\partial f(x) = \begin{cases} \{v \in \mathbb{R}^n : \exists x^k \to x, f(x^k) \to f(x), v^k \in \hat{\partial} f(x^k) \to v\}, & \text{if } x \in \text{dom}(f); \\ \emptyset, & \text{if } x \notin \text{dom}(f). \end{cases}$$

We denote by $\text{dom}\,\partial f = \{x \in \mathbb{R}^n : \partial f(x) \neq \emptyset\}$.

Remark 3.5.4   i) The limiting-subdifferential is closed and $\hat{\partial} f(x) \subset \partial f(x)$. If f is a lower semicontinuous and convex function, and $x \in \text{dom}(f)$, then $\hat{\partial} f(x)$ coincides with the classical subdifferential in the sense of convex analysis and it is a nonempty, closed and convex set.

ii) In our nonsmooth context, the well-known Fermat's rule remains barely unchanged and $\partial \Psi(u,v) = (\nabla_u S(u,v) + \partial h(u), \nabla_v S(u,v) + \partial k(v)) = (\partial_u \Psi(u,v), \partial_v \Psi(u,v))$.

iii) Points whose subdifferential contains 0 are called (limiting-)critical points.

Proposition 3.5.5 Let $\{z_n\} := \{(u_n, v_n)\}$ be a sequence generated by GPALM which is assumed to be bounded. For each $n \in \mathbb{N}$, take

$$A_u^n \in \nabla_u S(u_n, v_n) - \nabla_u S(u_{n-1}, v_{n-1}) - \lambda_{n-1} q_l(u_n, u_{n-1}) \partial q_l(u_{n-1}, u_n)$$

and

$$A_v^n \in \nabla_v S(u_n, v_n) - \nabla_v S(u_n, v_{n-1}) - \mu_{n-1} q_m(v_n, v_{n-1}) \partial q_m(v_{n-1}, v_n).$$

Then, $(A_u^n, A_v^n) \in \partial \Psi(u_n, v_n)$, and there exist constants $\omega_1 > 0$ and $\omega_2 > 0$ such that

$$\|(A_u^n, A_v^n)\| \leq \omega_1 q_l(u_{n-1}, u_n) + \omega_2 q_m(v_{n-1}, v_n).$$

Proof. Let n be an integer. Writing down the optimality condition of the iterative step (3.1), we have

$$0 \in \partial h(u_n) + \nabla_u S(u_{n-1}, v_{n-1}) + \lambda_n q_l(u_{n-1}, u_n) \partial q_l(u_{n-1}, u_n).$$

Hence,

$$\begin{aligned} A_u^n &\in \nabla_u S(u_n, v_n) - \nabla_u S(u_{n-1}, v_{n-1}) - \lambda_n q_l(u_{n-1}, u_n) \partial q_l(u_{n-1}, u_n) \\ &\subset \partial h(u_n) + \nabla_u S(u_n, v_n) = \partial_u \Psi(u_n, v_n). \end{aligned}$$

Similarly, from (3.2), we have $A_v^n \in \partial_v \Psi(u_n, v_n)$, and hence we obtain $(A_u^n, A_v^n) \in \partial \Psi(u_n, v_n)$.

We now have to estimate the norms of A_u^n and A_v^n. Note that $\nabla S(\cdot, \cdot)$ is Lipschitz continuous on bounded sets of $\mathbb{R}^l \times \mathbb{R}^m$ and $\partial q_l(u_{n-1}, \cdot)(u_n)$ is bounded on bounded sets of $\mathbb{R}^l$ and since $(z_n) = (u_n, v_n)$ is bounded, there exist constants $M, N_l > 0$ such that

$$\begin{aligned}
\|A_u^n\| &\leq M(\|u_n - u_{n-1}\| + \|v_n - v_{n-1}\|) + N_l \lambda_{n-1} q_l(u_{n-1}, u_n) \\
&\leq \frac{M}{\alpha_l} q_l(u_{n-1}, u_n) + \frac{M}{\alpha_m} q_m(v_{n-1}, v_n) + \frac{N_l \gamma_1 \kappa_1^+}{\alpha_l^2} q_l(u_{n-1}, u_n) \\
&= \frac{(M \alpha_l + N_l \gamma_1 \kappa_1^+)}{\alpha_l^2} q_l(u_{n-1}, u_n) + \frac{M}{\alpha_m} q_m(v_{n-1}, v_n).
\end{aligned}$$

On the other hand, the Lipschitz continuity of $\nabla_v(u, \cdot)$ and the fact that $\partial q_m(v_{n-1}, \cdot)(v_n)$ is bounded on bounded sets of $\mathbb{R}^m$ imply that there exists $N_m > 0$ such that

$$\begin{aligned}
\|A_v^n\| &\leq L_v(u_n) \|v_n - v_{n-1}\| + N_m \mu_{n-1} q_m(v_n, v_{n-1}) \\
&\leq \frac{\kappa_2^+}{\alpha_m} q_m(v_n, v_{n-1}) + \frac{N_m \gamma_2 \kappa_2^+}{\alpha_m^2} q_m(v_n, v_{n-1}) \\
&= \frac{\kappa_2^+}{\alpha_m} \left(1 + \frac{N_m \gamma_2}{\alpha_m}\right) q_m(v_n, v_{n-1}).
\end{aligned}$$

Summing up the above estimations we obtain the desired result with $\omega_1 = \frac{(M \alpha_l + N_l \gamma_1 \kappa_1^+)}{\alpha_l^2}$ and $\omega_2 = \frac{1}{\alpha_m}[M + \kappa_2^+(1 + \frac{N_m \gamma_2}{\alpha_m})]$. $\square$

Next, we refer to a limit point z^* of a sequence $\{z^k\}$ if there exists an increasing sequence of integers k_j such that $z^{k_j} \to z^*$ as $j \to \infty$. Clearly, if a sequence $\{z^k\}$ converges to a point z^*, then $\{z^k\}$ has an unique limit point z^*.

Theorem 3.5.6 *Let $\{z_n\} := \{(u_n, v_n)\}$ be a sequence generated by GPALM which is assumed to be bounded. The following assertions hold.*

i) The objective function Ψ is constant on the set of limit points of $\{z^k\}$.

ii) Every limit point of $\{z^k\}$ is a critical point of Ψ.

Proof. From Proposition 3.5.3 i) $\{\Psi(z_n)\}$ is a non increasing sequence which combined with the fact that Ψ is bounded from below implies that $\{\Psi(z_n)\}$ converges to Ψ^*. Let $z^* = (u^*, v^*)$ be a limit point of $z_n = (u_n, v_n)$ and let $\{(u_{n_q}, v_{n_q})\}$ be a subsequence such that $(u_{n_q}, v_{n_q}) \to (u^*, v^*)$ as $q \to \infty$. The lower semicontinuity of $h(\cdot)$ and $k(\cdot)$ means that

$$\liminf_{q \to \infty} h(u_{n_q}) \geq h(u^*) \quad \text{and} \quad \liminf_{q \to \infty} k(v_{n_q}) \geq k(v^*). \tag{3.6}$$

From the iterative step (3.1), we have

$$h(u_{n_q+1}) + \langle u_{n_q+1} - u_{n_q}, \nabla_u S(u_{n_q}, v_{n_q}) \rangle + \frac{\lambda_{n_q}}{2} q_l(u_{n_q}, u_{n_q+1})^2$$

$$\leq h(u^*) + \langle u^* - u_{n_q}, \nabla_u S(u_{n_q}, v_{n_q}) \rangle + \frac{\lambda_{n_q}}{2} q_l(u_{n_q}, u^*)^2.$$

From Proposition 3.5.3 ii) $q_l(u_{n_q}, u_{n_q+1})$ converges to 0 as $q \to \infty$, and hence, u_{n_q+1} converges to u^* as $q \to \infty$. Therefore, taking the lim sup in the above inequality, keeping in mind that $\{\lambda_n\}$ is bounded and ∇S is continuous, we obtain $\limsup_{q \to \infty} h(u_{n_q}) \leq h(u^*)$. Thus, in view of (3.6), $h(u_{n_q}) \to h(u^*)$ as $q \to \infty$. Arguing similarly with k and v_n we obtain

$$\Psi^* = \lim_{q \to \infty} \Psi(z_n) = \lim_{q \to \infty} \Psi(u_n, v_n) = \Psi(u^*, v^*) = \Psi(z^*),$$

and the first assertion is proved.

Now, from Proposition 3.5.5, we have $(A_u^n, A_v^n) \in \partial \Psi(u_n, v_n)$. We also have $(A_u^n, A_v^n) \to (0,0)$ as $n \to \infty$. Thus, the closedness property of $\partial \Psi$ implies that $(0,0) \in \partial \Psi(u^*, v^*)$ which means that $z^* = (u^*, v^*)$ is a critical point of Ψ. □

Dealing with descent methods for convex functions, we can expect that the algorithm provides globally convergent sequences, i.e., convergence of the whole sequence. When the functions under consideration are not convex, the method may provide sequences that exhibit highly oscillatory behaviors, and partial convergence results are obtained. The Kurdyka-Łojasiewicz property has been successfully applied in order to analyze various types of asymptotic behavior, in particular, proximal point methods; see for instance [7, 14, 22].

Let us recall the Kurdyka-Łojasiewicz property (KL property). A function $f : \mathbb{R}^n \to \mathbb{R}$ is said to have the Kurdyka-Łojasiewicz property at a point $x^* \in \text{dom}\,\partial f$ if there exist $\eta \in (0, +\infty]$, a neighborhood U of x^* and a continuous concave function $\varphi : [0, \eta) \to \mathbb{R}_+$ (called desingularizing function) such that

$$\varphi(0) = 0, \quad \varphi \text{ is } C^1 \text{ on } (0, \eta), \quad \varphi'(s) > 0, \forall s \in (0, \eta); \tag{3.7}$$

$$\varphi'^*))\text{dist}(0, \partial f(x)) \geq 1, \quad \forall x \in U \cap [f(x^*) < f < f(x^*) + \eta], \tag{3.8}$$

where $[\eta_1 < f < \eta_2] = \{x \in \mathbb{R}^n : \eta_1 < f(x) < \eta_2\}$ and C^1 means differentiable with continuous derivative; see the definition of the Kurdyka-Łojasiewicz property and other references about this subject in Attouch et al. [7]. We say that f is a KL function if f satisfies the Kurdyka-Łojasiewicz property at every point of its domain.

One can easily check that the Kurdyka-Łojasiewicz property is satisfied at any non-critical point $\hat{x} \in \text{dom}\,\partial f$. It follows from the Kurdyka-Łojasiewicz property that the critical points of f lying in $U \cap [f(x^*) < f < f(x^*) + \eta]$ have the same critical value $f(x^*)$. If f is differentiable and $f(x^*) = 0$, then (3.8) can be rewritten as

$$\nabla(\varphi \circ f)(x) \geq 1,$$

for each convenient $x \in \mathbb{R}^n$. This property basically expresses the fact that a function can be made sharp by a reparameterization of its values; see [7]. From Bolte et al. [19, Theorem 3.1], a subanalytic function f which is continuous when restricted to its closed domain satisfies the Kurdyka-Łojasiewicz property with desingularising function

$$\varphi(t) = Ct^{1-\theta} \text{ with } C > 0 \text{ and } \theta \in [0,1). \tag{3.9}$$

Theorem 3.5.7 *Suppose that Ψ is a KL function. Let $\{z^k\}$ be a sequence generated by GPALM which is assumed to be bounded. Then, $\{z^k\}$ converges to a critical point $z^* = (x^*, y^*)$ of Ψ.*

Proof. It is straightforward to check the assertion just applying Attouch et al. [7, Theorem 2.9] together with the following facts:

1. From Proposition 3.5.3
$$\Psi(z_{n+1}) + a\|z_{n+1} - z_n\|^2 \leq \Psi(z_n),$$
where $a := \min\{\theta_1 \alpha_l^2, \theta_2 \alpha_m^2\} > 0$;

2. From Proposition 3.5.5, we have $(A_u^n, A_v^n) \in \partial \Psi(z_n)$ and
$$\|(A_u^n, A_v^n)\| \leq b\|z_{n+1} - z_n\|,$$
where $b = \omega_1 \beta_l + \omega_2 \beta_m > 0$;

3. Since $\{z_n\}$ is bounded, let $\{z_{n_j}\}$ be a subsequence converging to z^*. From Theorem 3.5.6, $\{\Psi(z_n)\}$ converges to $\Psi(z^*)$.

This completes the proof. □

Remark 3.5.8 *An important and fundamental case of application of Theorem 3.5.7 is when the data functions h, k and S are semi-algebraic. This fact impacts the convergence rate of the method. If the desingularising function φ of Ψ is of the form (3.9), then, as in [3], the following estimations hold:*

1. *If $\theta = 0$, the sequence $\{z_n\}$ converges in a finite number of steps;*

2. *If $\theta \in (0, \frac{1}{2}]$, then there exist constants $\xi > 0$ and $\tau \in [0, 1)$ such that*
$$\|z_n - z^*\| \leq \xi \tau^n;$$

3. *If $\theta \in (\frac{1}{2}, 1)$, then there exists $\xi > 0$ such that*
$$\|z_n - z^*\| \leq \xi n^{-\frac{1-\theta}{2\theta - 1}}.$$

This result gives the speed of convergence of the potential Nash alternating game. This is a striking and unique finding in game theory.

References

[1] F. Acker and M. A. Prestel. Convergence d'un schema de minimisation alternée. *Ann. Fac. Sci. Toulouse V. Ser. Math.*, 2: 1–9, 1980.

[2] S. P. Anderson, J. K. Goeree and C. A. Holt. Noisy directional learning and the logit equilibrium. *The Scandinavian Journal of Economics*, 106(3): 581–602, 2004.

[3] H. Attouch and J. Bolte. On the convergence of the proximal algorithm for nonsmooth functions involving analytic features. *Math. Program*, 116: 5–16, 2009.

[4] H. Attouch, P. Redont and A. Soubeyran. A new class of alternating proximal minimization algorithms with costs to move. *SIAM J. Optim.*, 18(3): 1061–1081, 2007.

[5] H. Attouch, J. Bolte, P. Redont and A. Soubeyran. Alternating proximal algorithms for weakly coupled convex minimization problems. Applications to dynamical games and PDE's. *J. Convex Anal.*, 15(3): 485–506, 2008.

[6] H. Attouch, J. Bolte, P. Redont and A. Soubeyran. Proximal alternating minimization and projection methods for nonconvex problems: An approach based on the Kurdyka-Lojasiewicz inequality. *Math. Oper. Res.*, 35(2): 438–457, 2010.

[7] H. Attouch, J. Bolte and B. F. Svaiter. Convergence of descent methods for semi-algebraic and tame problems: Proximal algorithms, forward-backward splitting, and regularized Gauss-Seidel methods. *Math. Program. Ser. A.*, 137: 91–129, 2013.

[8] A. Auslender. Optimisation, méthodes numériques. Masson, Paris, 1976.

[9] A. Auslender. Asymptotic properties of the Fenchel dual functional and applications to decomposition problems. *J. Optim. Theory Appl.*, 73(3): 427–449, 1992.

[10] A. Auslender. Méthodes numériques pour la décomposition et la minimisation de fonctions non différentiables. *Numerische Mathematik*, 18: 213–223, 1971.

[11] B. Awerbuch, Y. Azar, A. Epstein, V. Mirrokni and A. Skopalik. Fast convergence to nearly optimal solutions in potential games. *Proc. of 9th EC*, 2008.

[12] H. H. Bauschke, P. L. Combettes and D. Noll. Joint minimization with alternating Bregman proximity operators. *Pac. J. Optim.*, 2: 401–424, 2006.

[13] A. Beck and M. Teboulle. Gradient-based algorithms with applications to signal recovery. *Convex Optimization in Signal Processing and Communications*, 42–88, 2009.

[14] G. C. Bento and A. Soubeyran. A generalized inexact proximal point method for nonsmooth functions that satisfies Kurdyka-Łojasiewicz inequality. *Set-Valued Var. Anal.*, 23: 501–517, 2015.

[15] D. P. Bertsekas. Nonlinear programming, Athena Scientic, Belmont, MA, 1995.

[16] D. P. Bertsekas and P. Tseng. Partial proximal minimization algorithms for convex programming. *SIAM J. Optim.*, 4(3):551–572, 1994.

[17] D. P. Bertsekas and J. N. Tsitsiklis. Parallel and distributed computation: Numerical methods. Prentice-Hall, New Jersey, 1989.

[18] J. Bolte, S. Sabach and M. Teboulle. Proximal alternating linearized minimization for nonconvex and nonsmooth problems. *Mathematical Programming*, 146(1-2): 459–494, 2014.

[19] J. Bolte, A. Daniliidis and A. Lewis. Łojasiewicz inequality for nonsmooth subanalytic functions with applications to subgradient dynamic systems. *SIAM Optim.*, 17(4): 1205–1223, 2007.

[20] Y. Chen and R. Gazzale. When does learning in games generate convergence to Nash equilibria? The role of supermodularity in an experimental setting. *American Economic Review*, 9(5): 1505–1535, 2004.

[21] P. L. Combettes and V. R. Wajs. Signal recovery by proximal forward-backward splitting. *Multiscale Modeling & Simulation*, 4(4): 1168–1200, 2005.

[22] J. X. Cruz Neto, P. R. Oliveira, A. Soubeyran and J. C. O. Souza. A generalized proximal linearized algorithm for DC functions with application to the optimal size of the firm problem. *Annals of Operations Research*, 1–27, 2019.

[23] J. X. Cruz Neto, P. R. Oliveira, P. A. Soares Junior and A. Soubeyran. Learning how to play Nash and Alternating minimization method for structured nonconvex problems on Riemannian manifolds. *Journal of Convex Analysis*, 20: 395–438, 2013.

[24] P. L. Combettes and J. C. Pesquet. Proximal splitting methods in signal processing. *In Fixed-point Algorithms for Inverse Problems in Science and Engineering* (pp. 185–212). Springer New York, 2011.

[25] I. Csiszar and G. Tusnady. Information geometry and alternating minimization procedures. *Statistics and Decisions* (supplement 1): 205–237, 1984.

[26] D. L. Donoho. Compressed sensing. *IEEE Trans. Inform. Theory*, 4: 1289–1306, 2006.

[27] P. Dubey, O. Haimanko and A. Zapechelnyuk. Strategic complements and substitutes, and potential games. *Games and Economic Behavior*, 54(1): 77–94, 2006.

[28] J. Falkinger, E. Fehr, S. Gächter and R. Winter-Ebmer. A simple mechanism for the efficient provision of public goods: Experimental evidence. *The American Economic Review*, 90(1): 247–264, 2000.

[29] F. Flores Bazán, D. T. Luc and A. Soubeyran. Maximal elements under reference-dependent preferences with applications to behavioral traps and games. *Journal of Optimization Theory and Applications,* 155(3): 883–901, 2012.

[30] J. K. Goeree and C. A. Holt. Coordination games. Encyclopedia of Cognitive Sciences. MacMillan, 2000.

[31] D. González-Sánchez and O. Hernández-Lerma. A survey of static and dynamic potential games. *Science China Mathematics,* 59(11): 2075–2102, 2016.

[32] L. Grippo and M. Sciandrone. On the convergence of the block nonlinear Gauss-Seidel method under convex constraints. *Oper. Res. Lett.,* 26: 127–136, 2000.

[33] I. Grubisic and R. Pietersz. Efficient rank reduction of correlation matrices. *Linear Algebra and its Applications,* 422: 629–653, 2007.

[34] J. Hofbauer and S. Sorin. Best response dynamics for continuous zero-sum games. *Discrete and Continuous Dynamical Systems. Ser B.6.,* 6(1): 215–224, 2006.

[35] M. Jensen and D. Oyama. Stability of pure strategy Nash equilibrium and best response potential games. Working paper, in progress. University of Birmingham JG Smith Building (Dep. of Economics). Birmingham, B15 2TT, United Kingdom, 2009.

[36] J. Levin. Supermodular games. Standford University, 2006.

[37] P. L. Lions. On the Schwarz alternating method. III. pp. 202–231. *In*: T. F. Chan, R. Glowinski, J. Periaux and O. Widlund (eds.). *A Variant for Non Overlapping Subdomains in Third International Symposium on Domain Decomposition Methods for Partial Differential Equations.* SIAM, 1990.

[38] P. Milgrom and J. Robert. Rationalizability, learning, and equilibrium in games with strategic complementarities. *Econometrica,* 58(6): 1255–1277, 1990.

[39] A. J. Monaco and T. Sabarwal. Games with strategic complements and substitutes. *Economic Theory,* 62(1-2): 65–91, 2016.

[40] D. Monderer and L. Shapley. Fictitious play property for games with identical players. *Journal of Economic Theory,* 68(14): 258–265, 1996.

[41] B. Mordukhovich. *Variational Analysis and Generalized Differentiation.* I. Basic Theory, Grundlehren der Mathematischen Wissenschaften, vol. 330. Springer, Berlin, 2006.

[42] F. G. Moreno, P. R. Oliveira and A. Soubeyran. A proximal point algorithm with quasi distance. Application to habit's formation. *Optimization,* 61, 1383–1403, 2012.

[43] M. Mohri, A. Rostamizadeh and A. Talwalkar. Foundations of machine learning. The MIT Press, 2012.

[44] J. M. Ortega and W. C. Rheinboldt. *Iterative solution of nonlinear equations in several variables.* Academic Press, New-York, 1970.

[45] M. J. D. Powell. On search directions for minimization algorithms. *Math. Program,* 4: 193–201, 1970.

[46] R. T. Rockafellar and R. Wets. *Variational Analysis Grundlehren der Mathematischen Wissenschaften,* vol. 317. Springer, Berlin, 1998.

[47] M. E. Slade. What does an oligopoly maximize? *Journal of Industrial Economics,* 42: 45–61, 1994.

[48] A. Soubeyran. Variational rationality, a theory of individual stability and change: Worthwhile and ambidextry behaviors. Preprint. GREQAM, Aix Marseille University, 2009.

[49] A. Soubeyran. Variational rationality and the unsatisfied man: A course pursuit between aspirations, capabilities, beliefs and routines. Preprint. GREQAM. Aix Marseille University, 2010.

[50] A. Soubeyran.*Variational rationality. Part 1. A theory of the unsatisfied men, making worthwhile moves to satisfy his needs.* Preprint. AMSE. Aix-Marseille University, 2019a.

[51] A. Soubeyran. *Variational rationality. Part 2. A unified theory of goal setting, intentions and goal striving.* Preprint. AMSE. Aix-Marseille University, 2019b.

[52] P. Tseng. Convergence of a block coordinate descent method for nondifferentiable minimization. *J. Optim. Theory Appl.,* 109(2001): 475–494, 2001.

[53] D. M. Topkis. Minimizing a submodular function on a lattice. *Operations Research,* 26(16): 305–321, 1979.

[54] G. Tullock. The economics of special privilege and rent seeking. Boston and Dordrecht: Kluwer Academic Publishers, 1989.

[55] X. Vives. Nash equilibrium with strategic complementarities. *Journal of Mathematical Economics,* 19: 305–321, 1990.

[56] J. Von Neumann. Functional operators. Annals of Mathematics Studies 22. Princeton University Press, 1950.

[57] M. Voorneveld. Best response potential games. *Economics Letters,* 66: 289–295, 2000.

[58] B. Widrow and E. Wallach. *Adaptive Inverse Control.* Prentice-Hall, New Jersey, 1996.

[59] Y. Xu and W. Yin. A block coordinate descent method for regularized multiconvex optimization with applications to nonnegative tensor factorization and completion. *SIAM Journal on Imaging Sciences,* 6(3): 1758–1789, 2013.

[60] W. I. Zangwill. Nonlinear programming: A unified approach. Prentice Hall, Englewood Cliffs, 1969.

Chapter 4

Sublinear-like Scalarization Scheme for Sets and its Applications to Set-valued Inequalities

Koichiro Ike
Niigata University, Niigata 950–2181, Japan.

Yuto Ogata
Niigata University, Niigata 950–2181, Japan.

Tamaki Tanaka
Niigata University, Niigata 950–2181, Japan.

Hui Yu
Niigata University, Niigata 950–2181, Japan.

4.1 Introduction

When we want to evaluate performance for sport teams or compare various factors in quality control, we should deal with families whose elements are some kinds of vectors, and we frequently encounter some difficult situations on the decision-making by a certain methodology for comparisons between sets.

Generally speaking, comparing two or more things is a basic and essential operation in various decision-making situations. It requires some comparison criterion, such as physical size, monetary value, future potential, or personal preference. In mathematics, this criterion is often expressed as an order relation or a binary relation satisfying good properties. For example, the set of all real numbers is totally ordered by the usual less-than-or-equal-to relation $\leq$, and vectors in a real vector space are compared on the basis of a preorder that is compatible with the linear structure. As for comparisons of sets, certain types of binary relations, called "set relations," have been vigorously researched; see [7, 11, 12]. These relations are considered to be natural criteria to represent relative superiority or inferiority of sets, holding a key position in the area of set optimization.

On the other hand, a linear functional on a real vector space is a bilinear form as a function of two variables of the original space and its dual space; it is an inner product of two vectors in the case of a finite-dimensional space. Also, it is one of the most useful tools for evaluation with respect to some index of the adequacy of efficiency in multi-objective programming or vector optimization. This is a typical approach to comparing vectors or sets, and it is referred to as "scalarization"; see [2, 23]. Each object is mapped to an (extended) real number so that any two objects become comparable as elements of the totally ordered space $\mathbb{R}$ (or $\overline{\mathbb{R}} := \mathbb{R} \cup \{\pm\infty\}$). Naturally, converting vectors or sets into scalars involves a decrease of the amount of information; however, a well-designed conversion makes the problem less significant and can be a powerful tool. A norm $\|\cdot\|$ and an inner product $\langle w, \cdot \rangle$ with a fixed vector w are familiar examples of a scalarization for vectors. In addition, Gerstewitz's (Tammer's) sublinear scalarizing functional has exerted a prominent presence in vector optimization (e.g., see [2, Sections 2.3 and 3.1]):

$$h_C(v;d) := \inf\{t \in \mathbb{R} \mid v \in td - C\} \tag{4.1}$$

where C is a convex cone and $d \in C$. This scalarizing functional $h_C(\cdot;d)$ is sublinear (i.e., $h_C(v_1 + v_2; d) \leq h_C(v_1) + h_C(v_2)$ and $h_C(tv;d) = th_C(v;d)$ for $t > 0$) and, hence, this conversion is called "sublinear scalarization", found early in [1, 20]. If convex cone C is a half space, that is, $C = \{v \mid \langle w, v \rangle \geq 0\}$ with w satisfying $\langle w, w \rangle = 1$, then $h_C(v;w) = \langle w, v \rangle$. Therefore, this special functional is a certain generalization of linear scalarization, including the notions of weighted sum and inner product. Accordingly, this idea has inspired some researchers to develop particular scalarization methods for sets, leading to several applicative results shown in [4, 14, 18, 27].

For example, we consider a scalarizing function $\psi : V \to \overline{\mathbb{R}}$ for vectors in a real vector space V. Let θ_V and C be the zero vector and a (nonempty) convex cone in V with $\theta_V \in C$. We define a preorder $\leq_C$ in V induced by C as follows: for $v_1, v_2 \in V$, $v_1 \leq_C v_2 :\iff v_2 - v_1 \in C$. This preorder is compatible with the linear structure:

$$v_1 \leq_C v_2 \implies v_1 + v_3 \leq_C v_2 + v_3 \text{ for all } v_1, v_2, v_3 \in V; \tag{4.2}$$

$$v_1 \leq_C v_2 \implies tv_1 \leq_C tv_2 \text{ for all } v_1, v_2 \in V \text{ and } t > 0. \tag{4.3}$$

For well-designed conversions in scalarization, we need the following "order-monotone" (that is, order preserving) property:

$$v_1 \leq_C v_2 \implies \psi(v_1) \leq \psi(v_2).$$

If the scalarizing function is an inner product, that is, $\psi(v) = \langle w, v \rangle$ with the weight vector w chosen as an element of the dual cone C^* in V^*, where

$$C^* = \{w \in V^* \mid \langle w, v \rangle \geq 0 \quad \forall v \in C\},$$

ψ has the order-monotone property. Owing to this property, any minimal or maximal element of a convex set in a vector optimization problem is characterized by optimal solutions of its scalarized problem with a certain nonzero weight vector in C^*, which is guaranteed by separation theorems for two convex sets.

In the case of sublinear scalarization, when $d \in C$, the lower level set of $h_C(\cdot; d)$ at each height t coincides with a parallel translation of $-C$ at offset td, that is,

$$\{v \mid h_C(v; d) \leq t\} = td - C,$$

and hence $h_C(\cdot; d)$ is the smallest strictly monotonic function with respect to C in the case that $d \in \text{int}\, C$, which is the topological interior of C; see page 21 in [15]. Hence, $h_C(\cdot; d)$ has the order-monotone property. Also this scalarizing functional has a dual form as follows:

$$-h_C(-v; d) = \sup\{t \in \mathbb{R} \mid v \in td + C\}. \tag{4.4}$$

When convex cone C is a half space, both sublinear functionals $h_C(v; d)$ and $-h_C(-v; d)$ are coincident with the value of $\langle d, v \rangle$. Nishizawa and Tanaka [17] study certain characterizations of set-valued mappings by using the inherited properties of the two sublinear functionals on cone-convexity and cone-continuity. Moreover, this kind of sublinear scalarization in (4.1) and (4.4) is a method used to measure how far the origin θ_V needs to be moved toward a specific direction d in order to fulfill $v \leq_C \theta_V + td$ or $\theta_V + td \leq_C v$ with respect to $\leq_C$.

These observations lead us into the work to study sublinear scalarization scheme for sets with respect to set relations in a general setting. The aim of this paper is to show a certain possibility to use such a scheme in order to establish generalizations of inequalities for set-valued cases as applications. We also mention an effort to apply the scalarization to fuzzy theoretical fields.

4.2 Set Relations and Scalarizing Functions for Sets

Let V be a real topological vector space unless otherwise specified. Let θ_V be the zero vector in V and $\mathcal{P}(V)$ denote the set of all subsets of V. The topological interior, topological closure, convex hull, and complement of a set $A \in \mathcal{P}(V)$ are denoted by $\text{int}\, A$, $\text{cl}\, A$, $\text{co}\, A$, and A^c, respectively.

For given $A, B \in \mathcal{P}(V)$ and $t \in \mathbb{R}$, the algebraic sum $A + B$ and the scalar multiplication tA are defined as follows:

$$A + B := \{a + b \mid a \in A,\ b \in B\}, \quad tA := \{ta \mid a \in A\}.$$

In particular, we denote $A + \{v\}$ by $A + v$ (or $v + A$) and $(-1)A$ by $-A$ for $A \in \mathcal{P}(V)$ and $v \in V$. We define $Tv := \{tv \mid t \in T\}$ for $v \in V$ and $T \subset \mathbb{R}$.

Also, we denote the composition of two functions f and g by $g \circ f$. Moreover, we recall some definitions of C-notions related to a convex cone $C \in \mathcal{P}(V)$ which are referred to in [15]. $A \in \mathcal{P}(V)$ is said to be C-convex (resp., C-closed) if $A + C$ is convex (resp., closed); C-proper if $A + C \neq V$. Besides, A is said to be C-bounded if for each open neighborhood U of θ_V there exists $t \geq 0$ such that $A \subset tU + C$. Furthermore, we say that set-valued map F has each C-notion mentioned above if the image set $F(x) \subset V$ for each x has the property of the corresponding C-notion.

Let X be a nonempty set and $\preccurlyeq$ a binary relation on X. The relation $\preccurlyeq$ is said to be

1. reflexive if $x \preccurlyeq x$ for all $x \in X$;

2. irreflexive if $x \not\preccurlyeq x$ for all $x \in X$;

3. transitive if $x \preccurlyeq y$ and $y \preccurlyeq z$ imply $x \preccurlyeq z$ for all $x, y, z \in X$;

4. antisymmetric if $x \preccurlyeq y$ and $y \preccurlyeq x$ imply $x = y$ for all $x, y \in X$;

5. complete if $x \preccurlyeq y$ or $y \preccurlyeq x$ for all $x, y \in X$.

The relation $\preccurlyeq$ is called

1. a preorder if it is reflexive and transitive;

2. a strict order if it is irreflexive and transitive;

3. a partial order if it is reflexive, transitive, and antisymmetric;

4. a total order if it is reflexive, transitive, antisymmetric, and complete.

A set $C \in \mathcal{P}(V)$ is called a cone if $tv \in C$ for every $v \in C$ and $t > 0$. Let us remark that a cone considered in this paper does not necessarily contain the zero vector θ_V. Let C be a convex cone in V. Then, $C + C = C$ holds, and $\text{int} C$ and $\text{cl} C$ are also convex cones. As mentioned in Section 4.1, if $\theta_V \in C$, then $\leq_C$ induced by C is reflexive and, hence, a preorder. If not, then $\leq_C$ is irreflexive and hence a strict order. In addition to $\theta_V \in C$, assuming that C is pointed (i.e., $C \cap (-C) = \{\theta_V\}$), one can check that $\leq_C$ is antisymmetric and becomes a partial order.

Proposition 4.2.1 *Let C, C' be convex cones in V and $d \in V$. Assume that $C + (0, +\infty) d \subset C'$. Then, for any $v_1, v_2 \in V$ and $t, t' \in \mathbb{R}$ with $t > t'$,*

$$v_1 + td \leq_C v_2 \implies v_1 + t'd \leq_{C'} v_2.$$

Proof. From the assumption, we get $C + td \subset C' + t'd$. Since $v_1 + td \leq_C v_2$ implies $v_2 - v_1 \in C + td$, we have $v_2 - v_1 \in C' + t'd$ and, thus, $v_1 + t'd \leq_{C'} v_2$. □

Next, we give a definition of certain binary relations between sets, called set relations, that are used as comparison criteria for sets.

Definition 4.2.2 *Let C be a convex cone in V. Eight types of set relations are defined by*

$$A \leq_C^{(1)} B :\iff \forall a \in A, \forall b \in B, a \leq_C b;$$
$$A \leq_C^{(2L)} B :\iff \exists a \in A, \forall b \in B, a \leq_C b;$$
$$A \leq_C^{(2U)} B :\iff \exists b \in B, \forall a \in A, a \leq_C b;$$
$$A \leq_C^{(2)} B :\iff A \leq_C^{(2L)} B \text{ and } A \leq_C^{(2U)} B;$$
$$A \leq_C^{(3L)} B :\iff \forall b \in B, \exists a \in A, a \leq_C b;$$
$$A \leq_C^{(3U)} B :\iff \forall a \in A, \exists b \in B, a \leq_C b;$$
$$A \leq_C^{(3)} B :\iff A \leq_C^{(3L)} B \text{ and } A \leq_C^{(3U)} B;$$
$$A \leq_C^{(4)} B :\iff \exists a \in A, \exists b \in B, a \leq_C b$$

for $A, B \in \mathcal{P}(V) \setminus \{\emptyset\}$.

Here, the letters L and U stand for "lower" and "upper," respectively.

Remark 4.2.3 *In the original definition given in [12], only six types of set relations are proposed, and they are expressed on the basis of the following equivalences: for $A, B \in \mathcal{P}(V) \setminus \{\emptyset\}$,*

$$A \leq_C^{(1)} B \iff A \subset \bigcap_{b \in B}(b-C) \iff B \subset \bigcap_{a \in A}(a+C);$$
$$A \leq_C^{(2L)} B \iff A \cap \left(\bigcap_{b \in B}(b-C)\right) \neq \emptyset;$$
$$A \leq_C^{(2U)} B \iff \left(\bigcap_{a \in A}(a+C)\right) \cap B \neq \emptyset;$$
$$A \leq_C^{(3L)} B \iff B \subset A+C;$$
$$A \leq_C^{(3U)} B \iff A \subset B-C;$$
$$A \leq_C^{(4)} B \iff A \cap (B-C) \neq \emptyset \iff (A+C) \cap B \neq \emptyset.$$

Definition 4.2.2 is a modified version of that original definition; more essential forms using quantifiers (suggested in [11]), two additional types of set relations, and a different numbering style are newly adopted.

Remark 4.2.4 *More set relations, including minmax-type ones, are dealt with in [7, 8]. Especially in the book [8], many useful concepts and results surrounding set relations and set-valued optimization are systematically summarized; some of them are closely related to our discussion in this paper.*

In general, the relation $\leq_C^{(j)}$ is transitive for $j = 1, 2L, 2U, 2, 3L, 3U, 3$ and not transitive for $j = 4$. If $\theta_V \in C$, $\leq_C^{(j)}$ is reflexive for $j = 3L, 3U, 3, 4$ and, hence, a preorder for $j = 3L, 3U, 3$. If $\theta_V \notin C$, $\leq_C^{(j)}$ is irreflexive and, hence, a strict order for $j = 1, 2L, 2U, 2$. For each $j = 1, 2L, 2U, 2, 3L, 3U, 3, 4$, the relation $\leq_C^{(j)}$ satisfies the following properties for all $A, B \in \mathcal{P}(V) \setminus \{\emptyset\}$:

1. $A \leq_C^{(j)} B \implies A+v \leq_C^{(j)} B+v$ for $v \in V$;

2. $A \leq_C^{(j)} B \implies tA \leq_C^{(j)} tB$ for $t > 0$.

From the definition, we easily obtain the following implications:

$$\begin{cases} A \leq_C^{(1)} B \implies A \leq_C^{(2L)} B \implies A \leq_C^{(3L)} B \implies A \leq_C^{(4)} B; \\ A \leq_C^{(1)} B \implies A \leq_C^{(2U)} B \implies A \leq_C^{(3U)} B \implies A \leq_C^{(4)} B; \\ A \leq_C^{(1)} B \implies A \leq_C^{(2)} B \implies A \leq_C^{(3)} B \implies A \leq_C^{(4)} B \end{cases} \quad (4.5)$$

for $A, B \in \mathcal{P}(V) \setminus \{\emptyset\}$. We also remark several singleton cases.

Remark 4.2.5 *Let C be a convex cone in V. For $A \in \mathcal{P}(V) \setminus \{\emptyset\}$ and $b \in V$, we have the following equivalences:*

1. $A \leq_C^{(1)} \{b\} \iff A \leq_C^{(2U)} \{b\} \iff A \leq_C^{(2)} \{b\}$
   $\iff A \leq_C^{(3U)} \{b\} \iff A \leq_C^{(3)} \{b\};$

2. $A \leq_C^{(2L)} \{b\} \iff A \leq_C^{(3L)} \{b\} \iff A \leq_C^{(4)} \{b\};$

3. $\{b\} \leq_C^{(1)} A \iff \{b\} \leq_C^{(2L)} A \iff \{b\} \leq_C^{(2)} A$
   $\iff \{b\} \leq_C^{(3L)} A \iff \{b\} \leq_C^{(3)} A;$

4. $\{b\} \leq_C^{(2U)} A \iff \{b\} \leq_C^{(3U)} A \iff \{b\} \leq_C^{(4)} A.$

Proposition 4.2.6 *Let C be a convex cone in V. For $A, B \in \mathcal{P}(V) \setminus \{\emptyset\}$, the following statements hold:*

$$A \leq_C^{(1)} B \iff B \leq_{-C}^{(1)} A; \quad A \leq_C^{(2L)} B \iff B \leq_{-C}^{(2U)} A;$$
$$A \leq_C^{(3L)} B \iff B \leq_{-C}^{(3U)} A; \quad A \leq_C^{(2U)} B \iff B \leq_{-C}^{(2L)} A;$$
$$A \leq_C^{(3U)} B \iff B \leq_{-C}^{(3L)} A; \quad A \leq_C^{(4)} B \iff B \leq_{-C}^{(4)} A.$$

Proof. The conclusion follows immediately from Definition 4.2.2. □

Proposition 4.2.7 *Let C, C' be convex cones in V and $d \in V$. Assume that $C + (0, +\infty)d \subset C'$. Then, for each $j = 1, 2L, 2U, 2, 3L, 3U, 3, 4$, any $A, B \in \mathcal{P}(V) \setminus \{\emptyset\}$, $s, s' \in \mathbb{R}$ with $s < s'$ and $t, t' \in \mathbb{R}$ with $t > t'$,*

$$A \leq_C^{(j)} B + sd \implies A \leq_{C'}^{(j)} B + s'd$$

and

$$A + td \leq_C^{(j)} B \implies A + t'd \leq_{C'}^{(j)} B.$$

Proof. The conclusion follows immediately from Definition 4.2.2 and Proposition 4.2.1. □

Remark 4.2.8 *If convex cone $C \in \mathcal{P}(V)$ contains θ_V, the following elementary properties hold for $A, B \in \mathcal{P}(V) \setminus \{\emptyset\}$:*

1. $\bigcap_{a \in A}(a+C) = \bigcap_{v \in A-C}(v+C)$ and $\bigcap_{b \in B}(b-C) = \bigcap_{v \in B+C}(v-C)$;
2. $A+C = A+C+C$ and $B-C = B-C-C$;
3. $B \subset (A+C)$ if and only if $(B+C) \subset (A+C)$;
4. $A \subset (B-C)$ if and only if $(A-C) \subset (B-C)$.

Moreover, for $A, B \in \mathcal{P}(V) \setminus \{\emptyset\}$, conditions $A \subset B$ and $A \cap B \neq \emptyset$ are invariant under translation and scalar multiplication; for $v \in V$ and $t > 0$,

5. $A \subset B \implies (A+v) \subset (B+v)$ and $(tA) \subset (tB)$;
6. $A \cap B \neq \emptyset \implies (A+v) \cap (B+v) \neq \emptyset$ and $(tA) \cap (tB) \neq \emptyset$.

Hence, for $A, B, D, E \in \mathcal{P}(V) \setminus \{\emptyset\}$, the following property holds:

7. $A \cap B \neq \emptyset$ and $D \cap E \neq \emptyset \implies (A+D) \cap (B+E) \neq \emptyset$.

Proposition 4.2.9 (Proposition 2.2 in [13]) *Assume that convex cone $C \in \mathcal{P}(V)$ contains θ_V, then the following equivalences hold for $A, B \in \mathcal{P}(V) \setminus \{\emptyset\}$:*

1. $A \leq_C^{(1)} B \iff A \leq_C^{(1)} (B+C) \iff (A-C) \leq_C^{(1)} B$
   $\iff (A-C) \leq_C^{(1)} (B+C)$;
2. $A \leq_C^{(2L)} B \iff A \leq_C^{(2L)} (B+C) \iff (A+C) \leq_C^{(2L)} B$
   $\iff (A+C) \leq_C^{(2L)} (B+C)$;
3. $A \leq_C^{(2U)} B \iff A \leq_C^{(2U)} (B-C) \iff (A-C) \leq_C^{(2U)} B$
   $\iff (A-C) \leq_C^{(2U)} (B-C)$;
4. $A \leq_C^{(3L)} B \iff A \leq_C^{(3L)} (B+C) \iff (A+C) \leq_C^{(3L)} B$
   $\iff (A+C) \leq_C^{(3L)} (B+C)$;
5. $A \leq_C^{(3U)} B \iff A \leq_C^{(3U)} (B-C) \iff (A-C) \leq_C^{(3U)} B$
   $\iff (A-C) \leq_C^{(3U)} (B-C)$;
6. $A \leq_C^{(4)} B \iff A \leq_C^{(4)} (B-C) \iff (A+C) \leq_C^{(4)} B$
   $\iff (A+C) \leq_C^{(4)} (B-C)$.

Based on these set relations, we introduce the following scalarizing functions for sets in a real vector space, which are certain generalizations as unification of several nonlinear scalarizations proposed in [5, 24].

Definition 4.2.10 *For each $j = 1, 2L, 2U, 2, 3L, 3U, 3, 4$, we define*

$$I_C^{(j)}(A; W, d) := \inf \left\{ t \in \mathbb{R} \,\middle|\, A \leq_C^{(j)} (W+td) \right\}; \tag{4.6}$$

$$S_C^{(j)}(A; W, d) := \sup \left\{ t \in \mathbb{R} \,\middle|\, (W+td) \leq_C^{(j)} A \right\} \tag{4.7}$$

for $A, W \in \mathcal{P}(V) \setminus \{\emptyset\}$ and $d \in C$; W and d are index parameters playing key roles as a reference set and a direction, respectively, on these sublinear-like scalarization for a given set A.

The reason why we refer to this type of scalarization as "sublinear-like" is mentioned later. The functions in Definition 4.2.10 are introduced in [13, 14], which originate from the idea of Gerstewitz's (Tammer's) sublinear scalarizing functional in [1]. These scalarizing functions measure how far a set needs to be moved toward a specific direction to fulfill each set relation. This idea is inspired by the scalarizing functions introduced by Hamel and Löhne [5]. Indeed, $z^l(\cdot)$ and $z^u(\cdot)$ in [5] can be represented in the form of (4.6) and (4.7) as follows:

$$z^l(A) := \inf\{t \in \mathbb{R} \mid td + W \subset A + \mathrm{cl}\,C\}$$
$$= \inf\left\{t \in \mathbb{R} \,\middle|\, A \leq_{\mathrm{cl}\,C}^{(3L)} (W+td)\right\} = I_{\mathrm{cl}\,C}^{(3L)}(A;W,d);$$
$$z^u(A) := \sup\{t \in \mathbb{R} \mid td + W \subset A - \mathrm{cl}\,C\}$$
$$= \sup\left\{t \in \mathbb{R} \,\middle|\, (W+td) \leq_{\mathrm{cl}\,C}^{(3U)} A\right\} = S_{\mathrm{cl}\,C}^{(3U)}(A;W,d).$$

Remark 4.2.11 *We consider singleton cases on reference set W in (4.6) and (4.7). If $W = \{w\}$ for $w \in V$, then we get the following reduced representation of the scalarization of $A \in \mathcal{P}(V) \setminus \{\emptyset\}$ by Remark 4.2.5:*

1. $I_C^{(1)}(A;\{w\},d) = I_C^{(2U)}(A;\{w\},d) = I_C^{(2)}(A;\{w\},d)$
$= I_C^{(3U)}(A;\{w\},d) = I_C^{(3)}(A;\{w\},d);$

2. $I_C^{(2L)}(A;\{w\},d) = I_C^{(3L)}(A;\{w\},d) = I_C^{(4)}(A;\{w\},d);$

3. $S_C^{(1)}(A;\{w\},d) = S_C^{(2L)}(A;\{w\},d) = S_C^{(2)}(A;\{w\},d)$
$= S_C^{(3L)}(A;\{w\},d) = S_C^{(3)}(A;\{w\},d);$

4. $S_C^{(2U)}(A;\{w\},d) = S_C^{(3U)}(A;\{w\},d) = S_C^{(4)}(A;\{w\},d).$

Especially, if $W = \{\theta_V\}$, we notice by (4.1) and (4.4) that for $A \in \mathcal{P}(V) \setminus \{\emptyset\}$,

$$I_C^{(j)}(A;\{\theta_V\},d) = \begin{cases} \sup_{v \in A} h_C(v;d) & \text{for } j = 1, 2U, 2, 3U, 3; \\ \inf_{v \in A} h_C(v;d) & \text{for } j = 2L, 3L, 4 \end{cases} \quad (4.8)$$

and

$$S_C^{(j)}(A;\{\theta_V\},d) = \begin{cases} \inf_{v \in A} (-h_C(-v;d)) & \text{for } j = 1, 2L, 2, 3L, 3; \\ \sup_{v \in A} (-h_C(-v;d)) & \text{for } j = 2U, 3U, 4. \end{cases} \quad (4.9)$$

They are certain generalizations of four types of scalarization for sets proposed in [16, 17]. These facts suggest a similar approach to characterize set-valued mappings in the same way used in [17] as mentioned in Section 4.1 and also to apply it to establish alternative theorems for set-valued maps without convexity assumptions in a similar way used in [16].

Proposition 4.2.12 *Let C be a convex cone in V. The following inequalities hold between each scalarizing function for sets:*

$$I_C^{(4)}(A;W,d) \leq I_C^{(3L)}(A;W,d) \leq I_C^{(2L)}(A;W,d) \leq I_C^{(1)}(A;W,d);$$
$$I_C^{(4)}(A;W,d) \leq I_C^{(3U)}(A;W,d) \leq I_C^{(2U)}(A;W,d) \leq I_C^{(1)}(A;W,d);$$
$$I_C^{(4)}(A;W,d) \leq I_C^{(3)}(A;W,d) \leq I_C^{(2)}(A;W,d) \leq I_C^{(1)}(A;W,d);$$
$$S_C^{(1)}(A;W,d) \leq S_C^{(2L)}(A;W,d) \leq S_C^{(3L)}(A;W,d) \leq S_C^{(4)}(A;W,d);$$
$$S_C^{(1)}(A;W,d) \leq S_C^{(2U)}(A;W,d) \leq S_C^{(3U)}(A;W,d) \leq S_C^{(4)}(A;W,d);$$
$$S_C^{(1)}(A;W,d) \leq S_C^{(2)}(A;W,d) \leq S_C^{(3)}(A;W,d) \leq S_C^{(4)}(A;W,d)$$

for $A, W \in \mathcal{P}(V) \setminus \{\emptyset\}$ *and* $d \in C$.

Proof. The conclusion follows immediately from (4.5). $\square$

Proposition 4.2.13 *Let C be a convex cone in V. There are certain relations among the scalarizations of types* (2L), (2U), (2) *as well as* (3L), (3U), (3):

1. $I_C^{(2)}(A;W,d) = \max\left\{I_C^{(2L)}(A;W,d), I_C^{(2U)}(A;W,d)\right\}$;

2. $I_C^{(3)}(A;W,d) = \max\left\{I_C^{(3L)}(A;W,d), I_C^{(3U)}(A;W,d)\right\}$;

3. $S_C^{(2)}(A;W,d) = \min\left\{S_C^{(2L)}(A;W,d), S_C^{(2U)}(A;W,d)\right\}$;

4. $S_C^{(3)}(A;W,d) = \min\left\{S_C^{(3L)}(A;W,d), S_C^{(3U)}(A;W,d)\right\}$

for $A, W \in \mathcal{P}(V) \setminus \{\emptyset\}$ *and* $d \in C$.

Proof. Let
$$I_1(j) := \left\{t \in \mathbb{R} \;\middle|\; A \leq_C^{(j)} (W+td)\right\}$$
and
$$I_2(j) := \left\{t \in \mathbb{R} \;\middle|\; (W+td) \leq_C^{(j)} A\right\}$$

for $j = 2L, 2U, 3L, 3U$. From Proposition 4.2.7, $I_1(j)$ ($j = 2L, 2U, 3L, 3U$) are either $\emptyset$ or $\mathbb{R}$ or a semi-infinite interval like $(a, +\infty)$ or $[a, +\infty)$ for some $a \in \mathbb{R}$. Hence, we obtain

$$\begin{aligned}
I_C^{(2)}(A;W,d) &= \inf\{t \in \mathbb{R} \mid t \in I_1(2L) \cap I_1(2U)\} \\
&= \max\{\inf\{t \in \mathbb{R} \mid t \in I_1(2L)\}, \inf\{t \in \mathbb{R} \mid t \in I_1(2U)\}\} \\
&= \max\left\{I_C^{(2L)}(A;W,d), I_C^{(2U)}(A;W,d)\right\}.
\end{aligned}$$

The others are derived similarly. $\square$

For each j without $j = 4$, scalarizing functions $I_C^{(j)}(\cdot;W,d)$ and $S_C^{(j)}(\cdot;W,d)$ with nonempty reference set W and direction d have the following monotonicity with respect to $\leq_C^{(j)}$, which is referred to as "j-monotonicity" in [9]:

$$\begin{cases} A \leq_C^{(j)} B \implies I_C^{(j)}(A;W,d) \leq I_C^{(j)}(B;W,d); \\ A \leq_C^{(j)} B \implies S_C^{(j)}(A;W,d) \leq S_C^{(j)}(B;W,d). \end{cases} \quad (4.10)$$

When $d \in \mathrm{int}\, C$, the scalarizing functions above keep the following properties, which are similar to convexity and concavity.

Proposition 4.2.14 (Propositions 2.14 and 2.15 in [9]) *For $A, B \in \mathcal{P}(V) \setminus \{\emptyset\}$, $\lambda \in (0,1)$ and $d \in \mathrm{int}\, C$, the following statements hold:*

1. *For $j = 1, 2L, 3L$,*

$$I_C^{(j)}(\lambda A + (1-\lambda)B; W, d) \leq \lambda I_C^{(j)}(A;W,d) + (1-\lambda)I_C^{(j)}(B;W,d);$$

2. *For $j = 2U, 3U, 4$, if W is $(-C)$-convex,*

$$I_C^{(j)}(\lambda A + (1-\lambda)B; W, d) \leq \lambda I_C^{(j)}(A;W,d) + (1-\lambda)I_C^{(j)}(B;W,d);$$

3. *For $j = 1, 2U, 3U$,*

$$\lambda S_C^{(j)}(A;W,d) + (1-\lambda)S_C^{(j)}(B;W,d) \leq S_C^{(j)}(\lambda A + (1-\lambda)B; W, d);$$

4. *For $j = 2L, 3L, 4$, if W is C-convex,*

$$\lambda S_C^{(j)}(A;W,d) + (1-\lambda)S_C^{(j)}(B;W,d) \leq S_C^{(j)}(\lambda A + (1-\lambda)B; W, d)$$

with the agreement that $+\infty - \infty = +\infty$ and $\alpha(+\infty) = +\infty$, $\alpha(-\infty) = -\infty$ for $\alpha > 0$.

Besides, it is easily seen that

$$\begin{cases} I_C^{(j)}(\alpha A; W, d) = \alpha I_C^{(j)}\left(A; \left(\frac{1}{\alpha}W\right), d\right) & \forall \alpha > 0; \\ S_C^{(j)}(\alpha A; W, d) = \alpha S_C^{(j)}\left(A; \left(\frac{1}{\alpha}W\right), d\right) & \forall \alpha > 0 \end{cases} \quad (4.11)$$

and then under some conditions like $(-C)$-convex and C-convex on W for some cases, the following inequalities hold:

$$I_C^{(j)}(A+B; W, d) \leq I_C^{(j)}\left(A; \left(\frac{1}{2}W\right), d\right) + I_C^{(j)}\left(B; \left(\frac{1}{2}W\right), d\right); \quad (4.12)$$

$$S_C^{(j)}(A+B; W, d) \geq S_C^{(j)}\left(A; \left(\frac{1}{2}W\right), d\right) + S_C^{(j)}\left(B; \left(\frac{1}{2}W\right), d\right). \quad (4.13)$$

If reference set W is a convex set, then it is $(-C)$-convex and C-convex. Therefore, properties (4.11)–(4.13) in case of convex cone W can be regarded as "positively homogeneous," "subadditive" and "superadditive," therefore, they suggest that $I_C^{(j)}(\cdot;W,d)$ and $S_C^{(j)}(\cdot;W,d)$ have "sublinear-like" and "superlinear-like" properties, respectively.

Moreover, they have monotonically increasing or decreasing properties with respect to the relation of inclusion.

Proposition 4.2.15 *Let C be a convex cone in V. For $A, B, W \in \mathcal{P}(V) \setminus \{\emptyset\}$ and $d \in C$, the following monotone properties hold:*

1. $A \subset B \implies I_C^{(j)}(A; W, d) \leq I_C^{(j)}(B; W, d)$ for $j = 1, 2U, 3U$;

2. $A \subset B \implies I_C^{(j)}(A; W, d) \geq I_C^{(j)}(B; W, d)$ for $j = 2L, 3L, 4$;

3. $A \subset B \implies S_C^{(j)}(A; W, d) \leq S_C^{(j)}(B; W, d)$ for $j = 1, 2U, 3U$;

4. $A \subset B \implies S_C^{(j)}(A; W, d) \geq S_C^{(j)}(B; W, d)$ for $j = 2L, 3L, 4$.

Proof. For $D \in \mathcal{P}(V) \setminus \{\emptyset\}$, let

$$I_1^{(j)}(D) := \left\{ t \in \mathbb{R} \mid D \leq_C^{(j)} (W + td) \right\}$$

and

$$I_2^{(j)}(D) := \left\{ t \in \mathbb{R} \mid (W + td) \leq_C^{(j)} D \right\}$$

for $j = 1, 2L, 2U, 3L, 3U, 4$. In the same way as the proof of Proposition 4.2.13, we obtain

$$A \subset B \implies I_1^{(j)}(A) \supset I_1^{(j)}(B) \text{ for } j = 1, 2U, 3U;$$
$$A \subset B \implies I_1^{(j)}(A) \subset I_1^{(j)}(B) \text{ for } j = 2L, 3L, 4;$$
$$A \subset B \implies I_2^{(j)}(A) \subset I_2^{(j)}(B) \text{ for } j = 1, 2U, 3U;$$
$$A \subset B \implies I_2^{(j)}(A) \supset I_2^{(j)}(B) \text{ for } j = 2L, 3L, 4.$$

Hence, the conclusion follows immediately. □

When we consider the possibility to apply the idea of set relations to classical mathematical inequalities, it is important to take into account when the sublinear scalarization for sets takes finite values.

Proposition 4.2.16 (Proposition 3.4 in [14]) *Let C be a convex cone in V. For $A, W \in \mathcal{P}(V) \setminus \{\emptyset\}$ and $d \in \text{int} C$, the following statements hold:*

1. *If A is $(-C)$-bounded and W is C-bounded, $I_C^{(1)}(A; W, d) \in \mathbb{R}$;*

2. *If A is C-proper and W is C-bounded, $I_C^{(j)}(A; W, d) \in \mathbb{R}$ for $j = 2L, 3L$;*

3. *If A is $(-C)$-bounded and W is $(-C)$-proper, $I_C^{(j)}(A; W, d) \in \mathbb{R}$ for $j = 2U, 3U$;*

4. *If A is C-proper and W is $(-C)$-bounded, $I_C^{(4)}(A; W, d) \in \mathbb{R}$.*

Proposition 4.2.17 (Proposition 3.5 in [14]) *Let C be a convex cone in V. For $A, W \in \mathcal{P}(V) \setminus \{\emptyset\}$ and $d \in \text{int} C$, the following statements hold:*

1. *If A is C-bounded and W is $(-C)$-bounded, $S_C^{(1)}(A; W, d) \in \mathbb{R}$;*

2. *If A is C-bounded and W is C-proper, $S_C^{(j)}(A; W, d) \in \mathbb{R}$ for $j = 2L, 3L$;*

3. *If A is $(-C)$-proper and W is $(-C)$-bounded, $S_C^{(j)}(A; W, d) \in \mathbb{R}$ for $j = 2U, 3U$;*

4. *If A is $(-C)$-proper and W is C-bounded, $S_C^{(4)}(A;W,d) \in \mathbb{R}$.*

Proposition 4.2.18 (Lemma 3.4 in [10], Propositions 3.1, 3.3 in [9]) *Let C be a convex cone in V. For $A \in \mathcal{P}(V) \setminus \{\emptyset\}$ and $d \in \text{int}\, C$, the following statements hold:*

1. *If A is C-proper, $I_C^{(3L)}(A;A,d) = 0$ and $S_C^{(3L)}(A;A,d) = 0$;*

2. *If A is $(-C)$-proper, $I_C^{(3U)}(A;A,d) = 0$ and $S_C^{(3U)}(A;A,d) = 0$.*

Remark 4.2.19 *In [27, 28], simple calculation algorithms of the scalarizing functions $I_C^{(j)}(\cdot;W,d)$ and $S_C^{(j)}(\cdot;W,d)$ are proposed in a finite-dimensional Euclidean space for certain polyhedral cases with a convex polyhedral cone inducing the ordering. As a result, it is shown that the problem to calculate each scalarizing function can be decomposed into finite numbers of linear programming subproblems.*

4.3 Inherited Properties of Scalarizing Functions

We would remark that each scalarizing function defined in Definition 4.2.10 inherits the properties of several types of "cone-convexity" (or "cone-concavity") and "cone-continuity" of a parent set-valued map. Given $F : X \to 2^V$, where X and V are real topological vector spaces, we consider composite functions of F with either scalarizing function $I_C^{(j)}(\cdot;W,d)$ or $S_C^{(j)}(\cdot;W,d)$ for each j as follows:

$$(I_{W,d}^{(j)} \circ F)(x) := I_C^{(j)}(F(x);W,d) \quad \text{and} \quad (S_{W,d}^{(j)} \circ F)(x) := S_C^{(j)}(F(x);W,d)$$

with fixed $W \in \mathcal{P}(V) \setminus \{\emptyset\}$, $d \in C$ and convex cone C in V. These compositions transmit several properties in an analogous fashion to linear scalarizing functions like inner product; general results for several types of convexity and quasiconvexity are summarized in [9] and useful results and examples for cone-continuity are in [25].

Definition 4.3.1 (cone-convexity, [12]) *For each $j = 1, 2L, 2U, 2, 3L, 3U, 3, 4$,*

1. *a map F is said to be type (j) C-convex if for each $x_1, x_2 \in X$ and $\lambda \in (0,1)$,*

$$F(\lambda x_1 + (1-\lambda)x_2) \leq_C^{(j)} \lambda F(x_1) + (1-\lambda)F(x_2);$$

2. *a map F is said to be type (j) properly quasi C-convex if for each $x_1, x_2 \in X$ and $\lambda \in (0,1)$,*

$$F(\lambda x_1 + (1-\lambda)x_2) \leq_C^{(j)} F(x_1) \quad \text{or} \quad F(\lambda x_1 + (1-\lambda)x_2) \leq_C^{(j)} F(x_2);$$

3. *a map F is said to be type (j) naturally quasi C-convex if for each $x_1, x_2 \in X$ and $\lambda \in (0,1)$, there exists $\mu \in [0,1]$ such that*

$$F(\lambda x_1 + (1-\lambda)x_2) \leq_C^{(j)} \mu F(x_1) + (1-\mu)F(x_2).$$

Definition 4.3.2 (cone-concavity, [14]) *For each $j = 1, 2L, 2U, 2, 3L, 3U, 3, 4$,*

Table 4.1: Inherited properties on convexity and concavity.

Assumptions ($j = 1, 2L, 2U, 3L, 3U, 4$) on		Conclusions on
ψ	F	$\psi \circ F$
j-monotone	cv type (j) C-convex	cv
	cv type (j) naturally quasi C-convex	
	qcv type (j) C-convex	qcv
	qcv type (j) naturally quasi C-convex	
	type (j) properly quasi C-convex	
	cc type (j) C-concave	cc
	cc type (j) naturally quasi C-concave	
	qcc type (j) C-concave	qcc
	qcc type (j) naturally quasi C-concave	
	type (j) properly quasi C-concave	

1. A map F is said to be type (j) C-concave if for each $x_1, x_2 \in X$ and $\lambda \in (0,1)$,

$$\lambda F(x_1) + (1-\lambda)F(x_2) \leq_C^{(j)} F(\lambda x_1 + (1-\lambda)x_2);$$

2. A map F is said to be type (j) properly quasi C-concave if for each $x_1, x_2 \in X$ and $\lambda \in (0,1)$,

$$F(x_1) \leq_C^{(j)} F(\lambda x_1 + (1-\lambda)x_2) \quad \text{or} \quad F(x_2) \leq_C^{(j)} F(\lambda x_1 + (1-\lambda)x_2);$$

3. A map F is said to be type (j) naturally quasi C-concave if for each $x_1, x_2 \in X$ and $\lambda \in (0,1)$, there exists $\mu \in [0,1]$ such that

$$\mu F(x_1) + (1-\mu)F(x_2) \leq_C^{(j)} F(\lambda x_1 + (1-\lambda)x_2).$$

Theorem 4.3.3 ([9]) *Let $\psi : \mathcal{P}(V) \setminus \{\emptyset\} \to \overline{\mathbb{R}}$ and $F : X \to \mathcal{P}(V)$. If ψ and F satisfy each assumption in Table 4.1, then its correspondent conclusion holds.*

We denote "convex," "quasiconvex," "concave," and "quasiconcave" by "cv," "qcv," "cc," and "qcc" for short in Table 4.1, respectively. By property (4.10) and Proposition 4.2.14, Theorem 4.3.3 guarantees that $I_{W,d}^{(j)} \circ F$ and $S_{W,d}^{(j)} \circ F$ ($j = 1, 2L, 2U, 3L, 3U$) possibly become to be cv or qcv / cc or qcc under some suitable cone-convexity / cone-concavity assumptions on F, respectively.

Remark 4.3.4 *Theorems 3.8 and 3.9 in [9] show some inverse results on the inherited properties, that is, certain convexity or concavity of parent set-valued map F can be derived from those of composite functions of F and its correspondent scalarizing functions $I_C^{(j)}(\cdot; W, d)$ or $S_C^{(j)}(\cdot; W, d)$ with respect to $j = 3L$ or $3U$.*

In addition, each scalarizing function defined in Definition 4.2.10 inherits several cone-continuity of parent set-valued map in Definitions 4.3.5 and 4.3.6, which are summarized in [18, 25]; cones C and $(-C)$ for cone-continuity of the set-valued map play certain roles for "l.s.c." and "u.s.c." of each composite function, respectively. For more details, see [18, 25].

Definition 4.3.5 (lower continuity and upper continuity, [2])

1. A map F is said to be lower continuous (l.c., for short) at x if for every open set $W \subset V$ with $F(x) \cap W \neq \emptyset$, there exists an open neighborhood U of x such that $F(x) \cap W \neq \emptyset$ for all $x \in U$. We say that F is lower continuous on X if F is l.c. at every point $x \in X$.

2. A map F is said to be upper continuous (u.c., for short) at x if for every open set $W \subset V$ with $F(x) \subset W$, there exists an open neighborhood U of x such that $F(x) \subset W$ for all $x \in U$. We say that F is upper continuous on X if F is u.c. at every point $x \in X$.

Definition 4.3.6 (cone-lower continuity and cone-upper continuity, [2])

1. A map F is said to be C-lower continuous (C-l.c., for short) at x if for every open set $W \subset V$ with $F(x) \cap W \neq \emptyset$, there exists an open neighborhood U of x such that $F(x) \cap (W+C) \neq \emptyset$ for all $x \in U$. We say that F is C-lower continuous on X if F is C-l.c. at every point $x \in X$.

2. A map F is said to be C-upper continuous (C-u.c., for short) at x if for every open set $W \subset V$ with $F(x) \subset W$, there exists an open neighborhood U of x such that $F(x) \subset (W+C)$ for all $x \in U$. We say that F is C-upper continuous on X if F is C-u.c. at every point $x \in X$.

4.4 Applications to Set-valued Inequality and Fuzzy Theory

We shall apply the idea of set relations and the sublinear-like scalarization scheme for sets in order to establish generalizations of inequalities for set-valued and fuzzy cases as applications.

4.4.1 Set-valued Fan-Takahashi Minimax Inequality

Theorem 4.4.1 ([26]) *Let X be a nonempty compact convex subset of a topological vector space and $f : X \times X \to \mathbb{R}$. If f satisfies the following conditions:*

1. *For each fixed $y \in X$, $f(\cdot, y)$ is lower semicontinuous,*

2. *For each fixed $x \in X$, $f(x, \cdot)$ is quasiconcave,*

3. *For all $x \in X$, $f(x, x) \leq 0$,*

then there exists $\bar{x} \in X$ such that $f(\bar{x}, y) \leq 0$ for all $y \in X$.

Based on the above theorem, we can get four kinds of Fan-Takahashi minimax inequality for set-valued maps as applications of sublinear-like scalarization scheme; it is shown that assumptions (i) and (ii) in Theorem 4.4.1 are fulfilled for each composition of a parent set-valued map and its correspondent scalarizing function by using several results in Sections 4.2 and 4.3.

Let X be a nonempty compact convex subset of a topological vector space, V a real topological vector space, C a proper closed convex cone in V with $\text{int}\, C \neq \emptyset$ and $F : X \times X \to \mathcal{P}(V) \setminus \{\emptyset\}$.

Theorem 4.4.2 ([14]) *If F satisfies the following conditions:*

1. *F is $(-C)$-bounded on $X \times X$,*

2. *For each fixed $y \in X$, $F(\cdot, y)$ is C-lower continuous,*

3. *For each fixed $x \in X$, $F(x, \cdot)$ is type $(3U)$ properly quasi C-concave,*

4. *For all $x \in X$, $F(x, x) \subset -C$ (that is, $F(x, x) \leq_C^{(3U)} \{\theta_V\}$),*

then there exists $\bar{x} \in X$ such that $F(\bar{x}, y) \subset -C$ (that is, $F(\bar{x}, y) \leq_C^{(3U)} \{\theta_V\}$) for all $y \in X$.

Theorem 4.4.3 ([14]) *If F satisfies the following conditions:*

1. *F is C-proper and C-closed on $X \times X$,*

2. *For each fixed $y \in X$, $F(\cdot, y)$ is C-upper continuous,*

3. *For each fixed $x \in X$, $F(x, \cdot)$ is type $(3L)$ properly quasi C-concave,*

4. *For all $x \in X$, $F(x, x) \cap (-C) \neq \emptyset$ (that is, $F(x, x) \leq_C^{(2L)} \{\theta_V\}$),*

then there exists $\bar{x} \in X$ such that $F(\bar{x}, y) \cap (-C) \neq \emptyset$ (that is, $F(\bar{x}, y) \leq_C^{(2L)} \{\theta_V\}$) for all $y \in X$.

Theorem 4.4.4 ([14]) *If F satisfies the following conditions:*

1. *F is $(-C)$-proper on $X \times X$,*

2. *For each fixed $y \in X$, $F(\cdot, y)$ is C-lower continuous,*

3. *For each fixed $x \in X$, $F(x, \cdot)$ is type $(3U)$ naturally quasi C-concave,*

4. *For all $x \in X$, $F(x, x) \cap \text{int}\, C = \emptyset$ (that is, $\{\theta_V\} \not\leq_{\text{int}\, C}^{(2U)} F(x, x)$),*

then there exists $\bar{x} \in X$ such that $F(\bar{x}, y) \cap \text{int}\, C = \emptyset$ (that is, $\{\theta_V\} \not\leq_{\text{int}\, C}^{(2U)} F(\bar{x}, y)$) for all $y \in X$.

Theorem 4.4.5 ([14]) *If F satisfies the following conditions:*

1. *F is compact-valued on $X \times X$,*

2. *For each fixed $y \in X$, $F(\cdot, y)$ is C-upper continuous,*

3. *For each fixed $x \in X$, $F(x, \cdot)$ is type $(3L)$ naturally quasi C-concave,*

4. *For all $x \in X$, $F(x, x) \not\subset \text{int}\, C$ (that is, $\{\theta_V\} \not\leq_{\text{int}\, C}^{(3L)} F(x, x)$),*

then there exists $\bar{x} \in X$ such that $F(\bar{x}, y) \not\subset \text{int}\, C$ (that is, $\{\theta_V\} \not\leq_{\text{int}\, C}^{(3L)} F(\bar{x}, y)$) for all $y \in X$.

Remark 4.4.6 *From Remark 4.2.5, it follows that set relations "3U" in Theorem 4.4.2, "2L" in Theorem 4.4.3, "2U" in Theorem 4.4.4 and "3L" in Theorem 4.4.5 can be replaced by "1, 2U, 2, 3," "3L, 4," "3U, 4," and "1, 2L, 2, 3," respectively.*

Remark 4.4.7 *In [21, 22], Ricceri type theorem on Fan-Takahashi minimax inequality with different assumptions from that of Theorem 4.4.1 is generalized into set-valued versions by using the sublinear-like scalarization scheme.*

Remark 4.4.8 *It is easy to check that if F is a single-valued function to the real numbers then Theorems 4.4.2–4.4.5 are reduced to Theorem 4.4.1.*

4.4.2 Set-valued Gordan-type Alternative Theorems

As another application of sublinear-like scalarization scheme for sets, we provide 12 kinds of Gordan-type alternative theorems. If the reference set W consists of the origin, they are reduced to those in [16], which are generalizations of the following classical one with non-negative orthant $\mathbb{R}_+^m$ as the ordering cone.

Theorem 4.4.9 ([3]) *Let A be an $m \times n$ matrix and then exactly one of the following systems has a solution:*

1. *There exists $x \in \mathbb{R}^n$ such that $Ax > \theta_{\mathbb{R}^m}$;*

2. *There exists $y \in \mathbb{R}^m$ such that $A^T y = \theta_{\mathbb{R}^n}$ and $y \geq \theta_{\mathbb{R}^m}, y \neq \theta_{\mathbb{R}^m}$,*

where $z_2 \geq z_1$ and $z_2 > z_1$ in $\mathbb{R}^i$ when $z_1 \leq_{\mathbb{R}_+^i} z_2$ and $z_1 \leq_{\text{int}\mathbb{R}_+^i} z_2$, that is, $z_2 - z_1 \in \mathbb{R}_+^i$ and $z_2 - z_1 \in \text{int}\mathbb{R}_+^i$, respectively for $i = m, n$.

This theorem focuses on geometry of a finite number of vectors and the origin. In [16], Nishizawa, Onodsuka, and Tanaka gave generalized forms by using sublinear scalarizaions in formulas (4.8) and (4.9) without any convexity assumption. At first, we show six kinds of generalization of Theorem 4.4.9 based on sublinear-like scalarization scheme.

Theorem 4.4.10 (Theorem 3.1 in [19]) *Let X be a nonempty set, $W \in \mathcal{P}(V) \setminus \{\emptyset\}$, C a convex solid pointed cone in V, $F : X \to \mathcal{P}(V) \setminus \{\emptyset\}$. If*

- *F is compact-valued on X and W is compact in case of $j = 1$;*

- *W is compact in case of $j = 2L, 3L$;*

- *F is compact-valued on X in case of $j = 2U, 3U$,*

then for $j = 1, 2L, 2U, 3L, 3U, 4$, exactly one of the following two systems is consistent:

1. *There exists $x \in X$ such that $F(x) \leq_{\text{int}C}^{(j)} W$;*

2. *There exists $d \in \text{int}C$ such that $I_C^{(j)}(F(x); W, d) \geq 0$ for all $x \in X$.*

Similarly, we have the following another six kinds of generalizations of Theorem 4.4.9.

Theorem 4.4.11 *Let X be a nonempty set, $W \in \mathcal{P}(V) \setminus \{\emptyset\}$, C a convex solid pointed cone in V, $F : X \to \mathcal{P}(V) \setminus \{\emptyset\}$. If*

- *F is compact-valued on X and W is compact in case of $j = 1$;*
- *W is compact in case of $j = 2L, 3L$;*
- *F is compact-valued on X in case of $j = 2U, 3U$,*

then for $j = 1, 2L, 2U, 3L, 3U, 4$, exactly one of the following two systems is consistent:

1. *There exists $x \in X$ such that $W \leq_{\mathrm{int}\, C}^{(j)} F(x)$;*
2. *There exists $d \in \mathrm{int}\, C$ such that $S_C^{(j)}(F(x); W, d) \leq 0$ for all $x \in X$.*

The set-valued alternative theorems in Theorems 4.4.10 and 4.4.11 are applied to considering robustness for perturbations, that is, the stability of mathematical programming models; see [19].

4.4.3 Application to Fuzzy Theory

The concept of fuzzy set (Zadeh [29]) is widely known as a natural extension of usual set. We provide a new methodology for comparing two fuzzy sets, using "fuzzy-set relations" induced by the set relations as comparison criteria of fuzzy sets. Also, by regarding the sublinear-like scalarization scheme as "difference evaluation functions" for fuzzy sets, we can show that these functions correspond well to the fuzzy-set relations. In the following, basic notions of fuzzy sets and the existing results are briefly presented. For more details, see [6].

A fuzzy set $\tilde{A}$ in V is a pair $(V, \mu_{\tilde{A}})$ where $\mu_{\tilde{A}}$ is a function from V to the unit interval $[0,1]$ and is called the membership function of $\tilde{A}$. For each $\alpha \in [0,1]$, the α-level set (α-cut) of $\tilde{A}$ is defined as

$$[\tilde{A}]_\alpha := \begin{cases} \{v \in V \mid \mu_{\tilde{A}}(v) \geq \alpha\} & (\alpha \in (0,1]) \\ \mathrm{cl}\{v \in V \mid \mu_{\tilde{A}}(v) > 0\} & (\alpha = 0). \end{cases}$$

A fuzzy set $\tilde{A}$ is said to be normal if $[\tilde{A}]_1 \neq \emptyset$. We denote by $\mathcal{F}_N(V)$ the set of all normal fuzzy sets in V. For given $d \in V$, the translation $\tilde{A} + d$ is defined by $\mu_{\tilde{A}+d}(v) := \mu_{\tilde{A}}(v - d)$, $v \in V$.

Let C be a convex cone in V, $d \in \mathrm{int}\, C$, and Ω a nonempty subset of $[0,1]$.

Definition 4.4.12 *For each $j = 1, 2L, 2U, 2, 3L, 3U, 3, 4$, a fuzzy-set relation $\leq_C^{\Omega(j)}$ is defined by*

$$\tilde{A} \leq_C^{\Omega(j)} \tilde{B} :\iff \forall \alpha \in \Omega,\ [\tilde{A}]_\alpha \leq_C^{(j)} [\tilde{B}]_\alpha$$

for $\tilde{A}, \tilde{B} \in \mathcal{F}_N(V)$.

Definition 4.4.13 *For each $j = 1, 2L, 2U, 2, 3L, 3U, 3, 4$, a difference evaluation function $D_{C,d}^{\Omega(j)} : \mathcal{F}_N(V) \times \mathcal{F}_N(V) \to \overline{\mathbb{R}}$ is defined by*

$$D_{C,d}^{\Omega(j)}(\tilde{A}, \tilde{B}) := \sup\left\{ t \in \mathbb{R} \,\middle|\, (\tilde{A} + td) \leq_C^{\Omega(j)} \tilde{B} \right\}$$

for $\tilde{A}, \tilde{B} \in \mathcal{F}_N(V)$.

The above functions are reduced to (4.7) of Definition 4.2.10 when $\tilde{A}$ and $\tilde{B}$ are crisp, that is, their membership functions coincide with the characteristic functions of some sets $A, B \in \mathcal{P}(V)$.

When V is a normed space, for each j under suitable assumptions of certain compactness and stability of $\tilde{A}, \tilde{B}$ and closedness of Ω, we have

$$\tilde{A} \leq_{\operatorname{cl} C}^{\Omega(j)} \tilde{B} \iff D_{C,d}^{\Omega(j)}(\tilde{A}, \tilde{B}) \geq 0 \tag{4.14}$$

and

$$\tilde{A} \leq_{\operatorname{int} C}^{\Omega(j)} \tilde{B} \iff D_{C,d}^{\Omega(j)}(\tilde{A}, \tilde{B}) > 0. \tag{4.15}$$

These equivalences can be considered as a generalization of the following obvious facts: For $a, b \in \mathbb{R}$,

$$a \leq b \iff b - a \geq 0 \quad \text{and} \quad a < b \iff b - a > 0.$$

In addition, numerical calculation methods of the difference evaluation functions are discussed in [6]. Thus, one can judge whether each fuzzy-set relation holds or not by using the calculation methods, (4.14) and (4.15).

Acknowledgments. The authors would like to express their sincere gratitude to the anonymous referee and the editors of the book for their helpful and sustainable encouragement on the paper.

References

[1] C. Gerstewitz and C. Tammer. Nichtkonvexe dualität in der vektoroptimierung. *Wiss. Z. Tech. Hochsch. Leuna-Merseburg,* 25(3): 357–364, 1983.

[2] A. Göpfert, H. Riahi, C. Tammer and C. Zălinescu. *Variational Methods in Partially Ordered Spaces.* Springer, New York, 2003.

[3] P. Gordan. Über die Auflösung linearer Gleichungen mit reellen coefficienten. *Mathematische Annalen,* 6: 23–28, 1873.

[4] C. Gutiérrez, B. Jiménez, E. Miglierina and E. Molho. Scalarization in set optimization with solid and nonsolid ordering cones. *J. Global Optim.,* 61: 525–552, 2015.

[5] A. H. Hamel and A. Löhne. Minimal element theorems and Ekeland's principle with set relations. *J. Nonlinear Convex Anal.,* 7: 19–37, 2006.

[6] K. Ike and T. Tanaka. Convex-cone-based comparisons of and difference evaluations for fuzzy sets. *Optimization,* 67: 1051–1066, 2018.

[7] J. Jahn and T. X. D. Ha. New order relations in set optimization. *J. Optim. Theory Appl.,* 148: 209–236, 2011.

[8] A. A. Khan, C. Tammer and C. Zălinescu. *Set-valued Optimization: An Introduction with Applications.* Springer, Berlin, 2015.

[9] S. Kobayashi, Y. Saito and T. Tanaka. Convexity for compositions of set-valued map and monotone scalarizing function. *Pacific J. Optim.,* 12: 43–54, 2016.

[10] T. Kubo, T. Tanaka and S. Yamada. Ekekand's variational principle for set-valued maps via scalarization. pp. 283–289. *In*: S. Akashi, W. Takahashi and T. Tanaka (eds.). *Proceedings of the 7th International Conference on Nonlinear Analysis and Convex Analysis -I,* Yokohama Publishers, Yokohama, 2012.

[11] D. Kuroiwa. On set-valued optimization. *Nonlinear Anal.,* 47: 1395–1400, 2001.

[12] D. Kuroiwa, T. Tanaka and T. X. D. Ha. On cone convexity of set-valued maps. *Nonlinear Anal.,* 30: 1487–1496, 1997.

[13] I. Kuwano, T. Tanaka and S. Yamada. Characterizaton of nonlinear scalarizing functions for set-valued maps. pp. 193–204. *In*: S. Akashi, W. Takahashi and T. Tanaka (eds.). *Proceedings of the Asian Conference on Nonlinear Analysis and Optimization*, Yokohama Publishers, Yokohama, 2009.

[14] I. Kuwano, T. Tanaka and S. Yamada. Unified scalarization for sets and set-valued Ky Fan minimax inequality. *J. Nonlinear Convex Anal.*, 11: 513–525, 2010.

[15] D. T. Luc. *Theory of Vector Optimization. Lecture Note in Econom. and Math. Systems*, 319, Springer, Berlin, 1989.

[16] S. Nishizawa, M. Onodsuka and T. Tanaka. Alternative theorems for set-valued maps based on a nonlinear scalarization. *Pacific J. Optim.*, 1: 147–159, 2005.

[17] S. Nishizawa and T. Tanaka. On inherited properties for set-Valued maps and existence theorems for generalized vector equilibrium problems. *J. Nonlinear Convex Anal.*, 5: 187–197, 2004.

[18] Y. Ogata, Y. Saito, T. Tanaka and S. Yamada. Sublinear scalarization methods for sets with respect to set-relations. *Linear Nonlinear Anal.*, 3: 121–132, 2017.

[19] Y. Ogata, T. Tanaka, Y. Saito, G. M. Lee and J. H. Lee. An alternative theorem for set-valued maps via set relations and its application to robustness of feasible sets. *Optimization*, 67: 1067–1075, 2018.

[20] A. M. Rubinov. Sublinear operators and their applications. *Russian Math. Surveys*, 32: 115–175, 1977.

[21] Y. Saito and T. Tanaka. A set-valued generalization of Ricceri's theorem related to Fan-Takahashi minimax inequality. *Nihonkai Math. J.*, 26: 135–144, 2015.

[22] Y. Saito, T. Tanaka and S. Yamada. On generalization of Ricceri's theorem for Fan-Takahashi minimax inequality into set-valued maps via scalarization. *J. Nonlinear Convex Anal.*, 16: 9–19, 2015.

[23] Y. Sawaragi, H. Nakayama and T. Tanino. *Theory of Multiobjective Optimization*. Academic Press, Orlando, Florida, 1985.

[24] A. Shimizu and T. Tanaka. Minimal element theorem with set-relations. *J. Nonlinear Convex Anal.*, 9: 249–253, 2008.

[25] Y. Sonda, I. Kuwano and T. Tanaka. Cone semicontinuity of set-valued maps by analogy with real-valued semicontinuity. *Nihonkai Math. J.*, 21: 91–103, 2010.

[26] W. Takahashi. Nonlinear variational inequalities and fixed point theorems. *J. Math. Soc. Japan*, 28: 168–181, 1976.

[27] H. Yu, K. Ike, Y. Ogata, Y. Saito and T. Tanaka. Computational methods for set-relation-based scalarizing functions. *Nihonkai Math. J.*, 28: 139–149, 2017.

[28] H. Yu, K. Ike, Y. Ogata and T. Tanaka. A calculation approach to scalarization for polyhedral sets by means of set relations. *Taiwanese J. Math.*, 23: 255–267, 2019.

[29] L. A. Zadeh. Fuzzy sets. *Inform. Control*, 8: 338–353, 1965.

Chapter 5

Functions with Uniform Sublevel Sets, Epigraphs and Continuity

Petra Weidner

HAWK Hochschule Hildesheim/Holzminden/Göttingen, University of Applied Sciences and Arts, Göttingen, Germany.

5.1 Introduction

In this chapter, we consider extended real-valued functions which are defined on some real vector space. We will study possibilities to construct functions with uniform sublevel sets having certain properties. The investigation focuses on translative functions. Moreover, any extended real-valued function will be completely described by some function with uniform sublevel sets. One effect is the characterization of continuity by the epigraph of the function.

Functions with uniform sublevel sets are an important tool for scalarization in multi-criteria optimization, decision theory, mathematical finance, production theory and operator theory, since they can represent orders, preference relations and other binary relations [19]. They were used in set-valued optimization, e.g. in [11, 12, 13, 14], [6] and [9].

Gerstewitz functionals are those functions with uniform sublevel sets which can be generated by a linear shift of some set along a line. The class of translative functions coincides with the class of Gerstewitz functionals [21]. Each extended real-valued function will be proved to be the restriction of some Gerstewitz functional to a hyperspace. A Gerstewitz functional is constructed by means of a set and a vector. We will characterize the class of sets and vectors needed to construct all translative functionals on the one hand and all of these functionals which are lower semicontinuous or continuous on the other hand. This investigation is based on statements about the directional closedness of sets. We will also discuss possibilities to generate nontranslative functionals with uniform sublevel sets and some properties of these functions.

In Section 5.2, we will introduce the notation used in this chapter. The directional closedness of sets is investigated in Section 5.3. Section 5.4 contains the general definition of functionals with uniform sublevel sets. The case in which these functions are translative and its relationship to Gerstewitz functionals are discussed in Section 5.5. It is shown in which way each of these translative functionals can be constructed. Corresponding statements are proved for all lower semicontinuous and for all continuous functionals of this type. In Section 5.6, we will study some properties of nontranslative functions with uniform sublevel sets. Finally, in Section 5.7, each extended real-valued function is described as the restriction of some translative function, and interdependencies between the properties of both functions are proved. Here, we will also characterize continuity by the epigraph of the function and apply this result to positively homogeneous functions.

5.2 Preliminaries

Throughout this chapter, Y is assumed to be a real vector space. $\mathbb{R}$ and $\mathbb{N}$ will denote the sets of real numbers and of nonnegative integers, respectively. We define $\mathbb{N}_> := \mathbb{N} \setminus \{0\}$, $\mathbb{R}_+ := \{x \in \mathbb{R}: x \geq 0\}$, $\mathbb{R}_> := \{x \in \mathbb{R}: x > 0\}$, $\mathbb{R}_+^n := \{(x_1,\ldots,x_n)^T \in \mathbb{R}^n: x_i \geq 0 \text{ for all } i \in \{1,\ldots,n\}\}$ for each $n \in \mathbb{N}_>$. Given any set $B \subseteq \mathbb{R}$ and some vector k in Y, we will use the notation $Bk := \{b \cdot k: b \in B\}$. A set $C \subseteq Y$ is a cone if $\lambda C \subseteq C$ for all $\lambda \in \mathbb{R}_+$. The cone C is called nontrivial if $C \neq \emptyset$, $C \neq \{0\}$ and $C \neq Y$ hold. For a subset A of Y, core$A$ will denote the algebraic interior (core) of A and $0^+A := \{u \in Y: A + \mathbb{R}_+ u \subseteq A\}$ the recession cone of A. This recession cone is always a convex cone. The recession cone of a convex cone A is A. Note that $0^+A = Y$ if $A = \emptyset$.

In a topological vector space Y, int$A$, cl$A$ and bd$A$ stand for the (topological) interior, the closure and the boundary, respectively, of A. Consider a functional $\varphi: Y \to \overline{\mathbb{R}}$, where $\overline{\mathbb{R}} := \mathbb{R} \cup \{-\infty, +\infty\}$. Its effective domain is defined as dom$\varphi := \{y \in Y: \varphi(y) \in \mathbb{R} \cup \{-\infty\}\}$. Its epigraph is epi$\varphi := \{(y,t) \in Y \times \mathbb{R}: \varphi(y) \leq t\}$. The sublevel sets of φ are given as sublev$_\varphi(t) := \{y \in Y: \varphi(y) \leq t\}$ with $t \in \mathbb{R}$. φ is said to be finite-valued on $F \subseteq Y$ if it attains only real values on F. It is said to be finite-valued if it attains only real values on Y. According to the rules of convex analysis, inf$\emptyset = +\infty$.

Moreover, we will deal with the following properties of functionals.

Definition 5.2.1 *A functional* $\varphi: Y \to \overline{\mathbb{R}}$ *is said to be*

(a) *convex if* epiφ *is convex,*

(b) *positively homogeneous if* epiφ *is a nonempty cone,*

(c) *subadditive if* epiφ + epi$\varphi \subseteq$ epiφ,

(d) *sublinear if* epiφ *is a nonempty convex cone.*

If φ is finite-valued, the defined properties are equivalent to those for real-valued functions [22].

Especially in vector optimization, monotonicity of functionals turns out to be essential for their usability in scalarizing problems.

Definition 5.2.2 *Assume $F, B \subseteq Y$. $\varphi : Y \to \overline{\mathbb{R}}$ is said to be*

(a) *B-monotone on F if $y^1, y^2 \in F$ and $y^2 - y^1 \in B$ imply $\varphi(y^1) \leq \varphi(y^2)$,*

(b) *strictly B-monotone on F if $y^1, y^2 \in F$ and $y^2 - y^1 \in B \setminus \{0\}$ imply $\varphi(y^1) < \varphi(y^2)$.*

Definition 5.2.3 *Assume that Y is a topological space. A function $\varphi : Y \to \overline{\mathbb{R}}$ is called lower semicontinuous if, for each $y^0 \in Y$, one of the following conditions holds:*

(i) $\varphi(y^0) = -\infty$.

(ii) *For each $h \in \mathbb{R}$ with $h < \varphi(y^0)$, there exists some neighborhood U of y^0 such that $\varphi(y) > h$ for all $y \in U$.*

Semicontinuity can be completely characterized by the sublevel sets of the function on the one hand and by the epigraph of the function on the other hand [15].

Lemma 5.2.4 *Let Y be a topological space, $\varphi : Y \to \overline{\mathbb{R}}$. Then the following statements are equivalent:*

(a) φ *is lower semicontinuous.*

(b) *The sublevel sets $\mathrm{sublev}_\varphi(t)$ are closed for all $t \in \mathbb{R}$.*

(c) *epi φ is closed in $Y \times \mathbb{R}$.*

Section 5.7 contains further properties of extended real-valued functions.

5.3 Directional Closedness of Sets

In this section, we introduce a directional closedness for sets. This property is closely related to the construction of functions with uniform sublevel sets.

Throughout this section, A will be a subset of Y and $k \in Y \setminus \{0\}$.

We define the directional closure according to [21].

Definition 5.3.1 *The k-directional closure $\mathrm{cl}_k(A)$ of A consists of all elements $y \in Y$ with the following property: For each $\lambda \in \mathbb{R}_>$, there exists some $t \in \mathbb{R}_+$ with $t < \lambda$ such that $y - tk \in A$.*
A is said to be k-directionally closed if $A = \mathrm{cl}_k(A)$.

Obviously, $\mathrm{cl}_k(Y) = Y$ and $\mathrm{cl}_k(\emptyset) = \emptyset$.

It is easy to verify that directional closedness can also be characterized in the following way.

Lemma 5.3.2

(a) A is k-directionally closed if and only if, for each $y \in Y$, we have $y \in A$ whenever there exists some sequence $(t_n)_{n \in \mathbb{N}}$ of real numbers with $t_n \searrow 0$ and $y - t_n k \in A$.

(b) $\text{cl}_k(A) = \{y \in Y : \exists (t_n)_{n \in \mathbb{N}} \text{ with } t_n \in \mathbb{R}_+ \text{ and } t_n \to 0 \text{ such that } y - t_n k \in A \text{ for all } n \in \mathbb{N}\}$.

Remark 5.3.3 *Definition 5.3.1 defines the k-directional closure in such a way that it is the closure into direction k. In [16], the k-vector closure of A is defined as $\text{vcl}_k(A) := \{y \in Y : \exists (t_n)_{n \in \mathbb{N}} \text{ with } t_n \to 0 \text{ such that } t_n \in \mathbb{R}_+ \text{ and } y + t_n k \in A \text{ for all } n \in \mathbb{N}\}$. Obviously, $\text{vcl}_k(A) = \text{cl}_{-k}(A)$. This k-vector closure has been used for the investigation of Gerstewitz functionals under the assumption that A is a convex cone and $k \in A \setminus (-A)$ in [16] and without this restrictive assumption in [7].*

Definition 5.3.1 implies immediately [21, Lemma 2.1]:

Lemma 5.3.4

(a) $A \subseteq \text{cl}_k(A)$.

(b) $\text{cl}_k(A) \subseteq \text{cl}_k(B)$ if $A \subseteq B \subseteq Y$.

(c) $\text{cl}_k(A + y) = \text{cl}_k(A) + y$ for each $y \in Y$.

Proposition 5.3.5 $\text{cl}_k(\text{cl}_k(A)) = \text{cl}_k(A)$.

Proof. For $A = \emptyset$, the assertion is true. Assume now $A \neq \emptyset$.

(i) $A \subseteq \text{cl}_k(A)$ implies $\text{cl}_k(A) \subseteq \text{cl}_k(\text{cl}_k(A))$ because of Lemma 5.3.4.

(ii) Take any $y \in \text{cl}_k(\text{cl}_k(A))$ and $\lambda \in \mathbb{R}_>$. By Definition 5.3.1, there exists some $t_0 \in \mathbb{R}_+$ with $t_0 < \frac{\lambda}{2}$ and $y - t_0 k \in \text{cl}_k(A)$. If $t_0 = 0$, then $y \in \text{cl}_k(A)$. Suppose now $t_0 \in \mathbb{R}_>$. By Definition 5.3.1, there exists some $t \in \mathbb{R}_+$ with $t < t_0$ and $(y - t_0 k) - tk \in A$, i.e., $y - (t_0 + t)k \in A$, where $0 < t + t_0 < 2t_0 < \lambda$. Hence $y \in \text{cl}_k(A)$. □

With Lemma 5.3.2(b), the following proposition can easily be verified.

Proposition 5.3.6

(a) *If A is a cone, then $\text{cl}_k(A)$ is a cone.*

(b) *If A is convex, then $\text{cl}_k(A)$ is convex.*

(c) *If $A + A \subseteq A$, then $\text{cl}_k(A) + \text{cl}_k(A) \subseteq \text{cl}_k(A)$.*

Remark 5.3.7 *We can derive from [16, Proposition 2.4]:*
If A is a nontrivial convex cone and $k \in -A \setminus A$, then $\text{cl}_k(A)$ is a convex cone with $k \in -\text{cl}_k(A) \setminus \text{cl}_k(A)$.
The statements of Lemma 5.3.4(a) and (c), Proposition 5.3.5 and Proposition 5.3.6(b) are mentioned for $\text{cl}_{-k}(A)$ in [7, p.2678f].

Lemma 5.3.2 yields:

Proposition 5.3.8 *A is k-directionally closed if $k \in 0^+A$.*

But in applications, we often have to deal with the case $k \in -0^+A$. Let us prove a statement from [21, Lemma 2.3].

Proposition 5.3.9 *If $k \in -0^+A$, then $\mathrm{cl}_k(A) = \{y \in Y : y - \mathbb{R}_> k \subseteq A\}$.*

Proof.

(i) Assume first $y \in \mathrm{cl}_k(A)$. If $y \in A$, then $y - tk \in A$ holds for all $t \in \mathbb{R}_>$. Suppose now $y \notin A$. Take any $t \in \mathbb{R}_>$. By Definition 5.3.1, there exists some $t_0 \in \mathbb{R}_>$ with $t_0 < t$ such that $y - t_0 k \in A$. Then $y - tk = y - t_0 k + (t_0 - t)k \in A + 0^+A \subseteq A$.

(ii) Take any $y \in Y$ for which $y - tk \in A$ holds for all $t \in \mathbb{R}_>$. Then $y \in \mathrm{cl}_k(A)$ by Definition 5.3.1. $\square$

Let us now investigate the relationship between directional and topological closedness in a topological vector space.

Proposition 5.3.10 *Assume that Y is a topological vector space.*

(a) *$\mathrm{cl}A$ is k-directionally closed, i.e., $\mathrm{cl}_k(\mathrm{cl}A) = \mathrm{cl}A$.*

(b) *$\mathrm{cl}_k(A) \subseteq \mathrm{cl}A$.*

(c) *If A is closed, then it is k-directionally closed, i.e., $\mathrm{cl}_k(A) = A$.*

(d) *$\mathrm{cl}_k(A)$ is closed if and only if $\mathrm{cl}_k(A) = \mathrm{cl}A$.*

(e) *$\mathrm{cl}A - \mathbb{R}_> k \subseteq A$ holds if and only if $k \in -0^+A$ and $\mathrm{cl}_k(A) = \mathrm{cl}A$.*

Proof.

(a) Take any $y \in Y$ for which there exists some sequence $(t_n)_{n \in \mathbb{N}}$ of real numbers with $t_n \searrow 0$ and $y - t_n k \in \mathrm{cl}A$. Since each neighborhood of y contains some $y - t_n k$, we get $y \in \mathrm{cl}(\mathrm{cl}A) = \mathrm{cl}A$. Hence, $\mathrm{cl}A$ is k-directionally closed.

(b) $A \subseteq \mathrm{cl}A$ yields $\mathrm{cl}_k(A) \subseteq \mathrm{cl}_k(\mathrm{cl}A) = \mathrm{cl}A$ by (a).

(c) results from (a).

(d) Assume first that $\mathrm{cl}_k(A)$ is closed. By Lemma 5.3.4, $A \subseteq \mathrm{cl}_k(A)$. This yields $\mathrm{cl}A \subseteq \mathrm{cl}(\mathrm{cl}_k(A)) = \mathrm{cl}_k(A)$. This implies $\mathrm{cl}A = \mathrm{cl}_k(A)$ by (b). The reverse direction of the assertion is obvious.

(e) $\mathrm{cl}A - \mathbb{R}_> k \subseteq A$ implies $k \in -0^+A$ and $\mathrm{cl}A \subseteq \mathrm{cl}_k(A)$ by Proposition 5.3.9; hence $\mathrm{cl}_k(A) = \mathrm{cl}A$ because of (b).
If $k \in -0^+A$ and $\mathrm{cl}_k(A) = \mathrm{cl}A$, then Proposition 5.3.9 yields $\mathrm{cl}_k(A) - \mathbb{R}_> k \subseteq A$ and, thus, $\mathrm{cl}A - \mathbb{R}_> k \subseteq A$. $\square$

Example 5.3.11 *The lexicographic order in* $\mathbb{R}^2$ *is represented by the convex, pointed cone* $A = \{(y_1,y_2)^T \in \mathbb{R}^2 : y_1 > 0\} \cup \{(y_1,y_2)^T \in \mathbb{R}^2 : y_1 = 0, y_2 \geq 0\}$. A *is not algebraically closed and, thus, not closed. Obviously,* $0^+A = A$ *and* $-0^+A \setminus \{0\} = -A \setminus \{0\} = Y \setminus A = -A \setminus A$.

(a) *For each* $k \in A \setminus \{0\}$, A *is k-directionally closed by Proposition 5.3.8.*

(b) *For each* $k \in -\operatorname{core} A = \{(y_1,y_2)^T \in \mathbb{R}^2 : y_1 < 0\}$, *we have* $\operatorname{cl} A - \mathbb{R}_> k \subseteq A$ *and* $\operatorname{cl}_k(A) = \operatorname{cl} A = \{(y_1,y_2)^T \in \mathbb{R}^2 : y_1 \geq 0\}$.

(c) *For each* $k \in -A \setminus ((-\operatorname{core} A) \cup \{0\}) = \{(y_1,y_2)^T \in \mathbb{R}^2 : y_1 = 0, y_2 < 0\}$, A *is k-directionally closed and* $\operatorname{cl} A - \mathbb{R}_> k \not\subseteq A$.

5.4 Definition of Functions with Uniform Sublevel Sets

If all sublevel sets of an extended real-valued function $\varphi : Y \to \overline{\mathbb{R}}$ have the same shape, then there exists some set $A \subseteq Y$ and some function $\zeta : \mathbb{R} \to Y$ with

$$\operatorname{sublev}_\varphi(t) = A + \zeta(t) \text{ for all } t \in \mathbb{R}. \tag{5.1}$$

Clearly, A and ζ have to fulfill the condition

$$A + \zeta(t_1) \subseteq A + \zeta(t_2) \text{ for all } t_1, t_2 \in \mathbb{R} \text{ with } t_1 < t_2.$$

Condition (5.1) coincides with

$$\operatorname{epi} \varphi = \{(y,t) \in Y \times \mathbb{R} : y \in A + \zeta(t)\}.$$

It implies

$$\varphi(y) = \inf\{t \in \mathbb{R} : y \in A + \zeta(t)\} \text{ for all } y \in Y \tag{5.2}$$

since $\varphi(y) = \inf\{t \in \mathbb{R} : \varphi(y) \leq t\}$.

Each function $\varphi : Y \to \{-\infty, +\infty\}$ is a function with uniform sublevel sets. In detail, $\varphi : Y \to \overline{\mathbb{R}}$ is a function that does not attain real values if and only if (5.1) holds with some set $A \subseteq Y$ and with the constant function ζ which attains the value 0_Y. Then

$$A = \{y \in Y : \varphi(y) = -\infty\}.$$

Hence, if (5.1) holds with some set $A \subseteq Y$ and some constant function ζ, then φ does not attain real values.

Moreover, (5.1) holds for $A = \emptyset$ if and only if φ is the constant function with the value $+\infty$. (5.1) holds for $A = Y$ if and only if φ is the constant function with the value $-\infty$. In both cases, the choice of ζ does not matter.

Consequently, if $\varphi : Y \to \overline{\mathbb{R}}$ is a function with uniform sublevel sets that attains at least one real value, then it fulfills (5.1) for some set A which is a proper subset of Y and some function ζ which is not constant. The reverse statement is not true in general.

Example 5.4.1 *Take* $Y = \mathbb{R}^2$, $A := \{(y_1,y_2)^T \in Y : 0 \leq y_1 \leq 1\}$ *and define* $\zeta(t) := t \cdot (0,1)^T$ *for each* $t \in \mathbb{R}$. *Then condition (5.1) corresponds to the functional* $\varphi : Y \to \overline{\mathbb{R}}$ *given by*

$$\varphi(y) := \begin{cases} -\infty & \text{if } y \in A, \\ +\infty & \text{if } y \in Y \setminus A. \end{cases}$$

5.5 Translative Functions

If there exists some fixed vector $k \in Y \setminus \{0\}$ such that $\zeta(t) = tk$ for each $t \in \mathbb{R}$, then (5.1) is equivalent to

$$\mathrm{sublev}_\varphi(t) = \mathrm{sublev}_\varphi(0) + tk \quad \text{for all } t \in \mathbb{R}, \tag{5.3}$$

and (5.2) coincides with the definition of a Gerstewitz functional.
A Gerstewitz functional $\varphi_{A,k} : Y \to \overline{\mathbb{R}}$ is given by

$$\varphi_{A,k}(y) := \inf\{t \in \mathbb{R} : y \in A + tk\} \quad \text{for all } y \in Y$$

with $A \subseteq Y$ and $k \in Y \setminus \{0\}$.

This formula was introduced by Gerstewitz (later Gerth, now Tammer) for convex sets A under more restrictive assumptions in the context of vector optimization [2]. Basic properties of this function on topological vector spaces Y have been proved in [3] and [18], later followed by [5], [17], [1] and [20]. For detailed bibliographical notes, see [20]. There, it is also pointed out that researchers from different fields of mathematics and economic theory have applied Gerstewitz functionals. Properties of Gerstewitz functionals on linear spaces were studied in [7], [19] and [21].

Example 5.5.1 *Assume that Y is the vector space of functions $f : X \to \mathbb{R}$, where X is some nonempty set. Choose $k \in Y$ as the function with the constant value 1 at each $x \in X$ and $A := \{f \in Y : f(x) \leq 0 \text{ for all } x \in X\}$. Then we get, for each $f \in Y$,*

$$\begin{aligned}
\varphi_{A,k}(f) &= \inf\{t \in \mathbb{R} : f \in A + tk\} \\
&= \inf\{t \in \mathbb{R} : f - tk \in A\} \\
&= \inf\{t \in \mathbb{R} : f(x) - t \leq 0 \text{ for all } x \in X\} \\
&= \inf\{t \in \mathbb{R} : f(x) \leq t \text{ for all } x \in X\} \\
&= \sup_{x \in X} f(x), \text{ where} \\
\mathrm{dom}\, \varphi_{A,k} &= \{f \in Y : \sup_{x \in X} f(x) \in \mathbb{R}\}.
\end{aligned}$$

Each Gerstewitz functional fulfills equation (5.3).

Proposition 5.5.2 *Consider a function $\varphi : Y \to \overline{\mathbb{R}}$ and $k \in Y \setminus \{0\}$. The following conditions are equivalent to each other:*

$$\mathrm{sublev}_\varphi(t) = \mathrm{sublev}_\varphi(0) + tk \quad \text{for all } t \in \mathbb{R}, \tag{5.4}$$

$$\mathrm{epi}\, \varphi = \{(y,t) \in Y \times \mathbb{R} : y \in \mathrm{sublev}_\varphi(0) + tk\}, \tag{5.5}$$

$$\varphi(y+tk) = \varphi(y) + t \quad \text{for all } y \in Y, t \in \mathbb{R}, \tag{5.6}$$

$$\varphi(y) = \inf\{t \in \mathbb{R} : y \in \mathrm{sublev}_\varphi(0) + tk\} \quad \text{for all } y \in Y. \tag{5.7}$$

φ is said to be k-translative if property (5.6) is satisfied. It is said to be translative if it is k-translative for some $k \in Y \setminus \{0\}$.

Remark 5.5.3 *Proposition 5.5.2 was proved in [21, Proposition 3.1]. Hamel proved the equivalence of (5.4)-(5.6) [8, Theorem 1] and that (5.6) implies (5.7) [10, Proposition 1].*

The k-translativity of $\varphi_{A,k}$ had already been mentioned in [4]. Hamel ([8], [10, Proposition 1]) studied relationships between Gerstewitz functionals and k-translative functions, where he also introduced a notion of directional closedness. However, his definition of k-directional closedness is different from the one we use. He defined A to be k-directionally closed if, for each $y \in Y$, one has $y \in A$ whenever there exists some sequence $(t_n)_{n \in \mathbb{N}}$ of real numbers with $t_n \to 0$ and $y \in A - t_n k$.

Lemma 5.5.4 *Assume $\varphi: Y \to \overline{\mathbb{R}}$ and $k \in Y \setminus \{0\}$. Then, the following conditions are equivalent to each other:*

$$\varphi(y+tk) = \varphi(y) + t \text{ for all } y \in Y, t \in \mathbb{R}, \quad (5.8)$$

$$\varphi(y+tk) \leq \varphi(y) + t \text{ for all } y \in Y, t \in \mathbb{R}, \quad (5.9)$$

$$\varphi(y+tk) \geq \varphi(y) + t \text{ for all } y \in Y, t \in \mathbb{R}, \quad (5.10)$$

$$\{\lambda(k,1) \in Y \times \mathbb{R}: \lambda \in \mathbb{R}\} \subseteq 0^+(\text{epi}\,\varphi). \quad (5.11)$$

Proof. Obviously, (5.8) implies (5.9). Assume now that (5.9) holds. Take any $y \in Y$ and $t \in \mathbb{R}$. Then $\varphi(y) = \varphi((y+tk)-tk) \leq \varphi(y+tk) - t. \Rightarrow \varphi(y+tk) \geq \varphi(y) + t$. By (5.9), $\varphi(y+tk) = \varphi(y) + t$. Hence, (5.8) is fulfilled.
The equivalence between (5.8) and (5.10) can be proved analogously.
Assume now that (5.8) holds. Take any $(y,t) \in \text{epi}\,\varphi$ and $\lambda \in \mathbb{R}$. Then $\varphi(y) \leq t$ implies $\varphi(y+\lambda k) = \varphi(y) + \lambda \leq t + \lambda$. Hence, $(y,t) + \lambda(k,1) \in \text{epi}\,\varphi$. Thus (5.11) is satisfied.
Assume now (5.11). Take first any $y \in Y$ with $\varphi(y) \in \mathbb{R}$. Then $(y, \varphi(y)) \in \text{epi}\,\varphi$. This implies $(y+tk, \varphi(y)+t) \in \text{epi}\,\varphi$ for each $t \in \mathbb{R}$ by (5.11). Hence, $\varphi(y+tk) \leq \varphi(y) + t$ holds for all $t \in \mathbb{R}$, where $\varphi(y+tk) < +\infty$ for each $t \in \mathbb{R}$. Take now any $y \in Y$ with $\varphi(y) = -\infty$. Then $(y, \lambda) \in \text{epi}\,\varphi$ for each $\lambda \in \mathbb{R}$. This yields $(y+tk, \lambda+t) \in \text{epi}\,\varphi$ for each $t, \lambda \in \mathbb{R}$ and, thus, $\varphi(y+tk) \leq \varphi(y) + t$ holds for all $t \in \mathbb{R}$, where $\varphi(y+tk) = -\infty$ for each $t \in \mathbb{R}$. Consider now any $y \in Y$ with $\varphi(y) = +\infty$. By the previous considerations, we get $\varphi(y+tk) = +\infty$ for each $t \in \mathbb{R}$. Hence (5.9) is fulfilled. □

Remark 5.5.5 *The equivalence between (5.8) and (5.9) was noted in [8, p.11] without proof. A proof of the equivalence between (5.8) and (5.11) was given in [8, Theorem 1].*

Note that the sublevel sets of a Gerstewitz functional $\varphi_{A,k}$ do not always have the shape of A. However, they can be described using A and k in the following way [21, Theorem 3.1].

Theorem 5.5.6 *Assume $A \subseteq Y$ and $k \in Y \setminus \{0\}$. Consider $\tilde{A} := \text{sublev}_{\varphi_{A,k}}(0)$.*

(a) *$\tilde{A}$ is the unique set for which*

$$\text{sublev}_{\varphi_{A,k}}(t) = \tilde{A} + tk \text{ for all } t \in \mathbb{R}$$

holds.

(b) $\tilde{A}$ is the unique set with the following properties:

   (i) $\tilde{A}$ is k-directionally closed,
   (ii) $k \in -0^+\tilde{A} \setminus \{0\}$ and
   (iii) $\varphi_{A,k}$ coincides with $\varphi_{\tilde{A},k}$ on Y.

(c) $\tilde{A}$ is the k-closure of $A - \mathbb{R}_+ k$. It consists of those points $y \in Y$ for which $y - tk \in A - \mathbb{R}_+ k$ holds for each $t \in \mathbb{R}_>$.

Proposition 5.5.2 and Theorem 5.5.6 completely characterize the class of translative functions. This is summarized in the following theorem [21, Theorem 3.2].

Theorem 5.5.7 *For each $k \in Y \setminus \{0\}$, the class of k-translative functions on Y coincides with the class of Gerstewitz functionals $\{\varphi_{A,k} : A \subseteq Y\}$ and with the class of functions $\varphi : Y \to \overline{\mathbb{R}}$ having uniform sublevel sets which fulfill the condition*

$$\mathrm{sublev}_\varphi(t) = \mathrm{sublev}_\varphi(0) + tk \text{ for all } t \in \mathbb{R}.$$

Hence, each translative functional can be constructed as a Gerstewitz functional. Moreover, the class of sets A and vectors k which have to be used in order to construct all translative functionals can be restricted. Theorem 5.5.6(b) implies:

Corollary 5.5.8 *For each $k \in Y \setminus \{0\}$, the class*

$$\{\varphi_{A,k} : A \subseteq Y\}$$

of all k-translative functions on Y coincides with

$$\{\varphi_{A,k} : A \subseteq Y \text{ is k-directionally closed}, k \in -0^+A \setminus \{0\}\}.$$

Note that $A - \mathbb{R}_+ k = A$ if and only if $k \in -0^+A$. Hence, we get from Theorem 5.5.6:

Corollary 5.5.9 *Assume $A \subseteq Y$ and $k \in -0^+A \setminus \{0\}$. Then*

$$\mathrm{sublev}_{\varphi_{A,k}}(t) = \mathrm{cl}_k(A) + tk \text{ for all } t \in \mathbb{R}.$$

Theorem 5.5.6 yields because of Lemma 5.2.4:

Proposition 5.5.10 *Assume that Y is a topological vector space, $A \subseteq Y$ and $k \in Y \setminus \{0\}$. Then $\varphi_{A,k}$ is lower semicontinuous if and only if $\mathrm{cl}_k(A - \mathbb{R}_+ k)$ is a closed set.*

Properties of lower semicontinuous Gerstewitz functionals have been studied in [20]. The proof of Theorem 2.9 from [20] contains the statement of the next lemma for proper subsets A of Y. If A is not a proper subset of Y, the statement is obvious.

Lemma 5.5.11 *Assume that Y is a topological vector space, $A \subseteq Y$ and $k \in Y \setminus \{0\}$ with*

$$\mathrm{cl} A - \mathbb{R}_> k \subseteq A.$$

$\varphi_{A,k}$ *is continuous if and only if* $\mathrm{cl} A - \mathbb{R}_> k \subseteq \mathrm{int} A$ *holds.*

Theorem 5.5.12 *Assume that Y is a topological vector space, $A \subseteq Y$ and $k \in -0^+A \setminus \{0\}$.*

(1) *The following conditions are equivalent to each other:*

  (a) $\varphi_{A,k}$ *is lower semicontinuous.*

  (b) $\mathrm{cl}_k(A) = \mathrm{cl}\,A$.

  (c) $\mathrm{sublev}_{\varphi_{A,k}}(t) = \mathrm{cl}\,A + tk$ *holds for all $t \in \mathbb{R}$.*

  (d) $\mathrm{cl}\,A - \mathbb{R}_> k \subseteq A$.

(2) $\varphi_{A,k}$ *is continuous if and only if $\mathrm{cl}\,A - \mathbb{R}_> k \subseteq \mathrm{int}\,A$.*

Proof.

(1) Because of Corollary 5.5.9 and Lemma 5.2.4, $\varphi_{A,k}$ is lower semicontinuous if and only if $\mathrm{cl}_k(A)$ is a closed set. This is equivalent to $\mathrm{cl}_k(A) = \mathrm{cl}\,A$ by Proposition 5.3.10(d).
(b) is equivalent to (c) because of Corollary 5.5.9.
(b) is equivalent to (d) by Proposition 5.3.10(e).

(2) Assume first that $\varphi_{A,k}$ is continuous. Then it is lower semicontinuous, and $\mathrm{cl}\,A - \mathbb{R}_> k \subseteq A$ by (1). Lemma 5.5.11 implies $\mathrm{cl}\,A - \mathbb{R}_> k \subseteq \mathrm{int}\,A$.
Assume now $\mathrm{cl}\,A - \mathbb{R}_> k \subseteq \mathrm{int}\,A$. Obviously, $\mathrm{cl}\,A - \mathbb{R}_> k \subseteq A$. Lemma 5.5.11 implies that $\varphi_{A,k}$ is continuous. □

We can restrict the class of sets A and vectors k which have to be used in order to construct all semicontinuous and continuous translative functions.

Theorem 5.5.13 *Assume Y to be a topological vector space, $k \in Y \setminus \{0\}$.*

(1) *The class*
$$\{\varphi_{A,k} : A \subseteq Y, \varphi_{A,k} \text{ lower semicontinuous}\} \tag{5.12}$$
of all lower semicontinuous k-translative functions on Y coincides with
$$\{\varphi_{A,k} : A \subseteq Y, \mathrm{cl}\,A - \mathbb{R}_> k \subseteq A\} \tag{5.13}$$
and with
$$\{\varphi_{A,k} : A \subseteq Y \text{ closed}, k \in -0^+A \setminus \{0\}\}. \tag{5.14}$$

(2) *The class*
$$\{\varphi_{A,k} : A \subseteq Y, \varphi_{A,k} \text{ continuous}\} \tag{5.15}$$
of all continuous k-translative functions on Y coincides with
$$\{\varphi_{A,k} : A \subseteq Y, \mathrm{cl}\,A - \mathbb{R}_> k \subseteq \mathrm{int}\,A\} \tag{5.16}$$
and with
$$\{\varphi_{A,k} : A \subseteq Y \text{ closed}, A - \mathbb{R}_> k \subseteq \mathrm{int}\,A\}. \tag{5.17}$$

Proof.

(1) (a) Take first any $A \subseteq Y$ for which $\varphi_{A,k}$ is lower semicontinuous. Because of Corollary 5.5.8, there exists $\tilde{A} \subseteq Y$ with $k \in -0^+\tilde{A} \setminus \{0\}$ and $\varphi_{A,k} = \varphi_{\tilde{A},k}$. Theorem 5.5.12 implies $\mathrm{cl}\tilde{A} - \mathbb{R}_> k \subseteq \tilde{A}$. Hence, set (5.12) is contained in set (5.13).

(b) Take now any $A \subseteq Y$ with $\mathrm{cl}A - \mathbb{R}_> k \subseteq A$. Consider $\varphi := \varphi_{A,k}$. $\mathrm{cl}A - \mathbb{R}_> k \subseteq A$ implies $k \in -0^+A \setminus \{0\}$ and $k \in -0^+(\mathrm{cl}A) \setminus \{0\}$. Theorem 5.5.12 implies $\mathrm{sublev}_\varphi(t) = \mathrm{cl}A + tk$ for all $t \in \mathbb{R}$. Proposition 5.5.2 yields $\varphi = \varphi_{\mathrm{cl}A,k}$. Hence, set (5.13) is contained in set (5.14).

(c) Take any closed set $A \subseteq Y$ with $k \in -0^+A \setminus \{0\}$. Then $\mathrm{cl}A - \mathbb{R}_> k = A - \mathbb{R}_> k \subseteq A$. Theorem 5.5.12 implies that $\varphi_{A,k}$ is lower semicontinuous. Hence, set (5.14) is contained in set (5.12).

(2) (a) Take first any $A \subseteq Y$ for which $\varphi_{A,k}$ is continuous. Because of Corollary 5.5.8, there exists $\tilde{A} \subseteq Y$ with $k \in -0^+\tilde{A} \setminus \{0\}$ and $\varphi_{A,k} = \varphi_{\tilde{A},k}$. Theorem 5.5.12 implies $\mathrm{cl}\tilde{A} - \mathbb{R}_> k \subseteq \mathrm{int}\tilde{A}$. Hence, set (5.15) is contained in set (5.16).

(b) Take now any $A \subseteq Y$ with $\mathrm{cl}A - \mathbb{R}_> k \subseteq \mathrm{int}A$. Consider $\varphi := \varphi_{A,k}$. $\mathrm{cl}A - \mathbb{R}_> k \subseteq \mathrm{int}A$ implies $k \in -0^+A \setminus \{0\}$. Theorem 5.5.12 yields $\mathrm{sublev}_\varphi(t) = \mathrm{cl}A + tk$ for all $t \in \mathbb{R}$. Hence, $\varphi = \varphi_{\mathrm{cl}A,k}$ by Proposition 5.5.2. Because of $\mathrm{cl}A - \mathbb{R}_> k \subseteq \mathrm{int}A \subseteq \mathrm{int}(\mathrm{cl}A)$, set (5.16) is contained in set (5.17).

(c) Take any closed set $A \subseteq Y$ with $A - \mathbb{R}_> k \subseteq \mathrm{int}A$. $\varphi_{A,k}$ is continuous by Theorem 5.5.12. Hence, set (5.17) is contained in set (5.15). □

Remark 5.5.14 *For topological vector spaces Y and $k \in Y \setminus \{0\}$, the equivalence between $\{\varphi: Y \to \overline{\mathbb{R}}: \varphi\ k\text{-translative and lower semicontinuous}\}$ and $\{\varphi_{A,k}: A \subseteq Y\ closed, k \in -0^+A \setminus \{0\}\}$ was proved in [8, Corollary 8], and the equivalence between $\{\varphi: Y \to \overline{\mathbb{R}}: \varphi\ k\text{-translative and continuous}\}$ and $\{\varphi_{A,k}: A \subseteq Y\ closed, A - \mathbb{R}_> k \subseteq \mathrm{int}A\}$ can be deduced from [8, Corollary 9].*

Let us mention some properties of Gerstewitz functionals which help to understand the examples in the following sections. We get from [19, Theorem 2.3]:

Proposition 5.5.15 *Assume that A is a proper, k-directionally closed subset of Y with $k \in -0^+A \setminus \{0\}$.*

(a) $\varphi_{A,k}$ *is convex if and only if A is convex.*

(b) $\varphi_{A,k}$ *is positively homogeneous if and only if A is a cone.*

(c) $\varphi_{A,k}$ *is subadditive if and only if $A + A \subseteq A$.*

(d) $\varphi_{A,k}$ *is sublinear if and only if A is a convex cone.*

The following statement was proved in [19, Theorem 2.2].

Proposition 5.5.16 *Assume that A is a proper subset of Y and $k \in -\mathrm{core}\,0^+A$. Then $\varphi_{A,k}$ is finite-valued.*

Moreover, we get from [19, Proposition 5.1(e)]:

Proposition 5.5.17 *Assume that A is a nontrivial convex cone in Y and $k \in -A \setminus \{0\}$. Then $\varphi_{A,k}$ is finite-valued if and only if $k \in -\operatorname{core} A$.*

We illustrate the statements of this section by an example.

Example 5.5.18 *In $Y = \mathbb{R}^2$, choose $A := \{(y_1, y_2)^T \in Y : 0 < y_1 \leq 2, 0 \leq y_2 \leq 2\}$ and $k := (-1, -1)^T$. Then $A - \mathbb{R}_+ k = \{(y_1, y_2)^T \in Y : 0 < y_1 < 2, 0 \leq y_2 < y_1 + 2\} \cup \{(y_1, y_2)^T \in Y : y_1 \geq 2, y_1 - 2 \leq y_2 < y_1 + 2\}$. $\tilde{A} := \operatorname{cl}_k(A - \mathbb{R}_+ k) = \{(y_1, y_2)^T \in Y : 0 \leq y_1 < 2, 0 \leq y_2 < y_1 + 2\} \cup \{(y_1, y_2)^T \in Y : y_1 \geq 2, y_1 - 2 \leq y_2 < y_1 + 2\}$, and $\operatorname{dom} \varphi_{\tilde{A},k} = \{(y_1, y_2)^T \in Y : y_1 - 2 \leq y_2 < y_1 + 2\}$. By Theorem 5.5.6, $\varphi_{A,k} = \varphi_{\tilde{A},k}$, $k \in -0^+ \tilde{A} \setminus \{0\}$ and*

$$\operatorname{sublev}_{\varphi_{A,k}}(t) = \tilde{A} + tk \text{ for all } t \in \mathbb{R}.$$

We have

$$\varphi_{A,k}(y) = \begin{cases} t & \text{if } y \in (\operatorname{bd} \mathbb{R}_+^2 + tk) \cap \operatorname{dom} \varphi_{A,k}, \\ +\infty & \text{if } y \in Y \setminus \operatorname{dom} \varphi_{A,k}. \end{cases}$$

$\varphi_{A,k}$ is convex by Proposition 5.5.15. Since $\operatorname{cl} \tilde{A} - \mathbb{R}_{>} k \not\subseteq \tilde{A}$, Theorem 5.5.12 implies that $\varphi_{A,k}$ is not lower semicontinuous.
Note that the restriction of $\varphi_{A,k}$ to its effective domain is a continuous function. For $\bar{A} := A \cup \{(0,2)^T\}$, we get a convex, lower semicontinuous functional $\varphi_{\bar{A},k}$, which coincides with $\varphi_{A,k}$ on $\operatorname{dom} \varphi_{A,k}$. Since $\operatorname{cl} \bar{A} - \mathbb{R}_{>} k \not\subseteq \operatorname{int} \bar{A}$, Theorem 5.5.12 implies that $\varphi_{\bar{A},k}$ is not continuous. But the restriction of $\varphi_{\bar{A},k}$ to its effective domain is a continuous function.

5.6 Nontranslative Functions with Uniform Sublevel Sets

In this section, we will give some examples for nontranslative functions with uniform sublevel sets and throw a glance at the sublevel sets of sublinear functions.

First, we show that a certain alternation in the definition of translativity does not really extend the class of translative functions.

Proposition 5.6.1 *Assume that $f : Y \to \overline{\mathbb{R}}$ is a function which attains at least one real value and has the property*

$$\forall y \in Y, t \in \mathbb{R} : f(y + tk) = f(y) + \xi(t)$$

*for some $k \in Y \setminus \{0\}$ and some continuous function $\xi : \mathbb{R} \to \mathbb{R}$.
Then f is a translative function, i.e., there exists some $\tilde{k} \in Y \setminus \{0\}$ such that*

$$\forall y \in Y, t \in \mathbb{R} : f(y + t\tilde{k}) = f(y) + t.$$

Proof. Take any $y^0 \in Y$ with $f(y^0) \in \mathbb{R}$.

$$\begin{aligned}
\forall t_1, t_2 \in \mathbb{R} : f(y^0 + (t_1 + t_2)k) &= f(y^0) + \xi(t_1 + t_2) \text{ and} \\
f(y^0 + (t_1 + t_2)k) &= f(y^0 + t_1 k) + \xi(t_2) \\
&= f(y^0) + \xi(t_1) + \xi(t_2). \\
\text{Hence, } \xi(t_1 + t_2) &= \xi(t_1) + \xi(t_2).
\end{aligned}$$

Thus, ξ is additive. Moreover, $f(y^0) = f(y^0 + 0 \cdot k) = f(y^0) + \xi(0)$. This yields $\xi(0) = 0$. Since ξ is additive and continuous and $\xi(0) = 0$, it has to be linear [22, Theorem 6.3]. Thus, there exists some $\lambda \in \mathbb{R}$ such that $\xi(t) = \lambda t$ for all $t \in \mathbb{R}$. Hence,

$$\forall y \in Y, t \in \mathbb{R}: \quad \begin{aligned} f(y+tk) &= f(y) + \lambda t, \\ f(y + \tfrac{t}{\lambda} \cdot k) &= f(y) + \lambda \cdot \tfrac{t}{\lambda} = f(y) + t, \\ f(y + t\tilde{k}) &= f(y) + t \text{ for } \tilde{k} := \tfrac{k}{\lambda}. \end{aligned}$$

$\square$

But the following proposition offers a possibility to construct nontranslative functions with uniform sublevel sets which are generated by shifting a set along a line.

Proposition 5.6.2 *Assume that A is a proper, k-directionally closed subset of Y with $k \in -0^+ A \setminus \{0\}$, and that $\xi: \mathbb{R} \to \mathbb{R}$ is a strictly monotonically increasing, continuous function. Define*

$$f(y) := \inf\{t \in \mathbb{R}: y \in A + \xi(t) \cdot k\} \text{ for each } y \in Y.$$

Then f is the function with the uniform sublevel sets

$$\mathrm{sublev}_f(t) = A + \xi(t) \cdot k \text{ for each } t \in \mathbb{R}.$$

Proof. Consider any $t \in \mathbb{R}$. By the definition of f, $y \in A + \xi(t) \cdot k$ implies $f(y) \leq t$. Thus, $A + \xi(t) \cdot k \subseteq \mathrm{sublev}_f(t)$.
Take first any $y \in Y$ with $f(y) < t$. Then there exists some $t_0 \in \mathbb{R}$ with $t_0 < t$ and $y \in A + \xi(t_0) \cdot k$. Since ξ is strictly monotonically increasing, we get $\xi(t_0) - \xi(t) < 0$. $k \in -0^+ A \setminus \{0\}$ implies $(\xi(t_0) - \xi(t)) \cdot k \in 0^+ A$ and $y \in A + \xi(t_0) \cdot k = A + \xi(t) \cdot k + (\xi(t_0) - \xi(t)) \cdot k \subseteq A + 0^+ A + \xi(t) \cdot k \subseteq A + \xi(t) \cdot k$.
Now take any $y \in Y$ with $f(y) = t$ and suppose $y \notin A + \xi(t) \cdot k$. The definition of f implies the existence of a sequence $(t_n)_{n \in \mathbb{N}}$ with $t_n \in \mathbb{R}$, $t_n \searrow t$ and $y \in A + \xi(t_n) \cdot k$ for all $n \in \mathbb{N}$. Since ξ is continuous and strictly monotonically increasing, $(\xi(t_n) - \xi(t))_{n \in \mathbb{N}}$ converges to zero from above. For each $n \in \mathbb{N}$, $y - \xi(t) \cdot k - (\xi(t_n) - \xi(t)) \cdot k = y - \xi(t_n) \cdot k \in A$. Since A is k-directionally closed, we get $y - \xi(t) \cdot k \in A$ by Lemma 5.3.2, a contradiction.
$\square$

Example 5.6.3 *Choose $Y = \mathbb{R}^2$, $A = -\mathbb{R}_+^2$, $k = (1,1)^T$. In (a)-(c), define*

$$f(y) := \inf\{t \in \mathbb{R}: y \in A + \xi(t) \cdot k\} \text{ for all } y \in Y,$$

depending on the given function ξ.

(a) $\xi(t) := t^3$ *for each $t \in \mathbb{R}$.*
We get, for each $t \in \mathbb{R}$, $\mathrm{sublev}_f(t) = A + t^3 \cdot k$ and

$$f(y) = t \text{ if } y \in \mathrm{bd} A + t^3 \cdot k.$$

(b) $\xi(t) := e^t$ *for each $t \in \mathbb{R}$.*
Then $\mathrm{sublev}_f(t) = A + e^t k$ for each $t \in \mathbb{R}$, and

$$f(y) = \begin{cases} t & \text{if } t \in \mathbb{R} \text{ and } y \in \mathrm{bd} A + e^t k, \\ -\infty & \text{if } y \in A. \end{cases}$$

(c) $\xi(t) := \arctan(t)$ *for each* $t \in \mathbb{R}$.
Then $\operatorname{sublev}_f(t) = A + \arctan(t) \cdot k$ *for each* $t \in \mathbb{R}$*, and*

$$f(y) = \begin{cases} t & \text{if } t \in \mathbb{R} \text{ and } y \in \operatorname{bd} A + \arctan(t) \cdot k, \\ -\infty & \text{if } y \in A - \frac{\pi}{2}k, \\ +\infty & \text{if } y \in Y \setminus (\operatorname{int} A + \frac{\pi}{2}k). \end{cases}$$

Consider any sublinear, continuous, finite-valued function $f \colon Y \to \overline{\mathbb{R}}$ with uniform sublevel sets. Can these sublevel sets be generated by shifting some set along a line, i.e., do there exist some set $A \subset Y$, some vector $k \in Y \setminus \{0\}$ and some function $\xi \colon \mathbb{R} \to \mathbb{R}$ such that $\operatorname{sublev}_f(t) = A + \xi(t)k$ holds for each $t \in \mathbb{R}$? This is not always the case.

Example 5.6.4 *Define* $Y := \mathbb{R}^2$, $A := -\mathbb{R}^2_+$, $k := (2,1)^T$, $\tilde{k} := (1,1)^T$,

$$f(y) := \begin{cases} \varphi_{A,k}(y) & \text{if } y \in A, \\ \varphi_{A,\tilde{k}}(y) & \text{if } y \in Y \setminus A. \end{cases}$$

Then $\operatorname{sublev}_f(t) = A + \tilde{\xi}(t)$ with

$$\tilde{\xi}(t) = \begin{cases} tk & \text{if } t \leq 0, \\ t\tilde{k} & \text{if } t > 0. \end{cases}$$

We get

$$f(y) = \begin{cases} \max\{\frac{y_1}{2}, y_2\} & \text{if } y \in A, \\ \max\{y_1, y_2\} & \text{if } y \in Y \setminus A. \end{cases}$$

f is continuous, finite-valued and positively homogeneous. In order to show that f is sublinear, we will now prove that it is subadditive.
f is subadditive on A and on $Y \setminus A$. Hence, f is subadditive on Y if

$$\forall a \in A, b \in Y \setminus A \colon f(a+b) \leq f(a) + f(b).$$

Take any $a \in A$, $b \in Y \setminus A$. We have

$$f(a+b) \leq \max\{\max\{\frac{a_1+b_1}{2}, a_2+b_2\}, \max\{a_1+b_1, a_2+b_2\}\}, \text{ and thus}$$

$$f(a+b) \leq \max\{\frac{a_1}{2} + \frac{b_1}{2}, a_1+b_1, a_2+b_2\}.$$

$$f(a) + f(b) = \max\{\frac{a_1}{2}, a_2\} + \max\{b_1, b_2\}, \text{ and thus}$$

$$f(a) + f(b) = \max\{\frac{a_1}{2} + b_1, \frac{a_1}{2} + b_2, a_2+b_1, a_2+b_2\}.$$

$\frac{a_1}{2} + \frac{b_1}{2} \leq \frac{a_1}{2} + b_1$ *if* $b_1 \geq 0$.
$\frac{a_1}{2} + \frac{b_1}{2} \leq \frac{a_1}{2} + b_2$ *if* $b_1 < 0$ *since then* $b_2 > 0$.
$a_1 + b_1 \leq \frac{a_1}{2} + b_1$ *since* $a_1 \leq 0$.
This implies $f(a+b) \leq f(a) + f(b)$.

A lower semicontinuous functional with uniform sublevel sets that are convex cones does not have to be convex, especially not sublinear.

Example 5.6.5 *In Example 5.6.4, replace k by $k := (1,0)^T$. Then the sublevel sets have again the shape of the convex cone A, and we have*

$$f(y) = \begin{cases} y_1 & \text{if } y \in A, \\ \max\{y_1, y_2\} & \text{if } y \in Y \setminus A. \end{cases}$$

f is lower semicontinuous, finite-valued and positively homogeneous. But it is not sublinear since we get, for $a := (-2,-1)^T$ and $b := (1,2)^T$, that $f(a+b) = f((-1,1)^T) = 1 > 0 = -2 + 2 = f(a) + f(b)$.
f cannot be convex since each convex, positively homogeneous functional is sublinear [22, Lemma 6.6].

5.7 Extension of Arbitrary Functionals to Translative Functions

In this section, each extended real-valued function will be shown to be the restriction of some translative function to a hyperspace. The results in this section are due to [23], where they have been applied to provide a formula for the epigraph of the inf-convolution.

Proposition 5.7.1 *Assume*

$$f: Y \to \overline{\mathbb{R}}, \; A := \operatorname{epi} f, \; k := (0_Y, -1) \in Y \times \mathbb{R}. \tag{H_{epi}}$$

Then $\operatorname{dom} \varphi_{A,k} = \operatorname{dom} f \times \mathbb{R}$ and

$$\varphi_{A,k}((y,s)) = f(y) - s \text{ for all } (y,s) \in Y \times \mathbb{R}. \tag{5.18}$$

Proof. Take any $(y,s) \in Y \times \mathbb{R}$.

$$\begin{aligned}
\varphi_{A,k}((y,s)) &= \inf\{t \in \mathbb{R} : (y,s) \in A + t(0_Y, -1)\} \\
&= \inf\{t \in \mathbb{R} : (y, s+t) \in A\} \\
&= \inf\{t \in \mathbb{R} : f(y) \leq s + t\} \\
&= \inf\{t \in \mathbb{R} : f(y) - s \leq t\} \\
&= f(y) - s
\end{aligned}$$

□

This implies:

Theorem 5.7.2 *Each extended real-valued function on a linear space is the restriction of some translative function to a hyperspace.*
In detail, assuming (H_{epi}), we get $\operatorname{dom} f = \{y \in Y : (y,0) \in \operatorname{dom} \varphi_{A,k}\}$ and

$$f(y) = \varphi_{A,k}((y,0)) \text{ for each } y \in Y.$$

We will now study interdependencies between the properties of f and $\varphi_{A,k}$.

Lemma 5.7.3 *Assume* (H_{epi}). *Then:*

(a) $k \in -0^+A \setminus \{0\}$.

(b) A *is k-directionally closed.*

(c) A *is a proper subset of $Y \times \mathbb{R}$ if and only if f is not a constant function with the function value $+\infty$ or $-\infty$.*

Proof.

(a) Take any $(y,t) \in A$, $\lambda \in \mathbb{R}_+$.
$\Rightarrow f(y) \leq t \leq t+\lambda$. $\Rightarrow (y,t) - \lambda k = (y, t+\lambda) \in A$. $\Rightarrow k \in -0^+A \setminus \{0\}$.

(b) Take any $(y,t) \in Y \times \mathbb{R}$ for which there exists some sequence $(t_n)_{n \in \mathbb{N}}$ of real numbers with $t_n \searrow 0$ and $(y,t) - t_n k \in A$ for all $n \in \mathbb{N}$. $(y, t+t_n) = (y,t) - t_n k \in A$.
$\Rightarrow f(y) \leq t + t_n$ for all $n \in \mathbb{N}$. $\Rightarrow f(y) \leq t$. $\Rightarrow (y,t) \in A$. Hence, A is k-directionally closed.

(c) $A = \emptyset$ if and only if $\text{dom} f = \emptyset$.
$A = Y \times \mathbb{R}$ if and only if $f(y) = -\infty$ for each $y \in Y$. □

Proposition 5.5.15 and Definition 5.2.1 yield:

Proposition 5.7.4 *Assume* (H_{epi}) *and that f is not a constant function with the function value $+\infty$ or $-\infty$. Then:*

(a) *f is convex if and only if $\varphi_{A,k}$ is convex.*

(b) *f is positively homogeneous if and only if $\varphi_{A,k}$ is positively homogeneous.*

(c) *f is subadditive if and only if $\varphi_{A,k}$ is subadditive.*

(d) *f is sublinear if and only if $\varphi_{A,k}$ is sublinear.*

Proposition 5.7.5 *Assume* (H_{epi}) *and $F, B \subseteq Y$ with $F \neq \emptyset$. Then:*

(1) *f is B-monotone on F if and only if $\varphi_{A,k}$ is $(B \times (-\mathbb{R}_+))$-monotone on $F \times \mathbb{R}$.*

(2) *$\varphi_{A,k}$ is strictly $(B \times (-\mathbb{R}_+))$-monotone on $F \times \mathbb{R}$ if and only if*

    (a) *f is strictly B-monotone on F, and*

    (b) *f is finite-valued on F or $0_Y \notin B$.*

Proof.

(1) Assume first that f is B-monotone on F. Take any $(y^1, s_1), (y^2, s_2) \in F \times \mathbb{R}$ with $(y^2, s_2) - (y^1, s_1) \in B \times (-\mathbb{R}_+)$. Then $f(y^1) \leq f(y^2)$ since f is B-monotone on F, and $s_1 \geq s_2$. Hence, $\varphi_{A,k}((y^1, s_1)) \leq \varphi_{A,k}((y^2, s_2))$ by (5.18). Thus, $\varphi_{A,k}$ is $(B \times (-\mathbb{R}_+))$-monotone on $F \times \mathbb{R}$.
Assume now that $\varphi_{A,k}$ is $(B \times (-\mathbb{R}_+))$-monotone on $F \times \mathbb{R}$. Take any $y^1, y^2 \in F$ with $y^2 - y^1 \in B$. Then, $\varphi_{A,k}((y^1, 0)) \leq \varphi_{A,k}((y^2, 0))$ by the monotonicity. Hence, $f(y^1) \leq f(y^2)$ by (5.18). Thus, f is B-monotone on F.

(2) (i) Assume first that $\varphi_{A,k}$ is strictly $(B \times (-\mathbb{R}_+))$-monotone on $F \times \mathbb{R}$. Take any $y^1, y^2 \in F$ with $y^2 - y^1 \in B \setminus \{0\}$. Then $\varphi_{A,k}((y^1,0)) < \varphi_{A,k}((y^2,0))$ by the assumed monotonicity. Hence, $f(y^1) < f(y^2)$ by (5.18). Thus, f is strictly B-monotone on F.
Suppose $0 \in B$. Take any $y \in F$. $(y,0) - (y,1) \in (B \times (-\mathbb{R}_+)) \setminus \{(0_Y, 0)\}$ implies $\varphi_{A,k}(y,1) < \varphi_{A,k}(y,0)$. By (5.18), $f(y) - 1 < f(y) - 0$. Hence, $f(y) \in \mathbb{R}$. Thus $\varphi_{A,k}$ is finite-valued on F.

(ii) Assume now that (a) and (b) hold. Take any $(y^1, s_1), (y^2, s_2) \in F \times \mathbb{R}$ with $(y^2, s_2) - (y^1, s_1) \in (B \times (-\mathbb{R}_+)) \setminus \{(0_Y, 0)\}$. If $y^1 = y^2$, then $s_1 > s_2$. Otherwise, $f(y^1) < f(y^2)$ since f is strictly B-monotone on F, and $s_1 \geq s_2$. If f is finite-valued on F, we get $f(y^1) - s_1 < f(y^2) - s_2$. If $0_Y \notin B$, then $y^1 \neq y^2$, and we get $f(y^1) - s_1 < f(y^2) - s_2$. This implies $\varphi_{A,k}((y^1, s_1)) < \varphi_{A,k}((y^2, s_2))$ by (5.18) Thus, $\varphi_{A,k}$ is strictly $(B \times (-\mathbb{R}_+))$-monotone on $F \times \mathbb{R}$. □

Proposition 5.7.6 *Assume* (H_{epi}) *and that Y is a topological vector space.*

(a) *f is lower semicontinuous if and only if $\varphi_{A,k}$ is lower semicontinuous.*

(b) *f is continuous if and only if $\varphi_{A,k}$ is continuous.*

Proof.

(a) Apply Lemma 5.2.4. f is lower semicontinuous if and only if A is closed. $\varphi_{A,k}$ is lower semicontinuous if and only if $\text{sublev}_{\varphi_{A,k}}(t)$ is closed for each $t \in \mathbb{R}$. By Lemma 5.7.3, A is k-directionally closed and $k \in -0^+ A \setminus \{0\}$. This implies $\text{sublev}_{\varphi_{A,k}}(t) = A + tk$ for each $t \in \mathbb{R}$ because of Theorem 5.5.6. Hence, A is closed if and only if $\text{sublev}_{\varphi_{A,k}}(t)$ is closed for each $t \in \mathbb{R}$.

(b) results from (5.18) by the properties of continuous functions. □

This implies a characterization of any continuous function by its epigraph.

Theorem 5.7.7 *Assume that Y is a topological vector space and $f: Y \to \overline{\mathbb{R}}$. f is continuous if and only if $\text{epi } f$ is closed and*

$$\text{epi } f + \mathbb{R}_> \cdot (0_Y, 1) \subseteq \text{int}(\text{epi } f). \tag{5.19}$$

Proof. Define $A := \text{epi } f$ and $k := (0_Y, -1) \in Y \times \mathbb{R}$. By Proposition 5.7.6, f is continuous if and only if $\varphi_{A,k}$ is continuous. $k \in -0^+ A \setminus \{0\}$ holds because of Lemma 5.7.3. We get from Theorem 5.5.12 that $\varphi_{A,k}$ is continuous if and only if $\text{cl} A - \mathbb{R}_> \cdot k \subseteq \text{int} A$. Hence, f is continuous if and only if

$$\text{cl} A + \mathbb{R}_> \cdot (0_Y, 1) \subseteq \text{int} A.$$

This yields the assertion. □

Let us apply this result to positively homogeneous functions.

Proposition 5.7.8 *Assume that Y is a topological vector space. Each continuous, positively homogeneous function $f\colon Y \to \overline{\mathbb{R}}$ is finite-valued or the constant function with the function value $-\infty$.*

Proof. Since f is positively homogeneous, $A := \operatorname{epi} f$ is a nonempty cone. By Theorem 5.7.7, $A - \mathbb{R}_> \cdot k \subseteq \operatorname{int} A$ for $k := (0_Y, -1)$. Hence, $k \in -\operatorname{int} A = -\operatorname{int}(0^+ A)$. If $A = Y$, then f is the constant function with the function value $-\infty$. Otherwise, $\varphi_{A,k}$ is finite-valued because of Proposition 5.5.16. Thus, f is finite-valued by (5.18). $\square$

A sublinear extended real-valued function which is neither finite-valued nor constant is given by $\varphi \colon \mathbb{R} \to \overline{\mathbb{R}}$ with

$$\varphi(x) := \begin{cases} -\infty & \text{if } x < 0, \\ 0 & \text{if } x = 0, \\ +\infty & \text{if } x > 0. \end{cases}$$

References

[1] M. Durea and C. Tammer. Fuzzy necessary optimality conditions for vector optimization problems. *Optimization*, 58: 449–467, 2009.

[2] C. Gerstewitz and E. Iwanow. Dualität für nichtkonvexe Vektoroptimierungsprobleme. *Wiss. Z. Tech. Hochsch. Ilmenau*, 2: 61–81, 1985.

[3] C. Gerth and P. Weidner. Nonconvex separation theorems and some applications in vector optimization. *J. Optim. Theory Appl.*, 67(2): 297–320, 1990.

[4] A. Göpfert, C. Tammer, and C. Zălinescu. On the vectorial Ekeland's variational principle and minimal points in product spaces. *Nonlinear Anal.*, 39(7, Ser. A: Theory Methods): 909–922, 2000.

[5] A. Göpfert, H. Riahi, C. Tammer, and C. Zălinescu. *Variational Methods in Partially Ordered Spaces*. Springer-Verlag, New York, 2003.

[6] C. Gutiérrez, B. Jiménez and V. Novo. Nonlinear scalarizations of set optimization problems with set orderings. *Springer Proc. Math. Stat.*, 151: 43–63, 2015.

[7] C. Gutiérrez, V. Novo, J.L. Ródenas-Pedregosa and T. Tanaka. Nonconvex separation functional in linear spaces with applications to vector equilibria. *SIAM J. Optim.*, 26(4): 2677–2695, 2016.

[8] A. H. Hamel. Translative sets and functions and their applications to risk measure theory and nonlinear separation. Preprint Serie D 21/2006, IMPA, Rio de Janeiro, 2006.

[9] A. Hamel, F. Heyde, and B. Rudloff. Set-valued risk measures for conical market models. *Math. Financ. Econ.*, 5: 1–28, 2011.

[10] A. Hamel. A very short history of directional translative functions. Manuscript, Yeshiva University, New York, 2012.

[11] E. Hernández and L. Rodríguez-Marín. Nonconvex scalarization in set optimization with set-valued maps. *J. Math. Anal. Appl.*, 325(1): 1–18, 2007.

[12] A.A. Khan, C. Tammer, and C. Zălinescu. *Set-valued Optimization.* Springer-Verlag, Berlin, 2015.

[13] E. Köbis and M. Köbis. Treatment of set order relations by means of a nonlinear scalarization functional: A full characterization. *Optimization*, 65: 1805–1827, 2016.

[14] S. J. Li, X. Q. Yang and G. Y. Chen. Nonconvex vector optimization of set-valued mappings. *J. Math. Anal. Appl.*, 283: 337–350, 2003.

[15] J. J. Moreau. Fonctionnelles convexes. Séminaire sur les équations aux dérivées partielles, Collège de France, 1967.

[16] J. H. Qiu and F. He. A general vectorial Ekeland's variational principle with a p-distance. *Acta Math. Sin. (Engl. Ser.)*, 29(9): 1655–1678, 2013.

[17] C. Tammer and C. Zălinescu. Lipschitz properties of the scalarization function and applications. *Optimization*, 59: 305–319, 2010.

[18] P. Weidner. Ein Trennungskonzept und seine Anwendung auf Vektoroptimierungsverfahren. Habilitation Thesis, Martin-Luther-Universität Halle-Wittenberg, 1990.

[19] P. Weidner. Gerstewitz functionals on linear spaces and functionals with uniform sublevel sets. *J. Optim. Theory Appl.*, 173(3): 812–827, 2017.

[20] P. Weidner. Lower semicontinuous functionals with uniform sublevel sets. *Optimization*, 66(4): 491–505, 2017.

[21] P. Weidner. Construction of functions with uniform sublevel sets. *Optim. Lett.*, 12(1): 35–41, 2018.

[22] P. Weidner. Extended real-valued functions—a unified approach. *Revista Investigación Operacional*, 39(3): 303–325, 2018.

[23] P. Weidner. A new tool for the investigation of extended real-valued functions. *Optim. Lett.*, DOI 10.1007/s11590-018-1370-7, 2018.

Chapter 6

Optimality and Viability Conditions for State-Constrained Optimal Control Problems

Robert Kipka
Lake Superior State University, 650 W Easterday Ave, Sault Ste. Marie, MI 49783.

6.1 Introduction

This chapter is about state constrained optimal control problems and provides a novel approach to the study of optimality conditions and viability through nonsmooth analysis and exact penalization. In this chapter, we focus on problems for which the state evolves in $\mathbb{R}^n$, leaving a possible extension to those for which the state evolves on a smooth manifold M for future work. Questions of viability are related to properties of a certain set-valued map and to the existence, or rather nonexistence, of adjoint arcs satisfying a degenerate form of the Pontryagin Maximum Principle.

There is a huge amount of literature relating to problems of optimal control with such state constraints. For an introduction, we recommend either the book by Vinter [26] or that by Clarke [5]. The paper [11] remains an excellent overview of mathematical results in this area and contains a considerable bibliography. Penalty methods are themselves well-established in the field of optimal control and inequalities such as in our Theorem 1.1.3 can also be found in work by Frankowska and Vinter, see [21, 22] and references therein for details. Our approach is entirely complementary to theirs and represents a natural generalization of work carried out in [13,14,15].

Applied problems of optimal control, in which the state is subject to a constraint of the form $g(t,x(t)) \le 0$, arise in a wide variety of fields. Theory of optimal control subject to such contraints has helped to solve mathematical problems related to the treatment of pathogenic diseases through theurapuetic agents [24]; epidemiology and design of vaccination schedules[18]; renewable resource harvesting [6]; chemical engineering, for example in the production of penicillin [20]; control of energy efficient vehicles [23]; aerospace engineering [25]; and many other fields.

In this chapter, we show how to use two theorems of nonsmooth analysis, namely, the *multidirectional mean value inequality* of Clarke and Ledyaev and a subgradient formula for maximum-type functions due to Ledyaev and Trieman, in order to derive the Pontryagin Maximum Principle for problems with state constraints and obtain viability results for state-constrained problems.

The approach through nonsmooth analysis has the benefit of providing information about the set-valued mapping $\mathcal{A} : \mathbb{R}^n \rightrightarrows \mathcal{U}$, with $\mathcal{U}$ the set of measurable controls, and $\mathcal{A}(x_a)$ the set of controls for which the state constraints are satisfied for a state trajectory beginning at x_a. If $x_a \in \text{dom}(\mathcal{A})$ then we say that x_a is *viable*. Our results in this direction take two forms, namely (i) conditions under which a given state x_a is in the interior of the domain of $\mathcal{A}$ and (ii) conditions under which this set-valued map is locally pseudo-Lipschitz.

6.1.1 Statement of Problem and Contributions

For the sake of exposition, this chapter deals with a relatively simple problem of optimal control: We are given a function $f : [a,b] \times \mathbb{R}^n \times \mathbb{R}^m \to \mathbb{R}^n$ and a compact set $\mathbb{U} \subset \mathbb{R}^m$. A *control* is a measurable function $u : [a,b] \to \mathbb{U}$ and a *state trajectory* $x(t;u,x_a)$ is the solution to the differential equation

$$x(a) = x_a \qquad \dot{x}(t) = f(t,x(t),u(t)) \qquad a.a.\ t \in [a,b]. \qquad (6.1)$$

We are asked to minimize a cost

$$J(x_a,u) := \ell(x(b;u,x_a)) \qquad (6.2)$$

subject to state constraints

$$g(t,x(t;u,x_a)) \le 0 \qquad \forall\, t \in [a,b]. \qquad (6.3)$$

Initial state x_a is said to be *viable* if there exists a control u for which state trajectory $x(t;u,x_a)$ satisfies constraints (6.3) and any such control u is said to be *admissible* for x_a. Associated with control system (6.1) is the *Hamiltonian* $H : [a,b] \times \mathbb{R}^n \times \mathbb{R}^n \times \mathbb{R}^m \to \mathbb{R}$ given by $H(t,x,p,u) := \langle p, f(t,x,u)\rangle$.

Our contributions to this problem include a new proof of the Pontryagin Maximum Principle:

Theorem 6.1.1 *Let x_a be a viable initial state, $\bar{u}$ a control which is admissible for x_a and optimal for (6.2), and $\bar{x}$ the state trajectory for $(x_a, \bar{u})$. If the standing hypotheses of section 6.1.2 hold near $(x_a, \bar{u})$ and ℓ is C^1-smooth, then there exist $\lambda \in \{0,1\}$, positive Borel measure μ satisfying*

$$\operatorname{supp} \mu \subseteq \{t : g(t, \bar{x}(t)) = 0\}, \tag{6.4}$$

and a function $p : [a,b] \to \mathbb{R}^n$ of bounded variation which satisfy the following conditions:

1. *The transversality condition:*

$$-p(b) = \lambda D\ell(\bar{x}(b)) + Dg(b, x(b)) \, \mu(\{b\}); \tag{6.5}$$

2. *The adjoint equations in the sense of distributions:*

$$\dot{p}(t) = -D_x H(t, \bar{x}(t), p(t), \bar{u}(t)) + D_x g(t, \bar{x}(t)) \mu; \tag{6.6}$$

3. *The maximum principle:*

$$H(t, \bar{x}(t), p(t), \bar{u}(t)) = \max_{u \in \mathbb{U}} H(t, \bar{x}(t), p(t), u) \quad a.a. \, t \in [a,b]; \tag{6.7}$$

4. *The nondegeneracy condition $\lambda + \|\mu\| \neq 0$ for all t.*

To say that p satisfies (6.6) in the *sense of distributions* is to say that for any function $\psi \in C_0^\infty(a, b; \mathbb{R})$ there holds

$$\int_a^b p(t) \psi'(t) \, dt = \int_a^b D_x H(t, \bar{x}(t), p(t), \bar{u}(t)) \psi(t) \, dt - \int_a^b D_x g(t, \bar{x}(t)) \psi(t) \, \mu(dt).$$

In order to state our results on viability and exact penalization, we first recall Ekeland's metric [8] for controls $u, v : [a, b] \to \mathbb{U}$. Let $\mathcal{U}$ denote the set of all measurable controls. With m the Lebesgue measure, the Ekeland metric

$$d(u, v) := m\{t : u(t) \neq v(t)\}$$

gives $\mathcal{U}$ the structure of a complete metric space. Given an initial state x_a, we write $\mathcal{A}(x_a) \subseteq \mathcal{U}$ for the set of controls which are admissible for x_a and set

$$d_{\mathcal{A}(x_a)}(u) := \inf \{d(u, v) : v \in \mathcal{A}(x_a)\}.$$

Definition 6.1.2 *Let x_a be a viable state, $\bar{u}$ a control which is admissible for x_a, and $\bar{x}$ the state trajectory for $(x_a, \bar{u})$. We say that $(x_a, \bar{u})$ is strictly normal if the only function $p : [a,b] \to \mathbb{R}^n$ which satisfies (6.4), (6.5), (6.6), (6.7), with $\lambda = 0$ is $p \equiv 0$.*

Both viability and exact penalization can be understood in part through the following theorem on metric regularity, which we prove in this chapter:

Theorem 6.1.3 *If (x_a, u) is strictly normal and the standing hypotheses hold near (x_a, u), then there exist real numbers $\varepsilon, \kappa > 0$ such that*

$$d_{\mathcal{A}(y_a)}(v) \leq \kappa \max_{t \in [a,b]} g_+(t, y(t; v, y_a))$$

for all (y_a, v) satisfying $|x_a - y_a| + d(u, v) < \varepsilon$.

Note that neither Theorem 6.1.3 nor the next corollary require the assumption of optimality.

Corollary 6.1.4 *If (x_a, u) is strictly normal and the standing hypotheses hold, then there is a neighborhood of x_a in which every initial state y_a is viable. That is, x_a is in the interior of the domain of $\mathcal{A}$. In addition, the map $\mathcal{A}$ is a locally pseudo-Lipschitz set-valued mapping.*

Corollary 6.1.5 *Let $(x_a, \bar{u})$ be locally optimal for (6.2) subject to constraints (6.3). If (x_a, u) is strictly normal then $(x_a, \bar{u})$ is a local unconstrained minimizer for the function*

$$\widehat{J}(y_a, u) := \ell(x(b; u, y_a)) + \kappa \max_{t \in [a,b]} g_+(t, x(t; u, y_a))$$

for κ sufficiently large.

6.1.2 Standing Hypotheses

Let x_a be a viable initial state, u a control which is admissible for x_a, and x the state trajectory for (x_a, u). The function f is assumed to be measurable in time, C^1-smooth in x, and continuous in u. In addition, we suppose there exists a number $\varepsilon > 0$ and constants m_f, k_f such that for any $x_1, x_2 \in x(t) + \varepsilon B_{\mathbb{R}^n}$, for almost all $t \in [a,b]$, and for all $u \in \mathbb{U}$ the inequalities

$$\|f(t, x, u)\| \leq m_f$$
$$\|D_x f(t, x, u)\| \leq m_f$$
$$\|D_x f(t, x, u) - D_x f(t, y, u)\| \leq k_f \|x - y\|$$

are satisfied. We suppose that g is jointly continuous in (t, x) and is C^1-smooth in x with locally Lipschitz derivative. Finally, we suppose that for each $t \in [a, b]$ we have either $g(t, \bar{x}(t)) \neq 0$ or $D_x g(t, \bar{x}(t)) \neq 0$.

6.2 Background

The main requirements for a careful reading of Sections 6.3 and 6.4 are in nonsmooth analysis and relaxed controls. For completeness, we provide here the necessary definitions and results from each. Readers may prefer to skip to Section 6.3, referring back to the current section as necessary.

6.2.1 Elements of Nonsmooth Analysis

Since the appearance of Clarke's subdifferential in 1973 [3], the field of nonsmooth analysis has seen considerable proliferation of subdifferentials, each with particular advantages and disadvantages. We will make use of the *Fréchet subgradient* in this chapter and we take a moment to explain why the definition which follows is a natural one. Recall that, for a Banach space X and a function $f : X \to \mathbb{R}$, one can define the *Fréchet derivative* at a point x as the unique functional p for which

$$\lim_{\|h\|\downarrow 0} \frac{f(x+h)-f(x)-\langle p,h\rangle}{\|h\|}=0. \quad (6.8)$$

In fact, one may define *a* Fréchet derivative to be *any* functional p satisfying (6.8) and the fact that such a covector is unique can be derived as a consequence of (6.8). A natural relaxation of Fréchet derivative is then the following:

Definition 6.2.1 *For a Banach space X, functional $p \in X^*$ is a* Fréchet subgradient *for lower semicontinuous function $f : X \to (-\infty, +\infty]$ provided that*

$$\liminf_{\|h\|\downarrow 0} \frac{f(x+h)-f(x)-\langle p,h\rangle}{\|h\|} \geq 0. \quad (6.9)$$

The possibly empty set $\partial_F f(x)$ of such functionals is the Fréchet subdifferential.

Thus, in defining $\partial_F f(x)$, one may interpret (6.8) as asking both that the quotient has a nonnegative limit infimum and nonpositive limit supremum and then asking that only (6.9) hold. In this case, one loses uniqueness of functional p but gains a powerful tool of nonconvex analysis.

The proofs in this chapter require only two theorems of nonsmooth analysis. The first is the *multidirectional mean value inequality* of Clarke and Ledyaev, which was first proved for Hilbert space in the early 1990s [4, Theorem 2.1]:

Theorem 6.2.2 *We are given a Hilbert space E; a point $x_0 \in E$; a lower semicontinuous $f : E \to \mathbb{R} \cup \{\infty\}$; and a closed, bounded, convex subset X of E. Let*

$$\rho = \sup_{\delta>0}\inf\{f(x)-f(x_0) : x \in X + \delta B_E\}.$$

For any $\varepsilon > 0$ there exists $x_\varepsilon \in \mathrm{co}(\{x_0\} \cup X) + \varepsilon B_E$ and $p_\varepsilon \in \partial_F f(x_\varepsilon)$ such that

$$\langle p_\varepsilon, x - x_0 \rangle \geq \rho - \varepsilon$$

for all $x \in X$.

The second proof is a formula for the computation of subgradients for max-type functions. This theorem is much more recent and is a result of Ledyaev and Treiman [19, Theorem 3.2]:

Theorem 6.2.3 *We are given a Hilbert space E and a collection of continuous functions $f_\gamma : E \to \mathbb{R}$, indexed by $\gamma \in \Gamma$, and we define*

$$f(x) := \sup_{\gamma \in \Gamma} f_\gamma(x).$$

If $p \in \partial_F f(x)$ then there exist convex coefficients $(\alpha_i)_{i=1}^k$, pairs $(x_i, \gamma_i)_{i=1}^k$ satisfying $x_i \in x + \varepsilon B_E$ and $f_{\gamma_i}(x_i) \geq f(x) - \varepsilon$, and functionals $p_i \in \partial_F f_{\gamma_i}(x_i)$ such that

$$\left\| p - \sum_{i=1}^k \alpha_i p_i \right\| < \varepsilon.$$

6.2.2 Relaxed Controls

The convexity assumption of Theorem 6.2.2 cannot be dropped, therefore, in order to apply this theorem in the context of control variations, we will eventually find ourselves in need of control variations with satisfactory convexity properties. This is done through the use of relaxed controls.

Informally, a relaxed control represents a kind of average of conventional controls through probability measures. Traditionally one takes a family of probability measures $(\rho_t)_{t \in [a,b]}$ on the set $\mathbb{U}$ and defines their effect on the control system (6.1) through

$$f(t,x,\rho_t) := \int_{\mathbb{U}} f(t,x,u) \rho_t(du).$$

Any conventional control u can be recovered through relaxed controls through a control of the type $\rho_t = \delta_{u(t)}$. We will always use the Greek ρ_t to denote a relaxed control and μ to denote a Borel measure on $[a,b]$, as appearing in Theorem 6.1.1.

It is interesting to note that the earliest use of relaxed controls appears to be in the work of L.C. Young [29] in context of the calculus of variations. Young's use of relaxed controls lay in the existence theory of the calculus of variations and his work is often related to the famous question of Hilbert: *"Has not every regular variation problem a solution ... provided also if need be that the notion of a solution shall be suitably extended?"* [12].

The extension of relaxed controls to problems of optimal controls was carried out by Warga during the 1960s [27] and the technique has been used extensively since, both to establish existence of optimal controls and to obtain necessary optimality conditions for such controls. Classical references include [2, 10, 28, 30] and, more recently, such controls have continued to play important roles both in infinite dimensions [9] and on manifolds [14, 15].

This chapter is not concerned with the existence theory of optimal control and our use of relaxed controls will be limited to certain variations of conventional controls. These variations are essentially equivalent to the *patch perturbations* in [9, §6.2]. We find it useful to use the formalism of relaxed controls, while forgoing the use of actual probability measures on $\mathbb{U}$. We therefore take the following as our definition of relaxed control:

Definition 6.2.4 *A relaxed control is a collection of convex coefficients* $(\alpha_j)_{j=1}^k$ *and conventional controls* $(u_j)_{j=1}^k$.

We then define the effect of a relaxed control on the system (6.1) by taking the corresponding convex combination of velocities $\sum_{j=1}^k \alpha_j f(t,x,u_j(t))$ and so we denote relaxed controls through

$$\rho_t = \sum_{j=1}^k \alpha_j \delta_{u_j(t)} \quad (6.10)$$

and then define

$$f(t,x,\rho_t) := \sum_{j=1}^k \alpha_j f(t,x,u_j(t)), \quad (6.11)$$

thus providing a connection to the classical approach through probability measures. We emphasize, however, that (6.10) is merely a formalism suggesting the definition (6.11).

We turn now to a short overview of technical lemmas required for our use of relaxed control variations:

Lemma 6.2.5 *Let u be a conventional control with state trajectory x and let $P_{s,t}$ denote the local flow of the vector field $(t,x) \mapsto f(t,x,u(t))$. For any relaxed control ρ_t and the state trajectory x^λ corresponding to control variation $(1-\lambda)\delta_{u(t)} + \lambda\rho_t$ satisfies*

$$\left.\frac{d}{d\lambda}\right|_{\lambda=0} x^\lambda(t) = \int_a^t DP_{s,t}(x(s))\left(f(s,x(s),\rho_s) - f(s,x(s),u(s))\right)ds$$

for any $t \in [a,b]$.

Proof. This is a standard result in the theory of nonlinear control systems, expressed here in terms of flows, applied to the perturbation of the vector field $(t,x) \mapsto f(t,x,u(t))$

$$f(t,x,u(t)) + \lambda W(t,x)$$

corresponding to the function $W(t,x) := \left(\sum_{j=1}^k \alpha_j f(t,x,u_j(t)) - f(t,x,u(t))\right)$. Proofs of this result in terms of flows can be found in [1, 16]. □

The set of functions $v(t)$ which can be expressed using the right-hand side of (6.2.5) is convex. Thus, by using relaxed control variations, we arrive at a convex set of variations of the state trajectory suitable for use with Theorem 6.2.2.

However, the use of relaxed controls introduces a problem, namely, that we are interested in optimality and viability results for *conventional controls*, not for relaxed controls, and we cannot guarantee that for a given relaxed control ρ_t there exists a conventional control u for which $f(t,x,\rho_t) = f(t,x,u(t))$. This problem can be overcome through a series of lemmas aimed at approximating relaxed controls through conventional controls. We provide next statements of these lemmas, along with sketches of their proofs.

Lemma 6.2.6 *Let u be a conventional control, ρ_t a relaxed control, and x^λ the state trajectory corresponding to relaxed control $(1-\lambda)\delta_{u(t)} + \lambda\rho_t$. If the standing assumptions hold, then for any $\varepsilon > 0$ and $\lambda \in [0,1]$ there exists a conventional control $u_{\varepsilon,\lambda}$ with $d(u,u_{\varepsilon,\lambda}) = \lambda(b-a)$ whose state trajectory $x^{\varepsilon,\lambda}$ satisfies*

$$\left\|x^{\varepsilon,\lambda} - x^\lambda\right\|_{L^\infty} < \varepsilon. \tag{6.12}$$

Proof. Control $u_{\varepsilon,\lambda}$ can be defined in the following way: Partition $[a,b]$ into subintervals $[t_j,t_{j+1}]$ with lengths $h_j := t_{j+1} - t_j$. Further partition each such interval into $k+1$ intervals of lengths $(1-\lambda)h_j$, $\alpha_1\lambda h_j$, $\alpha_2\lambda h_j$, and so on. On the first of these intervals, $[t_j, t_j + (1-\lambda)h_j)$, set $u_\varepsilon(t) = u(t)$. On the next, set $u_\varepsilon(t) = u_1(t)$, and so on.

Such a control is said to be a *chattering control* and for sufficiently small values of $\max_j h_j$ the rapid switching between controls has the desired effect of approximately averaging the controls in the sense of (6.12). Precise details in line with the current discussion appear in the papers [14, 15]. The idea of chattering controls is much older, dating at least to the works of Gamkrelidze [10] and Warga [28] from the 1970s. □

Given a conventional control u whose state trajectory is x, we may consider a variation x^λ through relaxed control $\rho_t^\lambda(t) = (1-\lambda)\delta_{u(t)} + \lambda\rho_t$ and set

$$v(t) = \int_a^t DP_{s,t}(x(s))\left(f(s,x(s),\rho_s) - f(s,x(s),u(s))\right)ds, \tag{6.13}$$

where $P_{s,t}$ denotes the local flow of vector field $(t,x) \mapsto f(t,x,u(t))$.

If we let $(\lambda_n)_{n=1}^{\infty}$ be a sequence of positive real numbers converging to zero, then Lemma 6.2.6 shows that we can choose a corresponding sequence of conventional controls with $d(u, u_n) = \lambda_n (b-a)$ for which $\lim_{n \to \infty} \frac{x_n - x}{\lambda_n} = v$ in the L^{∞} sense. This leads us to the final definition and technical lemma of this section:

Definition 6.2.7 A control variation *of a conventional control u is a sequence of pairs* $(u_n, \lambda_n)_{n=1}^{\infty}$, *with u_n a conventional control and λ_n a positive real number, with the property that* $\lim_{n \to \infty} \lambda_n = 0$ *and* $\limsup_{n \to \infty} \frac{d(u_n, u)}{\lambda_n} \leq b - a$.

Lemma 6.2.8 *Suppose the standing assumptions hold and let control $\bar{u}$ be given. There are numbers $c_0, \varepsilon_0, \lambda_0 > 0$ depending only on $\bar{u}$, f and $\mathbb{U}$ such that if u is a control with $d(u, \bar{u}) < \varepsilon_0$ whose state trajectory is x, ρ_t is an arbitrary relaxed control, and v is a function of the form (6.13), then there is a control variation $(u_n, \lambda_n)_{n=1}^{\infty}$ of u such that for all n satisfying $\lambda_n < \lambda_0$ we have*

$$\|x + \lambda_n v - x_n\|_{L^{\infty}} \leq c_0 \lambda_n^2,$$

where x_n is the state trajectory for control u_n.

We omit a complete proof of this lemma and refer to [14, 15] for a more careful presentation of Lemmas 6.2.5, 6.2.6, and 6.2.8. These lemmas constitute the necessary tools for the study of control variations of the system (6.1), with Lemma 6.2.8 playing a technical role in the proof of Theorem 6.1.3.

6.3 Strict Normality and the Decrease Condition

Equipped with an elementary background in nonsmooth analysis as outlined in Section 6.2.1 and with the technical lemmas of Section 6.2.2 we are now prepared for careful proofs of the main results of this chapter. These results rely, to a large extent, on the notion of *strict normality*, as given in Definition 6.1.2, and we provide next a high-level overview of the role played by strict normality and the main ideas behind the proofs.

6.3.1 Overview of the Approach Taken

We first show that if a pair (x_a, u) is strictly normal, then for all (y_a, v) sufficiently close to (x_a, u) the function

$$P(y_a, v) := \max_{t \in [a,b]} g_+(t, y(t; v, y_a))$$

can, whenever it takes on positive values, be strongly decreased through some control variation (Definition 6.2.7). That is, we first prove that if a pair (x_a, u) is strictly normal then for any (y_a, v) nearby for which the constraint (6.3) is violated, it is possible to strongly correct, in some sense, the *worst violation of the state constraints*. This will be made precise in the decrease condition of Theorem 6.3.2.

From this decrease condition, it is then possible to establish Theorem 6.1.3 using the Ekeland variational principle. This is done in Section 6.4.

Once Theorem 6.1.3 is established, we proceed to a proof of Theorem 6.1.1 using Clarke's technique of exact penalization. Here, the idea of strict normality plays the role of a constraint qualification: If strict normality fails, then Theorem 6.1.1 holds with $\lambda = 0$, as can be seen almost immediately from the definition of strict normality. In this case, the constraints are degenerate and the cost function ℓ does not appear in the necessary optimality conditions. On the other hand, if $(x_a, \bar{u})$ is strictly normal then we may penalize the constraints through Theorem 6.1.3 and so obtain Theorem 6.1.1 with $\lambda = 1$. This is the remaining content of Section 6.4.

We note that Section 6.4, while it contains the proofs of the main results of this chapter, is relatively short, relying on a careful reading of the proof of Theorem 6.3.2.

6.3.2 The Decrease Condition

Definition 6.3.1 *Let x_a be a viable initial state, u a control which is admissible for x_a, and x the state trajectory for the pair (x_a, u). We say that a nonnegative function $P : \mathbb{R}^n \times \mathcal{U} \to \mathbb{R}$ satisfies the* decrease condition *near (x_a, u) if there exist $\varepsilon, \delta > 0$ such that for all (y_a, v) for which $|x_a - y_a| + d(u, v) < \varepsilon$ and $P(y_a, u) > 0$ there exists a control variation $(v_n, \lambda_n)_{n=1}^{\infty}$ of v for which*

$$\liminf_{n \to \infty} \frac{P(y_a, v_n) - P(y_a, v)}{\lambda_n} \leq -\delta.$$

We emphasize that, while viability and admissibility assumptions are in place on (x_a, u) in Definition 6.3.1, they are not in place on (y_a, v).

Theorem 6.3.2 *Let x_a be a viable state, $\bar{u}$ a control which is admissible for x_a, and $\bar{x}$ the state trajectory for $(x_a, \bar{u})$. If $(x_a, \bar{u})$ is strictly normal and the standing hypotheses hold near this pair, then P satisfies the decrease condition near $(x_a, \bar{u})$.*

Proof. We will prove the contrapositive. If the decrease condition does not hold then we can find sequences of initial states $(x_i(a))_{i=1}^{\infty}$, conventional controls $(u_i)_{i=1}^{\infty}$, and positive real numbers $(\delta_i)_{i=1}^{\infty}$ for which $\|x_i(a) - x_a\| \to 0$, $d(u_i, \bar{u}) \to 0$, $P(x_i(a), u_i) > 0$, $\lim_{i \to \infty} \delta_i = 0$, and for every i

$$\liminf_{n \to \infty} \frac{P(x_i(a), u_{i,n}) - P(x_i(a), u_i)}{\lambda_{i,n}} \geq -\delta_i$$

for all control variations of $(u_{i,n}, \lambda_{i,n})_{n=1}^{\infty}$ of u_i.

Let x_i denote the state trajectory for initial state $x_i(a)$ and control u_i and let $P_{s,t}^i$ denote the flow of the vector field $(t, x) \mapsto f(t, x, u_i(t))$. Let $W^{1,2}$ denote the Hilbert space of absolutely continuous functions $y : [a, b] \to \mathbb{R}^n$ with inner-product

$$\langle y, z \rangle_{W^{1,2}} = \langle y(a), z(a) \rangle_{\mathbb{R}^n} + \int_a^b \langle \dot{y}(t), \dot{z}(t) \rangle_{\mathbb{R}^n} \, dt$$

and denote by $V_i \subset W^{1,2}$ the set of all functions which can be written as

$$v(t) = \int_a^t DP_{s,t}^i(x_i(s))\big(f(s, x_i(s), \rho_s) - f(s, x_i(s), u_i(s))\big) \, ds$$

for some relaxed control ρ_t. Let $\Lambda : W^{1,2}(a,b;\mathbb{R}^n) \to \mathbb{R}$ be the function $\Lambda(x) = \max_{t \in [a,b]} g_+(t,x(t))$.

For each i and n choose $v_{i,n} \in V_i$ for which

$$\inf_{v \in \overline{V_i}} \{\Lambda(x_i + \lambda_{i,n} v) - \Lambda(x_i)\} \geq \Lambda(x_i + \lambda_{i,n} v_{i,n}) - \Lambda(x_i) - \lambda_{i,n}^2.$$

Function Λ is locally Lipschitz, say with constant k_Λ in an appropriately chosen neighborhood of $\bar{x}$ and so, for a given i, we can use Lemma 6.2.8 in order to obtain control $u_{i,n}$ with $d(u_i, u_{i,n}) \leq \lambda_{i,n}(b-a)$ whose state trajectory $x_{i,n}$ satisfies

$$\Lambda(x_i + \lambda_{i,n} v_{i,n}) - \Lambda(x_i) - \lambda_{i,n}^2 \geq \Lambda(x_{i,n}) - \Lambda(x_i) - (c_0 k_\Lambda + 1)\lambda_{i,n}^2,$$

provided that n is sufficiently large in a sense that may depend on i.

The multidirectional mean value inequality, Theorem 6.2.2, now provides subgradient $\xi_{i,n} \in \partial_F \Lambda(y_i)$, $y_{i,n} \in (x_i + \lambda_{i,n} V_i) + \lambda_{i,n} B_{W^{1,2}}$, for which

$$\lambda_{i,n} \inf_{v \in V_i} \langle \xi_{i,n}, v \rangle \geq P(x_i(a), u_{i,n}) - P(x_i(a), u_i) - (c_0 k_\Lambda + 2)\lambda_{i,n}^2.$$

By Theorem 6.2.3, there are times $(t_{i,n,j})_{j=1}^k$ and convex coefficients $(\alpha_{i,n,j})_{j=1}^k$ such that $\|y_{i,n}(t_{i,n,j}) - x_i(t_{i,n,j})\| < \lambda_{i,n}^2$

$$\lambda_{i,n} \sum_{j=1}^k \alpha_{i,n,j} \langle D_x g(t_{i,n,j}, y_{i,n}(t_{i,n,j})), v(t_{i,n,j}) \rangle_{\mathbb{R}^n}$$

$$\geq P(x_i(a), u_{i,n}) - P(x_i(a), u_i) - (c_0 k_\Lambda + 2)\lambda_{i,n}^2$$

for all $v \in V_i$. Using the assumption that $D_x g$ is locally Lipschitz, we obtain

$$\lambda_{i,n} \sum_{j=1}^k \alpha_{i,n,j} \langle D_x g(t_{i,n,j}, x_{i,n}(t_{i,n,j})), v(t_{i,n,j}) \rangle_{\mathbb{R}^n} \qquad (6.14)$$

$$\geq P(x_i(a), u_{i,n}) - P(x_i(a), u_i) - o(\lambda_{i,n}),$$

where $o(\lambda_{i,n})$ denotes a quantity satisfying $o(\lambda_{i,n})/\lambda_{i,n} \to 0$ as $n \to \infty$.

We write (6.14) using probability measures on $[a,b]$ as

$$\lambda_{i,n} \sum_{j=1}^k \int_a^b \alpha_{i,n,j} \langle D_x g(t, x_{i,n}(t)), v(t) \rangle_{\mathbb{R}^n} \delta_{t_{i,n,j}}(dt) \qquad (6.15)$$

$$\geq P(x_i(a), u_{i,n}) - P(x_i(a), u_i) - o(\lambda_{i,n}).$$

We emphasize that $\delta_{t_{i,n,j}}$ is a true probability measure on $[a,b]$ and not a relaxed control. Since the set of probability measures on $[a,b]$ is sequentially weakly* compact, we may define

$$\mu_{i,n} := \sum_{j=1}^k \alpha_{i,n,j} \delta_{t_{i,n,j}}$$

and then choose, for each i, a measure μ_i and a subsequence of $(\mu_{i,n})_{n=1}^\infty$ which converges in the weak* sense to μ_i as n goes to infinity. We may further assume without loss of generality that the sequence $(\mu_i)_{i=1}^\infty$ converges in the weak* sense to a probability measure μ on $[a,b]$.

Dividing (6.15) by $\lambda_{i,n}$ and taking the limit as $n \to \infty$ we obtain

$$\int_a^b \langle D_x g(t, x_i(t)), v(t) \rangle \, \mu_i(dt) \geq -\delta_i.$$

This holds for any $v \in V_i$. Taking the limit as $i \to \infty$ we obtain

$$\int_a^b \langle D_x g(t, \bar{x}(t)), v(t) \rangle \, \mu(dt) \geq 0$$

for any v which is of the form

$$v(t) = \int_a^t DP_{s,t}(\bar{x}(s)) \big(f(s, \bar{x}(s), \rho_s) - f(s, \bar{x}(s), \bar{x}(s)) \big) \, ds,$$

where $P_{s,t}$ denotes the local flow of $(t,x) \mapsto f(t,x,\bar{u}(t))$ and ρ_t is an arbitrary relaxed control.

Define

$$p(t) = -\int_t^b DP_{t,s}(\bar{x}(t))^* D_x g(s, \bar{x}(s)) \mu(ds) \tag{6.16}$$

and use Fubini theorem to arrive at the inequality

$$\int_a^b \langle p(t), f(t, \bar{x}(t), \bar{u}(t)) - f(t, \bar{x}(t), u(t)) \rangle \, dt \geq 0, \tag{6.17}$$

which holds for any control u.

Function $p : [a,b] \to \mathbb{R}^n$ is of bounded variation and satisfies the adjoint equations (6.6) for control $\bar{u}$ and measure μ in a distributional sense. This fact can be checked using the derivative

$$\frac{d}{dt} DP_{t,s}(x(s))^* p_0 = -D_x H(t, x(t), p_0) \tag{6.18}$$

which holds for almost all $t \in [a,b]$ and all $p_0 \in \mathbb{R}^n$.

Using a standard measurable selection theorem we may then choose a control $u_{\max}$ such that $H(t, \bar{x}(t), p(t), u_{\max}(t)) = \max_{u \in \mathbb{U}} H(t, \bar{x}(t), p(t), u)$ for almost all t. For this control (6.17) becomes

$$\int_a^b \left(H(t, \bar{x}(t), p(t), \bar{u}(t)) - \max_{u \in \mathbb{U}} H(t, \bar{x}(t), p(t), u) \right) dt \geq 0.$$

Since the integrand is everywhere nonpositive and the integral is nonnegative, we obtain the pointwise maximum principle (6.7).

Equation (6.5) can be seen to hold, with $\lambda = 0$, from (6.16). Finally, (6.4) holds for each of the measures μ_i and so holds for μ by Theorem [17, Theorem 13.16].

Since μ is a probability measure and $D_x g(t, \bar{x}(t))$ is never zero on the support of μ by the standing assumptions we see that we cannot have $p \equiv 0$ and so $(x_a, \bar{u})$ is not strictly normal, completing the proof. $\square$

6.4 Metric Regularity, Viability, and the Maximum Principle

We are now in a position to prove Theorem 6.1.1, Theorem 6.1.3, and Corollary 6.1.4, beginning with a proof of Theorem 6.1.3.

In fact we will prove something slightly better: The constant κ can be taken to be $(b-a)/\delta$, where $\delta > 0$ is the constant from the decrease condition, which holds by Theorem 6.3.2. Thus, the harder it is to control the system in a way that corrects violations of state constraints the larger the constant that must be used for exact penalization.

Proof of Theorem 6.1.3: Since (x_a, u) is strictly normal and the standing hypotheses hold nearby, the decrease condition must hold for P near (x_a, u) by Theorem 6.3.2. Let $\varepsilon, \delta > 0$ be as in the decrease condition. Since $P(x_a, u) = 0$ we may choose $\varepsilon' \in (0, \frac{1}{2}\varepsilon)$ such that for all (y_a, v) with $\|x_a - y_a\| + d(u,v) < \varepsilon'$ there holds

$$\frac{4(b-a)}{\delta} P(y_a, v) \leq \frac{\varepsilon}{2}.$$

Fix any $\gamma \in (1,2)$ and suppose by way of contradiction that there exists (y_a, v) satisfying $\|x_a - y_a\| + d(u,v) < \varepsilon'$ such that

$$d_{\mathcal{A}(y_a)}(v) > \frac{\gamma(b-a)}{\delta} P(y_a, v).$$

In this case, we must have $P(y_a, v) > 0$ and we see that v is a $\gamma^{1/2} P(y_a, v)$-minimizer of the function $w \mapsto P(y_a, w)$. We apply the Ekeland variational principle with $\varepsilon = \gamma^{1/2} P(y_a, v)$ and $\sigma = (b-a)\gamma P(y_a, v)/\delta$ and obtain $\bar{v}$ with $d(v, \bar{v}) < (b-a)\gamma P(y_a, v)/\delta < d_{\mathcal{A}(y_a)}(v)$ and which minimizes

$$w \mapsto P(y_a, w) + \frac{\delta}{\gamma^{1/2}(b-a)} d(w, \bar{v}).$$

We can check that $d(v, \bar{v}) < \varepsilon$ and since $d_{\mathcal{A}(y_a)}(v) > d(v, \bar{v})$, we see that $\bar{v} \notin \mathcal{A}(y_a)$. It follows that $P(y_a, \bar{v}) > 0$.

By the decrease condition, we may obtain a control variation $(v_n, \lambda_n)_{n=1}^{\infty}$ of $\bar{v}$ for which

$$0 \leq \liminf_{n \to \infty} \frac{P(y_a, v_n) - P(y_a, \bar{v})}{\lambda_n} + \frac{\delta}{\gamma^{1/2}} \leq -\delta + \frac{\delta}{\gamma^{1/2}} < 0.$$

This contradiction means that for any $\gamma \in (1,2)$ and any (y_a, v) with $\|x_a - y_a\| + d(u,v) < \varepsilon'$ there must hold

$$d_{\mathcal{A}(y_a)}(v) \leq \frac{\gamma(b-a)}{\delta} P(y_a, v),$$

which completes the proof. $\square$

Proof of Corollary 6.1.4: Under the assumptions of Corollary 6.1.4, Theorem 6.1.3 implies that there exists $\varepsilon > 0$ such that for all $\|x_a - y_a\| + d(u,v) < \varepsilon$ there holds $d_{\mathcal{A}(y_a)}(v) \leq \frac{(b-a)}{\delta} P(y_a, v)$. In particular, $d_{\mathcal{A}(y_a)}(v) < \infty$ and so $\mathcal{A}(y_a) \neq \emptyset$ for all $y_a \in x_a + \varepsilon B$. Thus, if $\|x_a - y_a\| < \varepsilon$, then y_a is viable.

In addition, if $y_a, z_a \in x_a + \frac{1}{2}\varepsilon B$, $v \in \mathcal{A}(y_a)$, and $d(u,v) < \frac{1}{2}\varepsilon$ then

$$d_{\mathcal{A}(z_a)}(v) \leq \frac{(b-a)}{\delta} P(z_a, v) = \frac{(b-a)}{\delta} (P(z_a, v) - P(y_a, v)) \leq \frac{k_0(b-a)}{\delta} \|y_a - z_a\|$$

for some constant k_0, proving that set-valued mapping $\mathcal{A}$ is locally pseudo-Lipschitz. □

Proof of Corollary 6.1.5: By Corollary 6.1.4, there exists $\varepsilon > 0$ such that set-valued mapping $y_a \mapsto \mathcal{A}(y_a)$ is locally pseudo-Lipschitz on $x_a + \varepsilon B$, say with constant k_0. The function $(y_a, v) \mapsto d_{\mathcal{A}(y_a)}(v)$ is, therefore, locally Lipschitz for the metric space $\mathbb{R}^n \times \mathcal{U}$ with product metric.

Let $\mathrm{gr}(\mathcal{A}) \subset \mathbb{R}^n \times \mathcal{U}$ denote the set of all pairs (y_a, v) for which $v \in \mathcal{A}(y_a)$. Using a standard exact penalization result (c.f. [7, Theorem 1.6.3]), we have $J(x_a, \bar{u}) \leq J(y_a, v) + k_1 d_{\mathrm{gr}(\mathcal{A})}(y_a, v)$ for all (y_a, v) sufficiently close to $(x_a, \bar{u})$ and k_1 sufficiently large. We then have, by Theorem 6.1.3,

$$J(x_a, \bar{u}) \leq J(y_a, v) + k_1 d_{\mathcal{A}(y_a)}(v) \leq J(y_a, v) + k_1 \frac{b-a}{\delta} \max_{t \in [a,b]} g_+(t, y(t; v, y_a)),$$

completing the proof of Corollary 6.1.5. □

Proof of Theorem 6.1.1: If $(x_a, \bar{u})$ is not strictly normal, then the theorem is satisfied with $\lambda = 0$ as can be seen from Definition 6.1.2. We, therefore, suppose that $(x_a, \bar{u})$ is strictly normal.

Corollary 6.1.5 then implies that $(x_a, \bar{u})$ is a local minimizer for

$$\widetilde{J}(y_a, v) := \ell(x(b; v, y_a)) + \frac{c_0(b-a)}{\delta} \max_{t \in [a,b]} g_+(t, x(t; v, y_a)).$$

Thus, for any control variation $(v_n, \lambda_n)_{n=1}^{\infty}$ of $\bar{u}$ we have

$$\liminf_{n \to \infty} \frac{\widetilde{J}(v_n) - \widetilde{J}(\bar{u})}{\lambda_n} \geq 0.$$

The same type of argument carried out in the proof of Theorem 6.3.2, using mean value inequality, will now prove Theorem 6.1.1 with $\lambda = 1$ with only minor changes. In particular, one must replace function P with $\widetilde{J}$ and let $\Lambda : W^{1,2} \to \mathbb{R}$ be the function

$$\Lambda(x) := \ell(x(b)) + \kappa \max_{t \in [a,b]} g_+(t, x(t)).$$

The adjoint arc p will now be given by

$$p(t) = -DP_{t,b}(\bar{x}(t))^* d\ell(\bar{x}(b)) - \kappa \int_t^b DP_{t,s}(\bar{x}(t))^* D_x g(s, \bar{x}(s)) \mu(ds), \tag{6.19}$$

with the extra term coming from the new definition for Λ. Replacing $\kappa \mu$ with μ in (6.19) we arrive at Theorem 6.1.1. □

6.5 Closing Remarks

The essence of the ideas exposed above can be found in the following simple problem: For a given $a \in \mathbb{R}$, minimize $f(x) = -x$ subject to constraint $g(a,x) \leq 0$, where $g(a,x) = x^3 - a$. When $a = 0$, the constraint becomes $x^3 \leq 0$ and is degenerate, a fact which is reflected in the corresponding Karush-Kuhn-Tucker conditions. Indeed, there is no solution to the Lagrange multiplier problem

$$Df(0) + \lambda D_x g(0,0) = 0$$

because $D_x g(0,0)$ is zero. Indeed, for small positive values of x, we cannot bound $D_x g(0,x)$ away from zero and so we cannot "strongly correct" $g(0,x)$ for such values.

On the other hand, for nonzero values of a, we are able to "strongly correct" $g(a,x)$ for nonadmissible values of x, in the sense of Definition 1.3.1, and we can see not only that exact penalization is possible but that the Karush-Khun-Tucker conditions hold in a normal sense. Moreover, there is in this case no solution to the degenerate problem:

$$\lambda D_x g(a, a^{1/3}) = 0.$$

So it is with optimal control, as we have described in this paper. If, near a particular admissible control, nonadmissible controls can be varied in a way that strongly corrects violation of state constraints, then exact penalization is possible. What is required for a careful analysis are estimates for relaxed controls such as those in Lemma 1.2.8 and two theorems of nonsmooth analysis developed by Yuri Ledyaev, Frances Clarke, and Jay Trieman.

References

[1] A. Agrachev and Yu. Sachkov. *Control Theory from the Geometric Viewpoint.* Springer, New York, NY, 2004.

[2] L. Berkovitz. *Optimal Control Theory.* Springer, New York, NY, 1974.

[3] F. Clarke. *Necessary Conditions for Nonsmooth Problems in Optimal Control and the Calculus of Variations.* Ph.D. dissertation, Department of Mathematics, University of Washington, Seattle, 1973.

[4] F. Clarke and Yu. Ledyaev. Mean value inequalities in Hilbert space. *Transactions of the American Mathematical Society*, 344(1): 307–324, 1994.

[5] F. Clarke. *Functional Analysis, Calculus of Variations and Optimal Control.* Springer, New York, NY, 2013.

[6] W. Clark, F. Clarke and G. Munro. The optimal exploitation of renewable resource stocks: Problems of irreversible investment. *Econometrica: Journal of the Econometric Society*, 25–47, 1979.

[7] F. Clarke, Yu. Ledyaev, R. Stern and P. Wolenski. *Nonsmooth Analysis and Control Theory.* Springer, New York, NY, 1998.

[8] I. Ekeland. On the variational principle. *Journal of Mathematical Analysis and Applications*, 47(2): 324–353, 1974.

[9] Hector, Fattorini. *Infinite Dimensional Optimization and Control Theory.* Cambridge University Press, Cambridge, 1999.

[10] R. Gamkrelidze. *Principles of Optimal Control Theory.* Plenum Press, New York, NY, 1978.

[11] R. Hartl, S. Sethi and R. Vickson. A survey of the maximum principles for optimal control problems with state constraints. *SIAM Review*, 37(2): 181–218, 1995.

[12] D. Hilbert. Mathematical problems. *Bulletin of the American Mathematical Society*, 8(10): 437–479, 1902.

[13] R. Kipka. Mathematical methods of analysis for control and dynamic optimization problems on manifolds. Doctoral Dissertation, 2014.

[14] R. Kipka and Yu. Ledyaev. Optimal control on manifolds: Optimality conditions via nonsmooth analysis. *Communications in Applied Analysis*, 18: 563–590, 2014.

[15] R. Kipka and Yu. Ledyaev. Pontryagin maximum principle for control systems on infinite dimensional manifolds. *Set-Valued and Variational Analysis*, 23(1): 133 – 147, 2015.

[16] R. Kipka and Yu. Ledyaev. Extension of chronological calculus for dynamical systems on manifolds. *Journal of Differential Equations*, 258(5): 1765–1790, 2015.

[17] A. Klenke. *Probability Theory: A Comprehensive Course*. Springer, New York, NY, 2013.

[18] I. Kornienko, L. Paiva and M. De Pinho. Introducing state constraints in optimal control for health problems. *Procedia Technology*, 17: 415–422, 2014.

[19] Yu. Ledyaev and J. Treiman. Sub- and supergradients of envelopes, semicontinuous closures, and limits of sequences of functions. *Russian Mathematical Surveys*, 67(2): 2012.

[20] W. Mekarapiruk and R. Luus. Optimal control of inequality state constrained systems. *Industrial & Engineering Chemistry Research*, 36(5): 1686–1694, 1997.

[21] Piernicola, Bettiol and Frankowska, Helene. Regularity of solution maps of differential inclusions under state constraints. *Set-Valued Analysis*, 15.1: 21–45, 2007.

[22] Piernicola, Bettiol, Bressan, Alberto and Vinter, Richard. On trajectories satisfying a state constraint: $W^{1,1}$ estimates and counterexamples. *SIAM Journal on Control and Optimization*, 48.7: 4664–4679, 2010.

[23] A. Sciarretta, G. De Nunzio and L. Ojeda. Optimal ecodriving control: Energy-efficient driving of road vehicles as an optimal control problem. *IEEE Control Systems*, 35(5): 71–90, 2015.

[24] R. Stengel, R. Ghigliazza, N. Kulkarni and O. Laplace. Optimal control of innate immune response. *Optimal Control Applications and Methods*, 23(2): 91–104, 2002.

[25] E. Trélat. Optimal control and applications to aerospace: Some results and challenges. *Journal of Optimization Theory and Applications*, 154(3):713–758, 2012.

[26] R. Vinter. *Optimal Control*. Spring, New York, NY, 2010.

[27] J. Warga. Relaxed variational problems. *Journal of Mathematical Analysis and Applications*, 4: 111–128, 1962.

[28] J. Warga. *Optimal Control of Differential and Functional Equations*. Academic Press, New York, NY, 1972.

[29] L. C. Young. Generalized curves and the existence of an attained absolute minimum in the calculus of variations. *Comptes Rendus de la Societe des Sci. et des Lettres de Varsovie*, 30: 212–234, 1937.

[30] L. C. Young. *Lectures on the Calculus of Variations and Optimal Control Theory*. American Mathematical Society, Providence, RI, 2000.

Chapter 7

Lipschitz Properties of Cone-convex Set-valued Functions

Vu Anh Tuan

Martin-Luther-University Halle-Wittenberg, Institute of Mathematics, Faculty of Natural Sciences II, D-06126 Halle (Saale), Germany.
`anh.vu@mathematik.uni-halle.de`

Thanh Tam Le

Faculty of Basic Sciences, University of Transport and Communications, Hanoi, Vietnam.

Martin-Luther-University Halle-Wittenberg, Institute of Mathematics, Faculty of Natural Sciences II,D-06126 Halle (Saale), Germany.
`le.thanh-tam@mathematik.uni-halle.de`

7.1 Introduction

This manuscript is an overview of recent results concerning Lipschitz properties and convexities of set-valued functions. Both convexity and Lipschitz properties have various important and interesting applications. The convexity is a natural and powerful property of functions that plays a significant role in many areas of mathematics, not only in theoretical but also in applied problems. It connects notions from topology, algebra, geometry and analysis, and is an important tool in deriving optimality conditions in optimization. In optimization, in order to get sufficient conditions for optimal solutions, we need either a second order condition or a convexity assumption. The Lipschitz properties have been also known for a long time in applied sciences and optimization theory. For example, in order to show subdifferential chain rules or the relationships between the coderivative of a vector-valued function and the subdifferential of its scalarization, then this function should be strictly Lipschitzian; see [22, Theorem 3.28]. In particular, the Lipschitz properties for set-valued functions are used for deriving generalized differential calculus and necessary conditions for minimizers of the set-valued optimization problem; see Bao and Mordukhovich ([2]–[4]). A well-known result concerning the Lipschitz property of scalar functions states that: A convex function $f : \mathbb{R}^n \to \mathbb{R}$ is locally Lipschitz. This statement is generated by several authors for the case of vector-valued functions; see [6, 19, 25, 30, 33]. In this chapter, we study the Lipschitz property of a C-convex set-valued function $F : X \rightrightarrows Y$, where X, Y are topological vector spaces and C is a proper convex cone in Y. This problem has been investigated by different methods for the case of set-valued functions, since there exist several types of set relations and set differences. Minh and Tan [20], Kuwano and Tanaka [18] have studied Lipschitz properties of the function F, where X is a finite dimensional space. In addition, [18] assumed that the cone C is normal. Tuan et al. [31] have generated these results where X, Y are normed vector spaces, C is a proper convex cone and not necessarily normal. In this chapter, we present a selection of concepts and recent results on the Lipschitz property of a cone-convex set-valued functions. We not only derive some results on Lipschitz property in the sense of Kuwano et al. [18] but also relax the normal assumption on the cone C. Furthermore, by means of appropriate functionals, we study G-Lipschitzianity of F when it is $\mathfrak{C}_{\mathfrak{s}}$-convex (Definition 7.3.11).

This chapter is organized as follows: Section 7.2 presents basic concepts in topological and useful properties of the well-known noninear scalarizing functional. Section 7.3.1 defines several set relations and set differences. These concepts are beneficial for us to introduce many concepts of cone-convexity and Lipschitz property of set-valued functions in Sections 7.3.2 and 7.3.3. Section 7.4 derives many results on Lipschitz properties of cone-convex set-valued functions. In particular, Section 7.4.1 uses the well-known scalarizing functional in order to investigate the (C, e)-Lipschitz properties in the sense of Kuwano and Tanaka [18], in which e is taken in the interior of the proper, closed, pointed convex cone C. Section 7.4.2 proposes families of functionals with respect to cone-convex set-valued functions. Based on the equi-Lipschitz continuity of these families, we prove the C-Lipschitz properties of the original set-valued functions. Finally, we derive the G-Lipschitzianity for $\mathfrak{C}_{\mathfrak{s}}$-convex set-valued functions in Section 7.4.3.

7.2 Preliminaries

For the convenience of the reader, we collect some basic concepts in topological and some properties of functionals. These concepts and properties are presented in many classical references, so we omit their proofs. We shall be working in topological vector spaces X, whose elements are either vectors or points. The element 0_X is the origin of X. To simplify notation, we use the same symbol 0 for origin elements of all topological vector spaces if no confusion arises. In the case that X is a normed vector space, we shall denote the norm of x by $\|x\|_X$, and if there is no confusion, we omit the subscript X for brevity. In addition, the distance from an element $x \in X$ to a set $A \subset X$ is given by $d(x, A) := \inf\{\|x - a\|, a \in A\}$. We denote by X^* its dual space equipped with the weak* topology ω^*, while its dual norm is denoted by $\|\cdot\|_*$. We denote the closed unit ball and the unit sphere in X by U_X and S_X, respectively. The closed ball centered at $x_0 \in X$ with radius $r > 0$ is defined as

$$B(x_0, r) := x + rU_X = \{x \in X \mid \|x - x_0\|_X \leq r\}.$$

The symbol $*$ is used to indicate relations to dual spaces (dual elements, adjoint operators, dual cone, etc.). Furthermore, we use the notations $\mathbb{R}^n$ for the finite dimensional space of real vectors of dimension n, $\mathbb{R}^n_+$ for nonnegative orthant in $\mathbb{R}^n$, and $\overline{\mathbb{R}} := \mathbb{R} \cup \{\pm\infty\}$. The set of positive integers is denoted by $\mathbb{N}^* := \{1, 2, \ldots\}$. We use the convention $\inf \emptyset := +\infty$, $\sup \emptyset := -\infty$ and $(+\infty) + (-\infty) := +\infty$. We denote the inner product and the norm of $\mathbb{R}^n$ respectively by $\langle \cdot, \cdot \rangle$ and $\|\cdot\|$,

$$\langle x, y \rangle := \sum_{i=1}^n x_i y_i, \quad \|x\| := \langle x, x \rangle^{1/2}, \quad \text{for} \quad x = (x_1, \cdots, x_n), y = (y_1, \cdots, y_n).$$

Definition 7.2.1 *A nonempty set $C \subseteq X$ is said to be a cone if $tc \in C$ for every $c \in C$ and every $t \geq 0$. The cone C is called:*

(i) **convex** *if $C + C \subseteq C$,*

(ii) **proper** *if $C \neq \{0\}$ and $C \neq X$,*

(iii) **pointed** *if $C \cap (-C) = \{0\}$.*

Definition 7.2.2 *Let X be a topological vector space, and $C \subset X$ be a proper convex cone. Then C is called **normal** if there exists a neighborhood base of the origin $0 \in X$ formed by full sets w.r.t. C; a set $A \subset X$ is called full w.r.t. C if*

$$(A + C) \cap (A - C) = A.$$

Observe that if X is a finite-dimensional Hausdorff topological linear space and $C \subset X$ is a convex cone, then C is normal if and only if $\mathrm{cl}\, C$ is pointed. Furthermore, if C is a convex cone in a normed vector space X, the normal property of C is equivalent to the boundedness of the set $(U_X + C) \cap (U_X - C)$, see [12].

For a scalar function $f : X \to \overline{\mathbb{R}}$, the **domain** of f is given by

$$\mathrm{dom}\, f := \{x \in X \mid f(x) < +\infty\},$$

while its **graph** and **epigraph** are given, respectively, by

$$\mathrm{gph}\, f := \{(x, t) \in X \times \mathbb{R} \mid f(x) = t\},$$

$$\mathrm{epi}\, f := \{(x, t) \in X \times \mathbb{R} \mid f(x) \leq t\}.$$

Definition 7.2.3 *The function $f : X \to \overline{\mathbb{R}}$ is called:*

(i) **proper** *if $\mathrm{dom}\, f \neq \emptyset$ and $f(x) > -\infty$ for all $x \in X$;*

(ii) **upper (lower) semi-continuous** *around $x_0 \in X$ if for every $\varepsilon > 0$, there is a neighborhood U_{x_0} of x_0 such that for all $x \in U_{x_0}$,*

$$f(x) \leq f(x_0) + \varepsilon,$$

$$\left(f(x) \geq f(x_0) - \varepsilon, \text{respectively} \right).$$

(iii) **continuous** around $x_0 \in X$ if f is both upper semi-continuous and lower semi-continuous around $x_0 \in X$.

In the following definitions, we define the Lipschitz continuity and the convexity of scalar functions, then we recall their relationships. They will be generated in Section 7.4 for set-valued functions in general spaces.

Definition 7.2.4 *([7]) Let X be a normed vector space, $f : X \to \overline{\mathbb{R}}$ and $A \subseteq X$. The function f is said to be **Lipschitz** on A with a nonnegative constant ℓ provided that f is finite on A and*

$$|f(x) - f(x')| \leq \ell \|x - x'\|_X,$$

*for all points x, x' in A. This is also referred to as a Lipschitz condition of rank ℓ. We say that f is **Lipschitz at** x if there is a neighborhood U_x of x such that f is Lipschitz on U_x (in particular $x \in \text{int}(\text{dom} f)$). In addition, f is said to be **locally Lipschitz** on A, if f is Lipschitz at every point $x \in A$. Hence, $A \subseteq \text{int}(\text{dom} f)$.*

We introduce now the equi-Lipschitzianity of a family of functions that will be used in Section 7.4.2 to study the Lipschitz properties of convex set-valued functions.

Definition 7.2.5 *([26]) Let $\{f_\alpha\}_{\alpha \in I}$ be a family of functions $f_\alpha : X \to \overline{\mathbb{R}}$, where I is a nonempty index set. We say that the family $\{f_\alpha\}_{\alpha \in I}$ is **equi-Lipschitz** at $x_0 \in X$ if there are a neighborhood U_{x_0} of x_0 and a real number $L > 0$ such that for every $\alpha \in I$, f_α is finite and Lipschitz on U_{x_0} with the same Lipschitz constant L, i.e.,*

$$|f_\alpha(x) - f_\alpha(y)| \leq L\|x - y\|_X, \quad \text{for all} \quad x, y \in U_{x_0}, \alpha \in I.$$

For convenience, we will say that the function $f : X \to \overline{\mathbb{R}}$ is convex if f is convex on the whole space X. Obviously, if f is convex, then f is convex on every convex subset of X.

In the following lemma, Zălinescu [33, Corollary 2.2.12] proved the Lipschitz property of a proper convex function f, and estimated the Lipschitz constant.

Lemma 7.2.6 *Let $(X, \|\cdot\|_X)$ be a normed vector space, and $f : X \to \overline{\mathbb{R}}$ be a proper convex function. Suppose that $x_0 \in \text{dom} f$ and there exist $\theta > 0$, $m \geq 0$ such that*

$$\forall x \in B(x_0, \theta) : f(x) \leq f(x_0) + m.$$

Then

$$\forall \theta' \in (0, \theta), \forall x, x' \in B(x_0, \theta') : |f(x) - f(x')| \leq \frac{m}{\theta} \cdot \frac{\theta + \theta'}{\theta - \theta'} \cdot \|x - x'\|_X.$$

In the following, we recall "the nonlinear scalarizing functional" or "Gerstewitz scalarizing functional", which was first introduced in [11] by Tammer (Gerstewitz) and Weidner in order to prove separation theorems for nonconvex sets. This is a useful tool that we can use to study the Lipschitzianity of set-valued convex functions in Section 7.4.

Let A be a given proper closed subset of Y, and $e \in Y \setminus \{0\}$ such that

$$A + [0, +\infty) \cdot e \subseteq A. \tag{7.1}$$

The nonlinear scalarizing functional $\varphi_{A,e} : Y \to \overline{\mathbb{R}}$ is defined by

$$\varphi_{A,e}(y) := \inf\{\lambda \in \mathbb{R} \mid \lambda \cdot e \in y + A\}. \tag{7.2}$$

We present some important properties of $\varphi_{A,e}$ that will be used in the sequel. For the proofs and comments, we refer to [10, 12, 29].

Theorem 7.2.7 *([12, 29]) Let Y be a topological vector space, and $A \subset Y$ be a proper closed set. Let e be a given point in $Y \setminus \{0\}$ such that (7.1) holds, then the following properties hold for $\varphi := \varphi_{A,e}$:*

(a) *φ is lower semi-continuous, and $\mathrm{dom}\,\varphi = \mathbb{R}e - A$.*

(b) *$\forall y \in Y, \forall t \in \mathbb{R} : \varphi(y) \leq t$ if and ony if $y \in te - A$.*

(c) *$\forall y \in Y, \forall t \in \mathbb{R} : \varphi(y+te) = \varphi(y) + t$.*

(d) *φ is convex if and ony if A is convex; $\varphi(\lambda y) = \lambda \varphi(y)$ for all $\lambda > 0$ and $y \in Y$ if and ony if A is a cone.*

(e) *φ is proper if and ony if A does not contain lines parallel to e, i.e., $\forall y \in Y, \exists t \in \mathbb{R} : y + te \notin A$.*

(f) *φ takes finite values if and ony if A does not contain lines parallel to e and $\mathbb{R}e - A = Y$.*

The following corollary is a direct consequence of Theorem 7.2.7. For further properties of the function φ, we refer the reader to Hernández and Rodríguez-Marín [13].

Corollary 7.2.8 *[13] Let Y be a topological vector space, and C be a proper cone in Y with a nonempty interior. Let $B_1, B_2 \subset Y$ be proper sets such that $C - B_1$ is closed. The following assertion holds for every $e \in \mathrm{int}\,C$ and $t \in \mathbb{R}$,*

$$B_2 \subseteq te + B_1 - C \iff \sup_{b \in B_2} \varphi_{C-B_1,e}(b) \leq t.$$

Before stating the next result, we recall the D-monotonicity of a functional.

Definition 7.2.9 *Let Y be a topological vector space, and D be a nonempty subset of Y. A functional $\varphi : Y \to \overline{\mathbb{R}}$ is called D-monotone, if*

$$\forall y_1, y_2 \in Y : y_1 \in y_2 - D \implies \varphi(y_1) \leq \varphi(y_2).$$

Moreover, φ is said to be strictly D-monotone, if

$$\forall y_1, y_2 \in Y : y_1 \in y_2 - D \setminus \{0\} \implies \varphi(y_1) < \varphi(y_2).$$

The following results provide some monotonicity properties of the scalarizing functional φ. For the detailed proof, see [12, Theorem 2.3.1].

Theorem 7.2.10 *([12]) Assume that all the assumptions of Theorem 7.2.7 hold, and $\emptyset \neq D \subseteq Y$. The following properties hold:*

(a) *$\varphi_{A,e}$ is D-monotone if and only if $A + D \subseteq A$.*

(b) *$\varphi_{A,e}$ is subadditive if and only if $A + A \subseteq A$.*

7.3 Concepts on Convexity and Lipschitzianity of Set-valued Functions

7.3.1 Set Relations and Set Differences

This chapter works on convexity and Lipschitzianity of set-valued functions. For this aim, we recall many set relations and several types of set differences between two given sets $A, B \subseteq Y$. First, we define a *set less order relation*, which plays an important role in set optimization, and was first independently introduced by Nishnianidze [23] and Young [32].

Definition 7.3.1 *Let A, B be two nonempty subsets of Y, and C be a proper convex cone. The **set less order relation** $\preceq_C^s$ is defined by*

$$A \preceq_C^s B \iff B \subseteq A + C \text{ and } A \subseteq B - C.$$

Furthermore, there exist six kinds of set relations between two nonempty sets in the sense of Kuroiwa, Tanaka, and Ha [17] as follows:

Definition 7.3.2 *For two nonempty sets $A, B \subseteq Y$ and a proper convex cone C in Y, we introduce the following set relations*

(i) $A \preceq_C^{(i)} B \iff B \subseteq \bigcap_{a \in A} (a + C);$

(ii) $A \preceq_C^{(ii)} B \iff A \bigcap \left(\bigcap_{b \in B} b - C \right) \neq \emptyset;$

(iii) $A \preceq_C^{(iii)} B \iff B \subseteq A + C;$

(iv) $A \preceq_C^{(iv)} B \iff B \bigcap \left(\bigcap_{a \in A} a + C \right) \neq \emptyset;$

(v) $A \preceq_C^{(v)} B \iff A \subseteq B - C;$

(vi) $A \preceq_C^{(vi)} B \iff B \bigcap (A + C) \neq \emptyset.$

The set relations $\preceq_C^{(iii)}$ and $\preceq_C^{(v)}$ can be called by other names, the **lower** ($\preceq_C^l$) and the **upper set less order relation** ($\preceq_C^u$), respectively; see [15, 16] and the references therein. Note that the relation $\preceq_C^s$ is a combination of the relations $\preceq_C^l$ and $\preceq_C^u$.

The following proposition is directly verified from Definition 7.3.2.

Proposition 7.3.3 *([17]) Let $A, B \subseteq Y$ be nonempty sets, and C be a proper convex cone in Y. The following statements hold:*

$$\begin{array}{ccccc} A \preceq_C^{(i)} B & \Longrightarrow & A \preceq_C^{(ii)} B & \Longrightarrow & A \preceq_C^{(iii)} B \\ \Downarrow & & & & \Downarrow \\ A \preceq_C^{(iv)} B & \Longrightarrow & A \preceq_C^{(v)} B & \Longrightarrow & A \preceq_C^{(vi)} B \end{array}$$

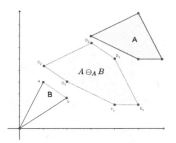 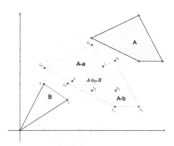

Figure 7.1: The algebraic difference and the geometric difference.

Next, we study several set differences which later motivate the corresponding Lipschitz properties in Section 7.3.3. These differences were investigated by Baier and Farkhi [1], Dempe and Pilecka [8], Rubinov and Akhundov [28] for finite-dimensional spaces, and by Jahn [14] for infinite-dimensional spaces.

We begin with the definition of the algebraic difference and geometric difference of two arbitrary sets. They were studied by Baier and Farkhi [1] in finite dimensional spaces. In this chapter, we consider them in topological vector spaces.

Definition 7.3.4 *([1]) Let Y be a topological vector space, and A, B be two subsets of Y. We define the*

(i) **algebraic difference** *as*

$$A \ominus_A B := A + (-1) \cdot B;$$

(ii) **geometric/star-shaped/Hadwiger-Pontryagin difference** *as*

$$A \ominus_G B := \{y \in Y : y + B \subseteq A\}.$$

The algebraic difference and geometric difference can be also presented as

$$A \ominus_A B = \bigcup_{b \in B} A^{-b}, \qquad A \ominus_G B = \bigcap_{b \in B} A^{-b},$$

where $A^{-b} := \{a - b \mid a \in A\} = A - b$ for $b \in B$. Of course, the second formulae for the algebraic difference and the geometric difference have more geometrical meaning than the first ones given in Definition 7.3.4. For an illustration, see Figure 7.1.

Note that the disadvantage of these two concepts is that the cardinality of the algebraic difference set is usually bigger than one of two original sets. Furthermore, the geometric difference sets in several cases can be empty, even if the spaces are finite or infinite dimensional.

Example 7.3.5 (i) *Let $Y := \mathbb{R}^n$, and two sets $A = B := U_Y$. Then, $A \ominus_A B = 2U_Y$; see Figure 7.2 (left).*

(ii) *Let $A, B \subset Y$ be two balls in $\mathbb{R}^n$ such that B has a radius bigger than A. Then, $A \ominus_G B = \emptyset$; see Figure 7.2 (right).*

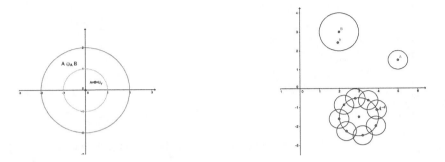

Figure 7.2: Illustration for Examples 7.3.5.

In the next definition, the l-difference is defined in a finite-dimensional $\mathbb{R}^n$ w.r.t. the relation $\preceq_C^{(iii)}$ given in Definition 7.3.2, where C is a proper convex cone in $\mathbb{R}^n$. This difference is an extension of the geometric difference; see Pilecka [24].

Definition 7.3.6 *Let A, B be arbitrary subsets of $\mathbb{R}^n$, and C be a proper convex cone in $\mathbb{R}^n$. The l-difference is defined as follows:*

$$A \ominus_l B := \{y \in \mathbb{R}^n : y + B \subseteq A + C\}, \tag{7.3}$$

or the equivalent formulation

$$A \ominus_l B = \{y \in \mathbb{R}^n : A \preceq_C^{(iii)} B + y\},$$

where $\preceq_C^{(iii)}$ is introduced in Definition 7.3.6.

Now, we recall the Demyanov differences which are studied in [1, 9, 28] on the class of nonempty compact subsets $\mathcal{K}(\mathbb{R}^n)$ and the class of nonempty convex compact subsets $\mathcal{C}(\mathbb{R}^n)$ in $\mathbb{R}^n$. They are essential to make Demyanov difference sets nonempty. For a given set $A \in \mathcal{K}(\mathbb{R}^n)$, the support function of A is given by

$$\sigma(\ell, A) := \max_{a \in A} \langle \ell, a \rangle \qquad (\ell \in \mathbb{R}^n),$$

and the supporting face of A is given by

$$M(\ell, A) := \{a \in A : \langle \ell, a \rangle = \sigma(\ell, A)\}.$$

Here, we denote by $\langle \ell, a \rangle$ the inner product of ℓ and a, and by $m(\ell, A)$ a point of the supporting face. $\mathcal{S}_A$ denotes the set of $\ell \in \mathbb{R}^n$ such that the supporting face $M(\ell, A)$ consists of only a single point $m(\ell, A)$. In general, one takes $\mathcal{S}_A$ in the unit sphere of $\mathbb{R}^n$.

Definition 7.3.7 *(Baier and Farkhi [1]) Let $A, B \in \mathcal{K}(\mathbb{R}^n)$. The Demyanov difference is defined as follows:*

$$A \ominus_D B := \mathrm{cl\,conv}\{m(\ell, A) - m(\ell, B) : \ell \in \mathcal{S}_A \cap \mathcal{S}_B\}. \tag{7.4}$$

There are several modifications of the Demyanov difference in the literature. In the following definitions, we introduce two approaches proposed by Dempe and Pilecka [8], and Jahn [14], which restrict the considered directions to vectors contained in the dual and negative dual cone of the ordering cone. To simplify notation, we use the same symbol $\ominus_D$ for all the Demyanov differences under consideration if no confusion arises.

Definition 7.3.8 *(Dempe and Pilecka [8]) Let C be a proper convex cone in $\mathbb{R}^n$, and A, B be two nonempty sets in $\mathbb{R}^n$. Then, the modified Demyanov difference is given by:*

$$A \ominus_D B := \operatorname{cl conv} \{ m(\ell, A) - m(\ell, B) : \ell \in \mathcal{S}_A \cap \mathcal{S}_B \cap (C^+ \cup (-C^+)) \}, \qquad (7.5)$$

where $C^+ := \{ y^* \in Y^* \mid y^*(y) \geq 0, \ \forall y \in C \}$.

In [14], Jahn derives new Demyanov differences for two arbitrary sets A, B in normed vector space Y which is partially ordered by a convex cone C. However, it is necessary to assume that the solutions of the following minimization and maximization problems are unique for every $\ell \in C_1^+ := C^+ \cap U_{Y^*}$:

$$\min_{a \in A} \langle \ell, a \rangle, \qquad (7.6)$$

$$\max_{a \in A} \langle \ell, a \rangle. \qquad (7.7)$$

The solutions of the problems (7.6) and (7.7) are denoted by $y_{\min}(\ell, A)$ and $y_{\max}(\ell, A)$, respectively. Note that if the constrained set A is weakly compact, then there exist solutions to these problems.

Definition 7.3.9 *(Jahn [14]) Let Y be a normed vector space, C be a proper convex cone in Y. Let $A, B \in Y$ be such that the solutions $y_{\min}(\ell, A), y_{\min}(\ell, B), y_{\max}(\ell, A)$ and $y_{\max}(\ell, B)$ of the problems (7.6), (7.7) are unique for every $\ell \in C_1^+$. The modified Demyanov difference is given by:*

$$A \ominus_D B := \bigcup_{\ell \in C_1^+} \{ y_{\min}(\ell, A) - y_{\min}(\ell, B), y_{\max}(\ell, A) - y_{\max}(\ell, B) \}. \qquad (7.8)$$

We end this part by defining the metric difference of two nonempty compact subsets of $\mathbb{R}^n$.

Definition 7.3.10 *([1]) Let $A, B \in \mathcal{K}(\mathbb{R}^n)$ be given. The metric difference is defined as:*

$$A \ominus_M B := \{ a - b : \|a - b\| = d(a, B) \quad \text{or} \quad \|b - a\| = d(b, A) \}. \qquad (7.9)$$

In the special case that $B = \{b\}$ is a singleton all the differences defined by Definitions 7.3.4-7.3.10 coincide and are exactly the set $A - b$.

For more details on properties and the calculus of the aforementioned set differences, as well as comparisons among them, we refer the reader to [1, 14].

7.3.2 Cone-convex Set-valued Functions

Throughout this section, X and Y are normed spaces, C is a proper convex cone in Y, and $F : X \rightrightarrows Y$ is a set-valued function. The **domain** of F is $\text{dom} F := \{x \in X \mid F(x) \neq \emptyset\}$. We define the **graph** of F by

$$\text{gph} F := \{(x,y) \in X \times Y \mid y \in F(x)\}.$$

The **epigraph** of F with respect to (w.r.t.) C is given by

$$\text{epi} F := \{(x,y) \in X \times Y \mid y \in F(x) + C\}. \tag{7.10}$$

We begin this part by introducing concepts of cone-convex set-valued functions corresponding to the relations $\preceq_C^s$ and $\preceq_C^{(k)}$, where $k \in \{i, ii, iii, iv, v, vi\}$ given in Definitions 7.3.1 and 7.3.2.

Definition 7.3.11 *Let $F : X \rightrightarrows Y$ with $\text{dom} F \neq \emptyset$, C be a proper convex cone, and $k \in \{i, ii, iii, iv, v, vi\}$. F is said to be:*

(i) **type-(k)-convex** *if*

$$F(\alpha x + (1-\alpha)y) \preceq_C^{(k)} \alpha F(x) + (1-\alpha) F(y).$$

holds for all $x, y \in \text{dom} F$ and $\alpha \in (0,1)$.

(ii) **$\mathfrak{C}$s-convex** *if*

$$F(\alpha x + (1-\alpha)y) \preceq_C^s \alpha F(x) + (1-\alpha) F(y),$$

holds for all $x, y \in \text{dom} F$ and $\alpha \in (0,1)$.

Remark 7.3.11 We say that F is **lower C-convex** (or **upper C-convex**) if F is type-(iii)-convex (type-(v)-convex, respectively). Obviously, if F is lower C-convex, then $\text{dom} F$ is convex, and $F(x) + C$ is a convex set for all $x \in \text{dom} F$. Furthermore, F is $\mathfrak{C}$s-convex if and only if F is lower C-convex and upper C-convex. ∎

By Proposition 7.3.3, we have some implications for the convexities above:

Proposition 7.3.12 *([17]) Let $F : X \rightrightarrows Y$ be a set-valued function. Then, the following statements hold:*

type-(i)-convex	$\Longrightarrow$	type-(ii)-convex	$\Longrightarrow$	type-(iii)-convex
$\Downarrow$				$\Downarrow$
type-(iv)-convex	$\Longrightarrow$	type-(v)-convex	$\Longrightarrow$	type-(vi)-convex.

As shown in the literature, the convexity of a vector function is equivalent to the convexity of its epigraph. However, the following result states that this is not always true for set-valued functions.

Proposition 7.3.13 *Let $F : X \rightrightarrows Y$ be a set-valued function and $k \in \{i, ii, \ldots, v\}$. If F is type-(k)-convex then*

$$\operatorname{epi}_{(k)} F := \{(x, V) \in X \times \mathcal{V} \mid F(x) \preceq_C^{(k)} V\}$$

is convex. Furthermore, the converse assertion holds if $k \in \{iii, iv, v\}$.

Proof. We will only prove the case $k = iii$, as the other cases can be proved by using similar arguments.

Take $(x_1; V_1), (x_2; V_2) \in \operatorname{epi}_{(iii)} F$. We have that

$$\begin{cases} V_1 \subseteq F(x_1) + C; \\ V_2 \subseteq F(x_2) + C. \end{cases}$$

These inclusions imply that

$$\lambda V_1 + (1-\lambda)V_2 \subseteq \lambda F(x_1) + (1-\lambda)F(x_2) + C \subseteq F(\lambda x_1 + (1-\lambda)x_2) + C$$

for all $\lambda \in (0, 1)$. This means $\operatorname{epi}_{(iii)} F$ is convex.
Conversely, since $\operatorname{epi}_{(iii)} F$ is convex, and $(x_1, F(x_1)), (x_2, F(x_2)) \in \operatorname{epi}_{(iii)} F$, we have

$$\big(\lambda x_1 + (1-\lambda)x_2, \lambda F(x_1) + (1-\lambda)F(x_2)\big) \in \operatorname{epi}_{(iii)} F.$$

Thus, F is type-(iii)-convex. □

In order to derive the equivalences between the convexities of set-valued functions and scalar functions, Minh and Tan [20] have used two useful scalarizing functionals $G_{y^*}, g_{y^*} : X \to \overline{\mathbb{R}}$, which are defined for each $y^* \in C^+$ as follows:

$$G_{y^*}(x) := \sup_{y \in F(x)} y^*(y), \quad x \in X, \qquad (7.11)$$

$$g_{y^*}(x) := \inf_{y \in F(x)} y^*(y), \quad x \in X, \qquad (7.12)$$

with the convention $\inf \emptyset := +\infty$, $\sup \emptyset := -\infty$.

Obviously, $\operatorname{dom} g_{y^*} = \operatorname{dom} F$ and $g_0 = \delta_{\operatorname{dom} F}$ (for $y^* = 0$), where δ_A is the indicator function of A defined by $\delta_A(x) = 0$ if $x \in A$, and $\delta_A(x) = +\infty$ otherwise.

The following propositions are stated in [21, Proposition 2.2] without proof. For convenience of the reader, we prove them in detail.

Proposition 7.3.14 *Let $F : X \rightrightarrows Y$ be a set-valued function such that $\operatorname{dom} F$ is convex and nonempty. Let C be a proper convex cone. The following implications hold:*

(i) *If F is an upper C-convex function, then G_{y^*} is convex on $\operatorname{dom} F$ for all $y^* \in C^+$.*

(ii) *Conversely, assume that $\operatorname{dom} F$ is convex, and $F(x) - C$ is closed and convex for all $x \in \operatorname{dom} F$. If G_{y^*} is convex for all $y^* \in C^+$, then F is upper C-convex.*

Proof. (i) Let F be upper C-convex, and $y^* \in C^+$ be arbitrarily chosen. For every $\lambda \in (0,1)$, $x_1, x_2 \in \text{dom}\, F$, we have

$$\begin{aligned}
G_{y^*}(\lambda x_1 + (1-\lambda)x_2) &= \sup_{y \in F(\lambda x_1 + (1-\lambda)x_2)} y^*(y) \\
&\leq \sup_{y \in \lambda F(x_1) + (1-\lambda)F(x_2) - C} y^*(y) \\
&\leq \sup_{y \in \lambda F(x_1) + (1-\lambda)F(x_2)} y^*(y) \\
&= \sup_{y \in \lambda F(x_1)} y^*(y) + \sup_{y \in (1-\lambda)F(x_2)} y^*(y) \\
&= \lambda \sup_{y \in F(x_1)} y^*(y) + (1-\lambda) \sup_{y \in F(x_2)} y^*(y) \\
&= \lambda G_{y^*}(x_1) + (1-\lambda)G_{y^*}(x_2).
\end{aligned}$$

Therefore, G_{y^*} is convex on $\text{dom}\, F$ for all $y^* \in C^+$.

(ii) Suppose by contradiction that F is not upper C-convex. Therefore, there exist $x_1, x_2 \in \text{dom}\, F$ and $\lambda \in (0,1)$ such that

$$F(\lambda x_1 + (1-\lambda)x_2) \not\subseteq \lambda F(x_1) + (1-\lambda)F(x_2) - C.$$

We can choose $\bar{y} \in F(\lambda x_1 + (1-\lambda)x_2)$ such that $\bar{y} \notin \lambda F(x_1) + (1-\lambda)F(x_2) - C$. Since $\lambda F(x_1) + (1-\lambda)F(x_2) - C$ is closed and convex, there exists $y^* \in Y^*$ such that

$$y^*(\bar{y}) > \sup\{y^*(y) \mid y \in \lambda F(x_1) + (1-\lambda)F(x_2) - C\}.$$

It follows that $y^* \in C^+ \setminus \{0\}$ and

$$\begin{aligned}
G_{y^*}(\lambda x_1 + (1-\lambda)x_2) &= \sup_{y \in F(\lambda x_1 + (1-\lambda)x_2)} y^*(y) \geq y^*(\bar{y}) \\
&> \sup_{y \in \lambda F(x_1) + (1-\lambda)F(x_2) - C} y^*(y) \\
&= \lambda G_{y^*}(x_1) + (1-\lambda)G_{y^*}(x_2).
\end{aligned}$$

This contradicts our assumption on the convexity of G_{y^*}. □

Proposition 7.3.15 *Let $F : X \rightrightarrows Y$ with $\text{dom}\, F \neq \emptyset$, and C be a proper convex cone. The following implications hold:*

(i) *If F is a lower C-convex function, then g_{y^*} is convex for all $y^* \in C^+$.*

(ii) *Conversely, if $F(x) + C$ is closed and convex for all $x \in \text{dom}\, F$, and g_{y^*} is convex for all $y^* \in C^+$, then F is lower C-convex.*

Proof. (i) Let $y^* \in C^+$ be arbitrarily chosen. For every $\lambda \in (0,1), x_1, x_2 \in \operatorname{dom} g_{y^*} = \operatorname{dom} F$, we have

$$\begin{aligned}
g_{y^*}(\lambda x_1 + (1-\lambda)x_2) &= \inf_{y \in F(\lambda x_1 + (1-\lambda)x_2)} y^*(y) = \inf_{y \in F(\lambda x_1 + (1-\lambda)x_2)+C} y^*(y) \\
&\leq \inf_{y \in \lambda F(x_1) + (1-\lambda)F(x_2)} y^*(y) \\
&= \inf_{y \in \lambda F(x_1)} y^*(y) + \inf_{y \in (1-\lambda)F(x_2)} y^*(y) \\
&= \lambda \inf_{y \in F(x_1)} y^*(y) + (1-\lambda) \inf_{y \in F(x_2)} y^*(y) \\
&= \lambda g_{y^*}(x_1) + (1-\lambda) g_{y^*}(x_2).
\end{aligned}$$

Therefore, g_{y^*} is convex for all $y^* \in C^+$.

(ii) Since $g_0 = \delta_{\operatorname{dom} F}$ is convex, so is $\operatorname{dom} F$. Suppose by contradiction that F is not lower C-convex, then there exist $x_1, x_2 \in \operatorname{dom} F$ and $\lambda \in (0,1)$ such that

$$\lambda F(x_1) + (1-\lambda) F(x_2) \not\subseteq F(\lambda x_1 + (1-\lambda)x_2) + C.$$

Take $\bar{y} \in \lambda F(x_1) + (1-\lambda) F(x_2)$ such that $\bar{y} \notin F(\lambda x_1 + (1-\lambda)x_2) + C$. Since $F(\lambda x_1 + (1-\lambda)x_2) + C$ is closed and convex, there exists $y^* \in Y^*$ such that

$$y^*(\bar{y}) < \inf\{y^*(y) \mid y \in F(\lambda x_1 + (1-\lambda)x_2) + C\}.$$

It follows that $y^* \in C^+ \setminus \{0\}$ and

$$\begin{aligned}
g_{y^*}(\lambda x_1 + (1-\lambda)x_2) &= \inf_{y \in F(\lambda x_1 + (1-\lambda)x_2)+C} y^*(y) > y^*(\bar{y}) \\
&\geq \inf_{y \in \lambda F(x_1) + (1-\lambda)F(x_2)} y^*(y) \quad (\text{as } \bar{y} \in \lambda F(x_1) + (1-\lambda)F(x_2)) \\
&= \lambda g_{y^*}(x_1) + (1-\lambda) g_{y^*}(x_2).
\end{aligned}$$

This contradicts our assumption on the convexity of g_{y^*}. □

7.3.3 Lipschitz Properties of Set-valued Functions

This section introduces several types of Lipschitz property for set-valued functions generated by a given proper convex cone C. We begin with the ***Lipschitz-like*** (also known as the ***Aubin property***, or the ***pseudo-Lipschitzian property***). This property has been used to derive necessary optimality conditions for vector optimization problems; see [2, 3, 5, 22].

Definition 7.3.16 *[22, Section 1.2.2] Let $F : X \rightrightarrows Y$ with $\operatorname{dom} F \neq \emptyset$, $U \subseteq X$ and $V \subseteq Y$ be two given nonempty sets. Let $(\bar{x}, \bar{y}) \in \operatorname{gph} F$ be given. We say that F is*

(i) ***Lipschitz-like*** *on U relative to V if there is $\ell \geq 0$ such that*

$$\forall x, x' \in U : \quad F(x) \cap V \subseteq F(x') + \ell \|x - x'\|_X U_Y. \quad (7.13)$$

(ii) ***Lipschitz-like*** *at $(\bar{x}, \bar{y})$ with modulus $\ell \geq 0$ if there are neighborhoods $U_{\bar{x}}$ of $\bar{x}$ and $V_{\bar{y}}$ of $\bar{y}$ such that (7.13) holds, hence, necessarily $\bar{x} \in \operatorname{int}(\operatorname{dom} F)$. The infimum of all such moduli ℓ is called the exact Lipschitz bound of F at $(\bar{x}, \bar{y})$ and is denoted by $\operatorname{lip} F(\bar{x}, \bar{y})$.*

(iii) **Lipschitz continuous** *on U if (7.13) holds with $V := Y$, in this case, the infimum of $\ell \geq 0$ is denoted by $\text{lip}_U F(\bar{x})$. Furthermore, F is **Lipschitz at** $\bar{x}$ if there is a neighborhood $U_{\bar{x}}$ of $\bar{x}$ such that F is Lipschitz continuous on $U_{\bar{x}}$.*

Remark 7.3.16

(i) If F is Lipschitz-like on U relative to V and $F(U) \cap V \neq \emptyset$, then $U \subseteq \text{dom} F$.

(ii) The Lipschitz-like property of F at $(\bar{x}, \bar{y})$ implies the Lipschitz-like property of F at $(x, y) \in \text{gph} F$ provided that it is close enough to $(\bar{x}, \bar{y})$.

(iii) If F is Lipschitz-like on U, then we have $U \cap \text{dom} F = \emptyset$ or $U \subseteq \text{dom} F$.

■

The *epigraphical multifunction* of $F : X \rightrightarrows Y$, $\mathcal{E}_F : X \rightrightarrows Y$, is defined by

$$\mathcal{E}_F(x) := F(x) + C. \tag{7.14}$$

From (7.10) and (7.14) it follows that $\text{gph} \mathcal{E}_F = \text{epi} F$.

Definition 7.3.17 *A set-valued function $F : X \rightrightarrows Y$ is **epigraphically Lipschitz-like** (ELL) at a given point $(\bar{x}, \bar{y}) \in \text{gph} F$ if $\mathcal{E}_F$ is Lipschitz- like at that point, i.e., there are neighborhoods $U_{\bar{x}}$ of $\bar{x}$ and $V_{\bar{y}}$ of $\bar{y}$ such that*

$$\forall x, x' \in U_{\bar{x}}: \quad \mathcal{E}_F(x) \cap V_{\bar{y}} \subseteq \mathcal{E}_F(x') + \ell \|x - x'\|_X U_Y. \tag{7.15}$$

Definition 7.3.18 *Let $F : X \rightrightarrows Y$ be a set-valued function such that $\text{dom} F \neq \emptyset$, and $C \subset Y$ be a proper convex cone.*

(i) *F is said to be **upper (lower) C-Lipschitz** at $x_0 \in X$ if there are a neighborhood U_{x_0} of x_0, and a constant $\ell \geq 0$ such that the following inclusions hold for all $x \in U_{x_0}$*

$$F(x) \subseteq F(x_0) + \ell \|x - x_0\|_X U_Y + C, \tag{7.16}$$

$$(F(x_0) \subseteq F(x) + \ell \|x - x_0\|_X U_Y + C, \text{respectively}). \tag{7.17}$$

(ii) *F is said to be **C-Lipschitz** at $x_0 \in X$ if there are a neighborhood U_{x_0} of x_0, and a constant $\ell \geq 0$ such that the following inclusions hold for all $x, x' \in U_{x_0}$*

$$F(x) \subseteq F(x') + \ell \|x - x'\|_X U_Y + C. \tag{7.18}$$

Remark 7.3.18

(i) If F is lower C-Lipschitz at x_0, then $x_0 \notin \text{cl}(\text{dom} F)$ or $x_0 \in \text{int}(\text{dom} F)$.

(ii) It is clear that if F is C-Lipschitz at $x_0 \in \text{dom} F$, then F is upper and lower C-Lipschitz at x_0. However, the converse assertion does not always hold true.

(iii) Obviously, if F is C-Lipschitz at $x_0 \in \text{dom}F$, from (7.18), we have that $\mathcal{E}_F$ is Lipschitz-like continuous at (x_0, y_0), for all $y_0 \in F(x_0)$. This also implies that F is (ELL) at (x_0, y_0).

(iv) Observe that the upper (lower) C-Lipschitz properties imply the Hausdorff C-upper (lower) continuities. Recall that F is **Hausdorff C-upper (lower) continuous** (H-C-u.c, H-C-l.c, respectively) at $x_0 \in X$ if there are a neighborhood U_{x_0} of x_0, and a constant $\rho \geq 0$ such that the following inclusions hold for all $x \in U_{x_0}$:

$$F(x) \subseteq F(x_0) + \rho U_Y + C, \tag{7.19}$$

$$(F(x_0) \subseteq F(x) + \rho U_Y + C, \text{respectively}), \tag{7.20}$$

see [16, Definition 3.1.16].

∎

We recall, in the following, other definitions of C-Lipschitz property of set-valued functions. They were introduced by Kuwano and Tanaka[18].

Definition 7.3.19 *Let X be a normed space, and Y be a linear topological space. Let $C \subset Y$ be a proper, closed, pointed, convex cone with a nonempty interior, $e \in \text{int}C$, and $F : X \rightrightarrows Y$. F is said to be **upper (lower) (C, e)-Lipschitz** at $x_0 \in X$ if there exist a positive constant L and a neighborhood U_{x_0} of x such that for any $x \in U_{x_0}$,*

$$F(x) \subseteq F(x_0) - L\|x - x_0\|e + C, \tag{7.21}$$

$$\big(F(x) \subseteq F(x_0) + L\|x - x_0\|e - C, \text{respectively}\big). \tag{7.22}$$

F is said to be (C,e)-**Lipschitz** at $x_0 \in X$ if F is upper (C,e)-Lipschitz and lower (C,e)-Lipschitz at this point.

Note that the upper C-Lipschitzianity in Definition 7.3.18(i) is more general than the upper (C,e)-Lipschitzianity in Definition 7.3.19, as it is not necessary to suppose additionally that the interior of C is nonempty. Moreover, in the case that $\text{int}C \neq \emptyset$, the upper (C,e)-Lipschitzianity is equivalent to upper C-Lipschitzianity. Indeed, the implication (7.21)$\Rightarrow$(7.16) is obvious. From $e \in \text{int}C$, there is $\mu' > 0$ such that $e + \mu' U_Y \subseteq C$, and so $U_Y \subseteq -\frac{1}{\mu'}e + C := -te + C$. It follows that $\ell\|x - x_0\|_X U_Y + C \subseteq -t\ell\|x - x_0\|_X e + C$. Therefore, the converse assertion is clear.

We now introduce several Lipschitz properties of set-valued maps w.r.t. the given set differences presented in Section 7.3.1. These Lipschitz properties were investigated for both finite-dimensional spaces in [1, 28], and infinite-dimensional spaces in [14]. Baier and Farkhi [1] presented a survey on Lipschitz properties of set-valued functions. Furthermore, they also studied the relationships between Lipschitz properties and existence of selections of set-valued functions. This matter has been attracting the attention of researchers for a long time.

We now define the Lipschitz properties of set-valued functions w.r.t. the algebraic difference $\ominus_A$ and the geometric difference $\ominus_G$ given in Definition 7.3.4. We use the notion Δ-*Lipschitz* standing for both A-*Lipschitz* (algebraic Lipschitz), and G-*Lipschitz* (geometric Lipschitz).

Definition 7.3.20 *([1],[24]) Let X,Y be two normed vector spaces, and $F: X \rightrightarrows Y$ be a set-valued function. F is called Δ-**Lipschitz at** $x_0 \in X$ w.r.t. the set difference $\ominus_\Delta$ (where $\Delta \in \{\mathcal{A}, G\}$) with a constant $L \geq 0$ if there is a neighborhood U_{x_0} of x_0 such that*

$$F(x) \ominus_\Delta F(y) \subseteq L\|x-y\|_X U_Y \quad \text{for all} \quad x,y \in U_{x_0}.$$

For other Lipschitz continuity concepts concerning the partial ordering relation, we derive the upper (lower) G-Lipschitz concepts at $x_0 \in X$. We will use these concepts to study the G-Lipschitz properties of $\mathfrak{C}_5$-convex set-valued functions in Section 7.4.3.

Definition 7.3.21 *Let X,Y be two normed vector spaces, $F : X \rightrightarrows Y$ be a set-valued function with $\operatorname{dom} F \neq \emptyset$, and $C \subset Y$ be a proper convex cone. F is said to be (C,G)-**Lipschitz at** $x_0 \in X$ if there is a neighborhood U_{x_0} of x_0, and a constant $\ell \geq 0$ such that for all $x, x' \in U_{x_0}$*

$$F(x) \ominus_G F(x') \subseteq \ell \|x-x'\|_X U_Y + C. \tag{7.23}$$

Remark 7.3.21

(i) If $F(x')$ is bounded (i.e., there exists $\rho > 0$ such that $F(x') \subseteq \rho B_Y$) then (7.18) implies (7.23). Indeed, if (7.18) holds, we utilize the definition of geometric difference (see Definition 7.3.4(ii)) to get that:

$$F(x') + F(x) \ominus_G F(x') \subseteq F(x) \subseteq F(x') + \ell \|x-x'\|_X U_Y + C.$$

By the law of cancellation for certain classes of convex sets ([27, Lemma 1]), the boundedness of $F(x')$, and the closedness and the convexity of $\ell \|x - x'\|_X U_Y + C$, it implies (7.23).

(ii) In the case C is a normal cone, F is G-Lipschitz at x_0 if and only if it is (C,G)-Lipschitz and $(-C,G)$-Lipschitz at x_0. Indeed, since C is normal, there is $\rho > 0$ such that

$$(\rho U_Y + C) \cap (\rho U_Y - C) \subseteq U_Y. \tag{7.24}$$

By (7.23), we get the following assertion for all $x, x' \in U, x \neq x'$

$$\frac{\rho(F(x) \ominus_G F(x'))}{\ell \|x-x'\|_X} \subseteq (\rho U_Y + C) \cap (\rho U_Y - C) \subseteq U_Y.$$

This means that F is G-Lipschitz at x_0.

■

To this end, we define the Lipschitzianity w.r.t. the metric difference $\ominus_M$ in (7.9), and the D-Lipschitzianity (or Demyanov Lipschitzianity) w.r.t. the Demyanov difference in (7.4).

Definition 7.3.22 *([1]) Let X be a vector space, and $F: X \rightrightarrows \mathcal{K}(\mathbb{R}^n)$ (or $F: X \rightrightarrows \mathcal{C}(\mathbb{R}^n)$) be a set-valued function. F is called M-**Lipschitz** (or D-**Lipschitz**) **at** $x_0 \in X$ with respect* to the metric difference $\ominus_M$ (the Demyanov difference $\ominus_D$, respectively) with a constant $L \geq 0$ if there is a neighborhood U_{x_0} of x_0 such that for all $x, y \in U_{x_0}$

$$F(x) \ominus_M F(y) \subseteq L\|x-y\|_X U_{\mathbb{R}^n},$$

$$(F(x) \ominus_D F(y) \subseteq L\|x-y\|_X U_{\mathbb{R}^n}, respectively).$$

Note that we can also use the formula in Definition 7.3.22 to define the Demyanov Lipschitz properties w.r.t. Dempe and Pilecka's Demyanov difference in (7.5), and Jahn's Demyanov difference in (7.8).

The following proposition presents the hierarchy of the Lipschitz notions above.

Proposition 7.3.23 ([1]) Let X be a vector space, and $F : X \rightrightarrows \mathbb{R}^n$ be a set-valued function with image in $\mathcal{K}(\mathbb{R}^n)$. Then, we have

$$F \text{ is } D\text{-Lipschitz} \implies F \text{ is } M\text{-Lipschitz} \implies F \text{ is } G\text{-Lipschitz}.$$

7.4 Lipschitz Properties of Cone-convex Set-valued Functions

7.4.1 (C, e)-Lipschitzianity

In [18], Kuwano and Tanaka introduced a new concept of the Lipschitz properties of set-valued functions. They used the nonlinear scalarizing functional to prove the Lipschitz continuity of convex set-valued functions with the assumption that C is a normal cone; see [18, Theorem 3.2]. In this section, we use another scalarization approach to prove these results without the aforementioned property of C.

Throughout this section, X is a normed space, Y is a linear topological space, $C \subset Y$ is a proper, closed, pointed, convex cone with a nonempty interior, and $e \in \text{int} C$. We begin by recalling the (C, e)-boundedness concepts in [18].

Definition 7.4.1 Let X be a normed space, Y be a linear topological space, and $F : X \rightrightarrows Y$ be a set-valued function. Let $C \subset Y$ be a proper, closed, pointed, convex cone with a nonempty interior, and $e \in \text{int} C$. F is said to be (C, e)-**bounded from above (resp. below)** around $x_0 \in X$ if there exist a positive t and a neighborhood U_{x_0} of x_0 such that

$$F(U_{x_0}) \subseteq te - C, \quad (resp.\ F(U_{x_0}) \subseteq -te + C).$$

Furthermore, F is called (C, e)-**bounded** around $x_0 \in X$ if it is C-bounded from above and C-bounded from below around x_0.

The following result proves that an upper C-convex set-valued function (see Definition 7.3.11) is lower (C, e)-Lipschitz in the sense of Definition 7.3.19.

Theorem 7.4.2 Let X be a normed space and Y be a linear topological space. Let $C \subset Y$ be a proper, closed, pointed, convex cone with a nonempty interior, and $e \in \text{int} C$. If $F : X \rightrightarrows Y$ is an upper C-convex set-valued function, F is (C, e)-bounded around a point $x_0 \in \text{int}(\text{dom} F)$, and $F(x_0) - C$ is closed and convex. Then F is lower (C, e)-Lipschitz at x_0.

Proof. Let $e \in \operatorname{int} C$. Because of the (C,e)-boundedness of F around x_0 (see Definition 7.4.1), there exist a positive t and a neighborhood $U_{x_0} := x_0 + \mu U_X \subseteq \operatorname{int}(\operatorname{dom} F)$ of x_0 such that $F(U_{x_0}) \subseteq te - C$ and $F(U_{x_0}) \subseteq -te + C$, which is equivalent to $-F(U_{x_0}) \subseteq te - C$. Thus, for every $x \in U_{x_0}$, we have $F(x) - F(x_0) \subseteq 2te - C$. This implies that

$$F(x) \subseteq 2te + F(x_0) - C. \tag{7.25}$$

Since $F(x_0) - C$ is closed, $A := C - F(x_0)$ is closed. Obviously, (7.1) is fullfilled. We consider a function $H : X \to \overline{\mathbb{R}}$ given by

$$H(x) := \sup_{a \in F(x)} \varphi_{A,e}(a),$$

in which the functional $\varphi_{A,e}$ is introduced in (7.2) with $A = C - F(x_0)$ and $e \in \operatorname{int} C$. Applying Theorem 7.2.10(a), we get that $\varphi_{A,e}$ is C-monotone. Taking into account the inclusion (7.25), for all $a \in F(x)$, we can choose $b \in F(x_0)$ such that $a \in 2te + b - C$. By Theorems 7.2.10(a) and 7.2.7(c),

$$\varphi_{A,e}(a) \le \varphi_{A,e}(2te + b) = 2t + \varphi_{A,e}(b) \le 2t + H(x_0).$$

By the definition of the functional H, it holds that

$$H(x) \le H(x_0) + 2t. \tag{7.26}$$

This implies that H is bounded from above around x_0.

Now, we prove that H is convex. Indeed, because of the upper C-convexity of F (see Definition 7.3.11) and the convexity of $\varphi_{A,e}$ (since $A = C - F(x_0)$ is convex), the following hold for every $x, y \in X$ and $\alpha \in (0,1)$,

$$\begin{aligned}
H(\alpha x + (1-\alpha)y) &= \sup_{a \in F(\alpha x + (1-\alpha)y)} \varphi_{A,e}(a) \\
&\le \sup_{a \in \alpha F(x) + (1-\alpha)F(y) - C} \varphi_{A,e}(a) \\
&= \sup_{a_1 \in F(x), a_2 \in F(y), c \in C} \varphi_{A,e}(\alpha a_1 + (1-\alpha)a_2 - C) \\
&= \sup_{a_1 \in F(x), a_2 \in F(y)} \varphi_{A,e}(\alpha a_1 + (1-\alpha)a_2) \\
&\le \alpha \sup_{a_1 \in F(x)} \varphi_{A,e}(a_1) + (1-\alpha) \sup_{a_2 \in F(y)} \varphi_{A,e}(a_2) \\
&= \alpha H(x) + (1-\alpha)H(y).
\end{aligned}$$

Applying Lemma 7.2.6 to the scalar proper convex function H, we get that H is Lipschitz on a neighborhood $U'_{x_0} := x_0 + \mu' U_X$ of x_0 ($0 < \mu' < \mu$) with the Lipschitz constant $L > 0$. This means that

$$|H(x) - H(x_0)| \le L\|x - x_0\| \quad \text{for all} \quad x \in U'_{x_0}.$$

Since $H(x_0) \le 0$, we get

$$H(x) \le L\|x - x_0\| \quad \text{for all} \quad x \in U'_{x_0},$$

which induces the conclusion due to Corollary 7.2.8. □

The following result can be proved similarly to that one of Theorem 7.4.2. For the purpose of the shortness, we omit its proof.

Theorem 7.4.3 *Let X be a normed space, and Y be a linear topological space. Let $C \subset Y$ be a proper, closed, pointed, convex cone with a nonempty interior, and $e \in \text{int}\, C$. Suppose that $F : X \rightrightarrows Y$ is a upper $(-C)$-convex set-valued function, F is (C,e)-bounded around a point $x_0 \in \text{int}\,(\text{dom}\, F)$, and $F(x_0) + C$ is closed and convex. Then F is locally upper (C,e)-Lipschitz at x_0.*

Observe that the assumptions in Theorem 7.4.2 and 7.4.3 are weaker than the ones in [18], since we do not need the conditions that C is a normal cone and X is a finite-dimensional space.

7.4.2 C-Lipschitzianity

In [20], Minh and Tan have studied the C-Lipschitzianity of C-convex set-valued functions $F : X \rightrightarrows Y$, where X is a finite dimensional space, and Y is a Banach space. In this section, we derive corresponding results where X, Y are general normed spaces.

We present in the following C-boundedness notions of a set-valued function $F : X \rightrightarrows Y$, where $C \subset Y$ is a proper convex cone.

Definition 7.4.4 *Let $F : X \rightrightarrows Y$ be a set-valued function, and C be a proper convex cone in Y.*

(i) *F is said to be **C-bounded from above** (resp. **below**) on a subset A of X if there exists a constant $\mu > 0$ such that*

$$F(A) \subseteq \mu U_Y - C \quad (\text{resp. } F(A) \subseteq \mu U_Y + C).$$

(ii) *F is said to be **C-bounded** on a subset A of X if it is C-bounded from above and C-bounded from below on A.*

Remark 7.4.4

(i) By [12, Proposition 2.6.2], if F is C-bounded from below on a neighborhood of $x_0 \in \text{int}\,(\text{dom}\, F)$ then F is C-bounded from below on a neighborhood of x for every $x \in \text{dom}\, F$.

(ii) In the case that $\text{int}\, C \neq \emptyset$, these concepts are equivalent to the (C,e)-boundedness in Definition 7.4.1. Indeed, since $e \in \text{int}\, C$, there is $\mu' > 0$ such that $-e + \mu' U_Y \subseteq -C$. It follows that $U_Y \subseteq \frac{1}{\mu'} e - C$, and so $\mu U_Y - C \subseteq \frac{\mu}{\mu'} e - C := te - C$, the converse assertion is clear. Similar arguments can be applied to the case F is C-bounded from below.

Definition 7.4.5 *([16, Definition 3.1.26])* We say that $F : X \rightrightarrows Y$ is **weakly C-upper (lower) bounded** on a set $A \subseteq X$ if there exists $\mu' > 0$ such that $F(x) \cap (\mu' U_Y - C) \neq \emptyset$ $(F(x) \cap (\mu' U_Y + C) \neq \emptyset$, respectively) for all $x \in A$.

Remark 7.4.5 If F is weakly C-upper bounded on a neighborhood of $x_0 \in \text{int}(\text{dom} F)$ then F is weakly C-upper bounded on a neighborhood of x for every $x \in \text{int}(\text{dom} F)$. Indeed, assume that $F(x) \cap (\mu U_Y - C) \neq \emptyset$, $\forall x \in B(x_0, r)$. Fix $\bar{x} \in \text{int}(\text{dom} F)$, then there exist $x_1 \in \text{dom} F$, $\lambda \in (0, 1)$ such that $\bar{x} = \lambda x_0 + (1-\lambda)x_1$. Fix $y_1 \in F(x_1)$ and take $u \in rU_X$, then $\bar{x} + \lambda u = \lambda(x_0 + u) + (1-\lambda)x_1$, and there exists $y_u \in F(x_0 + u) \cap (\mu U_Y - C)$. Hence, $y_u = \mu v_u - c_u$ with $\|v_u\| \leq 1$, $c_u \in C$. Then, $\lambda(\mu v_u - c_u) + (1-\lambda)y_1 \in F(\bar{x} + \lambda u) + C$, hence $\exists c'_u \in C$ and $\bar{y} \in F(\bar{x} + \lambda u)$ such that $\bar{y} = \lambda(\mu v_u - c_u) + (1-\lambda)y_1 - c'_u \in \bar{\mu} U_Y - C$ with $\bar{\mu} = \lambda\mu + (1-\lambda)\|y_1\|$. ∎

The following proposition states the equivalence between the C-Lipschitzianity of F (see Definition 7.3.18) and the equi-Lipschitzianity (see Definition 7.2.5) of the family $\{g_{y^*} | y^* \in C^+, \|y^*\|_* = 1\}$ corresponding to F.

Proposition 7.4.6 *Let X, Y be two normed vector spaces, $F : X \rightrightarrows Y$, and $F(x) + C$ be convex for all $x \in X$. Let x_0 be a given point in $\text{int}(\text{dom} F)$ such that $F(x_0)$ is C-bounded from below. Then, F is C-Lipschitz at x_0 if and only if the family $\{g_{y^*} | y^* \in C^+, \|y^*\|_* = 1\}$ is equi-Lipschitz at x_0.*

Proof. As F is C-Lipschitz at x_0, there exist a neighborhood $U_{x_0} \subseteq \text{dom} F$ of x_0 and a real number $\ell > 0$ such that

$$F(x) \subseteq F(x') + \ell\|x - x'\|_X U_Y + C, \quad \text{for all} \quad x, x' \in U_{x_0}. \tag{7.27}$$

As $F(x_0)$ is C-bounded from below, we assume that $F(x_0) \subseteq \mu U_Y + C$ for some $\mu > 0$; see Definition 7.4.4. Due to (7.27), we get

$$F(x) \subseteq (\mu + \ell\|x - x'\|_X)U_Y + C, \quad \text{for all} \quad x \in U_{x_0}. \tag{7.28}$$

Hence, for all $x \in U_{x_0}, y^* \in C^+, \|y^*\|_* = 1$, we have that

$$g_{y^*}(x) = \inf_{y \in F(x)} y^*(y) \geq -(\mu + \ell\|x - x'\|_X) > -\infty.$$

Thus, for all $y^* \in C^+$ that satisfies $\|y^*\|_* = 1$, g_{y^*} is finite on U_{x_0}.
Taking into account (7.27), we have the following estimation for all $x, x' \in U_{x_0} \subseteq \text{dom} F$

$$g_{y^*}(x) = \inf_{y \in F(x)} y^*(y) \geq \inf_{y \in F(x')} y^*(y) - \ell\|x - x'\|_X = g_{y^*}(x') - \ell\|x - x'\|_X.$$

Hence,

$$g_{y^*}(x') - g_{y^*}(x) \leq \ell\|x - x'\|_X, \quad \text{for all} \quad x, x' \in U_{x_0}, y^* \in C^+, \|y^*\|_* = 1.$$

By interchanging x and x', we get

$$|g_{y^*}(x) - g_{y^*}(x')| \leq \ell\|x - x'\|_X, \quad \text{for all} \quad x, x' \in U, y^* \in C^+, \|y^*\|_* = 1.$$

This shows that the family $\{g_{y^*} | y^* \in C^+, \|y^*\|_* = 1\}$ is equi-Lipschitz at x_0.

We prove the converse implication by contradiction that F is not C-Lipschitz at x_0. Then, there exist $x_n, x'_n \in B(x_0, \frac{1}{n})$ such that

$$F(x_n) \not\subseteq F(x'_n) + n\|x_n - x'_n\|_X U_Y + C \quad \text{for all} \quad n \in \mathbb{N}^*;$$

and hence $x_n \neq x'_n$ for all $n \in \mathbb{N}^*$.

Since $x_0 \in \text{int}(\text{dom} F)$, for n large enough, $B(x_0, \frac{1}{n}) \subseteq \text{dom} F$, we can take $y_n \in F(x_n)$ such that

$$y_n \notin B_n := F(x'_n) + n\|x_n - x'_n\|_X U_Y + C.$$

Since the set B_n is convex and $\text{int} B_n \neq \emptyset$, one can find $y_n^* \in Y^*$ with $\|y_n^*\|_* = 1$ such that

$$y_n^*(y_n) \leq y_n^*(v) \quad \text{for all} \quad v \in B_n.$$

Hence,

$$y_n^*(y_n) \leq \inf y_n^*(B_n) = \inf y_n^*(F(x'_n)) - n\|x_n - x'_n\|_X + \inf y_n^*(C).$$

It follows that $y_n^* \in C^+$ for large $n \in \mathbb{N}$ and

$$g_{y_n^*}(x_n) \leq g_{y_n^*}(x'_n) - n\|x_n - x'_n\|_X,$$

which yields that

$$n\|x_n - x'_n\|_X \leq g_{y_n^*}(x'_n) - g_{y_n^*}(x_n) \leq \ell\|x_n - x'_n\|_X,$$

and, hence, $n \leq \ell$, which could not hold true for n sufficiently large. $\square$

Remark 7.4.6 Proposition 7.4.6 is stated in [20, Theorem 2.5] without the assumption that $F(x_0)$ is C-bounded from below. Taking $F(x) = Y$ for all $x \in X$, it is clear that F is C-Lipschitz, but $\{g_{y^*} | y^* \in C^*, \|y^*\| = 1\}$ is not equi-Lipschitz. ∎

By using similar arguments presented in the proof of Proposition 7.4.6, we present in the next proposition an equivalence between the $(-C)$-Lipschitz property of a given set-valued function F (see Definition 7.3.18) and the equi-Lipschitz property (see Definition 7.2.5) of the family $\{G_{y^*} : X \to \overline{\mathbb{R}} \mid y^* \in C^+, \|y^*\|_* = 1\}$.

Proposition 7.4.7 *Let X, Y be two normed spaces, $F : X \rightrightarrows Y$, and $F(x) - C$ be convex for all $x \in X$. Let x_0 be a given point in $\text{int}(\text{dom} F)$ such that $F(x_0)$ is C-bounded from above. Then, F is $(-C)$-Lipschitz at x_0 if and only if the family $\{G_{y^*} | y^* \in C^+, \|y^*\|_* = 1\}$ is equi-Lipschitz at x_0.*

The next theorem is a relationship between the C-convexity and the C-Lipschitz property of set-valued functions.

Theorem 7.4.8 ([31, Theorem 3]) *Let X, Y be two normed spaces, C be a proper convex cone. Let $F : X \rightrightarrows Y$ be lower C-convex, C-bounded from below and weakly C-upper bounded on a neighborhood of $x_0 \in \text{int}(\text{dom} F)$, then F is C-Lipschitz at x_0.*

Proof. Without loss of generality, we suppose that $x_0 = 0$ and $0 \in F(0)$. As F is C-bounded from below and weakly C-upper bounded on a neighborhood $U = \theta U_X \subseteq \text{dom} F$ of 0 ($\theta > 0$), taking into account Definition 7.4.4 and Definition 7.4.5, there exists a real number $\mu > 0$ such that $F(U) \subseteq \mu U_Y + C$ and $F(x) \cap (\mu U_Y - C) \neq \emptyset$ for all $x \in U$. Take $y^* \in C^+$ with $\|y^*\|_* = 1$. Let $\bar{x} \in U$ be arbitrary, and take $\bar{y} \in F(\bar{x})$, $c \in C$, and $y' \in \mu U_Y$ such that $\bar{y} = y' - c$. We then have

$$g_{y^*}(\bar{x}) = \inf_{y \in F(\bar{x})} y^*(y) \leq y^*(\bar{y}) = y^*(y' - c) = y^*(y') - y^*(c)$$

$$\leq y^*(y') \leq \|y^*\|_* \|y'\|_Y = \|y'\|_Y \leq \mu, \quad \text{for all} \quad \bar{x} \in U.$$

Analogously, from $F(U) \subseteq \mu U_Y + C$, we get $g_{y^*}(x) = \inf_{y \in F(x)} y^*(y) \geq -\mu$ for every $x \in U$. It follows that g_{y^*} is finite on U and

$$g_{y^*}(x) \leq g_{y^*}(0) + 2\mu, \quad \text{for all} \quad x \in U = \theta U_X.$$

By Proposition 7.3.15, g_{y^*} is convex on U. Applying Lemma 7.2.6 to the convex function g_{y^*} and $\theta' \in (0, \theta)$, we get

$$|g_{y^*}(x) - g_{y^*}(x')| \leq L\|x - x'\|_X, \quad \text{for all} \quad x, x' \in \theta' U_X,$$

where $L := 2\mu(\theta + \theta')/[\theta(\theta - \theta')]$, which does not depend on y^*.
So $\{g_{y^*} \mid y^* \in C^+, \|y^*\|_* = 1\}$ is equi-Lipschitz at x_0 with the Lipschitz constant L. Because of the convexity of F, $F(x) + C$ is convex for all $x \in X$. Applying Proposition 7.4.6, we have that F is C-Lipschitz at x_0. □

Remark 7.4.8   It follows from Remark 7.3.18(ii) that, under the assumptions given in Theorem 7.4.8, the set-valued function F is upper and lower C-Lipschitz at x_0. Furthermore, it is both H-C-u.c and H-C-l.c (Remark 7.3.18(iv)). These continuity properties are also given by Khan et al. [16, Corollary 3.2.7]. ■

The following result illustrates the relationship between the upper C-convexity and the $(-C)$-Lipschitz property of set-valued functions.

Theorem 7.4.9 Let X, Y be two normed spaces, C be a proper convex cone. Let $F: X \rightrightarrows Y$ be upper C-convex, and $F(x) - C$ be convex for all $x \in X$. If F is C-bounded from above on a neighborhood of $x_0 \in \text{int}(\text{dom} F)$, then F is $(-C)$-Lipschitz at x_0.

Proof. We follow the same lines in the proof of Theorem 7.4.8, where the family $\{g_{y^*} \mid y^* \in C^+, \|y^*\|_* = 1\}$ is replaced by $\{G_{y^*} \mid y^* \in C^+, \|y^*\|_* = 1\}$. □

The next theorem is a consequence of Theorem 7.4.8 and Remarks 7.4.4 and 7.4.5 above.

Theorem 7.4.10 Let X, Y be two normed vector spaces, $C \subset Y$ be a proper convex cone, and let $F: X \rightrightarrows Y$ be lower C-convex. If F is C-bounded from below and weakly C-upper bounded on a neighborhood of some point $x \in \text{int}(\text{dom} F)$, then F is C-Lipschitz at every point in $\text{int}(\text{dom} F)$.

Remark 7.4.10 It is clear that the assumptions in Theorem 7.4.10 are much weaker than those in [20, Theorem 2.9]; while the assumption that X is finite-dimensional is not necessary. We even do not need any additional conditions for the cone C such as $C^+ = \text{cone}(\text{conv}\{y_1^*, \ldots, y_n^*\})$ for some $y_1^*, \ldots, y_n^* \in Y^*$ and $0 \notin \text{conv}\{y_1^*, \ldots, y_n^*\}$ as in [20]. ∎

In the following, we consider Theorem 7.4.10 for the case that $F : X \rightrightarrows Y$ is an at most single-valued function (i.e., $F(x)$ is a single point in Y for every $x \in \text{dom} F$). We assume that C is a normal cone in Y and $\leq_C$ is binary relation given by $x \leq_C y$ if and only if $y - x \in C$. We consider the corresponding vector-valued function $f : X \to Y^\bullet$ determined by

$$f(x) := \begin{cases} y & \text{if } x \in \text{dom} F \text{ and } F(x) = \{y\}, \\ +\infty & \text{if } x \notin \text{dom} F, \end{cases} \tag{7.29}$$

where $y \leq_C +\infty$ for all $y \in Y$ and $Y^\bullet := Y \cup \{+\infty\}$. We recall that f is Lipschitz at x_0 if there are a neighborhood U_{x_0} of x_0 and $L > 0$ such that for every $x, x' \in U_{x_0}$, we have

$$\|f(x) - f(x')\|_Y \leq L\|x - x'\|_X.$$

The next result presents the Lipschitz continuity of cone-convex vector-valued function $f : X \to Y^\bullet$.

Theorem 7.4.11 *Under the hypotheses of Theorem 7.4.8, if $F : X \rightrightarrows Y$ is an at most single-valued function and C is a normal cone, then the function $f : X \to Y^\bullet$ defined in (7.29) is Lipschitz at x_0.*

Proof. Theorem 7.4.8 implies that F is lower C-Lipschitz at $x_0 \in \text{int}(\text{dom} F)$. Thus, there is a neighborhood U_{x_0} of x_0 in $\text{dom} F$ and a constant $\ell > 0$ such that

$$F(x) \subseteq F(x') + \ell\|x - x'\|_X U_Y + C, \quad \text{for all} \quad x, x' \in U_{x_0},$$

or equivalently

$$f(x) \in f(x') + \ell\|x - x'\|_X U_Y + C, \quad \text{for all} \quad x, x' \in U_{x_0}. \tag{7.30}$$

Since C is normal, there is $\rho > 0$ such that

$$(\rho U_Y + C) \cap (\rho U_Y - C) \subseteq U_Y.$$

From (7.30), we have

$$\frac{\rho(f(x) - f(x'))}{\ell\|x - x'\|_X} \subseteq \rho U_Y + C, \quad \text{for all} \quad x, x' \in U_{x_0}, x \neq x'.$$

By interchanging x and x', we also have

$$\frac{\rho(f(x) - f(x'))}{\ell\|x - x'\|_X} \subseteq \rho U_Y - C, \quad \text{for all} \quad x, x' \in U_{x_0}, x \neq x'.$$

Therefore,

$$\frac{\rho(f(x) - f(x'))}{\ell\|x - x'\|_X} \subseteq (\rho U_Y + C) \cap (\rho U_Y - C) \subseteq U_Y,$$

for all $x, x' \in U_{x_0}, x \neq x'$. This shows that f is Lipschitzian at x_0. □

When C is normal, by [31, Proposition 2], the C-boundedness from below and weakly C-upper boundedness of f in Theorem 7.4.11 can be replaced by the C-boundedness from above, and then we obtain the Lipschitz property of convex vector-valued functions; compare [31, Theorem 2].

7.4.3 G-Lipschitzianity

In this section, we prove that a C-bounded, $\mathfrak{C}$s-convex function $F: X \rightrightarrows Y$ on a neighborhood of $x_0 \in \text{int}(\text{dom} F)$ is G-Lipschitz (see Definition 7.3.20) at x_0. For this aim, we use the equi-Lipschitzianity of a functional family in the sense of Definition 7.2.5, and the (C, G)-Lipschitzianity of F (see Definition 7.3.21).

We assume that X, Y are two normed spaces, C is a proper convex cone and $F: X \rightrightarrows Y$ is a set-valued function. We consider the functions G_{y^*}, g_{y^*} given by (7.11), (7.12) w.r.t. F.

The following propositions will be used for the main results of this section.

Proposition 7.4.12 *Let X, Y be two normed spaces, $F: X \rightrightarrows Y$, and $x_0 \in \text{int}(\text{dom} F)$. If the family $\{G_{y^*} | y^* \in C^+, \|y^*\|_* = 1\}$ is equi-Lipschitz at x_0, then F is $(-C, G)$-Lipschitz at x_0.*

Proof. Suppose by contradiction that F is not $(-C, G)$-Lipschitz at x_0, it follows that for any $n \in \mathbb{N}^*$, there are $x_n, x_n' \in B(x_0, \frac{1}{n})$ such that

$$F(x_n) \ominus_G F(x_n') \not\subseteq n\|x_n - x_n'\|_X U_Y - C.$$

This implies that $x_n \neq x_n'$ for all $n \in \mathbb{N}^*$.
Since $x_0 \in \text{int}(\text{dom} F)$, for n large enough, $B(x_0, \frac{1}{n}) \subseteq \text{dom} F$, we can take $y_n \in F(x_n) \ominus_G F(x_n')$ such that $y_n + F(x_n') \subseteq F(x_n)$ and

$$y_n \notin B_n := n\|x_n - x_n'\|_X U_Y - C.$$

Since B_n is convex and $\text{int} B_n \neq \emptyset$, one can find $y_n^* \in Y^*$ such that $\|y_n^*\|_* = 1$ and

$$y_n^*(y_n) \geq y_n^*(v) \quad \text{for all} \quad v \in B_n.$$

This implies that

$$y_n^*(y_n) \geq \sup y_n^*(B_n) = n\|x_n - x_n'\|_X - \sup y_n^*(C).$$

It follows that $y_n^* \in C^+$ for all $n \in \mathbb{N}$ and

$$G_{y_n^*}(x_n) = \sup y_n^*(F(x_n)) \geq \sup y_n^*(F(x_n')) + y_n^*(y_n) \geq G_{y_n^*}(x_n') + n\|x_n - x_n'\|_X.$$

Therefore,

$$n\|x_n - x_n'\|_X \leq G_{y_n^*}(x_n) - G_{y_n^*}(x_n') \leq \ell\|x_n - x_n'\|_X.$$

This yields that $n \leq \ell$, which could not hold true for arbitrarily large n. □

Proposition 7.4.13 *Let X,Y be two normed spaces, $F: X \rightrightarrows Y$, and x_0 be a given point in $\mathrm{int}(\mathrm{dom} F)$. If the family $\{g_{y^*} | y^* \in C^+, ||y^*||_* = 1\}$ is equi-Lipschitz at x_0, then F is (C,G)-Lipschitz at x_0.*

Proof. We follow the same lines presented in Proposition 7.4.12. □

The main result of this section is illustrated by the following theorem.

Theorem 7.4.14 *Let X,Y be two normed spaces, C be a proper convex cone, and $F: X \rightrightarrows Y$ be $\mathfrak{C}$s-convex. If F is C-bounded on a neighborhood of $x_0 \in \mathrm{int}(\mathrm{dom} F)$, then F is (C,G)-Lipschitz and $(-C,G)$-Lipschitz at x_0. Moreover, if C is normal cone, then F is G-Lipschitz at x_0.*

Proof. Without loss of generality, we suppose that $x_0 = 0$ and $0 \in F(0)$. As F is C-bounded on a neighborhood $U = \theta U_X \subseteq \mathrm{dom} F$ of 0 ($\theta > 0$), and taking into account Definition 7.4.4, there exist real numbers $\mu, \mu' > 0$ such that $F(U) \subseteq \mu U_Y + C$ and $F(x) \subseteq \mu' U_Y - C$ for all $x \in U$. Take $y^* \in C^+$ with $||y^*||_* = 1$. Take $\bar{x} \in U$ be arbitrary, $\bar{y} \in F(\bar{x})$, $c \in C$, and $y' \in \mu' U_Y$ such that $\bar{y} = y' - c$. It follows from the definition of g_{y^*} that

$$g_{y^*}(\bar{x}) = \inf_{y \in F(\bar{x})} y^*(y) \leq y^*(\bar{y}) = y^*(y' - c) \leq y^*(y') - y^*(c)$$
$$\leq y^*(y') \leq ||y^*||_* ||y'||_Y = ||y'||_Y \leq \mu', \quad \text{for all} \quad \bar{x} \in U.$$

Analogously, from $F(U) \subseteq \mu U_Y + C$, we get $g_{y^*}(x) = \inf_{y \in F(x)} y^*(y) \geq -\mu$ for every $x \in U$. Therefore, g_{y^*} is finite on U and

$$g_{y^*}(x) \leq g_{y^*}(0) + \mu + \mu', \quad \text{for all} \quad x \in U = \theta U_X.$$

By Proposition 7.3.15, g_{y^*} is convex. Applying Lemma 7.2.6 to g_{y^*} and $\theta' \in (0, \theta)$, we get

$$|g_{y^*}(x) - g_{y^*}(x')| \leq L||x - x'||_X, \quad \text{for all} \quad x, x' \in \theta' U_X,$$

where $L := (\mu + \mu')(\theta + \theta')/[\theta(\theta - \theta')]$, which clearly does not depend on y^*.
So, $\{g_{y^*} | y^* \in C^+, ||y^*||_* = 1\}$ is equi-Lipschitz at x_0 with the Lipschitz constant L. By Proposition 7.4.13, we have that F is (C,G)-Lipschitz at x_0.
Similarly, $\{G_{y^*} | y^* \in C^+, ||y^*||_* = 1\}$ is equi-Lipschitz at x_0. Applying Proposition 7.4.12, F is $(-C,G)$-Lipschitz at x_0.
By Remark 7.3.21, if C is a normal cone, then F is G-Lipschitz at x_0. □

The following example is an illustration of Theorem 7.4.14.

Example 7.4.15 *Set $F: [0, +\infty) \times \mathbb{R} \rightrightarrows \mathbb{R}^2$ assuming*

$$F(x_1, x_2) = \mathrm{conv}\left\{\begin{pmatrix}0\\0\end{pmatrix}, \begin{pmatrix}x_1\\-\sqrt{x_1}\end{pmatrix}\right\}.$$

Obviously, F is C-bounded around every point $x \in (0, +\infty) \times \mathbb{R}$, and $\mathfrak{C}$s-convex with $C := \mathbb{R}^2_+$. Hence F is G-Lipschitz. However, F is not M-Lipschitz.

7.5 Conclusions

This manuscript studies six kinds of cone-convexity, as well as several types of Lipschitz property of set-valued functions. Section 7.4 indicates that, under appropriate assumptions, upper (lower) C-convex functions are (C, e)-Lipschitz, C-Lipschitz or G-Lipschitz. Futhermore, all these results also hold for set-valued functions which are type-(k)-convex, $k \in \{i, ii, iv\}$ by taking into account Proposition 7.3.12.

Acknowledgements. The authors would like to offer special thanks to Prof. Christiane Tammer, Prof. Constantin Zălinescu and the referees for their support and constructive remarks. Their valuable comments help us to improve not only the content but also the structure of this manuscript. This research is supported by Vietnamese government project MOET 911, the PhD Finalization Grant and the Ministry Federal for Education and Research (Martin-Luther-University Halle-Wittenberg).

References

[1] R. Baier and E. Farkhi. Regularity of set-valued maps and their selections through set differences. *Part 1: Lipschitz continuity. Serdica Math. J.,* 39(3-4): 365–390, 2013.

[2] T. Q. Bao and B. S. Mordukhovich. Existence of minimizers and necessary conditions in set-valued optimization with equilibrium constraints. *Appl. Math.,* 52(6): 453–472, 2007.

[3] T. Q. Bao and B. S. Mordukhovich. Relative Pareto minimizers for multiobjective problems: Existence and optimality conditions. *Math. Program.,* 122-2, Ser. A): 301–347, 2010.

[4] T. Q. Bao and B. S. Mordukhovich. Necessary nondomination conditions in set and vector optimization with variable ordering structures. *J. Optim. Theory Appl.,* 162(2): 350–370, 2014.

[5] T. Q. Bao and C. Tammer. Lagrange necessary conditions for Pareto minimizers in Asplund spaces and applications. *Nonlinear Anal.,* 75(3): 1089–1103, 2012.

[6] J. M. Borwein. Continuity and differentiability properties of convex operators. *Proc. London Math. Soc. (3),* 44(3): 420–444, 1982.

[7] F. H. Clarke. Optimization and nonsmooth analysis, volume 5 of Classics in Applied Mathematics. Society for Industrial and Applied Mathematics (SIAM), Philadelphia, PA, second edition, 1990.

[8] S. Dempe and M. Pilecka. Optimality conditions for set-valued optimisation problems using a modified Demyanov difference. *Journal of Optimization Theory and Applications,* 1–20, 2015.

[9] P. Diamond, P. Kloeden, A. Rubinov and A. Vladimirov. Comparative properties of three metrics in the space of compact convex sets. *Set-Valued Anal.,* 5(3): 267–289, 1997.

[10] Christiane Gerstewitz. Nichtkonvexe Dualität in der Vektoroptimierung. *Wiss. Z. Tech. Hochsch. Leuna-Merseburg,* 25(3): 357–364, 1983.

[11] C. Gerth and P. Weidner. Nonconvex separation theorems and some applications in vector optimization. *J. Optim. Theory Appl.,* 67(2): 297–320, 1990.

[12] A. Gopfert, H. Riahi, C. Tammer and C. Zalinescu. Variational methods in partially ordered spaces. *CMS Books in Mathematics/Ouvrages de Mathematiques de la SMC, 17.* Springer-Verlag, New York, 2003.

[13] E. Hernández and L. Rodríguez-Marín. Nonconvex scalarization in set optimization with set-valued maps. *J. Math. Anal. Appl.,* 325(1): 1–18, 2007.

[14] J. Jahn. Directional derivatives in set optimization with the set less order relation. *Taiwanese J. Math.,* 19(3): 737–757, 2015.

[15] J. Jahn and T. X. D. Ha. New order relations in set optimization. *J. Optim. Theory Appl.,* 148(2): 209–236, 2011.

[16] A. A. Khan, C. Tammer and C. Zalinescu. Set-valued optimization. Vector Optimization. Springer, Heidelberg, 2015. An introduction with applications.

[17] D. Kuroiwa, T. Tanaka and T. X. D. Ha. On cone convexity of set-valued maps. *In Proceedings of the Second World Congress of Nonlinear Analysts, Part 3 (Athens, 1996),* 30: 1487–1496, 1997.

[18] I. Kuwano and T. Tanaka. Continuity of cone-convex functions. *Optim. Lett.,* 6(8): 1847–1853, 2012.

[19] D. T. Luc, N. X. Tan and P. N. Tinh. Convex vector functions and their subdifferential. *Acta Math. Vietnam.,* 23(1): 107–127, 1998.

[20] N. B. Minh and N. X. Tan. On the C-Lipschitz continuities and C-approximations of multivalued mappings. *Vietnam J. Math.,* 30(4): 343–363, 2002.

[21] N. B. Minh and N. X. Tan. On the continuity of vector convex multivalued functions. *Acta Math. Vietnam.,* 27(1): 13–25, 2002.

[22] B. S. Mordukhovich. Variational analysis and generalized differentiation. I, volume 330 of Grundlehren der Mathematischen Wissenschaften [Fundamental Principles of Mathematical Sciences]. Springer-Verlag, Berlin, 2006. Basic theory.

[23] Z. G. Nishnianidze. Fixed points of monotone multivalued operators. *Soobshch. Akad. Nauk Gruzin. SSR,* 114(3): 489–491, 1984.

[24] M. Pilecka. Set-valued optimization and its application to bilevel optimization. Dissertation, Technische Universitat Bergakademie Freiberg, 2016.

[25] T. W. Reiland. Nonsmooth analysis and optimization on partially ordered vector spaces. *Internat. J. Math. Math. Sci.,* 15(1): 65–81, 1992.

[26] R. T. Rockafellar. *Convex Analysis.* Princeton Mathematical Series, No. 28. Princeton University Press, Princeton, N.J., 1970.

[27] Hans Rådström. An embedding theorem for spaces of convex sets. *Proc. Amer. Math. Soc.,* 3: 165–169, 1952.

[28] A. M. Rubinov and I. S. Akhundov. Difference of compact sets in the sense of Demyanov and its application to nonsmooth analysis.*Optimization,* 23(3): 179–188, 1992.

[29] C. Tammer and C. Zălinescu. Lipschitz properties of the scalarization function and applications. *Optimization,* 59(2): 305–319, 2010.

[30] L. Thibault. Subdifferentials of compactly Lipschitzian vector-valued functions. *Ann. Mat. Pura Appl. (4),* 125: 157–192, 1980.

[31] V. A. Tuan, C. Tammer and C. Zalinescu. The Lipschitzianity of convex vector and set-valued functions. *TOP,* 24(1): 273–299, 2016.

[32] R. C. Young. The algebra of many-valued quantities. *Math. Ann.,* 104(1): 260–290, 1931.

[33] C. Zaalinescu. Convex analysis in general vector spaces. World Scientific Publishing Co. Inc., River Edge, NJ, 2002.

Chapter 8

Efficiencies and Optimality Conditions in Vector Optimization with Variable Ordering Structures

Marius Durea
Faculty of Mathematics, "Alexandru Ioan Cuza" University, Bd. Carol I, nr. 11, 700506 – Iaşi, Romania and "Octav Mayer" Institute of Mathematics of the Romanian Academy, Iaşi, Romania. E-mail: durea@uaic.ro

Elena-Andreea Florea
Department of Research, Faculty of Mathematics, "Alexandru Ioan Cuza" University, Bd. Carol I, nr. 11, 700506, Iaşi, Romania. E-mail: andreea_acsinte@yahoo.com

Radu Strugariu
Department of Mathematics, "Gheorghe Asachi" Technical University, Bd. Carol I, nr. 11, 700506 – Iaşi, Romania. E-mail: rstrugariu@tuiasi.ro

8.1 Introduction

The main purpose of this work is to present some new developments for the theory of vector optimization with variable ordering structures under the paradigm introduced in

[22], that is in the natural setting when the objective and the ordering mappings share the same input and output space. This corresponds, intuitively, to the situation when the preferences of the decision maker may vary with respect to the same parameters as the objective itself. Moreover, a suitable treatment of the problems coming from the proposed paradigm requires the strong tools and methods of Variational Analysis and, for this reason, one can say that this work fits into the broadest topics of nonlinear analysis and mathematical programming governed by set-valued maps.

The extension of the classical case of vector optimization we propose to survey and to enrich in this work is a natural one, given the evolution over the last 60–70 years of the mathematical objects and techniques used to study many concrete problems for which the use of smooth functions is not sufficient. For the study of many issues, including optimal control problems, abstract or vector optimization problems, an important aspect was the introduction of concepts of differentiability for multivalued applications, along the generalized subdifferentials objects introduced before for single-valued mappings (that is, functions). For set-valued mappings, the generalized differentiation objects can be defined either on primary spaces using tangent cones, or on dual spaces using normal cones, defined in turn by subdifferentials of some specific functions. Many important mathematicians have offered contributions in this area, which have subsequently received the name of Variational Analysis; a detailed account is to be found in the two-volume comprehensive monograph [41], [42].

If we only refer to the case of vector optimization, which is the main subject of the current work, we will find that, within the already classical case of Pareto-type minima, introduced by means of a partial order induced by a convex, closed and pointed cone, many of the generalized differentiation objects have been successfully used, firstly by using scalarization techniques (which reduce the vector minimization problems to scalar minimization problems), and then into the general framework, where scalarization no longer works.

The present work addresses, through the tools and techniques of Variational Analysis, the wider framework of the situation where the cone is not fixed, but varies with the objective mapping argument. This expansion of the usual vector case has been known since the 1970s (see [52] and [53]), but has gained a new impetus over the past 10 years with the description by Eichfelder (see [25] and [26]), and Bao, Mordukhovich, Soubeyran (see [5]) of possible applications in medical imaging and behavioral sciences. Thus, on the basis of this impetus, the theoretical progress has multiplied, and this work is part of the effort to theoretically substantiate new facts arising from the investigation of the situation addressed, by using a mathematical apparatus based on the main ideas of Variational Analysis.

As mentioned, the results proposed in this work survey, from a wider perspective, some results obtained by the authors in the last years in the papers [22], [31], [12], [13]. However, there are several novelties and generalizations that we propose and present systematically as follows: (i) *a more complete comparison of the efficiency concepts involved in our study;* (ii) *a more general technique to penalize constrained problems in the proposed setting;* (iii) *a better way of getting openness properties for a finite family of set-valued mappings that allows one to drop, to unify, and to improve some technical assumptions;* (iv) *enhanced necessary optimality conditions for the problems under study.*

160 ■ *Variational Analysis and Set Optimization*

Now, we briefly describe the structure of this chapter.

In the second section we introduce the notation and the main tools of nonlinear analysis that we need in this work. Starting from the end of this section, the presentation is self-contained, excepting some well-known facts on nonlinear analysis that cannot be fully introduced here because of the lack of space (e.g., Ekeland Variational Principle, and some results concerning the Mordukhovich generalized differentiation calculus).

The third section deals with the efficiency concepts we study in the framework of vector optimization with the variable ordering structures for the case of the same input and output spaces, both for objective and ordering multifunctions. Moreover, we present several links and differences between the concepts we deal with from different perspectives: Direct comparisons, scalarization and penalization results that reduce certain constrained efficiencies to other unconstrained ones. We end this section by emphasizing a way to achieve necessary optimality conditions for our notions by means of openness results for set-valued maps, and, in this sense, we devise two different kinds of openness that are studied, from the perspective of derivative and coderivative sufficient conditions, in the next section.

Therefore, Section 4 deals with two openness results for a finite sum of set-valued maps and our aim here is to get sufficient conditions in primal spaces (in terms of derivatives) and then in dual spaces (in terms of coderivatives) for an openness property that combines the two types of openness discussed before.

Based on the results developed in Sections 3 and 4, in Section 5 we present several optimality conditions for the efficiency concepts introduced in Section 2 and, again, both primal spaces and dual spaces settings are envisaged.

In order to make the presentation as readable as possible, we avoid stopping too often to compare the presented results with other existing ones in literature. For this reason, we end this work with a section containing bibliographic notes and comments. Moreover, several conclusions and possible developments are presented.

8.2 Preliminaries

In order to develop the main ideas of this chapter, we need several tools from variational analysis, nonsmooth analysis and optimization theory. The general setting is that of Banach spaces and the notation is standard. If X is a Banach space, then we denote by $B_X(x,r)$, $D_X(x,r)$ and $S_X(x,r)$ the open ball, the closed ball and the sphere of center $x \in X$ and radius $r > 0$, respectively. In the case where $x = 0$ and $r = 1$, we use the shorter notation B_X, D_X and S_X, respectively. For $x \in X$, the symbol $\mathcal{V}(x)$ stands for the system of neighborhoods of x with respect to the norm topology. For a set $A \subset X$, we denote by int$A$, cl$A$, bd$A$ its topological interior, closure and boundary, respectively. The cone generated by A is designated by cone$A$, and the convex hull of A is conv$A$. The indicator function of A is $\delta_A : X \to \mathbb{R} \cup \{+\infty\}$, defined as $\delta_A(x) = 0$ if $x \in A$ and $\delta_A(x) = \infty$ if $x \notin A$. The distance from a point x to a nonempty set $A \subset X$ is $d(x,A) = \inf\{\|x-a\| \mid a \in A\}$, and the distance function to A is $d_A : X \to \mathbb{R}$, given by $d_A(x) = d(x,A)$. Based on the distance function, the oriented distance function $\Delta : X \to \mathbb{R}$, given by

$$\Delta(x,A) = d(x,A) - d(x, X \setminus A) \text{ for all } x \in X,$$

was defined by Hiriart-Urruty in [38]. The notation X^* stands for the topological dual of X, while w^* denotes the weak-star topology on X^*. On a product space, we consider the sum norm, unless otherwise stated.

We begin by introducing, in a synthetic manner, the main concepts and results that we need in the sequel.

One of the key ingredients that we use in this work is a penalization technique, which allows us to look at a local minimum point of a constrained problem as a local minimum point of an unconstrained problem, by adding to the objective mapping a penalty term built on the constraint system. In fact, the penalty term we use here is based on special form of a minimal time function (see [15]), which, in fact, is one of the key concepts that we use in this work.

Consider two nonempty sets $S, M \subset X$. Then, one defines the minimal time function associated to the target set S, with respect to dynamics set M as $T_M(\cdot, S) : X \to [0, \infty]$,

$$T_M(x, S) = \inf\{t \geq 0 \mid \exists u \in M : x + tu \in S\}.$$

Note that this function naturally appears from differential inclusions of the type

$$\dot{x}(t) \in M, \quad x(0) = x_0 \qquad (8.1)$$

as the infimum of positive t (time), for which there is a trajectory x satisfying (8.1) such that $x(t) \in S$.

The amount of literature dedicated to minimal time function, under different assumptions on the underlying space X, and on the sets S and M, is important and is due to various instances and to different purposes this function can be applied for. However, it should be mentioned that in the majority of the papers in literature, the set M is taken as a closed convex neighborhood of 0, and this allows for the use of associated Minkovski functional (gauge) and tools from convex analysis in order to study and use the minimal time function. For more details, see the last section.

For our purposes, we propose to work with a set of directions $M \subset S_X$, as done in [15], i.e., M is a subset of the unit sphere, so M is not always convex, is never a neighborhood of 0, and this significantly changes both the use and the study of this function.

We list now some basic properties for $T_M(\cdot, S)$ as proved in [15], but before that, we recall that, for $\Omega \subset X$, its recession cone is

$$\Omega_\infty = \{u \in X \mid \omega + tu \in \Omega, \forall t \geq 0, \forall \omega \in \Omega\}.$$

Further, a set $\Omega \subset X$ is said to be epi-Lipschitz at $\bar{x}$ in a direction $v \neq 0$ if there exists $\delta > 0$ such that

$$\Omega \cap B_X(\bar{x}, \delta) + [0, \delta] \cdot B_X(v, \delta) \subset \Omega. \qquad (8.2)$$

Proposition 8.2.1 *Let $S \subset X$, $M \subset S_X$ be two nonempty sets. The following assertions hold:*

(i) The domain of the directional minimal time function with respect to M is

$$\mathrm{dom}\, T_M(\cdot, S) = S - \mathrm{cone}\, M.$$

(ii) One has
$$T_M(x,S) = \inf_{u \in M} T_{\{u\}}(x,S).$$

(iii) One has, for any $x \in X$,
$$d(x,S) \leq T_M(x,S).$$

If $M = S_X$, then $\operatorname{dom} T_M(\cdot, S) = X$ and
$$T_M(x,S) = d(x,S), \ \forall x \in X. \tag{8.3}$$

(iv) If one of the sets S and M is compact, and the other one is closed, $T_M(\cdot, S)$ is lower semicontinuous.

(v) If $\operatorname{cone} M$ is convex and either
$$\operatorname{cone} M \cap S_\infty \neq \emptyset \ \text{or} \ \operatorname{int}(\operatorname{cone} M) \cap S_\infty \neq \emptyset,$$
then $T_M(\cdot, S)$ is globally Lipschitz.

(vi) If $\operatorname{cone} M$ is convex and S is epi-Lipschitz at $\bar{x} \in S$ in a direction $v \in M$, then $T_M(\cdot, S)$ is Lipschitz around $\bar{x}$.

Let Y be a Banach space. In order to cover all the situations that we face here, we make some natural conventions related to the multiplication between $\pm \infty$ and the nonzero elements of a convex cone $C \subset Y$. Moreover, aiming at covering the case of functions with extended values, we add to Y two abstract and distinct elements $-\infty_C, +\infty_C$, which do not belong to Y, and we set $\overline{Y} = Y \cup \{-\infty_C, +\infty_C\}$. The ordering and algebraic rules we need for the new elements are the following ones

$$-(+\infty_C) = -\infty_C; \ \forall y \in \overline{Y}: -\infty_C \leq_C y \leq_C +\infty_C,$$
$$\forall y \in Y \cup \{+\infty_C\}: y + (+\infty_C) = (+\infty_C) + y = +\infty_C,$$
$$\forall y \in Y \cup \{-\infty_C\}: y + (-\infty_C) = (-\infty_C) + y = -\infty_C$$

and if $\operatorname{int} C \neq \emptyset$,
$$-\infty_C <_C y <_C +\infty_C,$$

where $\leq_C$ is the partial order relation induced by C, namely
$$y_1 \leq_C y_2 \Leftrightarrow y_2 - y_1 \in C,$$

and $<_C$ is the associated strict order given by
$$y_1 <_C y_2 \Leftrightarrow y_2 - y_1 \in \operatorname{int} C.$$

When needed, a neighborhood of $+\infty_C$ is understood in the following sense: A set $V \subset \overline{Y}$ is a neighborhood of $+\infty_C$ if there exists $\theta > 0$ such that $(C \cap Y \setminus D_Y(0, \theta)) \cup \{+\infty_C\} \subset V$. Similarly, a set $V \subset \overline{Y}$ is a neighborhood of $-\infty_C$ if there exists $\theta > 0$ such that $(-C \cap Y \setminus D_Y(0, \theta)) \cup \{-\infty_C\} \subset V$. We denote by τ the norm topology on Y and we endow the space $\overline{Y}$ with the following topology:

$$\overline{\tau} = \tau \cup \{D \subset \overline{Y} \mid D \ \text{neighborhood for} \ -\infty_C \ \text{or} \ +\infty_C \ \text{and} \ D \setminus \{-\infty_C, +\infty_C\} \in \tau\}.$$

We also use the convention that $-\infty \cdot e = -\infty_C$ and $+\infty \cdot e = +\infty_C$ for any $e \in C \setminus \{0\}$.

Other key ingredients in deriving optimality conditions are, of course, the generalized differentiation objects (for set-valued maps) defined on primal spaces or on dual spaces. We briefly recall some main parts of the theory.

Consider a set-valued map (or multifunction) $F : X \rightrightarrows Y$. The graph of F is $\mathrm{Gr}\, F = \{(x,y) \mid y \in F(x)\}$, and the inverse set-valued map is $F^{-1} : Y \rightrightarrows X$ given by $x \in F^{-1}(y)$ iff $y \in F(x)$. If $A \subset X$, then $F(A)$ is the union of all $F(x)$ with $x \in A$. The set-valued mapping F is said to have the Aubin property around a point $(\bar{x}, \bar{y}) \in \mathrm{Gr}\, F$ if there exist a constant $L > 0$ and two neighborhoods, U of $\bar{x}$, and V of $\bar{y}$, respectively, such that, for every $x, u \in U$,

$$F(x) \cap V \subset F(u) + L\|x-u\| D_Y.$$

The construction of the generalized differentiation objects on primal spaces is based on the concepts of tangent cones. The next definitions are well-known (see, for instance, [2]).

Definition 8.2.2 *Let S be a nonempty subset of X and $\bar{x} \in X$.*
(i) The Bouligand (contingent) tangent cone to S at $\bar{x}$ is the set

$$T_B(S, \bar{x}) = \{u \in X \mid \exists (t_n) \downarrow 0, \exists (u_n) \to u, \exists n_0 \in \mathbb{N}, \forall n \geq n_0, \bar{x} + t_n u_n \in S\},$$

where $(t_n) \downarrow 0$ means $(t_n) \subset]0, \infty[$ and $(t_n) \to 0$.
(ii) The Ursescu (adjacent) tangent cone to S at $\bar{x}$ is the set

$$T_U(S, \bar{x}) = \{u \in X \mid \forall (t_n) \downarrow 0, \exists (u_n) \to u, \exists n_0 \in \mathbb{N}, \forall n \geq n_0, \bar{x} + t_n u_n \in S\}.$$

Some remarks are in order: The sets $T_B(S, \bar{x})$, $T_U(S, \bar{x})$ are closed cones (not necessarily convex) and $T_B(S, \bar{x}) \neq \emptyset$ if and only if $\bar{x} \in \mathrm{cl}\, S$, and $T_U(S, \bar{x}) \subset T_B(S, \bar{x})$. It is a well-known fact that if S is a convex set, then $T_U(S, \bar{x}) = T_B(S, \bar{x}) = \mathrm{cl}\, \mathrm{cone}(S - \bar{x})$.

Based on these concepts, one defines the associated derivatives for set-valued maps.

Definition 8.2.3 *Let $(\bar{x}, \bar{y}) \in \mathrm{Gr}\, F$. The Bouligand derivative of F at $(\bar{x}, \bar{y})$ is the set valued map $D_B F(\bar{x}, \bar{y}) : X \rightrightarrows Y$ defined by*

$$\mathrm{Gr}\, D_B F(\bar{x}, \bar{y}) = T_B(\mathrm{Gr}\, F, (\bar{x}, \bar{y})).$$

The definition of Ursescu derivative, denoted by $D_U F(\bar{x}, \bar{y})$, is similar.

An important concept for getting calculus rules for above objects under mild conditions is the proto-differentiability, as introduced by Rockafellar (see [48]): One says that a set-valued map F is proto-differentiable at $\bar{x}$ relative to $\bar{y} \in F(\bar{x})$ if $D_U F(\bar{x}, \bar{y}) = D_B F(\bar{x}, \bar{y})$.

Now we recall some elements from Mordukhovich's generalized differentiability theory (see [41]) on dual spaces. The construction of these objects is based on various concepts of normal cone and their main features hold on Asplund spaces, which represent a special class of Banach spaces: X is Asplund if, and only if, every continuous convex function on any open convex set $U \subset X$ is Fréchet differentiable on a dense G_δ−subset of U. However, one of the main properties of Asplund spaces we use is that every bounded sequence from the topological dual admits a w^*−convergent subsequence.

Take a nonempty subset S of the Banach space X and $x \in S$. For every $\varepsilon \geq 0$, the set of ε−normals to S at x is

$$\widehat{N}_\varepsilon(S,x) = \left\{ x^* \in X^* \mid \limsup_{u \xrightarrow{S} x} \frac{x^*(u-x)}{\|u-x\|} \leq \varepsilon \right\},$$

where $u \xrightarrow{S} x$ means that $u \to x$ and $u \in S$. If $\varepsilon = 0$, then $\widehat{N}_0(S,x)$ is denoted by $\widehat{N}(S,x)$ and it is called the Fréchet normal cone to S at x.

Let $\bar{x} \in S$. The Mordukhovich normal cone to S at $\bar{x}$ is

$$N(S,\bar{x}) = \{ x^* \in X^* \mid \exists \varepsilon_n \downarrow 0, x_n \xrightarrow{S} \bar{x}, x_n^* \xrightarrow{w^*} x^*, x_n^* \in \widehat{N}_{\varepsilon_n}(S,x_n), \forall n \in \mathbb{N} \}.$$

If X is an Asplund space, and S is closed around $\bar{x}$, the formula for the Mordukhovich normal cone takes a simpler form, namely:

$$N(S,\bar{x}) = \{ x^* \in X^* \mid \exists x_n \xrightarrow{S} \bar{x}, x_n^* \xrightarrow{w^*} x^*, x_n^* \in \widehat{N}(S,x_n), \forall n \in \mathbb{N} \}.$$

Let $F : X \rightrightarrows Y$ be a set-valued map and $(\bar{x},\bar{y}) \in \operatorname{Gr} F$. Then the Fréchet coderivative at $(\bar{x},\bar{y})$ is the set-valued map $\widehat{D}^* F(\bar{x},\bar{y}) : Y^* \rightrightarrows X^*$ given by

$$\widehat{D}^* F(\bar{x},\bar{y})(y^*) = \{ x^* \in X^* \mid (x^*, -y^*) \in \widehat{N}(\operatorname{Gr} F, (\bar{x},\bar{y})) \}.$$

Similarly, the Mordukhovich coderivative of F at $(\bar{x},\bar{y})$ is the set-valued map $D^* F(\bar{x},\bar{y}) : Y^* \rightrightarrows X^*$ given by

$$D^* F(\bar{x},\bar{y})(y^*) = \{ x^* \in X^* \mid (x^*, -y^*) \in N(\operatorname{Gr} F, (\bar{x},\bar{y})) \}.$$

As usual, when $F = f$ is a function, since $\bar{y} \in F(\bar{x})$ means $\bar{y} = f(\bar{x})$, we write $\widehat{D}^* f(\bar{x})$ for $\widehat{D}^* f(\bar{x},\bar{y})$, and similarly for D^*.

If $S \subset X$ is a convex set, then

$$N(S,\bar{x}) = \{ x^* \in X^* \mid x^*(x-\bar{x}) \leq 0, \forall x \in S \}$$

and this coincides with the negative polar of $T_B(S,\bar{x})$. Moreover, if F has convex graph,

$$D^* F(\bar{x},\bar{y})(y^*) = \{ x^* \in X^* \mid x^*(\bar{x}) - y^*(\bar{y}) \geq x^*(x) - y^*(y), \forall (x,y) \in \operatorname{Gr} F \}.$$

Let X, Y be Asplund spaces, $S \subset Y$ be a set closed around $\bar{y} \in S$ and suppose that $F : X \rightrightarrows Y$ is a set-valued map closed around $(\bar{x},\bar{y}) \in \operatorname{Gr} F$. Following [41, pages 27, 76, 266], one defines the following compactness notions.

One says that S is sequentially normally compact ((SNC), for short) at $\bar{y}$ if

$$\left[y_n \xrightarrow{S} \bar{y}, y_n^* \xrightarrow{w^*} 0, y_n^* \in \widehat{N}(S,y_n) \right] \Rightarrow y_n^* \to 0.$$

Remark that, in the case where $S = C$ is a closed convex cone, the (SNC) property at 0 is equivalent to

$$\left[(y_n^*) \subset C^+, y_n^* \xrightarrow{w^*} 0 \right] \Rightarrow y_n^* \to 0,$$

where
$$C^+ = \{y^* \in Y^* \mid y^*(y) \geq 0 \text{ for all } y \in C\}.$$
In particular, if $\text{int} C \neq \emptyset$, then C is (SNC) at 0.

One says that F is partially sequentially normally compact at $(\bar{x}, \bar{y})$ ((PSNC), for short) if
$$\left[(x_n, y_n) \xrightarrow{\text{Gr} F} (\bar{x}, \bar{y}), x_n^* \xrightarrow{w^*} 0, y_n^* \to 0, (x_n^*, y_n^*) \in \widehat{N}(\text{Gr} F, (x_n, y_n))\right] \Rightarrow x_n^* \to 0.$$

Let $f : X \to \mathbb{R} \cup \{+\infty\}$ be finite at $\bar{x} \in X$ and lower semicontinuous around $\bar{x}$; the Fréchet subdifferential of f at $\bar{x}$ is the set
$$\widehat{\partial} f(\bar{x}) = \{x^* \in X^* \mid (x^*, -1) \in \widehat{N}(\text{epi} f, (\bar{x}, f(\bar{x})))\},$$
where $\text{epi} f$ denotes the epigraph of f. The Mordukhovich subdifferential of f at $\bar{x}$ is given, according to [41], by
$$\partial f(\bar{x}) = \{x^* \in X^* \mid (x^*, -1) \in N(\text{epi} f, (\bar{x}, f(\bar{x})))\}.$$

It is well-known that if f is a convex function, then $\widehat{\partial} f(\bar{x})$ and $\partial f(\bar{x})$ coincide with the Fenchel subdifferential. However, in general, $\widehat{\partial} f(\bar{x}) \subset \partial f(\bar{x})$, and the following generalized Fermat rule holds: If $\bar{x} \in X$ is a local minimum point for $f : X \to \mathbb{R} \cup \{+\infty\}$, then $0 \in \widehat{\partial} f(\bar{x})$.

Consider now some closed subsets $C_1, ..., C_k$ of a normed vector space X ($k \in \mathbb{N} \setminus \{0, 1\}$). One says that $C_1, ..., C_k$ are allied at $\bar{x} \in C_1 \cap ... \cap C_k$ if for every $(x_{in}) \xrightarrow{C_i} \bar{x}$, $x_{in}^* \in \widehat{N}(C_i, x_{in}), i \in \overline{1, k}$, the relation $(x_{1n}^* + ... + x_{kn}^*) \to 0$ implies $(x_{in}^*) \to 0$ for every $i \in \overline{1, k}$. This concept was introduced and used by Penot and his coauthors in [47] and [40] in order to get a calculus rule for the Fréchet normal cone to the intersection of sets. More precisely, if the subsets $C_1, ..., C_k$ are allied at $\bar{x} \in C_1 \cap ... \cap C_k$ and X is an Asplund space, then there exists $r > 0$ such that, for every $\varepsilon > 0$ and every $x \in [C_1 \cap ... \cap C_k] \cap B_X(\bar{x}, r)$, there exist $x_i \in C_i \cap B_X(x, \varepsilon), i \in \overline{1, k}$ such that
$$\widehat{N}(C_1 \cap ... \cap C_k, x) \subset \widehat{N}(C_1, x_1) + ... + \widehat{N}(C_k, x_k) + \varepsilon D_{X^*}.$$

For this reason, the alliedness property is a corner stone for deriving optimality conditions, as we make precise in several results in what follows.

As an example of computation of some of these generalized differentiation objects we recall the form of Fréchet and Mordukhovich subdifferentials for the minimal time function under the paradigm introduced before (see [15]).

Proposition 8.2.4 *Let X be a normed vector space, $M \subset S_X$ and $\Omega \subset X$.*
(i) If $\bar{x} \in \Omega$, then
$$\widehat{\partial} T_M(\cdot, \Omega)(\bar{x}) = \{x^* \in X^* \mid x^*(-u) \leq 1, \forall u \in M\} \cap \widehat{N}(\Omega, \bar{x}).$$

(ii) Suppose that one of the sets Ω and M is compact and the other one is closed. Take $\bar{x} \in (\Omega - \text{cone} M) \setminus \Omega$. Then for every $u \in M$ and $\omega \in \Omega$ with $\bar{x} + T_M(\bar{x}, \Omega)u = \omega$, one has
$$\widehat{\partial} T_M(\cdot, \Omega)(\bar{x}) \subset \{x^* \in X^* \mid x^*(u) = -1\} \cap \widehat{N}(\Omega, \omega).$$

Proposition 8.2.5 *Let X be an Asplund space, $M \subset S_X$ and $\Omega \subset X$. Suppose that M is compact and Ω closed.*
  (i) If $\bar{x} \in \Omega$, then
$$\partial T_M(\cdot, \Omega)(\bar{x}) \subset \{x^* \in X^* \mid \exists u \in M : x^*(-u) \leq 1\} \cap N(\Omega, \bar{x}).$$

  (ii) If $\bar{x} \in (\Omega - \text{cone} M) \setminus \Omega$, then there exists $u \in \{v \in M \mid x + T_M(x,\Omega)v \in \Omega\}$ such that
$$\partial T_M(\cdot, \Omega)(\bar{x}) \subset \{x^* \in X^* \mid x^*(u) = -1\} \cap N(\Omega, \bar{x} + T_M(\bar{x}, \Omega)u).$$

Also, in the next lemma, we collect from [36, Theorem 2.3.1, Corollary 2.3.5] and [23, Lemma 2.4] the properties of the Gerstewitz's (Tammer's) scalarizing functional, which was defined in [33], including the calculation for its subdifferential.

Lemma 8.2.6 *Let $K \subset Y$ be a closed convex cone with nonempty interior. Then for every $e \in \text{int} K$, the functional $s_{e,K} : Y \to \mathbb{R}$ given by*
$$s_{e,K}(y) = \inf\{\lambda \in \mathbb{R} \mid \lambda e \in y + K\} \tag{8.4}$$

is continuous, and for every $\lambda \in \mathbb{R}$
$$\{y \in Y \mid s_{e,K}(y) < \lambda\} = \lambda e - \text{int} K, \tag{8.5}$$
$$\{y \in Y \mid s_{e,K}(y) \leq \lambda\} = \lambda e - K. \tag{8.6}$$

Moreover, $s_{e,K}$ is sublinear, and for every $u \in Y$, $\partial s_{e,K}(u)$ is nonempty and
$$\partial s_{e,K}(u) = \{v^* \in K^+ \mid v^*(e) = 1, v^*(u) = s_{e,K}(u)\}.$$

In addition, $s_{e,K}$ is $(d(e, \text{bd} K))^{-1}$-Lipschitz, and for every $u \in Y$ and $v^* \in \partial s_{e,K}(u)$, one has
$$\|e\|^{-1} \leq \|v^*\| \leq (d(e, \text{bd} K))^{-1}.$$

We close this preliminary part by defining another important notion for the next developments. Consider $C \subset X$ as a closed convex pointed cone. A convex set B is said to be a base for the cone C if $0 \notin \text{cl} B$ and $C = \text{cone} B$. A cone which admits a base is called based. For a based cone it is possible to define some dilating cones enjoying nice properties in view of applications to vector optimization problems. This theoretical construction and its properties are presented in the next result (see [36, Lemma 3.2.51]).

Proposition 8.2.7 *Let $C \subset Y$ be a closed convex cone with a base B and take $\delta = d(0,B) > 0$. For $\varepsilon \in]0,\delta[$, consider $B_\varepsilon = \{y \in Y \mid d(y,B) \leq \varepsilon\}$ and $C_\varepsilon = \text{cone} B_\varepsilon$. Then*
  (i) C_ε is a closed convex cone for every $\varepsilon \in]0,\delta[$;
  (ii) if $0 < \gamma < \varepsilon < \delta$, then $C \setminus \{0\} \subset C_\gamma \setminus \{0\} \subset \text{int} C_\varepsilon$;
  (iii) $C = \bigcap_{\varepsilon \in]0,\delta[} C_\varepsilon = \bigcap_{n \in \mathbb{N}} C_{\varepsilon_n}$, where $(\varepsilon_n) \subset]0,\delta[$ converges to 0.

The cones C_ε constructed in this way are termed as Henig dilating cones. One of the main features that make these cones attractive from an optimization point of view is the property that for every $\varepsilon > 0$, C_ε has nonempty interior.

8.3 Efficiency Concepts

First, we introduce the efficiency notions we deal with using the paradigm introduced in [22], that is the case where the order is expressed by means of a set-valued map acting between the same spaces as the objective mapping. Note that this approach is different from the one taken by other authors (see [24], [28], [29], [4], [27], [30]).

Let X, Y be Banach spaces, $F : X \rightrightarrows Y$ be a multifunction, and $\emptyset \neq S \subset X$ be a closed set. Consider $K : X \rightrightarrows Y$ a multifunction whose values are proper pointed closed convex cones. This leads us, for every $x \in X$, to the following order relation on Y:

$$y_1 \leq_{K(x)} y_2 \Leftrightarrow y_2 - y_1 \in K(x).$$

We consider the following optimization problem

$$(P_S) \text{ minimize } F(x), \text{ subject to } x \in S,$$

where the minimality is understood according to different efficiency notions, given in what follows. If $S = X$, we denote the associated (unconstrained) problem by (P).

The following efficiency notions were defined in [22] and represent extensions of proper efficiency from fixed order setting to variable order setting.

Definition 8.3.1 *Let* $(\bar{x}, \bar{y}) \in \operatorname{Gr} F \cap (S \times Y)$.

(i) One says that $(\bar{x}, \bar{y})$ *is a local nondominated point for F on S with respect to K if there is a neighborhood U of $\bar{x}$ such that, for every $x \in U \cap S$,*

$$(F(x) - \bar{y}) \cap (-K(x)) \subset \{0\}.$$

(ii) If $\operatorname{int} K(x) \neq \emptyset$ *for every x in a neighborhood V of $\bar{x}$, then one says that $(\bar{x}, \bar{y})$ is a local weak nondominated point for F on S with respect to K if there is a neighborhood $U \subset V$ of $\bar{x}$ such that, for every $x \in U \cap S$,*

$$(F(x) - \bar{y}) \cap (-\operatorname{int} K(x)) = \emptyset.$$

In the above definition, the argument x of the objective and ordering set-valued maps is the same, which means that the decision maker takes into account only the preferences at the moment of the decision. However, in the real world, we often meet optimization problems with uncertain data. Thus, as a natural extension of the above notions, the following concepts of robustness were introduced in [31].

Definition 8.3.2 *Let* $(\bar{x}, \bar{y}) \in \operatorname{Gr} F \cap (S \times Y)$.

(i) One says that $(\bar{x}, \bar{y})$ *is a local robust efficient point for F on S with respect to K if there is a neighborhood U of $\bar{x}$ such that, for every $x, z \in U \cap S$,*

$$(F(x) - \bar{y}) \cap (-K(z)) \subset \{0\}. \tag{8.7}$$

Moreover, if relation (8.7) holds for every $x \in U \cap S$ and $z \in U$, then one says that $(\bar{x}, \bar{y})$ is a local strong robust efficient point for F on S with respect to K.

(ii) If $\operatorname{int} K(z) \neq \emptyset$ *for every z in a neighborhood V of* $\bar{x}$*, then one says that* $(\bar{x}, \bar{y})$ *is a local robust weak efficient point for F on S with respect to K if there is a neighborhood* $U \subset V$ *of* $\bar{x}$ *such that, for every* $x, z \in U \cap S$,

$$(F(x) - \bar{y}) \cap (-\operatorname{int} K(z)) = \emptyset. \tag{8.8}$$

Similarly, if relation (8.8) holds for every $x \in U \cap S$ *and* $z \in U$*, then one says that* $(\bar{x}, \bar{y})$ *is a local strong robust weak efficient point for F on S with respect to K.*

We mention that the global versions of all the efficiency concepts introduced in this section are obtained by taking U and V as the whole space X. Also, when $S = X$, then we speak about unconstrained efficiencies.

Observe that if $K(x) = C$ for every $x \in U$, where by C we have denoted a closed, convex, proper and pointed cone in Y, then the notions given in Definitions 8.3.1 and 8.3.2 do coincide and both reduce to the classical ones: The Pareto minimality and the weak Pareto minimality of F with respect to C, respectively.

Of course, if $(\bar{x}, \bar{y})$ is a local strong robust (weak) efficient point for F on S with respect to K, then $(\bar{x}, \bar{y})$ is a local robust (weak) efficient point for F on S with respect to K. Moreover, it is easy to see that if $(\bar{x}, \bar{y})$ is a local robust (weak) efficient point for F on S with respect to K, then $(\bar{x}, \bar{y})$ is a local (weak) nondominated point for F on S with respect to K. However, in general, the reverse implications are not true. In this sense, we have the following example, which points out, as well, that robustness does not imply strong robustness.

Example 8.3.3 *Let* $S_1 = [-1, 0]$, $S_2 = [0, 1]$, *and the set-valued maps* $F, K : \mathbb{R} \rightrightarrows \mathbb{R}^2$ *given by*

$$F(x) = \begin{cases} \{0\} \times [0, 1], & \text{if } x = 0, \\ [(0,0), (x, \frac{1}{x})], & \text{if } x \in]0, 1], \\ \emptyset, & \text{otherwise}, \end{cases}$$

and

$$K(x) = \begin{cases} \{(0, z) \mid z \leq 0\}, & \text{if } x \in [-1, 0[, \\ \mathbb{R}_+^2, & \text{otherwise}, \end{cases}$$

where $[(a,b), (c,d)]$ *is the line segment joining* (a,b) *and* (c,d)*. It is easy to see that* $(\bar{x}, \bar{y}) = (0, (0, 0))$ *is a global nondominated point for F with respect to K, but it is not even a local robust efficient point for F on* S_1 *with respect to K. In addition,* $(\bar{x}, \bar{y})$ *is a local robust efficient point for F on* S_2 *with respect to K, but it is not a local strong robust efficient point for F on* S_2 *with respect to K.*

Further, we define a Henig type proper efficiency using the dilating cones $K(x)_\varepsilon$. In order to do this, we use the extension from fixed to variable order setting given in [13] for a such minimum point, based on the reduction to Henig dilating cones obtained in the previously cited paper for the usual definition of the Henig proper efficiency in the fixed order case governed by a based cone.

Definition 8.3.4 *Let* $(\bar{x}, \bar{y}) \in \operatorname{Gr} F \cap (S \times Y)$. *If there exists a neighborhood V of* $\bar{x}$ *such that* $K(x)$ *is based for every* $x \in V \cap S$*, then one says that* $(\bar{x}, \bar{y})$ *is a Henig proper nondominated point for F on S with respect to K if there exist* $\varepsilon > 0$ *and a neighborhood* $U \subset V$ *of* $\bar{x}$ *such that for every* $x \in U \cap S$,

$$(F(x) - \bar{y}) \cap (-K(x)_\varepsilon) \subset \{0\}.$$

Observe that the Henig efficiency concept is defined on similar lines as the notion of nondominated point presented above. Moreover, since $K(x) \subset K(x)_\varepsilon$ for every $x \in X$, every Henig proper nondominated point is also a nondominated point. However, the converse implication does not hold. To show this, we give the following example.

Example 8.3.5 *Let* $S = [-1, 0]$ *and the set-valued maps* $F, K : \mathbb{R} \rightrightarrows \mathbb{R}^2$ *given by*

$$F(x) = \begin{cases} \{(-a,a) \mid a \in [0,x]\}, & \text{if } x > 0, \\ \{(0,a) \mid a \in [x,0]\}, & \text{if } x \leq 0 \end{cases}$$

and

$$K(x) = \operatorname{cone conv} \{(-1,0), (x,1)\}.$$

Observe first that $K(x)$ *admits a base for every* $x \in S$. *Then, it is easy to see that* $(0,(0,0))$ *is a local nondominated point for F on S with respect to K, but* $(0,(0,0))$ *is not a Henig proper nondominated point for F on S with respect to K.*

Concerning the Henig proper efficiency and the robust efficiency, they are independent, in the sense that, in general, none of these concepts implies the other one. In order to illustrate this, we consider the following example.

Example 8.3.6 *Let* $S_1 = [0,1]$, $S_2 = [-1,0]$ *and the set-valued maps* $F, K : \mathbb{R} \rightrightarrows \mathbb{R}^2$ *given by*

$$F(x) = \begin{cases} \{(a,b) \in \mathbb{R}^2 \mid 0 \leq a \leq -b \leq x\}, & \text{if } x \geq 0, \\ \{(a,b) \in \mathbb{R}^2 \mid 0 \geq a \geq -b \geq x\}, & \text{if } x < 0 \end{cases}$$

and

$$K(x) = \begin{cases} \operatorname{cone conv} \{(1,0),(x,1)\}, & \text{if } x \neq 0, \\ \{(a,-a) \mid a \geq 0\}, & \text{if } x = 0. \end{cases}$$

Observe first that $K(x)$ *admits a base for every* $x \in S_1 \cup S_2$. *Then, it is easy to see that* $(0,(0,0))$ *is not a Henig proper nondominated point for F on* S_1 *with respect to K, but it is a local robust efficient point for F on* S_1 *with respect to K. Moreover,* $(0,(0,0))$ *is a Henig proper nondominated point for F on* S_2 *with respect to K, but it is not a local robust efficient point for F on* S_2 *with respect to K.*

We continue with a notion of sharp efficiency for set-valued mappings, which, in the fixed order case, was investigated under several versions in [32] and [16].

Definition 8.3.7 *Let* $\psi : [0, \infty[\to \mathbb{R}$ *be a nondecreasing function with the property that* $\psi(t) = 0$ *if and only if* $t = 0$. *One says that a point* $(\bar{x}, \bar{y}) \in \operatorname{Gr} F \cap (S \times Y)$ *is a weak* ψ-*sharp local nondominated point for* (P_S) *if there exist* $\alpha > 0$ *and a neighborhood U of* $\bar{x}$ *such that for every* $x \in U \cap S$, $y \in F(x)$ *one has*

$$\Delta(y - \bar{y}, -K(x)) \geq \alpha \psi(d(x, W)), \tag{8.9}$$

where $W = \{x \in S \mid \bar{y} \in F(x)\}.$

170 ■ *Variational Analysis and Set Optimization*

This notion is a vectorial version of the notion of weak sharp minimum in scalar optimization, in the sense made precise, for instance, in [54, Section 3.10]. The term "weak" refers here to the fact that the set of argmin points (that is, W) may be not a singleton. In the case where $W = \{\bar{x}\}$, we drop the word "weak" and then we get the concept of ψ-sharp local nondominated point. Some examples of functions ψ which can be considered are: $\psi_1 : [0, +\infty[\to \mathbb{R}$, $\psi_1(t) = t^m$, $m \in \mathbb{N} \setminus \{0\}$, $\psi_2 : [0, +\infty[\to \mathbb{R}$, $\psi_2(t) = \ln(1+t)$, $\psi_3 : [0, +\infty[\to \mathbb{R}$, $\psi_3(t) = \frac{t}{1+t}$, $\psi_4 : [0, +\infty[\to \mathbb{R}$, $\psi_4(t) = \arctan t$.

Remark 8.3.8 *If we suppose that there exists a neighborhood U of $\bar{x}$ such that $\mathrm{int}\, K(x) \neq \emptyset$ for all $x \in U \cap S$, and if $(\bar{x}, \bar{y})$ is a weak ψ-sharp local nondominated point for (P_S), then $(\bar{x}, \bar{y})$ is also a local weak nondominated point for F on S with respect to K.*

Next, we prove two results which show that the Gerstewitz's scalarizing functional is a suitable tool for investigating sharp nondominated points when some interiority assumptions hold.

Lemma 8.3.9 *Let $(\bar{x}, \bar{y}) \in \mathrm{Gr}\, F$ be a weak ψ-sharp local nondominated point (with the neighborhood U and with the constant $\alpha > 0$) for the problem (P_S) and take $\overline{K} = \bigcap_{x \in U \cap S} K(x)$. If $\mathrm{int}\, \overline{K} \neq \emptyset$, then for every $e \in \mathrm{int}\, \overline{K}$, there exists $\gamma > 0$ such that for every $x \in U \cap S$, $y \in F(x)$ and $z \in K(x)$, one has*

$$s_{e, \overline{K}}(y + z - \bar{y}) \geq \gamma \psi(d(x, W)). \tag{8.10}$$

Proof. Take $e \in \mathrm{int}\, \overline{K}$ with $\|e\| = 1$ and fix $x \in U \cap S$, $y \in F(x)$ and $z \in K(x)$. There are three possible cases.

First, if $d(x, W) = 0$ and $d(y - \bar{y}, -K(x)) = 0$ then $\psi(d(x, W)) = 0$ and, using (8.9) we have $d(y - \bar{y}, Y \setminus -K(x)) = 0$. Therefore $y - \bar{y} \in \mathrm{cl}(Y \setminus -K(x))$, from which we get $y - \bar{y} \notin -\mathrm{int}\, K(x)$. Then, since $z \in K(x)$, we get that $y + z - \bar{y} \notin -\mathrm{int}\, K(x)$ and as $\overline{K} \subset K(x)$ for every $x \in U \cap S$, we obtain that $y + z - \bar{y} \notin -\mathrm{int}\, \overline{K}$. Using relation (8.5), we get that $s_{e, \overline{K}}(y + z - \bar{y}) \geq 0$, which shows the desired inequality for any $\gamma > 0$.

Now, suppose that $d(x, W) = 0$ and $d(y - \bar{y}, -K(x)) > 0$. It follows that $y - \bar{y} \notin -\mathrm{cl}\, K(x) = -K(x)$, so $y + z - \bar{y} \notin -K(x)$, from which we get $y + z - \bar{y} \notin -\overline{K}$. So, in virtue of formula (8.6) we obtain that $s_{e, \overline{K}}(y + z - \bar{y}) > 0 = \gamma \psi(d(x, W))$ for every $\gamma > 0$.

In the third case we suppose that $d(x, W) > 0$. In this situation, we have $y \neq \bar{y}$ because otherwise we would have $-d(0, Y \setminus -K(x)) \geq \gamma \psi(d(x, W)) > 0$, which is impossible. We want to prove that for any $\mu \in]0, \alpha[$ one has

$$y + z - \bar{y} \notin \mu \psi(d(x, W)) e - \mathrm{int}\, \overline{K}.$$

Suppose that there exists $\mu \in]0, \alpha[$ such that $y + z - \bar{y} \in \mu \psi(d(x, W)) e - \mathrm{int}\, \overline{K}$. Since $z \in K(x)$, we obtain that $y - \bar{y} \in \mu \psi(d(x, W)) e - \mathrm{int}\, K(x)$. Then, using the fact that $\psi(d(x, W)) > 0$, we have

$$\Delta(y - \bar{y}, -K(x)) \leq d(y - \bar{y}, -K(x)) \leq \|\mu \psi(d(x, W)) e\| < \alpha \psi(d(x, W)),$$

which is a contradiction with the inequality (8.9). Using (8.5) again, we find that $s_{e, \overline{K}}(y + z - \bar{y}) \geq \mu \psi(d(x, W))$.

Finally, take $e_1 \in \operatorname{int} \overline{K}$ arbitrarily. It's easy to see that $s_{e_1,\overline{K}}(\cdot) = \frac{1}{\|e_1\|} s_{e,\overline{K}}(\cdot)$, where $e = \frac{e_1}{\|e_1\|}$. It follows that we have the conclusion in every of the previous cases for x, y and z fixed before. Consequently, as the obtained constant does not depend on x, y or z, we obtain the conclusion. □

Lemma 8.3.10 *Let $(\overline{x},\overline{y}) \in \operatorname{Gr} F \cap (S \times Y)$ a local weak nondominated point for F on S with respect to K (with the neighborhood U) and suppose that $\operatorname{int} \overline{K} \neq \emptyset$, where $\overline{K} = \bigcap_{x \in U \cap S} K(x)$. If for every $e \in \operatorname{int} \overline{K}$, there exists $\gamma > 0$ such that for every $x \in U \cap S$, $y \in F(x)$ and $z \in K(x)$ relation (8.10) holds, then $(\overline{x},\overline{y})$ is a weak ψ-sharp local nondominated point for (P_S).*

Proof. From the hypothesis, there exists a neighborhood U of $\overline{x}$ such that $\operatorname{int} \overline{K} \neq \emptyset$, and for every $x \in U \cap S$

$$(F(x) - \overline{y}) \cap (-\operatorname{int} K(x)) = \emptyset.$$

We prove that the conclusion of this lemma holds for the neighborhood U and a constant $\alpha > 0$.

Fix now $x \in U \cap S$ and $y \in F(x)$ such that $y - \overline{y} \notin -\operatorname{int} K(x)$. It follows that $d(y - \overline{y}, Y \setminus -K(x)) = 0$, so $\Delta(y - \overline{y}, -K(x)) = d(y - \overline{y}, -K(x))$. If $d(x, W) = 0$, then $\psi(d(x,W)) = 0$, and since $\Delta(y - \overline{y}, -K(x)) \geq 0$, we get the conclusion for every constant $\alpha > 0$. Now, if $d(x,W) > 0$, then $\psi(d(x,W)) > 0$. As $e \in \operatorname{int} \overline{K}$, there exists $r > 0$ such that $B_Y(e,r) = e - B_Y(0,r) \subset \operatorname{int} \overline{K}$. Since $\psi(d(x,W)) > 0$, we obtain that $B_Y(0, r\psi(d(x,W))) \subset \psi(d(x,W)) e - \operatorname{int} \overline{K}$. Suppose, by contradiction, that the conclusion does not hold. Taking $\alpha = \gamma r$, there exists $z \in K(x)$ such that $\|y + z - \overline{y}\| < \gamma r \psi(d(x,W))$, i.e.,

$$y + z - \overline{y} \in B_Y(0, \gamma r \psi(d(x,W))) \subset \gamma \psi(d(x,W)) e - \operatorname{int} \overline{K}.$$

Using again the relation (8.5), one gets the contradiction $s_{e,\overline{K}}(y + z - \overline{y}) < \gamma \psi(d(x,W))$. The conclusion follows. □

If in the previous lemma we do not suppose that $(\overline{x},\overline{y})$ is a local weak nondominated point for F on S with respect to K, then, according to Remark 8.3.8, $(\overline{x},\overline{y})$ is not a weak ψ-sharp local nondominated point for (P_S). Also, if we suppose that relation (8.10) holds only for some $z \in K(x)$, then the previous lemma does not necessarily hold. In this sense, we give the following example.

Example 8.3.11 *Let $X = Y = S = \mathbb{R}^2$, the set-valued maps $F, K : X \rightrightarrows Y$ given by $F(x_1, x_2) = [\|(x_1, x_2)\|, +\infty[\times \{0\}$ and*

$$K(x_1, x_2) = \begin{cases} \mathbb{R}_+^2, & \text{if } (x_1, x_2) = (0, 0), \\ \operatorname{cone conv}\{(-1, 0), (1, 1)\}, & \text{otherwise,} \end{cases}$$

the point $(\overline{x},\overline{y}) = ((0,0),(0,0)) \in \operatorname{Gr} F$ and $\psi(t) = t$ for every $t \in [0, +\infty[$. Then $W = \{(0,0)\}$ and $\psi(d((x_1,x_2), W)) = \|(x_1,x_2)\|$ for every $(x_1,x_2) \in \mathbb{R}^2$. It is easy to see that $(\overline{x},\overline{y})$ is a local weak nondominated point for F with respect to K. Also, the inequality (8.10) is true for some z. For this, we take U a neighborhood of $\overline{x}$, $(e_1, e_2) \in \operatorname{int} \overline{K}$,

$(x_1,x_2) \in U$, $(y_1,y_2) \in F(x_1,x_2)$, $z = (0,z_2) \in K(x_1,x_2)$ and $\gamma = \frac{1}{e_1} > 0$. Then

$$(\gamma e_1 \|(x_1,x_2)\| - y_1, \gamma e_2 \|(x_1,x_2)\| - y_2 - z_2)$$
$$= \left(\|(x_1,x_2)\| - y_1, \frac{e_2}{e_1} \|(x_1,x_2)\| - z_2 \right) \notin \operatorname{int} \overline{K},$$

because $y_1 \geq \|(x_1,x_2)\|$. Using the property (8.5) of the scalarization function, it follows that $s_{e,\overline{K}}((y_1,y_2) + (0,z_2)) \geq \gamma \|(x_1,x_2)\|$. However, $(\bar{x}, \bar{y})$ is not a weak ψ-sharp local nondominated point for (P) since, if we take $U \in \mathcal{V}(\bar{x})$ and $\alpha > 0$ arbitrarily, then there exist $(x_1,x_2) \in U \setminus \{(0,0)\}$, $(y_1,y_2) = (\|(x_1,x_2)\|, 0) \in F(x)$ such that

$$\Delta((y_1,y_2), -K(x_1,x_2)) = 0 < \alpha \|(x_1,x_2)\|.$$

In the following, we give some penalization results for the efficiency notions introduced above, where the efficiency is taken with respect to a variable ordering structure, and the penalization term is the minimal time function with respect to directions of a closed subset of the unit sphere.

In the first penalization result, we reduce the global notion of weak robust efficiency point of a constrained problem to the notion of nondominated efficiency point to an unconstrained problem, by adding to the objective map the minimal time function associated to the target set S, with respect to directions of M. Similar to the Clarke penalization method, this task is accomplished by assuming some Lipschitz behavior of the involved mappings. Note that, even in the conclusion of the next results, extended-valued multifunctions do appear, in view of the calculus rules with infinity terms, the efficiency notions are defined in the exactly same way as for multifunctions with values in Y.

Theorem 8.3.12 *Let $S \subset X$ be a nonempty closed set, $M \subset S_X$ be a nonempty set and $(\bar{x}, \bar{y}) \in \operatorname{Gr} F \cap (S \times Y)$. Take $C = \left(\bigcap_{x \in X} K(x) \right)$ and suppose that $\operatorname{int} K(x) \neq \emptyset$ for every $x \in X$, and that there exist $e \in C \setminus \{0\}$ and $L > 0$ such that for all $u, v \in X$,*

$$F(u) \subset F(v) - L \|u - v\| e + K(u).$$

If $(\bar{x}, \bar{y})$ is a strong robust weak efficient point for F on S with respect to K, then $(\bar{x}, \bar{y})$ is a weak nondominated point for $F(\cdot) + LT_M(\cdot, S)e$ with respect to K.

Proof. Suppose that there exist $x \in X$, $y \in F(x)$ such that

$$\bar{y} - y - LT_M(x,S)e \in \operatorname{int} K(x). \tag{8.11}$$

Note that, in view of the calculus rules for $\pm\infty_C$, $T_M(x,S) \in \mathbb{R}$. Further, from (8.11), we get that for every $\delta > 0$ small enough,

$$\bar{y} - y - L(1+\delta)T_M(x,S)e \in \operatorname{int} K(x). \tag{8.12}$$

Taking into account that $(\bar{x}, \bar{y})$ is a strong robust weak efficient point for F on S with respect to K, we get that $x \notin S$ and so there exists $x_1 \in S$ such that

$$\|x - x_1\| < (1+\delta) d(x,S).$$

Now, from the Lipschitz property of F we get that there exists $y_1 \in F(x_1)$ such that
$$y - y_1 + L\|x - x_1\|e \in K(x).$$
Then, taking into account that $e \in \left(\bigcap_{x \in X} K(x)\right) \setminus \{0\}$ and using the last two relations, one gets that
$$y - y_1 + L(1+\delta)d(x,S)e \in K(x),$$
from which we get
$$y - y_1 + L(1+\delta)T_M(x,S)e \in K(x). \tag{8.13}$$
Further, adding relations (8.12) and (8.13), and taking into account that $K(x)$ is a convex cone, we obtain that
$$\bar{y} - y_1 \in \operatorname{int} K(x),$$
which contradicts the efficiency hypothesis. The conclusion follows. $\square$

Now we do not suppose that $\operatorname{int} K(x) \neq \emptyset$, and we get a result in the setting known in vector optimization as the strong case.

Theorem 8.3.13 *Let $S \subset X$ be a nonempty closed set, $M \subset S_X$ be a nonempty set and $(\bar{x}, \bar{y}) \in \operatorname{Gr} F \cap (S \times Y)$. Take $C = \left(\bigcap_{x \in X} K(x)\right)$ and suppose that there exist $e \in C \setminus \{0\}$ and $L > 0$ such that for all $x, z \in X$,*
$$F(x) \subset F(z) - L\|x-z\|e + K(x).$$
If $(\bar{x}, \bar{y})$ is a strong robust efficient point for F on S with respect to K, then for all $\beta > L$, $(\bar{x}, \bar{y})$ is a nondominated point for $F(\cdot) + \beta T_M(\cdot, S)e$ with respect to K.

Proof. Suppose that there exists $\beta > L$ such that $(\bar{x}, \bar{y})$ is not a nondominated point for $F(\cdot) + \beta T_M(\cdot, S)e$ with respect to K, so there exist $x \in X$, $y \in F(x)$ such that
$$\bar{y} - y - \beta T_M(x,S)e \in K(x) \setminus \{0\}. \tag{8.14}$$
Note that, in view of the calculus rules for $\pm \infty_C$, $T_M(x,S) \in \mathbb{R}$. Taking into account that $(\bar{x}, \bar{y})$ is a strong robust efficient point for F on S with respect to K, we get that $x \notin S$ and since $\frac{\beta}{L} - 1 > 0$, it follows that there exists $x_1 \in S$ such that
$$\|x - x_1\| < \frac{\beta}{L}d(x,S).$$
Now, from the Lipschitz property of F we get that there exists $y_1 \in F(x_1)$ such that
$$y - y_1 + L\|x - x_1\|e \in K(x).$$
Then, taking into account that $e \in \left(\bigcap_{x \in X} K(x)\right) \setminus \{0\}$ and using the last two relations, one gets that
$$y - y_1 + \beta d(x,S)e \in K(x),$$

from which we get
$$y - y_1 + \beta T_M(x,S)e \in K(x). \tag{8.15}$$
Further, adding relations (8.14) and (8.15), and taking into account that $K(x)$ is a pointed convex cone, we obtain that
$$\bar{y} - y_1 \in K(x) \setminus \{0\},$$
from which we get contradicts the efficiency hypothesis. The conclusion follows. □

In order to illustrate the above penalization result, we consider the following example.

Example 8.3.14 *Let $X = \mathbb{R}$, $Y = \mathbb{R}^2$, $S = [-1,1]$ and the set-valued maps $F, K : \mathbb{R} \rightrightarrows \mathbb{R}^2$ given by*
$$F(x) = \begin{cases} \{(u,v) \in \mathbb{R}^2 \mid 0 \le v \le u|x|, \, u \ge 0\}, & \text{if } x \in [-1,1], \\ \{(u,0) \in \mathbb{R}^2 \mid u \in \mathbb{R}\}, & \text{if } x \in]-\infty, -1[\cup]1, \infty[, \end{cases}$$

and
$$K(x) = \begin{cases} \operatorname{cone conv}\{(0,1),(1,1)\}, & \text{if } x \in [-1,1], \\ \mathbb{R}^2_+, & \text{if } x \in]-\infty, -1[\cup]1, \infty[. \end{cases}$$

We prove that all the assumptions from Theorem 8.3.13 are satisfied for $(\bar{x}, \bar{y}) = (0,(0,0)) \in \operatorname{Gr} F \cap (S \times Y)$, $L = 1$ and $e = (1,1) \in \left(\bigcap_{x \in X} K(x) \right) \setminus \{(0,0)\}$. Firstly, it is easy to see that $(\bar{x}, \bar{y})$ is a strong robust efficient point for F on S with respect to K, i.e., for all $x \in [-1,1]$ and $z \in \mathbb{R}$, one has
$$F(x) \cap (-K(z)) \subset \{0\}. \tag{8.16}$$

Observe that if relation (8.16) would be satisfied for every $x \in \mathbb{R}$, then the conclusion of Theorem 8.3.13 would hold as well for every $e \in \bigcap_{x \in X} K(x)$, without any additional assumptions. However, it is not the case, since for $x, z \notin [-1,1]$, relation (8.16) does not hold. We take now $x, z \in \mathbb{R}$ arbitrarily, and we prove that
$$F(x) \subset F(z) - L \|x - z\| e + K(x).$$

Let $(u,v) \in F(x)$. It follows that there exists $(u,0) \in F(z)$ such that
$$(u,v) - (u,0) + (|x-z|, |x-z|) = (|x-z|, v + |x-z|) \in K(x).$$

So the multifunction F satisfies the Lipschitz property from Theorem 8.3.13, and then all the assumptions from the above theorem hold, hence, $(\bar{x}, \bar{y})$ is a nondominated point for $F(\cdot) + \beta T_M(\cdot, S)e$ with respect to K, for all $\beta > L$ and for every nonempty set $M \subset S_X$.

We mention that local versions of Theorems 8.3.12 and 8.3.13 can be proved as well. Now, proceeding similarly to the proof of Theorems 8.3.12 and 8.3.13, we give in the following two local penalization results in which a Henig proper nondominated point for F on S with respect to K is reduced to an unconstrained local robust weak efficient point, and to a local robust efficient point, respectively. However, the structure of the involved objects is different from the one used in the previous theorems. Thus, due to specific differences that appear, we give the complete proofs of the penalization results.

First, we consider the situation when the ordering cones have a nonempty topological interior.

Theorem 8.3.15 *Let $S \subset X$ be a nonempty closed set, $M \subset S_X$ be a nonempty set and $(\bar{x}, \bar{y}) \in \mathrm{Gr}\, F \cap (S \times Y)$. Suppose that there exists a neighborhood U of $\bar{x}$ such that:*
(i) $\mathrm{int}\, K(x) \neq \emptyset$ for all $x \in U$, and there exists $e \in Y \setminus \{0\}$ such that $e \in C = \bigcap_{x \in U} K(x)$;
(ii) there exists a constant $L > 0$ such that for all $u, v \in U$,

$$F(u) \subset F(v) - L\|u - v\|e + K(v);$$

(iii) $K(x)$ is based for all $x \in U$ (the base of $K(x)$ is denoted by $B(x)$);
(iv) there exists a constant $l > 0$ such that for all $u, v \in U$,

$$B(u) \subset B(v) + l\|u - v\|D_Y.$$

If $(\bar{x}, \bar{y})$ is a Henig proper nondominated point for F on S with respect to K, then $(\bar{x}, \bar{y})$ is a local robust weak efficient point for $F(\cdot) + LT_M(\cdot, S)e$ with respect to K.

Proof. Since $(\bar{x}, \bar{y})$ is a Henig proper nondominated point for F on S with respect to K, by the definition, there exist $\varepsilon > 0$ and $r > 0$ such that for every $x \in B_X(\bar{x}, r) \cap S$,

$$(F(x) - \bar{y}) \cap (-K(x)_e) \subset \{0\}.$$

Without loss of generality we can suppose that $U = B_X(\bar{x}, r)$.

Suppose that $(\bar{x}, \bar{y})$ is not a local robust weak efficient point for $F(\cdot) + LT_M(\cdot, S)e$ with respect to K. Take $\alpha > 0$ such that $\alpha < \min\{3^{-1}r, (3l)^{-1}\varepsilon\}$, where $l > 0$ is given by the hypothesis (iv). Hence there are $x, z \in B_X(\bar{x}, \alpha)$ and $y \in F(x)$ such that

$$\bar{y} - y - LT_M(x, S)e \in \mathrm{int}\, K(z).$$

Again, $T_M(x, S) \in \mathbb{R}$. Without loss of generality, we can write

$$\bar{y} - y - L(1 + \delta)T_M(x, S)e \in \mathrm{int}\, K(z), \tag{8.17}$$

for every positive δ small enough. Fix δ such that $l(3 + \delta)\alpha < \varepsilon$ and $\delta < 1$. Then there exists $u_\delta \in S$ such that

$$\|x - u_\delta\| \leq (1 + \delta)d(x, S) \leq (1 + \delta)\|x - \bar{x}\|,$$

from which we get

$$\|u_\delta - \bar{x}\| \leq (2 + \delta)\|x - \bar{x}\| < r.$$

Therefore, from the Lipschitz property of F with respect to K (that is, assumption (*ii*)),

$$F(x) \subset F(u_\delta) - L\|x - u_\delta\|e + K(u_\delta),$$

from which we get

$$F(x) + L(1 + \delta)d(x, S)e \subset F(u_\delta) + K(u_\delta).$$

In particular, for $y \in F(x)$, there exists $v_\delta \in F(u_\delta)$ such that

$$y - v_\delta + L(1 + \delta)d(x, S)e \in K(u_\delta),$$

from which we get
$$y - v_\delta + L(1+\delta)T_M(x,S)e \in K(u_\delta).$$
Adding this relation with (8.17), we get by (*iii*) and (*iv*) that
$$\begin{aligned}\bar{y} - v_\delta &\in K(u_\delta) + \operatorname{int} K(z) \subset K(u_\delta) + \operatorname{int}(\operatorname{cone} B(z))\\ &\subset K(u_\delta) + \operatorname{int}(\operatorname{cone}(B(u_\delta) + l\|z - u_\delta\|D_Y)).\end{aligned}$$
Since
$$\begin{aligned}l\|z - u_\delta\| &\leq l(\|z - \bar{x}\| + \|\bar{x} - x\| + \|x - u_\delta\|)\\ &< l(3+\delta)\alpha < \varepsilon,\end{aligned}$$
we have that
$$B(u_\delta) + l\|z - u_\delta\|D_Y \subset B(u_\delta)_\varepsilon,$$
so
$$\bar{y} - v_\delta \in K(u_\delta) + \operatorname{int} K(u_\delta)_\varepsilon \subset \operatorname{int} K(u_\delta)_\varepsilon \subset K(u_\delta)_\varepsilon \setminus \{0\},$$
which is a contradiction. Then the conclusion follows. □

In the next result, the case where the ordering cones are not necessarily solid is considered.

Theorem 8.3.16 *Let $S \subset X$ be a nonempty closed set, $M \subset S_X$ be a nonempty set and $(\bar{x}, \bar{y}) \in \operatorname{Gr} F \cap (S \times Y)$. Suppose that there exists a neighborhood U of $\bar{x}$ such that:*
(i) there exists $e \in Y \setminus \{0\}$ such that $e \in C = \bigcap_{x \in U} K(x)$;
(ii) there exists a constant $L > 0$ such that for all $u, v \in U$,
$$F(u) \subset F(v) - L\|u - v\|e + K(v);$$
(iii) $K(x)$ is based for all $x \in U$ (the base of $K(x)$ is denoted by $B(x)$);
(iv) there exists a constant $l > 0$ such that for all $u, v \in U$,
$$B(u) \subset B(v) + l\|u - v\|D_Y.$$
If $(\bar{x}, \bar{y})$ is a Henig proper nondominated point for F on S with respect to K, then for all $\beta > L$, $(\bar{x}, \bar{y})$ is a local robust efficient point for $F(\cdot) + \beta T_M(\cdot, S)e$ with respect to K.

Proof. Since $(\bar{x}, \bar{y})$ is a Henig proper nondominated point for F on S with respect to K, by the definition, there exist $\varepsilon > 0$ and $r > 0$ such that for every $x \in B_X(\bar{x}, r) \cap S$,
$$(F(x) - \bar{y}) \cap (-K(x)_\varepsilon) \subset \{0\}.$$

Again, without loss of generality we can suppose that $U = B_X(\bar{x}, r)$. Taking $l > 0$ from (iv), we have that there exists $\gamma > 0$ such that $\gamma l < \varepsilon$.

Suppose that there exists $\beta > L$ such that $(\bar{x}, \bar{y})$ is not a local robust efficient point for $F(\cdot) + \beta T_M(\cdot, S)e$ with respect to K. Take $\alpha = \min\left\{\left(1 + \frac{\beta}{L}\right)^{-1} r, \left(2 + \frac{\beta}{L}\right)^{-1}\gamma\right\}$. Whence there exist $x, z \in B_X(\bar{x}, \alpha)$ and $y \in F(x)$ such that
$$\bar{y} - y - \beta T_M(x,S)e \in K(z) \setminus \{0\}. \tag{8.18}$$

Again, $T_M(x,S) \in \mathbb{R}$. As $\frac{\beta}{L} - 1 > 0$, there exists $u \in S$ such that

$$\|x-u\| \leq \frac{\beta}{L} d(x,S) \leq \frac{\beta}{L} \|x - \bar{x}\|,$$

from which we get

$$\|u - \bar{x}\| \leq \left(1 + \frac{\beta}{L}\right) \|x - \bar{x}\| < r.$$

Therefore, from the Lipschitz property of F with respect to K we obtain that,

$$F(x) \subset F(u) - L\|x-u\|e + K(u),$$

from which we get

$$F(x) + \beta d(x,S)e \subset F(u) + K(u).$$

In particular, for $y \in F(x)$, there exists $v \in F(u)$ such that

$$y - v + \beta d(x,S)e \in K(u),$$

from which we get

$$y - v + \beta T_M(x,S)e \in K(u).$$

Adding this relation with (8.18), we get

$$\begin{aligned}\bar{y} - v &\in K(z) \setminus \{0\} + K(u) = (\operatorname{cone} B(z)) \setminus \{0\} + K(u) \\ &\subset (\operatorname{cone}(B(u) + l\|z-u\|D_Y)) \setminus \{0\} + K(u).\end{aligned}$$

Since

$$\begin{aligned}l\|z-u\| &\leq l(\|z-\bar{x}\| + \|\bar{x}-x\| + \|x-u\|) \\ &< l\left(2\alpha + \frac{\beta}{L}\|x-\bar{x}\|\right) \leq l\left(2 + \frac{\beta}{L}\right)\alpha \leq l\gamma < \varepsilon,\end{aligned}$$

we have that

$$B(u) + l\|z-u\|D_Y \subset (B(u))_\varepsilon,$$

so

$$\bar{y} - v \in K(u)_\varepsilon \setminus \{0\} + K(u).$$

Now, since $K(u)_\varepsilon$ is a pointed convex cone, we get that

$$\bar{y} - v \in K(u)_\varepsilon \setminus \{0\},$$

which is a contradiction. Then the conclusion follows. $\square$

In order to illustrate Theorem 8.3.16, we consider the following example.

Example 8.3.17 *Let $X = \mathbb{R}$, $S = [0,1]$ and the set-valued maps $F,K : \mathbb{R} \rightrightarrows \mathbb{R}^2$ from Example 8.3.5. We prove that all the assumptions of Theorem 8.3.16 are satisfied for $(\bar{x}, \bar{y}) = (0,(0,0))$ and $U =]-1,1[\subset \mathbb{R}$.*

It is easy to see that the hypothesis (i) from the previous theorem holds for $e = (-1,1) \in \mathbb{R}^2$.

Further, we prove that for every $u, v \in U$ *the following inclusion holds*

$$F(u) + |u-v| \cdot e \subset F(v) + K(v). \tag{8.19}$$

There are two possible cases.

First, if $u \in]0,1[$ *and* $v \in]-1,1[$, *or if* $u \in]-1,0]$ *and* $v \in]0,1[$, *since* $(0,0) \in F(v)$, *one has the inclusion.*

In the second case we take $u \leq 0$, $v \leq 0$ *and* $y \in F(u)$. *It follows that there exists* $a \in [u,0]$ *such that* $y = (0,a)$. *Now, if* $v \leq u$, *or if* $v > u$ *and* $a \in [v,0]$, *since* $(0,a) \in F(v)$, *the inclusion (8.19) holds. Finally, if* $v > u$ *and* $a \in [u,v[$, *because* $(0,v) \in F(v)$, *one has*

$$(0, a-v) + (u-v, v-u) = (u-v, a-u) \in K(v).$$

Therefore, the multifunction F *satisfies the assumption (ii) of Theorem 8.3.16, with* $L = 1$ *and* $e = (-1,1)$.

Next, we define the multifunction $B :]-1,1[\rightrightarrows \mathbb{R}^2$ *as*

$$B(x) = \left\{ \left(a, \frac{a+1}{x+1}\right) \mid a \in [-1,x] \right\}.$$

It follows immediately that for every $x \in]-1,1[$, *the convex set* $B(x)$ *is a base for* $K(x)$, *and thus, the hypothesis (iii) holds.*

In order to verify the Lipschitz property for the base, we take $u, v \in U$ *and* $y \in B(u)$. *It follows that there is* $\bar{u} \in [-1, u]$ *such that* $y = \left(\bar{u}, \frac{\bar{u}+1}{u+1}\right)$. *Take* $l = 1$ *and* $\bar{v} = \frac{\bar{u}+1}{u+1}(v+1) - 1$. *Then* $\frac{\bar{v}+1}{v+1} = \frac{\bar{u}+1}{u+1}$, *and an easy calculation shows that* $\bar{v}$ *belongs to* $[-1, v]$ *and*

$$\left\| \left(\bar{u}, \frac{\bar{u}+1}{u+1}\right) - \left(\bar{v}, \frac{\bar{v}+1}{v+1}\right) \right\| \leq |u-v|.$$

Moreover, it is easy to see that $(0,(0,0))$ *is a Henig proper nondominated point for* F *on* S *with respect to* K. *Therefore, all the assumptions of Theorem 8.3.16 are satisfied, so there exists* $U \in \mathcal{V}(0)$ *such that for every* $x, z \in U$,

$$(F(x) + \beta T_M(x,S)e) \cap (-K(z)) \subset \{0\}$$

for all $\beta > 1$ *and for every nonempty set* $M \subset S_X$.

Remark 8.3.18 *Notice that if in the previous example we take* $S = [-1,0]$, *the hypotheses (i)-(iv) from Theorem 8.3.16 are satisfied, because their fulfillment does not depend on the set of restrictions (that is, S). In this situation, the conclusion of Theorem 8.3.16 does not hold. To prove this we can take* $U \in \mathcal{V}(0)$ *and* $\beta > L$ *arbitrary; then there exist* $x \in U \cap S \setminus \{0\}$, $z = 0$ *and* $y = (0,x) \in F(x)$ *such that* $y \in -K(z) \setminus \{0\}$. *However, there exists* $U \in \mathcal{V}(0)$ *such that for every* $x \in U \cap S$ *there exists* $\varepsilon_x > 0$ *such that*

$$(F(x) - \bar{y}) \cap (-K(x)_{\varepsilon_x}) \subset \{0\}, \tag{8.20}$$

but $(0,(0,0))$ *is not a Henig proper nondominated point for* F *on* S *with respect to* K. *These facts allow us to conclude that, in Definition 8.3.4, it is essential to have the same* $\varepsilon > 0$ *for every* $x \in U \cap S$.

In order to deal with the above efficiency concepts from the point of view of necessary optimality conditions, we have to discuss some openness concepts for set-valued maps, having in mind the incompatibility between openness and efficiency, which we will emphasize in the sequel.

Consider the multifunctions $F_1, F_2 : X \rightrightarrows Y$. One says that F_1 is open (in the standard sense) at $(\bar{x}, \bar{y}) \in \operatorname{Gr} F_1$ if for every $U \in \mathcal{V}(\bar{x})$, $F_1(U) \in \mathcal{V}(\bar{y})$. One says that a pair (F_1, F_2) is *weakly open* at $(\bar{x}, \bar{y}) \in \operatorname{Gr}(F_1 + F_2)$, or $F_1 + F_2$ is *weakly open* at $(\bar{x}, \bar{y}) \in \operatorname{Gr}(F_1 + F_2)$, if for every $U \in \mathcal{V}(\bar{x})$, $F_1(U) + F_2(U) \in \mathcal{V}(\bar{y})$. Since, in general, $(F+K)(U) \subset F(U) + K(U)$, if $F+K$ is open in the standard sense at $(\bar{x}, \bar{y})$, then $F+K$ is also weakly open at the same point. However, the reverse implication is not true. In this sense, we give the following example.

Example 8.3.19 Let $F, K : \mathbb{R} \rightrightarrows \mathbb{R}$,

$$F(x) = \begin{cases} x, & \text{if } x \in \mathbb{Q}, \\ \emptyset, & \text{if } x \notin \mathbb{Q}, \end{cases}$$

and

$$K(x) = x, \ \forall x \in \mathbb{R},$$

where $\mathbb{Q}$ denotes the set of rational numbers. It is easy to see that $F + K$ is not open in the standard sense at $(0,0)$, but (F, K) is weakly open at $(0,0)$.

We end this section by giving two extensions in our setting of a well-known fact in scalar optimization, pointed out in several recent works (see, e.g., [14], [34], [31], [13], [22]): A function cannot be open at an extremum point.

Lemma 8.3.20 *Let $F : X \rightrightarrows Y$ be the objective multifunction and $K : X \rightrightarrows Y$ be the ordering multifunction. Suppose that there is a neighborhood U of $\bar{x}$ such that $\bigcap_{x \in U} K(x) \neq \{0\}$. If $(\bar{x}, \bar{y}) \in \operatorname{Gr} F$ is a local nondominated point for F with respect to K, then the multifunction $F + K$ is not open at $(\bar{x}, \bar{y})$.*

Proof. We may assume that for every $x \in U$,

$$(F(x) - \bar{y}) \cap (-K(x)) \subset \{0\},$$

that is

$$(F(x) + K(x) - \bar{y}) \cap (-K(x)) \subset \{0\}.$$

Suppose, by contradiction, that $F + K$ is open at $(\bar{x}, \bar{y})$. Then, for the neighborhood U chosen before, there is an open set V such that $\bar{y} \in V \subset (F+K)(U)$. Choose $y \in V$. Then there is $x \in U$ such that $y \in F(x) + K(x)$, hence

$$y - \bar{y} \in (F(x) + K(x) - \bar{y}) \subset \{0\} \cup (Y \setminus -K(x)) \subset \{0\} \cup \left(Y \setminus \bigcap_{x \in U} -K(x)\right).$$

However, this means that $V - \bar{y} \subset \{0\} \cup \left(Y \setminus \bigcap_{x \in U} -K(x) \right)$, and, by the fact that the first set is absorbing and the second one is a cone, it follows that

$$Y = \{0\} \cup \left(Y \setminus \bigcap_{x \in U} -K(x) \right),$$

in contradiction with $\bigcap_{x \in U} K(x) \neq \{0\}$. $\square$

Lemma 8.3.21 *Let $F : X \rightrightarrows Y$ be the objective multifunction and $K : X \rightrightarrows Y$ be the ordering multifunction. Suppose that there exists a neighborhood U of $\bar{x}$ such that $\bigcap_{x \in U} K(x) \neq \{0\}$. If $(\bar{x}, \bar{y}) \in \mathrm{Gr}\, F$ is a local robust efficient point for F with respect to K, then (F, K) is not weakly open at $(\bar{x}, \bar{y})$.*

Proof. As above, one can suppose that if $(\bar{x}, \bar{y})$ is a local robust efficient point for F with respect to K, then for every $x, z \in U$,

$$(F(x) + K(z) - \bar{y}) \cap (-K(z)) \subset \{0\}. \tag{8.21}$$

If the conclusion does not hold, then there exists V, a neighborhood of $\bar{y}$, such that $V \subset F(U) + K(U)$. Let $y \in V$. So, there exist $z_1, z_2 \in U$ such that $y \in F(z_1) + K(z_2)$. From (8.21) it follows that

$$y - \bar{y} \in \{0\} \cup (Y \setminus -K(z_2)) \subset \{0\} \cup \left(Y \setminus \bigcap_{x \in U} -K(x) \right).$$

But y was chosen arbitrarily in V, so $V - \bar{y} \subset \{0\} \cup \left(Y \setminus \bigcap_{x \in U} -K(x) \right)$. From the same arguments as in the previous proof, we deduce that $Y = \{0\} \cup \left(Y \setminus \bigcap_{x \in U} -K(x) \right)$, which contradicts the fact that $\bigcap_{x \in U} K(x) \neq \{0\}$. $\square$

Therefore, it becomes clear that in order to investigate the efficiency concepts introduced in Definitions 8.3.1, 8.3.2, 8.3.4 from the point of view of necessary optimality conditions, we need to provide sufficient conditions for the (weak) openness of a sum of set-valued maps. To this aim, we generalize some results on sufficient conditions for openness in literature, and, along this process, we use some slightly weaker assumptions when compared to those already used in other works (see Section 6 for details).

8.4 Sufficient Conditions for Mixed Openness

In this section, we give some sufficient conditions for a mixed type openness of a finite family of set-valued maps, since the conclusions of our results involve both standard and weak openness mentioned in the previous section. To achieve this goal, we are going to follow two paths, by obtaining such conditions on the primal spaces, working with derivatives, and on the dual spaces, using coderivatives. The method employed here for both paths relies on Ekeland's Variational Principle. However, the techniques and the assumptions used in the two directions of study are different.

We begin with the result on primal spaces; the method employed here was first implemented in [49, Theorem 1], and gives the possibility to decouple the properties of proto-differentiability and the Aubin property, which could be satisfied separately by the two multifunctions.

Consider p a nonzero natural number and the multifunctions $F_1, F_2, ..., F_p : X \rightrightarrows Y$. We denote by $(F_1, F_2, ..., F_p)$ the multifunction $(F_1, F_2, ..., F_p) : X \rightrightarrows Y^p$ given by $(F_1, F_2, ..., F_p)(x) = \prod_{i=1}^{p} F_i(x)$ for every $x \in X$. Also, in what follows we use the notation $(D_B F(x,y) \cap B_Y(0,1))(B_X(0,1))$ for $D_B F(x,y)(B_X(0,1)) \cap B_Y(0,1)$, where $F : X \rightrightarrows Y$ is a multifunction and (x,y) is a point from its graph.

Theorem 8.4.1 *Let X, Y be Banach spaces, p, k be natural numbers such that p is nonzero and $k \leq p$, and $F_1, F_2, ..., F_p : X \rightrightarrows Y$ be multifunctions such that $\mathrm{Gr}\, F_i$ is closed around $(\overline{x}, \overline{y}_i) \in \mathrm{Gr}\, F_i$ for every $i \in \overline{1,p}$. Suppose that there exists $\lambda > 0$ such that for every*

$$(x, y_1, y_2, ..., y_k, x_{k+1}, y_{k+1}, ..., x_p, y_p) \in \mathrm{Gr}(F_1, F_2, ..., F_k) \times \prod_{i=k+1}^{p} \mathrm{Gr}\, F_i$$

around $(\overline{x}, \overline{y}_1, \overline{y}_2, ..., \overline{y}_k, \overline{x}, \overline{y}_{k+1}, ..., \overline{x}, \overline{y}_p)$, the following assumptions are satisfied:
(i) the next inclusion holds

$$B_Y(0,\lambda) \subset \mathrm{cl}\left[\left(D_B F_1(x,y_1) + \sum_{i=2}^{k}(D_B F_i(x,y_i) \cap B_Y(0,1))\right)(B_X(0,1)) \right.$$
$$\left. + \sum_{i=k+1}^{p}(D_B F_i(x_i,y_i) \cap B_Y(0,1))(B_X(0,1))\right]; \quad (8.22)$$

(ii) either F_i is proto-differentiable at x relative to y_i for every $i \in \overline{1,k} \setminus \{i_0\}$, where $i_0 \in \overline{1,k}$ and F_i is proto-differentiable at x_i relative to y_i for every $i \in \overline{k+1,p}$, or F_i is proto-differentiable at x relative to y_i for every $i \in \overline{1,k}$ and F_i is proto-differentiable at x_i relative to y_i for every $i \in \overline{k+1,p} \setminus \{i_0\}$, where $i_0 \in \overline{k+1,p}$;
(iii) F_i has the Aubin property around the point (x, y_i) for every $i \in \overline{1,k} \setminus \{i_0\}$, where $i_0 \in \overline{1,k}$.

Then there exists $\varepsilon > 0$ such that for every $(x, y_1, y_2, ..., y_p) \in \mathrm{Gr}(F_1, F_2, ..., F_p)$ around $(\overline{x}, \overline{y}_1, \overline{y}_2, ..., \overline{y}_p)$, and for every $\rho \in]0, \varepsilon[$,

$$B_Y\left(\sum_{i=1}^{p} y_i, \lambda\rho\right) \subset \left(\sum_{i=1}^{k} F_i\right)(B_X(x,\rho)) + \sum_{i=k+1}^{p} F_i(B_X(x,\rho)).$$

Proof. The assumptions of the theorem allow us to find $r > 0$ such that for every

$$(x, y_1, y_2, ..., y_k, x_{k+1}, y_{k+1}, ..., x_p, y_p) \in \left[\text{Gr}(F_1, F_2, ..., F_k) \times \prod_{i=k+1}^{p} \text{Gr} F_i \right] \cap W,$$

the assumptions (i)-(iii) hold and the set $\left[\text{Gr}(F_1, F_2, ..., F_k) \times \prod_{i=k+1}^{p} \text{Gr} F_i \right] \cap \text{cl} W$ is closed, where

$$W = \left(B_X(\bar{x}, r) \times \prod_{i=1}^{k} B_Y(\bar{y}_i, r) \right) \times \prod_{i=k+1}^{p} \left(B_X(\bar{x}, r) \times B_Y(\bar{y}_i, r) \right).$$

Consider $\varepsilon = 2^{-1}(p + \lambda)^{-1} r$ and fix arbitrarily

$$(x, y_1, y_2, ..., y_p) \in \text{Gr}(F_1, F_2, ..., F_p)$$
$$\cap \left[B_X(\bar{x}, \varepsilon) \times B_Y(\bar{y}_1, (p-1+\lambda)\varepsilon) \times \prod_{i=2}^{p} B_Y(\bar{y}_i, \varepsilon) \right]$$

and $\rho \in]0, \varepsilon[$.

Take then $v \in B_Y\left(\sum_{i=1}^{p} y_i, \lambda \rho \right)$. One finds $\mu > 1$ such that

$$\mu \left\| v - \sum_{i=1}^{p} y_i \right\| < \lambda \rho.$$

Endow the product space $X \times Y^k \times (X \times Y)^{p-k}$ with the norm

$$\| (a, b_1, b_2, ..., b_k, a_{k+1}, b_{k+1}, ..., a_p, b_p) \|_0$$
$$= \max \left\{ \lambda \|a\|, \left\| \sum_{i=1}^{p} b_i \right\|, \max_{j \in \overline{k+1, p}} \lambda \|a_j\|, \max_{i \in \overline{2, p}} \lambda \|b_i\| \right\}$$

and apply the Ekeland Variational Principle to the function

$$\left[\text{Gr}(F_1, F_2, ..., F_k) \times \prod_{i=k+1}^{p} \text{Gr} F_i \right] \cap \text{cl} W \ni (a, b_1, b_2, ..., b_k, a_{k+1}, b_{k+1}, ..., a_p, b_p)$$

$$\mapsto \left\| v - \sum_{i=1}^{p} b_i \right\| \in \mathbb{R}_+,$$

in order to get a point

$$(a, b_1, b_2, ..., b_k, a_{k+1}, b_{k+1}, ..., a_p, b_p) \in \left[\text{Gr}(F_1, F_2, ..., F_k) \times \prod_{i=k+1}^{p} \text{Gr} F_i \right] \cap \text{cl} W$$

such that

$$\mu \left\| v - \sum_{i=1}^{p} b_i \right\| \leq \mu \left\| v - \sum_{i=1}^{p} y_i \right\|$$
$$- \|(a, b_1, b_2, \ldots, b_k, a_{k+1}, b_{k+1}, \ldots, a_p, b_p)$$
$$- (x, y_1, y_2, \ldots, y_k, x, y_{k+1}, \ldots, x, y_p)\|_0, \quad (8.23)$$

and for every

$$(c, q_1, q_2, \ldots, q_k, c_{k+1}, q_{k+1}, \ldots, c_p, q_p) \in \left[\mathrm{Gr}(F_1, F_2, \ldots, F_k) \times \prod_{i=k+1}^{p} \mathrm{Gr}\, F_i \right] \cap \mathrm{cl}\, W$$

$$\mu \left\| v - \sum_{i=1}^{p} b_i \right\| \leq \mu \left\| v - \sum_{i=1}^{p} q_i \right\|$$
$$+ \|(a, b_1, b_2, \ldots, b_k, a_{k+1}, b_{k+1}, \ldots, a_p, b_p)$$
$$- (c, q_1, q_2, \ldots, q_k, c_{k+1}, q_{k+1}, \ldots, c_p, q_p)\|_0. \quad (8.24)$$

We deduce from (8.23) that

$$\|(a, b_1, b_2, \ldots, b_k, a_{k+1}, b_{k+1}, \ldots, a_p, b_p) - (x, y_1, y_2, \ldots, y_k, x, y_{k+1}, \ldots, x, y_p)\|_0$$
$$\leq \mu \left\| v - \sum_{i=1}^{p} y_i \right\| < \lambda \rho < \lambda \varepsilon,$$

hence

$$\max \left\{ \lambda \|a - x\|, \left\| \sum_{i=1}^{p} (b_i - y_i) \right\|, \max_{i \in \overline{k+1, p}} \lambda \|a_i - x\|, \max_{i \in \overline{2, p}} \lambda \|b_i - y_i\| \right\} < \lambda \varepsilon,$$

and, moreover,

$$a, a_j \in B_X(x, \varepsilon) \subset B_X(\bar{x}, 2\varepsilon) \subset B_X(\bar{x}, r), \ \forall j \in \overline{k+1, p},$$
$$b_i \in B_Y(y_i, \varepsilon) \subset B_Y(\bar{y}_i, 2\varepsilon) \subset B_Y(\bar{y}_i, r), \ \forall i \in \overline{2, p}.$$

Also, $\sum_{i=1}^{p} b_i \in B_Y \left(\sum_{i=1}^{p} y_i, \lambda \varepsilon \right)$, and because $b_i \in B_Y(y_i, \varepsilon)$ for every $i \in \overline{2, p}$, we deduce that $b_1 \in B_Y(y_1, (p-1+\lambda)\varepsilon) \subset B_Y(\bar{y}_1, 2(p-1+\lambda)\varepsilon) \subset B_Y(\bar{y}_1, r)$.
In conclusion,

$$(a, b_1, b_2, \ldots, b_k, a_{k+1}, b_{k+1}, \ldots, a_p, b_p) \in \left[\mathrm{Gr}(F_1, F_2, \ldots, F_k) \times \prod_{i=k+1}^{p} \mathrm{Gr}\, F_i \right] \cap W,$$

so assumptions (i)-(iii) hold for $(a, b_1, b_2, \ldots, b_k, a_{k+1}, b_{k+1}, \ldots, a_p, b_p)$.

According to (8.22), it follows that

$$B_Y(0,\lambda) \subset \operatorname{cl}\left[\left(D_B F_1(a,b_1) + \sum_{i=2}^{k}(D_B F_i(a,b_i) \cap B_Y(0,1))\right)(B_X(0,1)) \right.$$
$$\left. + \sum_{i=k+1}^{p}(D_B F_i(a_i,b_i) \cap B_Y(0,1))(B_X(0,1))\right].$$

We want to prove that $v = \sum_{i=1}^{p} b_i$. Denote $w = v - \sum_{i=1}^{p} b_i$ and consider a sequence $(r_n)_{n\in\mathbb{N}} \subset]0,\infty[$ such that $r_n \to 0$. Then, for every $n \in \mathbb{N}$,

$$B_Y(0,\|w\|+r_n) \subset \operatorname{cl}\left[\left(D_B F_1(a,b_1) + \sum_{i=2}^{k}(D_B F_i(a,b_i) \cap B_Y(0,s_n))\right)(B_X(0,s_n)) \right.$$
$$\left. + \sum_{i=k+1}^{p} D_B F_i(a_i,b_i)(B_X(0,s_n)) \cap B_Y(0,s_n)\right],$$

where $s_n = \frac{\|w\|+r_n}{\lambda}$. Using this relation, since $w \in B_Y(0,\|w\|+r_n)$, one can find a sequence $(a_n, b_n^1, b_n^2, \ldots, b_n^k, a_n^{k+1}, b_n^{k+1}, \ldots, a_n^p, b_n^p)_{n\in\mathbb{N}}$ in $X \times Y^k \times (X \times Y)^{p-k}$ such that $\sum_{i=1}^{p} b_n^i \to w$ and, for every $n \in \mathbb{N}$,

$$a_n, a_n^i \in B_X(0,s_n), \ \forall i \in \overline{k+1,p},$$
$$b_n^i \in D_B F_i(a,b_i)(a_n), \ \forall i \in \overline{1,k} \text{ with } b_n^j \in B_Y(0,s_n), \forall j \in \overline{2,k},$$
$$b_n^i \in D_B F_i(a_i,b_i)(a_n^i) \cap B_Y(0,s_n), \ \forall i \in \overline{k+1,p}. \tag{8.25}$$

Suppose now that F_i is proto-differentiable at a relative to b_i for every $i \in \overline{2,k}$, F_i is proto-differentiable at a_i relative to b_i for every $i \in \overline{k+1,p}$ and F_i has the Aubin property around the point (a,b_i) for every $i \in \overline{2,k}$.

Further, consider a sequence $(t_n, \alpha_n, \lambda_n^1) \in]0,\infty[\times X \times Y$ such that $(t_n, \alpha_n, \lambda_n^1) \to (0,0,0)$ and for every $n \in \mathbb{N}$,

$$(a,b_1) + t_n((\alpha_n, \lambda_n^1) + (a_n, b_n^1)) \in \operatorname{Gr} F_1.$$

Using the proto-differentiability assumptions and relation (8.25) for the sequence (t_n) from before, we can find $(\alpha_n^i, \beta_n^i) \to (0,0)$ and $(\alpha_n^j, \lambda_n^j) \to (0,0)$ with $i \in \overline{2,k}$, $j \in \overline{k+1,p}$ such that for every n,

$$(a,b_i) + t_n((\alpha_n^i, \beta_n^i) + (a_n, b_n^i)) \in \operatorname{Gr} F_i, \ \forall i \in \overline{2,k}$$

and

$$(a_i,b_i) + t_n((\alpha_n^i, \lambda_n^i) + (a_n^i, b_n^i)) \in \operatorname{Gr} F_i, \ \forall i \in \overline{k+1,p}.$$

Using now the Aubin property of F_i around (a,b_i) (with constant $L_{F_i} > 0$) with $i \in \overline{2,k}$, one gets that for a neighborhood V of b_i and for every n sufficiently large

$$b_i + t_n(\beta_n^i + b_n^i) \in F_i(a + t_n(\alpha_n^i + a_n)) \cap V \subset F_i(a + t_n(\alpha_n + a_n)) + L_{F_i} t_n \left\| \alpha_n^i - \alpha_n \right\| D_Y,$$

for all $i \in \overline{2,k}$.

In conclusion, for every $i \in \overline{2,k}$, there exists $(f_n^i) \subset D_Y$ such that $\lambda_n^i = \beta_n^i + L_{F_i} \left\| \alpha_n^i - \alpha_n \right\| f_n^i \to 0$ and

$$b_i + t_n(b_n^i + \lambda_n^i) \in F_i(a + t_n(\alpha_n + a_n)).$$

Since $\sum_{i=1}^{p} b_n^i \to w$, using (8.25), it follows that all the sequences $(a_n), (a_n^i), (b_n^j)$ are bounded with $i \in \overline{k+1, p}$, $j \in \overline{1,p}$. Hence, for every n sufficiently large,

$$\begin{aligned}
t_n \Big((\alpha_n, \lambda_n^1, \lambda_n^2, ..., \lambda_n^k, \alpha_n^{k+1}, \lambda_n^{k+1}, ..., \alpha_n^p, \lambda_n^p) \\
+ (a_n, b_n^1, b_n^2, ..., b_n^k, a_n^{k+1}, b_n^{k+1}, ..., a_n^p, b_n^p) \Big) \\
+ (a, b_1, b_2, ..., b_k, a_{k+1}, b_{k+1}, ..., a_p, b_p) \\
\in \left[\operatorname{Gr}(F_1, F_2, ..., F_k) \times \prod_{i=k+1}^{p} \operatorname{Gr} F_i \right] \cap W,
\end{aligned} \qquad (8.26)$$

as the above sequence converges to $(a, b_1, b_2, ..., b_k, a_{k+1}, b_{k+1}, ..., a_p, b_p) \in W$.

Notice that for all the combinations of the assumptions (ii) and (iii), the proof of a relation of the type (8.26) is similar.

One now uses (8.24) to get that

$$\begin{aligned}
\mu \|w\| &\leq t_n \left\| (\alpha_n, \lambda_n^1, \lambda_n^2, ..., \lambda_n^k, \alpha_n^{k+1}, \lambda_n^{k+1}, ..., \alpha_n^p, \lambda_n^p) \right. \\
&\quad + \left. (a_n, b_n^1, b_n^2, ..., b_n^k, a_n^{k+1}, b_n^{k+1}, ..., a_n^p, b_n^p) \right\|_0 + \mu \left\| w - t_n \sum_{i=1}^{p} (\lambda_n^i + b_n^i) \right\| \\
&\leq t_n \left\| (\alpha_n, \lambda_n^1, \lambda_n^2, ..., \lambda_n^k, \alpha_n^{k+1}, \lambda_n^{k+1}, ..., \alpha_n^p, \lambda_n^p) \right. \\
&\quad + \left. (a_n, b_n^1, b_n^2, ..., b_n^k, a_n^{k+1}, b_n^{k+1}, ..., a_n^p, b_n^p) \right\|_0 \\
&\quad + \mu \left[t_n \left(\left\| \left(\sum_{i=1}^{p} b_n^i \right) - w \right\| + \left\| \sum_{i=1}^{p} \lambda_n^i \right\| \right) + (1 - t_n) \|w\| \right]
\end{aligned}$$

and, consequently,

$$\begin{aligned}
\mu \|w\| &\leq \left\|(\alpha_n, \lambda_n^1, \lambda_n^2, ..., \lambda_n^k, \alpha_n^{k+1}, \lambda_n^{k+1}, ..., \alpha_n^p, \lambda_n^p)\right. \\
&+ \left.(a_n, b_n^1, b_n^2, ..., b_n^k, a_n^{k+1}, b_n^{k+1}, ..., a_n^p, b_n^p)\right\|_0 \\
&+ \mu \left(\left\|\left(\sum_{i=1}^p b_n^i\right) - w\right\| + \left\|\sum_{i=1}^p \lambda_n^i\right\| \right) \\
&\leq \mu \left(\left\|\left(\sum_{i=1}^p b_n^i\right) - w\right\| + \left\|\sum_{i=1}^p \lambda_n^i\right\| \right) \\
&+ \left\|(\alpha_n, \lambda_n^1, \lambda_n^2, ..., \lambda_n^k, \alpha_n^{k+1}, \lambda_n^{k+1}, ..., \alpha_n^p, \lambda_n^p)\right\|_0 \\
&+ \max\left\{ \lambda \|a_n\|, \left\|\sum_{i=1}^p b_n^i\right\|, \max_{i\in \overline{k+1,p}} \lambda \|a_n^i\|, \max_{i\in \overline{2,p}} \lambda \|b_n^i\| \right\} \\
&\leq \mu \left(\left\|\left(\sum_{i=1}^p b_n^i\right) - w\right\| + \left\|\sum_{i=1}^p \lambda_n^i\right\| \right) \\
&+ \left\|(\alpha_n, \lambda_n^1, \lambda_n^2, ..., \lambda_n^k, \alpha_n^{k+1}, \lambda_n^{k+1}, ..., \alpha_n^p, \lambda_n^p)\right\|_0 \\
&+ \max\left\{ \lambda \lambda^{-1}(\|w\| + r_n), \left\|\sum_{i=1}^p b_n^i\right\| \right\}.
\end{aligned}$$

By passing to the limit above, one gets that

$$\mu \|w\| \leq \|w\|,$$

hence $w = 0$. Therefore,

$$v = \sum_{i=1}^p b_i \in \left(\sum_{i=1}^k F_i\right)(a) + \sum_{i=k+1}^p F_i(a_i)$$

$$\subset \left(\sum_{i=1}^k F_i\right)(B_X(x,\rho)) + \sum_{i=k+1}^p F_i(B_X(x,\rho)).$$

The proof is now complete. □

Further, using an approach based on Ekeland's Variational Principle implemented in [46, Theorem 2.3] and [41, Theorem 4.1], we present some sufficient conditions on dual spaces for a mixed type openness of a finite family of set-valued maps.

Let p, k be natural numbers such that p is nonzero and $k \leq p$, and $F_1, F_2, ..., F_p : X \rightrightarrows Y$ be some multifunctions. Consider the following sets

$$C_i = \left\{ (x, y_1, y_2, ..., y_k) \in X \times Y^k \mid y_i \in F_i(x) \right\}, \tag{8.27}$$

for $i \in \overline{1, k}$.

Theorem 8.4.2 *Let X,Y be Asplund spaces, p,k be natural numbers such that p is nonzero and $k \leq p$, and $F_1, F_2, ..., F_p : X \rightrightarrows Y$ be multifunctions with $(\bar{x}, \bar{y}_i) \in \operatorname{Gr} F_i$ for all $i \in \overline{1,p}$. Assume that the following assumptions are satisfied:*
(i) $\operatorname{Gr} F_i$ are closed around $(\bar{x}, \bar{y}_i)$ for all $i \in \overline{1,p}$;
(ii) the sets $C_1, C_2, ..., C_k$ are allied at $(\bar{x}, \bar{y}_1, \bar{y}_2, ..., \bar{y}_k)$;
(iii) there exist $c > 0$, $r > 0$ such that for every $i \in \overline{1,p}$, $(x_i, y_i) \in \operatorname{Gr} F_i \cap [B_X(\bar{x},r) \times B_Y(\bar{y}_i,r)]$, $y^* \in S_{Y^*}$, $y_i^* \in cD_{Y^*}$, $x_i^* \in \widehat{D}^* F_i(x_i, y_i)(y^* + y_i^*)$ with $x_j^* \in B_{X^*}$ for every $j \in \overline{k+1,p}$ we have

$$c \left\| py^* + \sum_{i=1}^{p} y_i^* \right\| \leq \left\| \sum_{i=1}^{p} x_i^* \right\|.$$

Then for every $a \in]0,c[$ there exists $\varepsilon > 0$ such that for every $\rho \in]0,\varepsilon]$,

$$B_Y \left(\sum_{i=1}^{p} \bar{y}_i, \rho a \right) \subset \left(\sum_{i=1}^{k} F_i \right)(B_X(\bar{x}, \rho)) + \sum_{i=k+1}^{p} F_i(B_X(\bar{x}, \rho)).$$

Proof. Fix $a \in]0,c[$ and choose $b \in]0,1[$ such that $\frac{a}{a+1} < b < \frac{c}{c+1}$. There exists $\varepsilon > 0$ such that the following are true:

$b^{-1} a\varepsilon < r$;

$\frac{a}{a+1} < b + \varepsilon < \frac{c}{c+1}$;

$\left(\bigcap_{i=1}^{k} C_i \right) \cap \operatorname{cl} W$ is closed, where $W = B_X\left(\bar{x}, b^{-1}a\varepsilon\right) \times \left[\prod_{i=1}^{k} B_Y\left(\bar{y}_i, b^{-1}a\varepsilon\right) \right]$;

$\operatorname{Gr} F_i \cap \operatorname{cl} W_i$ is closed, where $W_i = B_X\left(\bar{x}, b^{-1}a\varepsilon\right) \times B_Y\left(\bar{y}_i, b^{-1}a\varepsilon\right)$,

for all $i \in \overline{k+1,p}$.

Fix $\rho \in]0,\varepsilon]$, and take $v \in B_Y\left(\sum_{i=1}^{p} \bar{y}_i, \rho a \right)$. Let us denote

$$V = \left[\left(\bigcap_{i=1}^{k} C_i \right) \cap \operatorname{cl} W \right] \times \left[\prod_{i=k+1}^{p} (\operatorname{Gr} F_i \cap \operatorname{cl} W_i) \right] \subset X \times Y^k \times (X \times Y)^{p-k}.$$

We endow the space $X \times Y^k \times (X \times Y)^{p-k}$ with the sum norm and define the function $f : V \to \mathbb{R}_+$,

$$f(x, y_1, y_2, ..., y_k, x_{k+1}, y_{k+1}, ..., x_p, y_p) = \left\| \sum_{i=1}^{p} y_i - v \right\|.$$

Taking into account that the set V is closed, we can apply the Ekeland Variational Principle for f and $(\bar{x}, \bar{y}_1, \bar{y}_2, ..., \bar{y}_k, \bar{x}, \bar{y}_{k+1}, ..., \bar{x}, \bar{y}_p) \in V = \operatorname{dom} f$, where $\operatorname{dom} f$ denotes the domain of f. So, there exists $(x^b, y_1^b, y_2^b, ..., y_k^b, x_{k+1}^b, y_{k+1}^b, ..., x_p^b, y_p^b) \in V$ such that

$$\left\| \sum_{i=1}^{p} y_i^b - v \right\| \leq \left\| \sum_{i=1}^{p} \bar{y}_i - v \right\| - b \left(\left\| \bar{x} - x^b \right\| + \sum_{i=1}^{p} \left\| \bar{y}_i - y_i^b \right\| + \sum_{i=k+1}^{p} \left\| \bar{x} - x_i^b \right\| \right) \quad (8.28)$$

and

$$\left\|\sum_{i=1}^{p} y_i^b - v\right\| \le \left\|\sum_{i=1}^{p} y_i - v\right\| + b\left(\left\|x - x^b\right\| + \sum_{i=1}^{p}\left\|y_i - y_i^b\right\| + \sum_{i=k+1}^{p}\left\|x_i - x_i^b\right\|\right), \quad (8.29)$$

for every $(x, y_1, y_2, \ldots, y_k, x_{k+1}, y_{k+1}, \ldots, x_p, y_p) \in V$.

From (8.28) we obtain that

$$\left\|\bar{x} - x^b\right\| + \sum_{i=1}^{p}\left\|\bar{y}_i - y_i^b\right\| + \sum_{i=k+1}^{p}\left\|\bar{x} - x_i^b\right\| \le b^{-1}\left\|\sum_{i=1}^{p}\bar{y}_i - v\right\| < b^{-1}\rho a \le b^{-1}\varepsilon a,$$

from which we get $(x^b, y_1^b, y_2^b, \ldots, y_k^b) \in W$ and $(x_i^b, y_i^b) \in W_i$ for all $i \in \overline{k+1, p}$. Again from (8.28), if $v = \sum_{i=1}^{p} y_i^b \in B_Y\left(\sum_{i=1}^{p}\bar{y}_i, \rho a\right)$ we obtain that

$$\begin{aligned} b\left\|\bar{x} - x^b\right\| &\le \left\|\sum_{i=1}^{p}\bar{y}_i - v\right\| - b\left(\sum_{i=1}^{p}\left\|\bar{y}_i - y_i^b\right\| + \sum_{i=k+1}^{p}\left\|\bar{x} - x_i^b\right\|\right) \\ &\le \left\|\sum_{i=1}^{p}\bar{y}_i - v\right\| - b\left\|\sum_{i=1}^{p}(\bar{y}_i - y_i^b)\right\| = (1-b)\left\|\sum_{i=1}^{p}\bar{y}_i - v\right\| \\ &\le (1-b)\rho a < b\rho, \end{aligned}$$

and analogously

$$b\left\|\bar{x} - x_i^b\right\| < b\rho, \quad \forall i \in \overline{k+1, p},$$

therefore, $x^b, x_i^b \in B_X(\bar{x}, \rho)$ for all $i \in \overline{k+1, p}$.

Thus,

$$v = \sum_{i=1}^{p} y_i^b \in \left(\sum_{i=1}^{k} F_i\right)(x^b) + \sum_{i=k+1}^{p} F_i(x_i^b) \subset \left(\sum_{i=1}^{k} F_i\right)(B_X(\bar{x}, \rho)) + \sum_{i=k+1}^{p} F_i(B_X(\bar{x}, \rho)).$$

Let us prove that $v = \sum_{i=1}^{p} y_i^b$ is the only possible situation. For this, suppose that $v \ne \sum_{i=1}^{p} y_i^b$ and define the function $h: X \times Y^k \times (X \times Y)^{p-k} \to \mathbb{R}_+$,

$$h(x, y_1, y_2, \ldots, y_k, x_{k+1}, y_{k+1}, \ldots, x_p, y_p)$$
$$= \left\|\sum_{i=1}^{p} y_i - v\right\| + b\left(\left\|x - x^b\right\| + \sum_{i=1}^{p}\left\|y_i - y_i^b\right\| + \sum_{i=k+1}^{p}\left\|x_i - x_i^b\right\|\right).$$

From relation (8.29) we obtain that $(x^b, y_1^b, y_2^b, \ldots, y_k^b, x_{k+1}^b, y_{k+1}^b, \ldots, x_p^b, y_p^b)$ is a minimum point for h on V. Then, $(x^b, y_1^b, y_2^b, \ldots, y_k^b, x_{k+1}^b, y_{k+1}^b, \ldots, x_p^b, y_p^b)$ is a global minimum point for $h + \delta_V$, where δ_V denotes the indicator function of the set V. By means of the generalized Fermat rule, we have that

$$(0, \ldots, 0) \in \widehat{\partial}\left(h(\cdot, \ldots, \cdot) + \delta_V(\cdot, \ldots, \cdot)\right)\left(x^b, y_1^b, y_2^b, \ldots, y_k^b, x_{k+1}^b, y_{k+1}^b, \ldots, x_p^b, y_p^b\right).$$

Since $\left(x^b, y_1^b, y_2^b, ..., y_k^b\right) \in W$ and $\left(x_i^b, y_i^b\right) \in W_i$ for all $i \in \overline{k+1, p}$, we can choose $\gamma \in]0, \rho[$ such that

$$D_X\left(x^b, \gamma\right) \times \left[\prod_{i=1}^k D_Y\left(y_i^b, \gamma\right)\right] \times \left[\prod_{i=k+1}^p \left(D_X\left(x_i^b, \gamma\right) \times D_Y\left(y_i^b, \gamma\right)\right)\right] \subset W \times \left[\prod_{i=k+1}^p W_i\right]$$

and

$$v \notin D_Y\left(\sum_{i=1}^p y_i^b, \frac{p}{2}\gamma\right).$$

Taking into account that h is Lipschitz and δ_V is lower semicontinuous, we can apply the approximate calculus rule for the Fréchet subdifferential given in Theorem 2.33 from [41]. Thus, it follows that there exist

$$\left(x^{\gamma_1}, y_1^{\gamma_1}, y_2^{\gamma_1}, ..., y_k^{\gamma_1}, x_{k+1}^{\gamma_1}, y_{k+1}^{\gamma_1}, ..., x_p^{\gamma_1}, y_p^{\gamma_1}\right) \in D_X\left(x^b, \frac{\gamma}{2}\right) \times \left[\prod_{i=1}^k D_Y\left(y_i^b, \frac{\gamma}{2}\right)\right]$$
$$\times \left[\prod_{i=k+1}^p \left(D_X\left(x_i^b, \frac{\gamma}{2}\right) \times D_Y\left(y_i^b, \frac{\gamma}{2}\right)\right)\right],$$

and

$$\left(x^{\gamma_2}, y_1^{\gamma_2}, y_2^{\gamma_2}, ..., y_k^{\gamma_2}\right) \in \left(\bigcap_{i=1}^k C_i\right) \cap \left[D_X\left(x^b, \frac{\gamma}{2}\right) \times \prod_{i=1}^k D_Y\left(y_i^b, \frac{\gamma}{2}\right)\right],$$
$$\left(x_i^{\gamma_2}, y_i^{\gamma_2}\right) \in \mathrm{Gr}\, F_i \cap \left[D_X\left(x_i^b, \frac{\gamma}{2}\right) \times D_Y\left(y_i^b, \frac{\gamma}{2}\right)\right], \forall i \in \overline{k+1, p}$$

such that

$$\begin{aligned}(0, ..., 0) \in\ & \hat{\partial} h\left(x^{\gamma_1}, y_1^{\gamma_1}, y_2^{\gamma_1}, ..., y_k^{\gamma_1}, x_{k+1}^{\gamma_1}, y_{k+1}^{\gamma_1}, ..., x_p^{\gamma_1}, y_p^{\gamma_1}\right) \quad (8.30)\\ +\ & \hat{\partial} \delta_V\left(x^{\gamma_2}, y_1^{\gamma_2}, y_2^{\gamma_2}, ..., y_k^{\gamma_2}, x_{k+1}^{\gamma_2}, y_{k+1}^{\gamma_2}, ..., x_p^{\gamma_2}, y_p^{\gamma_2}\right)\\ +\ & \frac{\gamma}{2}\left[D_{X^*} \times (D_{Y^*})^k \times (D_{X^*} \times D_{Y^*})^{p-k}\right].\end{aligned}$$

Since

$$\left(x^{\gamma_2}, y_1^{\gamma_2}, y_2^{\gamma_2}, ..., y_k^{\gamma_2}, x_{k+1}^{\gamma_2}, y_{k+1}^{\gamma_2}, ..., x_p^{\gamma_2}, y_p^{\gamma_2}\right) \in \left[\left(\bigcap_{i=1}^k C_i\right) \cap W\right] \times \left[\prod_{i=k+1}^p (\mathrm{Gr}\, F_i \cap W_i)\right],$$

we have that

$$\hat{\partial} \delta_V\left(x^{\gamma_2}, y_1^{\gamma_2}, y_2^{\gamma_2}, ..., y_k^{\gamma_2}, x_{k+1}^{\gamma_2}, y_{k+1}^{\gamma_2}, ..., x_p^{\gamma_2}, y_p^{\gamma_2}\right) = \hat{N}\left(\bigcap_{i=1}^k C_i, \left(x^{\gamma_2}, y_1^{\gamma_2}, y_2^{\gamma_2}, ..., y_k^{\gamma_2}\right)\right)$$
$$\times \prod_{i=k+1}^p \hat{N}\left(\mathrm{Gr}\, F_i, \left(x_i^{\gamma_2}, y_i^{\gamma_2}\right)\right).$$

Notice that the closedness assumption of graphs of F_i for all $i \in \overline{1,p}$ implies that $C_1, C_2, ..., C_k$ are closed around $(\bar{x}, \bar{y}_1, \bar{y}_2, ..., \bar{y}_k)$. Using the hypothesis (ii), we obtain that

$$\widehat{N}\left(\bigcap_{i=1}^{k} C_i, \left(x^{\gamma 2}, y_1^{\gamma 2}, y_2^{\gamma 2}, ..., y_k^{\gamma 2}\right)\right) \subset \sum_{i=1}^{k} \widehat{N}\left(C_i, \left(x^{\gamma 2 i}, y_1^{\gamma 2 i}, y_2^{\gamma 2 i}, ..., y_k^{\gamma 2 i}\right)\right) + \frac{\gamma}{2}\left(D_{X^*} \times (D_{Y^*})^k\right), \qquad (8.31)$$

where

$$\left(x^{\gamma 2 i}, y_1^{\gamma 2 i}, y_2^{\gamma 2 i}, ..., y_k^{\gamma 2 i}\right) \in \left[B_X\left(x^{\gamma 2}, \frac{\gamma}{2}\right) \times \prod_{j=1}^{k} B_Y\left(y_j^{\gamma 2}, \frac{\gamma}{2}\right)\right] \cap C_i, \forall i \in \overline{1,k}.$$

As h is a sum of convex continuous functions, the Fréchet subdifferential $\widehat{\partial} h$ coincides with the sum of the Fenchel subdifferentials.

Now, defining the linear operator $A : Y^p \to Y$ by $A(y_1, y_2, ..., y_p) = \sum_{i=1}^{p} y_i$, the function $(y_1, y_2, ..., y_p) \mapsto \left\|\sum_{i=1}^{p} y_i - v\right\|$ can be expressed as the composition between the convex function $y \mapsto \|y - v\|$ and A. Applying [54, Theorem 2.8.6], we have that

$$\partial \|\cdot + ... + \cdot - v\|\left(y_1^{\gamma 1}, y_2^{\gamma 1}, ..., y_p^{\gamma 1}\right) = A^*(\partial \|\cdot - v\|)\left(\sum_{i=1}^{p} y_i^{\gamma 1}\right),$$

where $A^* : Y^* \to (Y^*)^p$ denotes the adjoint of A. Remarking also that $v \neq \sum_{i=1}^{p} y_i^{\gamma 1} \in D_Y\left(\sum_{i=1}^{p} y_i^p, \frac{p}{2}\gamma\right)$ and as $A^*(y^*) = (y^*, ..., y^*)$ for every $y^* \in Y^*$, we obtain that

$$\partial \|\cdot + ... + \cdot - v\|\left(y_1^{\gamma 1}, y_2^{\gamma 1}, ..., y_p^{\gamma 1}\right) = \left\{(y^*, ..., y^*) \mid y^* \in S_{Y^*}, y^*\left(\sum_{i=1}^{p} y_i^{\gamma 1} - v\right)\right.$$
$$= \left.\left\|\sum_{i=1}^{p} y_i^{\gamma 1} - v\right\|\right\}.$$

Consequently, using (8.30) and (8.31) we have that there exists $y^* \in S_{Y^*}$ such that

$$
\begin{aligned}
(0,\ldots,0) \quad &\in \quad \{0\} \times \{y^*\}^k \times \{(0,y^*)\}^{p-k} + b \left(D_{X^*} \times (D_{Y^*})^k \times (D_{X^*} \times D_{Y^*})^{p-k} \right) \\
&\quad + \left[\sum_{i=1}^{k} \widehat{N}\left(C_i, \left(x^{\gamma 2i}, y_1^{\gamma 2i}, y_2^{\gamma 2i}, \ldots, y_k^{\gamma 2i} \right) \right) + \frac{\gamma}{2} \left(D_{X^*} \times (D_{Y^*})^k \right) \right] \\
&\quad \times \left[\prod_{i=k+1}^{p} \widehat{N}\left(\operatorname{Gr} F_i, \left(x_i^{\gamma 2}, y_i^{\gamma 2} \right) \right) \right] \\
&\quad + \frac{\gamma}{2} \left[D_{X^*} \times (D_{Y^*})^k \times (D_{X^*} \times D_{Y^*})^{p-k} \right] \\
&\subset \left[\sum_{i=1}^{k} \widehat{N}\left(C_i, \left(x^{\gamma 2i}, y_1^{\gamma 2i}, y_2^{\gamma 2i}, \ldots, y_k^{\gamma 2i} \right) \right) \right] \times \left[\prod_{i=k+1}^{p} \widehat{N}\left(\operatorname{Gr} F_i, \left(x_i^{\gamma 2}, y_i^{\gamma 2} \right) \right) \right] \\
&\quad + \{0\} \times \{y^*\}^k \times \{(0,y^*)\}^{p-k} + (b+\rho) \left[D_{X^*} \times (D_{Y^*})^k \times (D_{X^*} \times D_{Y^*})^{p-k} \right].
\end{aligned}
$$

It follows that for every $i \in \overline{1,p}$, there exist $y_i^* \in D_{Y^*}$, $x_i^* \in X^*$ with $\left\| \sum_{i=1}^{k} x_i^* \right\| \leq b+\rho$ and $x_i^* \in D_{X^*}$ for $i \in \overline{k+1,p}$ such that

$$
\begin{aligned}
(x_i^*, -y^* - (b+\rho) y_i^*) &\in \widehat{N}\left(\operatorname{Gr} F_i, \left(x^{\gamma 2i}, y_i^{\gamma 2i} \right) \right), \forall i \in \overline{1,k}, \\
(-(b+\rho) x_i^*, -y^* - (b+\rho) y_i^*) &\in \widehat{N}\left(\operatorname{Gr} F_i, \left(x_i^{\gamma 2}, y_i^{\gamma 2} \right) \right), \forall i \in \overline{k+1,p},
\end{aligned}
$$

i.e.,

$$
\begin{aligned}
x_i^* &\in \widehat{D}^* F_i \left(x^{\gamma 2i}, y_i^{\gamma 2i} \right) \left(y^* + (b+\rho) y_i^* \right), \forall i \in \overline{1,k}, \quad (8.32)\\
-(b+\rho) x_i^* &\in \widehat{D}^* F_i \left(x_i^{\gamma 2}, y_i^{\gamma 2} \right) \left(y^* + (b+\rho) y_i^* \right), \forall i \in \overline{k+1,p}.
\end{aligned}
$$

Furthermore, we observe that $\|-(b+\rho) x_i^*\| \leq b + \varepsilon < \frac{c}{c+1} < 1$ for every $i \in \overline{k+1,p}$,

$$
\begin{aligned}
\left(x^{\gamma 2i}, y_i^{\gamma 2i} \right) &\in \operatorname{Gr} F_i \cap \left[B_X \left(x^{\gamma 2}, \frac{\gamma}{2} \right) \times B_Y \left(y_i^{\gamma 2}, \frac{\gamma}{2} \right) \right] \\
&\subset \operatorname{Gr} F_i \cap \left[B_X \left(\overline{x}, b^{-1} a\varepsilon \right) \times B_Y \left(\overline{y}_i, b^{-1} a\varepsilon \right) \right], \forall i \in \overline{1,k},
\end{aligned}
$$

and

$$
\begin{aligned}
\left(x_i^{\gamma 2}, y_i^{\gamma 2} \right) &\in \operatorname{Gr} F_i \cap \left[D_X \left(x_i^b, \frac{\gamma}{2} \right) \times D_Y \left(y_i^b, \frac{\gamma}{2} \right) \right] \\
&\subset \operatorname{Gr} F_i \cap \left[B_X \left(\overline{x}, b^{-1} a\varepsilon \right) \times B_Y \left(\overline{y}_i, b^{-1} a\varepsilon \right) \right], \forall i \in \overline{k+1,p}.
\end{aligned}
$$

Therefore, using the hypothesis (iii) and relation (8.32), we obtain

$$
c \left\| p y^* + (b+\rho) \sum_{i=1}^{p} y_i^* \right\| \leq \left\| \sum_{i=1}^{k} x_i^* - (b+\rho) \sum_{i=k+1}^{p} x_i^* \right\| \leq p(b+\rho).
$$

However,

$$c \left\| py^* + (b+\rho)\sum_{i=1}^{p} y_i^* \right\| \geq c \left(\|py^*\| - (b+\rho) \left\| \sum_{i=1}^{p} y_i^* \right\| \right) \geq cp(1-(b+\rho)),$$

so we have

$$c(1-(b+\rho)) \leq (b+\rho).$$

This contradicts the inequality

$$b+\rho \leq b+\varepsilon < \frac{c}{c+1}.$$

The proof is complete. □

Remark 8.4.3 *Remark that the alliedness of the sets $C_1, C_2, ..., C_k$ in the particular form (8.27) at $(\bar{x}, \bar{y}_1, \bar{y}_2, ..., \bar{y}_k) \in \bigcap_{i=1}^{k} C_i$ means that for every sequences $(x_n^i, y_n^i) \xrightarrow{\text{Gr } F_i} (\bar{x}, \bar{y}_i)$ and for every $x_n^{i*} \in \widehat{D}^* F_i (x_n^i, y_n^i)(y_n^{i*})$, with $i \in \overline{1,k}$ one has the implication:*

$$\sum_{i=1}^{k} x_n^{i*} \to 0, \ (y_n^{i*}) \to 0 \Rightarrow (x_n^{i*}) \to 0, \ \forall i \in \overline{1,k}.$$

Thus, based on Theorem 1.43 from [41], if there exists $i \in \overline{1,k}$ such that F_i has the Aubin property around $(\bar{x}, \bar{y}_i)$, then we obtain that the alliedness property of the sets $C_1, C_2, ..., C_k \subset X \times Y^k$ reduces to the alliedness property of the sets $C_1, ..., C_{i-1}, C_{i+1}, ..., C_k$.

Finally, we mention that Theorems 8.4.2 and 8.4.1 give sufficient conditions for both weak openness of pairs of the form $(F_1 + F_2 + ... + F_k, F_i)$ with $i \in \overline{k+1,p}$ and for openness in the standard sense for $F_1 + F_2 + ... + F_k$, so now we are ready to give some necessary optimality conditions for the efficiency notions introduced in Section 3.

8.5 Necessary Optimality Conditions

In the following, we investigate the efficiency notions introduced in Section 3 from the point of view of necessary optimality conditions. To achieve this goal, we are going to obtain such conditions on primal spaces, working with derivatives, and on the dual spaces, using coderivatives. For the minimum points introduced in Definitions 8.3.1, 8.3.4 and 8.3.2, we employ a method that was used in several papers (e.g., [22], [31], [13], [18], [12], [19]), and is based on three main ingredients, namely, on the incompatibility between (weak) openness and optimality, on sufficient conditions for the (weak) openness for a finite family of set-valued maps, and on the penalty results. Concerning the efficiency concept introduced in Definition 8.3.7, we have divided the necessary optimality conditions obtained into two parts, as follows. On one hand, we obtained such conditions in the solid case, by means of Gerstewitz's scalarizing functional, and, on the other hand, we tackle the general situation, where we have no additional assumptions on the ordering map. In general, all the results included in this section generalize theorems from the fixed order case, a tool for their control being the fact that they can be reduced to the known corresponding results.

We begin by providing necessary optimality conditions in terms of Bouligand derivatives for the constrained robust efficiency notion in the variable ordering structures setting.

Theorem 8.5.1 *Let X,Y be Banach spaces, $S \subset X$, $M \subset S_X$ be nonempty sets such that S is closed, $\operatorname{cone} M$ is convex and S is epi-Lipschitz at $\bar{x} \in S$ in a direction $v \in M$, and $F,K : X \rightrightarrows Y$ be multifunctions such that $\operatorname{Gr} F$ and $\operatorname{Gr} K$ are closed around $(\bar{x}, \bar{y}) \in \operatorname{Gr} F$ and $(\bar{x}, 0) \in \operatorname{Gr} K$, respectively. Suppose that there exist $e \in \left(\bigcap_{x \in X} K(x) \right) \setminus \{0\}$ and $L > 0$ such that for all $x, z \in X$,*

$$F(x) \subset F(z) - L\|x-z\|e + K(x),$$

and that for every $(x,y,z) \in \operatorname{Gr}(F,K)$ around $(\bar{x},\bar{y},0)$ the following assertions hold:

(i) F is proto-differentiable at x relative to y and K is proto-differentiable at x relative to z;

(ii) either F has the Aubin property around (x,y), or K has the Aubin property around (x,z).

If $(\bar{x},\bar{y})$ is a strong robust efficient point for F on S with respect to K, then for every $\varepsilon > 0$ and $\beta > L$, there exist $(x_\varepsilon, y_\varepsilon, z_\varepsilon) \in [\operatorname{Gr}(F,K)] \cap [B_X(\bar{x},\varepsilon) \times B_Y(\bar{y},\varepsilon) \times B_Y(0,\varepsilon)]$ and $w_\varepsilon \in B_Y(0,\varepsilon) \setminus \{0\}$ such that

$$\begin{aligned} w_\varepsilon &\notin \operatorname{cl}[(D_B F(x_\varepsilon, y_\varepsilon) + D_B(\beta T_M(\cdot, S)e)(x_\varepsilon) \cap B_Y(0,1) \\ &+ D_B K(x_\varepsilon, z_\varepsilon) \cap B_Y(0,1))(B_X(0,1))]. \end{aligned}$$

Proof. Since all the assumptions of Theorem 8.3.13 are satisfied, we obtain that $(\bar{x}, \bar{y})$ is a nondominated point for $F(\cdot) + \beta T_M(\cdot, S)e$ with respect to K, for all $\beta > L$. Using Lemma 8.3.20, we get that $F(\cdot) + \beta T_M(\cdot, S)e + K(\cdot)$ is not open at $(\bar{x}, \bar{y})$. Let $G : X \to Y$, $G(x) = \beta T_M(x, S)e$. Using Proposition 8.2.1, we get that G is Lipschitz around $\bar{x}$. Since the conclusion of Theorem 8.4.1 does not hold, we have that the first assumption is not satisfied (the others do). Therefore, for every $\varepsilon > 0$, there exist $(x_\varepsilon, y_\varepsilon, t_\varepsilon, z_\varepsilon) \in [\operatorname{Gr}(F,G,K)] \cap [B_X(\bar{x},\varepsilon) \times B_Y(\bar{y},\varepsilon) \times B_Y(0,\varepsilon) \times B_Y(0,\varepsilon)]$ and $w_\varepsilon \in B_Y(0,\varepsilon)$ such that

$$\begin{aligned} w_\varepsilon &\notin \operatorname{cl}[(D_B F(x_\varepsilon, y_\varepsilon) + D_B G(x_\varepsilon, \beta T_M(x_\varepsilon, S)e) \cap B_Y(0,1) \\ &+ D_B K(x_\varepsilon, z_\varepsilon) \cap B_Y(0,1))(B_X(0,1))]. \end{aligned}$$

Since 0 clearly belongs to the right-hand set, we infer that $w_\varepsilon \neq 0$. The conclusion follows. $\square$

In the next result we get necessary optimality conditions for constrained Henig proper nondominated points on primal spaces.

Theorem 8.5.2 *Let X,Y be Banach spaces, $S \subset X$, $M \subset S_X$ be nonempty sets such that S is closed, $\mathrm{cone}\, M$ is convex and S is epi-Lipschitz at $\bar{x} \in S$ in a direction $v \in M$, and $F, K : X \rightrightarrows Y$ be multifunctions such that $\mathrm{Gr}\, F$ and $\mathrm{Gr}\, K$ are closed around $(\bar{x}, \bar{y}) \in \mathrm{Gr}\, F$ and $(\bar{x}, 0) \in \mathrm{Gr}\, K$, respectively. Suppose that there exists a neighborhood U of $\bar{x}$ such that:*
  (i) there exists $e \in Y \setminus \{0\}$ such that $e \in K(x)$ for all $x \in U$;
  (ii) there exists a constant $L > 0$ such that for all $u, v \in U$,

$$F(u) \subset F(v) - L\|u - v\|e + K(v);$$

  (iii) $K(x)$ is based for all $x \in U$ (the base of $K(x)$ is denoted by $B(x)$);
  (iv) there exists a constant $l > 0$ such that for all $u, v \in U$

$$B(u) \subset B(v) + l\|u - v\|D_Y.$$

Moreover, suppose that for every $(x, y, t, z) \in \mathrm{Gr}\, F \times \mathrm{Gr}\, K$ around $(\bar{x}, \bar{y}, \bar{x}, 0) \in \mathrm{Gr}\, F \times \mathrm{Gr}\, K$, K is proto-differentiable at t relative to z and F is proto-differentiable at x relative to y.

If $(\bar{x}, \bar{y})$ is a Henig proper nondominated point for F on S with respect to K, then for every $\varepsilon > 0$ and $\beta > L$, there exist $(x_\varepsilon, y_\varepsilon, t_\varepsilon, z_\varepsilon) \in [\mathrm{Gr}\, F \times \mathrm{Gr}\, K] \cap [B_X(\bar{x}, \varepsilon) \times B_Y(\bar{y}, \varepsilon) \times B_X(\bar{x}, \varepsilon) \times B_Y(0, \varepsilon)]$ and $w_\varepsilon \in B_Y(0, \varepsilon) \setminus \{0\}$ such that

$$\begin{aligned} w_\varepsilon \notin\ & \mathrm{cl}[(D_B F(x_\varepsilon, y_\varepsilon) + D_B(\beta T_M(\cdot, S)e)(x_\varepsilon) \cap B_Y(0,1))(B_X(0,1)) \\ & + D_B K(t_\varepsilon, z_\varepsilon)(B_X(0,1)) \cap B_Y(0,1)]. \end{aligned}$$

Proof. As all the assumptions of Theorem 8.3.16 are satisfied, we obtain that $(\bar{x}, \bar{y})$ is a local robust efficient point for $F(\cdot) + \beta T_M(\cdot, S)e$ with respect to K, for all $\beta > L$. Using Lemma 8.3.21, we get that $(F(\cdot) + \beta T_M(\cdot, S)e, K(\cdot))$ is not weakly open at $(\bar{x}, \bar{y})$. Further, using the same arguments as in Theorem 8.5.1, the conclusion follows. □

We mention that similar results can be proved in the unconstrained case. In this sense, in the sequel we give necessary optimality conditions in terms of Bouligand derivatives for the local nondominated points.

Theorem 8.5.3 *Let X, Y be Banach spaces, $F : X \rightrightarrows Y$, $K : X \rightrightarrows Y$ be multifunctions such that $\mathrm{Gr}\, F$ and $\mathrm{Gr}\, K$ are closed around $(\bar{x}, \bar{y}) \in \mathrm{Gr}\, F$ and $(\bar{x}, 0) \in \mathrm{Gr}\, K$, respectively. Suppose that there is a neighborhood U of $\bar{x}$ such that $\bigcap_{x \in U} K(x) \neq \{0\}$ and that for every $(x, y, z) \in \mathrm{Gr}(F, K)$ around $(\bar{x}, \bar{y}, 0)$, the next assumptions hold:*
  (i) either F is proto-differentiable at x relative to y, or K is proto-differentiable at x relative to z;
  (ii) either F has the Aubin property around (x, y), or K has the Aubin property around (x, z).

If $(\bar{x}, \bar{y})$ is a local nondominated point for F with respect to K, then, for every $\varepsilon > 0$, there are $(x_\varepsilon, y_\varepsilon, z_\varepsilon) \in \mathrm{Gr}(F, K) \cap [B_X(\bar{x}, \varepsilon) \times B_Y(\bar{y}, \varepsilon) \times B_Y(0, \varepsilon)]$ and $w_\varepsilon \in B_Y(0, \varepsilon) \setminus \{0\}$ such that

$$w_\varepsilon \notin \mathrm{cl}\{[D_B F(x_\varepsilon, y_\varepsilon) + D_B K(x_\varepsilon, z_\varepsilon) \cap B_Y(0,1)](B_X(0,1))\}. \tag{8.33}$$

Proof. Observe that, in view of Lemma 8.3.20, the multifunction $F + K$ is not open at $(\bar{x}, \bar{y})$. As all the other assumptions of Theorem 8.4.1, except (i), are satisfied, the conclusion easily follows. □

Concerning the approach on dual spaces, in order to get nontrivial Lagrange multipliers for the objective, we need the following mixed qualification condition for k sets in a product of arbitrary Asplund spaces.

Let X and Y be Asplund spaces and $C_1, C_2, ..., C_k$ be closed subsets of $X \times Y$. We say that the system $(C_1, C_2, ..., C_k)$ satisfies the mixed qualification condition at $(\bar{x}, \bar{y}) \in \bigcap_{i=1}^{k} C_i$ if for every $i \in \overline{1,k}$, $(x_{in}, y_{in}) \xrightarrow{C_i} (\bar{x}, \bar{y})$ and $(x_{in}^*, y_{in}^*) \xrightarrow{w^*} (x_i^*, y_i^*)$ with $(x_{in}^*, y_{in}^*) \in \widehat{N}(C_i, (x_{in}, y_{in}))$ the following implication holds:

$$\sum_{i=1}^{k} x_{in}^* \to 0, \ \sum_{i=1}^{k} y_{in}^* \xrightarrow{w^*} 0 \Rightarrow \sum_{i=1}^{k} y_{in}^* \to 0.$$

Remark 8.5.4 *Observe that if we consider the sets C_i given in relation (8.27), the system $(C_1, C_2, ..., C_k)$ satisfies the mixed qualification condition at $(\bar{x}, \bar{y}_1, ..., \bar{y}_k) \in \bigcap_{i=1}^{k} C_i$ if for every $i \in \overline{1,k}$, $(x_{in}, y_{in}) \xrightarrow{\mathrm{Gr}\, F_i} (\bar{x}, \bar{y}_i)$, $(y_{in}^*) \subset Y^*$ and $x_{in}^* \xrightarrow{w^*} x_i^*$ with $x_{in}^* \in \widehat{D}^* F_i(x_{in}, y_{in})(y_n^{i*})$, one has*

$$\sum_{i=1}^{k} x_{in}^* \to 0, y_{in}^* \xrightarrow{w^*} 0 \Rightarrow y_{in}^* \to 0.$$

Moreover, if $k = 1$, then the mixed qualification condition means that F^{-1} is (PSNC) at $(\bar{y}, \bar{x})$. In addition, observe that if Y is a finite-dimensional space, then the mixed qualification condition always holds.

In the next results, we present necessary optimality conditions for the efficiency concepts introduced in Section 3, by means of Mordukhovich generalized differentiation objects. More precisely, in the case of unconstrained problems, the optimality conditions are formulated in terms of a sum of Mordukhovich coderivatives for both the objective and ordering set-valued mappings, and when the optimality conditions are obtained in the constrained case, to the above mentioned sum is added the Mordukhovich subdifferential of the minimal time function associated to the restriction set $S \subset X$, with respect to the set of directions $M \subset S_X$. We mention that such optimality conditions can be also obtained using the Fréchet coderivatives, that are, in fact, to be found in the proofs of the next results.

The first result, which provides necessary optimality conditions on dual spaces, concerns the local nondomination property and it reads as follows.

Theorem 8.5.5 *Let X, Y be Asplund spaces, $F, K : X \rightrightarrows Y$ be multifunctions such that $\mathrm{Gr}\, F$ and $\mathrm{Gr}\, K$ are closed around $(\bar{x}, \bar{y}) \in \mathrm{Gr}\, F$ and $(\bar{x}, 0) \in \mathrm{Gr}\, K$, respectively. Moreover, assume that the following assumptions are satisfied:*

(i) the sets C_1, C_2 are allied at $(\bar{x}, \bar{y}, 0)$ and the system (C_1, C_2) satisfies the mixed qualification condition at $(\bar{x}, \bar{y}, 0)$, where the sets C_1, C_2 are given in (8.27) for $F_1 = F$, $F_2 = K$ and $k = 2$;

(ii) there is a neighborhood U of $\bar{x}$ such that $\bigcap_{x \in U} K(x) \neq \{0\}$;

(iii) K is lower semicontinuous at $\bar{x}$.

If $(\bar{x}, \bar{y})$ is a local nondominated point for F with respect to K, then there exists $y^* \in K(\bar{x})^+ \setminus \{0\}$ such that

$$0 \in D^* F(\bar{x}, \bar{y})(y^*) + D^* K(\bar{x}, 0)(y^*).$$

Proof. We know from Lemma 8.3.20 that the multifunction $F + K$ is not open at $(\bar{x}, \bar{y})$. As the other conditions from Theorem 8.4.2 are satisfied, it means that condition (iii) is not. Consequently, for every $n \in \mathbb{N} \setminus \{0\}$, there exist $(x_n^1, y_n^1) \in \operatorname{Gr} F \cap \left(B_X\left(\bar{x}, \frac{1}{n}\right) \times B_Y\left(\bar{y}, \frac{1}{n}\right) \right)$, $(x_n^2, y_n^2) \in \operatorname{Gr} K \cap \left(B_X\left(\bar{x}, \frac{1}{n}\right) \times B_Y\left(0, \frac{1}{n}\right) \right)$, $(y_n^*) \subset S_{Y^*}$, $(z_n^{1*}) \subset \frac{1}{n} B_{Y^*}$, $(z_n^{2*}) \subset \frac{1}{n} B_{Y^*}$, $x_n^{1*} \in \widehat{D}^* F(x_n^1, y_n^1)(y_n^* + z_n^{1*})$, $x_n^{2*} \in \widehat{D}^* K(x_n^2, y_n^2)(y_n^* + z_n^{2*})$ such that

$$\left\| x_n^{1*} + x_n^{2*} \right\| < \frac{1}{n} \left\| 2y_n^* + z_n^{1*} + z_n^{2*} \right\| < \frac{1}{n}\left(2 + \frac{2}{n}\right).$$

Therefore, in particular,

$$x_n^{1*} + x_n^{2*} \to 0.$$

As $(y_n^*) \subset S_{Y^*}$ and Y is an Asplund space, (y_n^*) contains a weak* convergent subsequence (denoted also (y_n^*) for simplicity) to an element y^*. Then, it follows that $y_n^* + z_n^{1*} \xrightarrow{w^*} y^*$ and $y_n^* + z_n^{2*} \xrightarrow{w^*} y^*$. Next we want to prove that the sequences (x_n^{1*}) and (x_n^{2*}) are bounded. In virtue of $x_n^{1*} + x_n^{2*} \to 0$, it is sufficient to prove that one of these two sequences is bounded.

By contradiction, suppose that both sequences are unbounded. Then, for every n, there exists $k_n \in \mathbb{N}$ sufficiently large such that

$$n < \min\left\{ \left\| x_{k_n}^{1*} \right\|, \left\| x_{k_n}^{2*} \right\| \right\}. \tag{8.34}$$

Again, in order to keep the notation simple, we denote the subsequences $(x_{k_n}^{1*}), (x_{k_n}^{2*})$ by $(x_n^{1*}), (x_n^{2*})$, respectively. Remark that $\frac{1}{n}(y_n^* + z_n^{1*}) \to 0$ and $\frac{1}{n}(y_n^* + z_n^{2*}) \to 0$. Also, because of positive homogeneity of the Fréchet coderivatives, we have that

$$\frac{1}{n} x_n^{1*} \in \widehat{D}^* F(x_n^1, y_n^1)\left(\frac{1}{n}(y_n^* + z_n^{1*})\right),$$

$$\frac{1}{n} x_n^{2*} \in \widehat{D}^* K(x_n^2, y_n^2)\left(\frac{1}{n}(y_n^* + z_n^{2*})\right),$$

and also

$$\frac{1}{n} x_n^{1*} + \frac{1}{n} x_n^{2*} \to 0.$$

Then, one has from the alliedness of the sets C_1, C_2 that $\frac{1}{n} x_n^{1*} \to 0$ and $\frac{1}{n} x_n^{2*} \to 0$, which is impossible in virtue of relation (8.34). Consequently, the sequences $(x_n^{1*}), (x_n^{2*})$ are bounded, and because X is Asplund, we can suppose again, without loosing the generality, that $(x_n^{1*}), (x_n^{2*})$ are weak* convergent to some $x^{1*}, x^{2*} \in X^*$, respectively. Using again that $x_n^{1*} + x_n^{2*} \to 0$, it follows that $x^{1*} = -x^{2*}$. In conclusion, $x^{1*} \in D^* F(\bar{x}, \bar{y})(y^*)$ and $-x^{1*} \in D^* K(\bar{x}, 0)(y^*)$. Further, using the fact that K has convex cones values and taking into account the semicontinuity hypothesis on K, we obtain from [22, Lemma 4.9] that $y^* \in K(\bar{x})^+$.

To complete the proof, we only need to prove that $y^* \neq 0$. For this, we suppose that $y^* = 0$. Since $x_n^{1*} \in \widehat{D}^*F(x_n^1, y_n^1)(y_n^* + z_n^{1*})$, $x_n^{2*} \in \widehat{D}^*K(x_n^2, y_n^2)(y_n^* + z_n^{2*})$, $y_n^* + z_n^{1*} \xrightarrow{w^*} 0$, $y_n^* + z_n^{2*} \xrightarrow{w^*} 0$, $x_n^{1*} + x_n^{2*} \to 0$, $(x_n^1, y_n^1) \xrightarrow{\mathrm{Gr}F} (\bar{x}, \bar{y})$ and $(x_n^2, y_n^2) \xrightarrow{\mathrm{Gr}K} (\bar{x}, 0)$ it follows from the mixed qualification condition that $y_n^* + z_n^{1*} \to 0$. As $z_n^{1*} \to 0$, we obtain that $y_n^* \to 0$, but since $(y_n^*) \subset S_{Y^*}$ we arrive at a contradiction, so $y^* \neq 0$ and complete the proof of the theorem. □

Now, we obtain the following result that concerns the robust efficiency concept in the unconstrained case.

Theorem 8.5.6 *Let X, Y be Asplund spaces, $F, K : X \rightrightarrows Y$ be set-valued maps with $\mathrm{Gr}\, F$ and $\mathrm{Gr}\, K$ closed around $(\bar{x}, \bar{y}) \in \mathrm{Gr}\, F$ and $(\bar{x}, 0) \in \mathrm{Gr}\, K$, respectively. Suppose that the following assumptions are satisfied:*

(i) the system (C_1, C_2) satisfies the mixed qualification condition at $(\bar{x}, \bar{y}, 0) \in C_1 \cap C_2$, where the sets C_1, C_2 are given in (8.27) for $F_1 = F$, $F_2 = K$ and $k = 2$;

(ii) there exists a neighborhood U of $\bar{x}$ such that $\bigcap_{x \in U} K(x) \neq \{0\}$;

(iii) K is lower semicontinuous at $\bar{x}$.

If $(\bar{x}, \bar{y})$ is a local robust efficient point for F with respect to K, then there exists $y^* \in K(\bar{x})^+ \setminus \{0\}$ such that

$$0 \in D^*F(\bar{x}, \bar{y})(y^*) + D^*K(\bar{x}, 0)(y^*). \tag{8.35}$$

Proof. Using Lemma 8.3.21 and taking into account the hypotheses, it follows that the third assumption of Theorem 8.4.2 is not satisfied. Thus, there exist the sequences $(x_n^1, y_n^1) \xrightarrow{\mathrm{Gr}F} (\bar{x}, \bar{y})$, $(x_n^2, y_n^2) \xrightarrow{\mathrm{Gr}K} (\bar{x}, 0)$, $(y_n^*) \subset S_{Y^*}$, $(z_n^{1*}) \to 0$, $(z_n^{2*}) \to 0$, $(x_n^{1*}) \subset X^*$, $(x_n^{2*}) \subset X^*$ such that $x_n^{1*} \in \widehat{D}^*F(x_n^1, y_n^1)(y_n^* + z_n^{1*}) \cap B_{X^*}$ and $x_n^{2*} \in \widehat{D}^*K(x_n^2, y_n^2)(y_n^* + z_n^{2*}) \cap B_{X^*}$ for every n and

$$x_n^{1*} + x_n^{2*} \to 0. \tag{8.36}$$

As X and Y are Asplund spaces, $(y_n^*) \subset S_{Y^*}$, $x_n^{1*} + x_n^{2*} \to 0$, $x_n^{1*} \in B_{X^*}$, $x_n^{2*} \in B_{X^*}$ we obtain that there exist $y^* \in Y^*$, $x^{1*} \in X^*$ and $x^{2*} \in X^*$ such that $y_n^* \xrightarrow{w^*} y^*$, $x_n^{1*} \xrightarrow{w^*} x^{1*}$, $x_n^{2*} \xrightarrow{w^*} x^{2*}$ and $x^{1*} + x^{2*} = 0$ (without relabelling). Using the definition of the Mordukhovich coderivative, it follows that $0 \in D^*F(\bar{x}, \bar{y})(y^*) + D^*K(\bar{x}, 0)(y^*)$. Further, in order to get nontrivial positive Lagrange multipliers for the objective, we use the same arguments as in Theorem 8.5.5. The proof is now complete. □

The penalization results given in Theorems 8.3.12 and 8.3.13 allow us to treat the robust efficiency concept in the constrained case. We include here only the (global) strong case.

Theorem 8.5.7 *Let X, Y be Asplund spaces, $S \subset X$, $M \subset S_X$ be nonempty sets such that S is closed, $\mathrm{cone}\, M$ is convex and S is epi-Lipschitz at $\bar{x} \in S$ in a direction $v \in M$ and F, $K : X \rightrightarrows Y$ be multifunctions such that $\mathrm{Gr}\, F$ and $\mathrm{Gr}\, K$ are closed around $(\bar{x}, \bar{y}) \in \mathrm{Gr}\, F$ and $(\bar{x}, 0) \in \mathrm{Gr}\, K$, respectively. Suppose that the following assumptions are satisfied:*

(i) there exist $e \in \left(\bigcap_{x \in X} K(x)\right) \setminus \{0\}$ and $L > 0$ such that for all $u, v \in X$

$$F(u) \subset F(v) - L\|u - v\|e + K(u);$$

(ii) K is lower semicontinuous at $\bar{x}$;

(iii) the sets C_1, C_2 are allied at $(\bar{x}, \bar{y}, 0)$ and the system (C_1, C_2) satisfies the mixed qualification condition at $(\bar{x}, \bar{y}, 0) \in C_1 \cap C_2$, where the sets C_1, C_2 are given in (8.27) for $F_1 = F$, $F_2 = K$ and $k = 2$.

If $(\bar{x}, \bar{y})$ is a strong robust efficient point for F on S with respect to K, then for all $\beta > L$, there exists $y^* \in K(\bar{x})^+ \setminus \{0\}$ such that

$$0 \in D^*F(\bar{x}, \bar{y})(y^*) + D^*K(\bar{x}, 0)(y^*) + \beta y^*(e)\partial T_M(\bar{x}, S). \tag{8.37}$$

Proof. Since $(\bar{x}, \bar{y})$ is a strong robust efficient point for F on S with respect to K and all the assumptions from Theorem 8.3.13 are satisfied, we obtain that $(\bar{x}, \bar{y})$ is a nondominated point for $F(\cdot) + \beta T_M(\cdot, S)e$ with respect to K, for all $\beta > L$, and from Lemma 8.3.20 we get that $F(\cdot) + \beta T_M(\cdot, S)e + K(\cdot)$ is not open at $(\bar{x}, \bar{y})$. Let $G : X \to Y$, $G(x) = \beta T_M(x, S)e$. Taking into account that G is Lipschitz around $\bar{x}$ and using Theorem 1.43 from [41] and the alliedness hypothesis, we obtain that the sets C_1, C_2 and C_3 are allied at $(\bar{x}, \bar{y}, 0, 0)$, where the sets C_1, C_2, C_3 are defined in (8.27) for $F_1 = F$, $F_2 = K$, $F_3 = G$ and $k = 3$. Therefore, taking $p = 3$, all the other assumptions of Theorem 8.4.2, except (iii), are satisfied, we get that there exist the sequences $(x_n^1, y_n^1) \xrightarrow{\text{Gr }F} (\bar{x}, \bar{y})$, $(x_n^2, y_n^2) \xrightarrow{\text{Gr }K} (\bar{x}, 0)$, $x_n^3 \to \bar{x}$, $(y_n^*) \subset S_{Y^*}$, $(z_n^{1*}) \to 0$, $(z_n^{2*}) \to 0$, $(z_n^{3*}) \to 0$ such that $x_n^{1*} \in \hat{D}^*F(x_n^1, y_n^1)(y_n^* + z_n^{1*})$, $x_n^{2*} \in \hat{D}^*K(x_n^2, y_n^2)(y_n^* + z_n^{2*})$ and $x_n^{3*} \in \hat{D}^*G(x_n^3)(y_n^* + z_n^{3*})$ for every n and

$$x_n^{1*} + x_n^{2*} + x_n^{3*} \to 0.$$

Again, since, (y_n^*) is bounded, we can suppose, without loss of generality, that it weakly* converges to an element y^*. Therefore for every $i \in \{1, 2, 3\}$, $y_n^* + z_n^{i*} \xrightarrow{w^*} y^*$, so $(y_n^* + z_n^{i*})$ is bounded. Now, since G is Lipschitz and $x_n^{3*} \in \hat{D}^*G(x_n^3)(y_n^* + z_n^{3*})$, we get from [41, Theorem 1.43] that (x_n^{3*}) is bounded, whence there exists $x^{3*} \in X^*$ such that $x_n^{3*} \xrightarrow{w^*} x^{3*}$. Further, the boundedness of the sequences (x_n^{1*}) and (x_n^{2*}) follows analogously as in Theorem 8.5.5, whence there exist $x^{1*}, x^{2*} \in X^*$ such that $x_n^{1*} \xrightarrow{w^*} x^{1*}$ and $x_n^{2*} \xrightarrow{w^*} x^{2*}$. Therefore, since $x_n^{1*} + x_n^{2*} + x_n^{3*} \to 0$, we obtain that $x^{1*} + x^{2*} + x^{3*} = 0$, with $x^{1*} \in D^*F(\bar{x}, \bar{y})(y^*)$, $x^{2*} \in D^*K(\bar{x}, 0)(y^*)$, $x^{3*} \in D^*G(\bar{x})(y^*)$, whence, using again [22, Lemma 4.9], we have that $y^* \in K(\bar{x})^+$.

Further, in order to scalarize the normal coderivative we use [41, Theorem 3.28], and for this we have to prove that G is w^*-strictly Lipschitzian at $\bar{x}$ (see [41, Definition 3.25]). In fact, we will prove that G belongs to a subclass of strictly Lipschitzian mappings. More precisely, we will prove that G is compactly strictly Lipschitzian at $\bar{x}$ (see [41, Definition 3.32]). For this we consider $u_n \to \bar{x}$, $h_n \to 0$ with $h_n \in X \setminus \{0\}$. Using the Lipschitz property of the minimal time function, we obtain that there exists $l > 0$ such that

$$\left|\frac{T_M(u_n + h_n, S) - T_M(u_n, S)}{\|h_n\|}\right| \leq l,$$

from which we get that $\left(\frac{T_M(u_n+h_n,S)-T_M(u_n,S)}{\|h_n\|}\right)_n$ is a bounded sequence of real numbers. Therefore, we get that the sequence $\left(\frac{G(u_n+h_n)-G(u_n)}{\|h_n\|}\right)_n$ has a norm-convergent subsequence, so according to [41, Definition 3.32], G is compactly strictly Lipschitzian at $\bar{x}$. Thus, G is also strictly Lipschitzian at $\bar{x}$ which, according to [41, Proposition 3.26], implies the w^*-strictly Lipschitzian property at $\bar{x}$. In this way we obtain the following representation of the normal coderivative of G:

$$D^*G(\bar{x})(y^*) = \partial\left(\beta y^*(e)T_M(\cdot,S)\right)(\bar{x}).$$

Now, taking into account the positivity of the involved multiplier, we get that

$$D^*G(\bar{x})(y^*) = \beta y^*(e)\partial T_M(\bar{x},S).$$

It remains to prove that $y^* \neq 0$. For this, we suppose that $y^* = 0$. As G is compactly strictly Lipschitzian at $\bar{x}$, $x_n^{3*} \in \widehat{D}^*G(x_n^3)(y_n^* + z_n^{3*})$, $x_n^3 \to \bar{x}$ and $y_n^* + z_n^{3*} \xrightarrow{w^*} 0$, we get from [41, Lemma 3.33] that $x_n^{3*} \to 0$. Therefore, since $x_n^{1*} + x_n^{2*} + x_n^{3*} \to 0$, we get that $x_n^{1*} + x_n^{2*} \to 0$. Now, using the same arguments as in Theorem 8.5.5 we get that $y^* \neq 0$ and the conclusion follows. □

Further, we derive necessary optimality conditions for the constrained Henig proper nondominated minimum.

Theorem 8.5.8 *Let X,Y be Asplund spaces, $S \subset X$, $M \subset S_X$ be nonempty sets such that S is closed, cone$M$ is convex and S is epi-Lipschitz at $\bar{x} \in S$ in a direction $v \in M$ and $F,K : X \rightrightarrows Y$ be multifunctions such that $\mathrm{Gr}F$ and $\mathrm{Gr}K$ are closed around $(\bar{x},\bar{y}) \in \mathrm{Gr}F$ and $(\bar{x},0) \in \mathrm{Gr}K$, respectively. Suppose that there exists a neighborhood U of $\bar{x}$ such that:*
  (i) *there exists $e \in C \setminus \{0\}$, where $C = \bigcap\limits_{x \in U} K(x)$;*
  (ii) *there exists a constant $L > 0$ such that for all $u,v \in U$,*

$$F(u) \subset F(v) - L\|u-v\|e + K(v);$$

  (iii) *$K(x)$ is based for all $x \in U$ (the base of $K(x)$ is denoted by $B(x)$);*
  (iv) *there exists a constant $l > 0$ such that for all $u,v \in U$*

$$B(u) \subset B(v) + l\|u-v\|D_Y.$$

Moreover, suppose that K is lower semicontinuous at $\bar{x}$ and the system (C_1,C_2) satisfies the mixed qualification condition at $(\bar{x},\bar{y},0) \in C_1 \cap C_2$, where the sets C_1,C_2 are given in (8.27) for $F_1 = F$, $F_2 = K$ and $k = 2$.

If $(\bar{x},\bar{y})$ is a Henig proper nondominated point for F on S with respect to K, then for all $\beta > L$, there exists $y^* \in K(\bar{x})^+ \setminus \{0\}$ such that

$$0 \in D^*F(\bar{x},\bar{y})(y^*) + D^*K(\bar{x},0)(y^*) + \beta y^*(e)\partial T_M(\bar{x},S). \tag{8.38}$$

Proof. Again, as all the assumptions of Theorem 8.3.16 are satisfied, we obtain that $(\bar{x},\bar{y})$ is a local robust efficient point for $F(\cdot) + \beta T_M(\cdot,S)e$ with respect to K for all $\beta > L$, and from Lemma 8.3.21 we get that $(F(\cdot) + \beta T_M(\cdot,S)e, K(\cdot))$ is not weakly open at $(\bar{x},\bar{y})$.

Taking into account that $G: X \to Y$, $G(x) = \beta T_M(x,S))e$ is Lipschitz around $\bar{x}$, we obtain, as above, that the sets C_1 and C_2 are allied at $(\bar{x},\bar{y},0)$, where the sets C_1, C_2 are given in 8.27 for $F_1 = F$, $F_2 = G$ and $k = 2$. Therefore, as all the other assumptions of Theorem 8.4.2, except (iii), are satisfied, we get that there exist the sequences $(x_n^1, y_n^1) \stackrel{\mathrm{Gr}F}{\to} (\bar{x},\bar{y})$, $(x_n^2, y_n^2) \stackrel{\mathrm{Gr}K}{\to} (\bar{x}, 0)$, $x_n^3 \to \bar{x}$, $(y_n^*) \subset S_{Y^*}$, $(z_n^{1*}) \to 0$, $(z_n^{2*}) \to 0$, $(z_n^{3*}) \to 0$ such that $x_n^{1*} \in \widehat{D}^* F(x_n^1, y_n^1)(y_n^* + z_n^{1*})$, $x_n^{2*} \in \widehat{D}^* K(x_n^2, y_n^2)(y_n^* + z_n^{2*}) \cap B_{X^*}$ and $x_n^{3*} \in \widehat{D}^* G(x_n^3)(y_n^* + z_n^{3*})$ for every n and

$$x_n^{1*} + x_n^{2*} + x_n^{3*} \to 0. \tag{8.39}$$

Observe first that from above (x_n^{2*}) and (y_n^*) are bounded. Then, using the Lipschitz property of G we have that (x_n^{3*}) is bounded, and using relation (8.39) the sequence (x_n^{1*}) is also bounded. Further, using the same arguments as in Theorem 8.5.7, we find that there exist $y^* \in K(\bar{x})^+ \setminus \{0\}$ and $x^{1*}, x^{2*}, x^{3*} \in X^*$ such that $x^{1*} \in D^* F(\bar{x}, \bar{y})(y^*)$, $x^{2*} \in D^* K(\bar{x}, 0)(y^*)$, $x^{3*} \in D^* G(\bar{x})(y^*) = \beta y^*(e) \partial T_M(\bar{x},S)$, and $x^{1*} + x^{2*} + x^{3*} = 0$, i.e., the conclusion. $\square$

Remark 8.5.9 *Remark that corresponding results to Theorems 8.5.8 and 8.5.7 can be formulated and proved by similar arguments in the weak case, using Theorems 8.3.15 and 8.3.12 instead of Theorems 8.3.16 and 8.3.13. Note that, in Theorem 8.5.5, when compared to Theorem 8.5.6, a supplementary alliedness condition is needed, which is natural taking into account the established relation between the nondominated and the robust efficient points.*

Remark 8.5.10 *Note that in the above theorems, if instead of supposing the lower semi-continuity of K, we would assume that there exists $U \in \mathcal{V}(\bar{x})$ such that $\overline{K} \neq \{0\}$, where $\overline{K} = \bigcap_{x \in U} K(x)$, then the multiplier y^* would belong to the set $\overline{K}^+$, which is bigger than $K(\bar{x})^+$. Moreover, if one supposes that $\overline{K}$ is (SNC) at 0, then the mixed qualification condition is not needed.*

Remark 8.5.11 *In the proofs of Theorems 8.5.3, 8.5.1 and 8.5.2 we have used Theorem 8.4.1 for two and three multifunctions. Also, in the proofs of Theorems 8.5.5, 8.5.6, 8.5.7 and 8.5.8, we have used Theorem 8.4.2 for two and three multifunctions, respectively. However, on the same techniques and by use of the mentioned theorems for more multifunctions we can tackle problems where the objective map is given as a sum of mappings.*

Remark 8.5.12 *According to Proposition 8.2.1 (iii), if $M = S_X$, then $T_M(x,S) = d(x,S)$ for every $x \in X$. Thus, in the penalization results proved in Section 3, instead of $T_M(\cdot, S)$ we have $d(\cdot, S)$, which is a Lipschitz single-valued map. Therefore, in the above theorems, in which we treated the constrained case, we drop the assumption on the restrictions set S, and the conclusions should be written slightly different.*

Further, we formulate two results, in which we provide necessary optimality conditions in terms of Mordukhovich coderivatives for the concept of sharp efficiency.

Theorem 8.5.13 *Let X, Y be Asplund spaces, $F, K : X \rightrightarrows Y$ be two multifunctions such that $\operatorname{Gr} F$ and $\operatorname{Gr} K$ are closed around $(\bar{x}, \bar{y}) \in \operatorname{Gr} F$ and $(\bar{x}, 0) \in \operatorname{Gr} K$, respectively. Assume that the following assumptions are satisfied:*

(i) $(\bar{x}, \bar{y})$ is a weak ψ-sharp local nondominated point (with the constant $\mu > 0$ and the neighborhood U) for the problem (P);

(ii) the sets C_1 and C_2 are allied at $(\bar{x}, \bar{y}, 0)$, where the sets C_1, C_2 are given in (8.27) for $F_1 = F$, $F_2 = K$ and $k = 2$;

(iii) ψ is differentiable at 0 with $\psi'(0) > 0$;

(iv) $\operatorname{int} \overline{K} \neq \emptyset$, where $\overline{K} = \bigcap_{x \in U} K(x)$.

Then for every $t^* \in \mu \psi'(0) D_{X^*} \cap \widehat{N}(W, \bar{x})$, and every $e \in \operatorname{int} \overline{K}$, there exists $z^* \in \overline{K}^+$ such that $z^*(e) = 1$, and

$$t^* \in D^* F(\bar{x}, \bar{y})(z^*) + D^* K(\bar{x}, 0)(z^*). \tag{8.40}$$

Proof. Using Lemma 8.3.9, we obtain that for every $e \in \operatorname{int} \overline{K}$, there exist $\mu > 0$ and a neighborhood U of $\bar{x}$ such that for every $x \in U$, $y \in F(x)$, $z \in K(x)$

$$s_{e, \overline{K}}(y + z - \bar{y}) \geq \mu \psi(d(x, W)),$$

from which we get that $(\bar{x}, \bar{y}, 0)$ is a minimum point for the function

$$X \times Y \times Y \ni (x, y, z) \mapsto s_{e, \overline{K}}(y + z - \bar{y}) - \mu \psi(d(x, W)) \in \mathbb{R}$$

on $(U \times Y \times Y) \cap (C_1 \cap C_2)$. Using the infinite penalization, we obtain that $(\bar{x}, \bar{y}, 0)$ is a local minimum point (without constraints) for the function $g : X \times Y \times Y \to \mathbb{R}$ given by

$$g(x, y, z) = s_{e, \overline{K}}(y + z - \bar{y}) - \mu \psi(d(x, W)) + \delta_{C_1 \cap C_2}(x, y, z).$$

It follows, from the generalized Fermat rule, that

$$(0, 0, 0) \in \widehat{\partial} g(\bar{x}, \bar{y}, 0).$$

Let $g_1, g_2 : X \times Y \times Y \to \mathbb{R}$ be the functions given by $g_1(x, y, z) = \mu \psi(d(x, W))$ and $g_2(x, y, z) = s_{e, \overline{K}}(y + z - \bar{y}) + \delta_{C_1 \cap C_2}(x, y, z)$, respectively. Using Corollary 1.96 from [41] and Corollary 3.8 from [44], one has that

$$\widehat{\partial} g_1(\bar{x}, \bar{y}, 0) = \left[\mu \psi'(0) D_{X^*} \cap \widehat{N}(W, \bar{x}) \right] \times \{0\} \times \{0\} \supset \{(0, 0, 0)\},$$

so $\widehat{\partial} g_1(\bar{x}, \bar{y}, 0) \neq \emptyset$. It follows from [41, Theorem 3.1] that

$$(0, 0, 0) \in \bigcap_{t^* \in \mu \psi'(0) D_{X^*} \cap \widehat{N}(W, \bar{x})} \left[\widehat{\partial} g_2(\bar{x}, \bar{y}, 0) - (t^*, 0, 0) \right]. \tag{8.41}$$

Taking into account that g_2 is the sum between a Lipschitz function and a lower semicontinuous one around $(\bar{x}, \bar{y}, 0)$, we can apply the fuzzy calculus rule for the Fréchet subdifferential given in Theorem 2.33 from [41]. Thus, for every $\varepsilon > 0$, there exist

$$(x_1, y_1, z_1) \in B_X\left(\bar{x}, \frac{\varepsilon}{2}\right) \times B_Y\left(\bar{y}, \frac{\varepsilon}{2}\right) \times B_Y\left(0, \frac{\varepsilon}{2}\right)$$

and
$$(x_2, y_2, z_2) \in \left[B_X\left(\bar{x}, \frac{\varepsilon}{2}\right) \times B_Y\left(\bar{y}, \frac{\varepsilon}{2}\right) \times B_Y\left(0, \frac{\varepsilon}{2}\right) \right] \cap (C_1 \cap C_2)$$

such that
$$\widehat{\partial} g_2(\bar{x},\bar{y},0) \subset \{0\} \times \partial s_{e,\overline{K}}(\cdot + \cdot - \bar{y})(y_1, z_1) + \widehat{\partial} \delta_{C_1 \cap C_2}(x_2, y_2, z_2) + \frac{\varepsilon}{2} B_{X^*} \times B_{Y^*} \times B_{Y^*}. \tag{8.42}$$

Note that the closedness assumption of the graphs of F and K implies that C_1 and C_2 are closed around $(\bar{x},\bar{y},0)$. Using the hypothesis (ii), we obtain that

$$\widehat{\partial} \delta_{C_1 \cap C_2}(x_2, y_2, z_2) = \widehat{N}(C_1 \cap C_2, (x_2, y_2, z_2))$$
$$\subset \widehat{N}(C_1, (x_{21}, y_{21}, z_{21})) + \widehat{N}(C_2, (x_{22}, y_{22}, z_{22})) + \frac{\varepsilon}{2} D_{X^*} \times D_{Y^*} \times D_{Y^*}, \tag{8.43}$$

where
$$(x_{21}, y_{21}, z_{21}) \in \left[B_X\left(x_2, \frac{\varepsilon}{2}\right) \times B_Y\left(y_2, \frac{\varepsilon}{2}\right) \times B_Y\left(z_2, \frac{\varepsilon}{2}\right) \right] \cap C_1$$

and
$$(x_{22}, y_{22}, z_{22}) \in \left[B_X\left(x_2, \frac{\varepsilon}{2}\right) \times B_Y\left(y_2, \frac{\varepsilon}{2}\right) \times B_Y\left(z_2, \frac{\varepsilon}{2}\right) \right] \cap C_2.$$

Now, defining the linear operator $A : Y \times Y \mapsto Y$ by $A(y,z) = y + z$, the function $(y,z) \mapsto s_{e,\overline{K}}(y+z-\bar{y})$ can be expressed as the composition between the convex function $y \mapsto s_{e,\overline{K}}(y-\bar{y})$ and A. Applying [54, Theorem 2.8.6], we have that

$$\partial s_{e,\overline{K}}(\cdot + \cdot - \bar{y})(y_1, z_1) = A^* \left(\partial s_{e,\overline{K}}(\cdot - \bar{y}) \right)(y_1 + z_1),$$

where $A^* : Y^* \mapsto Y^* \times Y^*$ denotes the adjoint of A. As $A^*(y^*) = (y^*, y^*)$ for every $y^* \in Y^*$, we obtain that

$$\partial s_{e,\overline{K}}(\cdot + \cdot - \bar{y})(y_1, z_1) = \left\{ (y^*, y^*) \mid y^* \in \partial s_{e,\overline{K}}(y_1 + z_1 - \bar{y}) \right\}.$$

Fix $t^* \in \mu \psi'(0) D_{X^*} \cap \widehat{N}(W,\bar{x})$. Thus, from (8.41), (8.42) and (8.43) we obtain that there exist $y \in B_Y(0,\varepsilon)$, $y^* \in \partial s_{e,\overline{K}}(y)$,

$$(x_1^*, -y_1^*, 0) \in \widehat{N}(C_1, (x_{21}, y_{21}, z_{21})) \Leftrightarrow x_1^* \in \widehat{D}^* F(x_{21}, y_{21})(y_1^*),$$
$$(x_2^*, 0, -z_2^*) \in \widehat{N}(C_2, (x_{22}, y_{22}, z_{22})) \Leftrightarrow x_2^* \in \widehat{D}^* K(x_{22}, z_{22})(z_2^*),$$

such that
$$t^* \in x_1^* + x_2^* + \varepsilon D_{X^*},$$
$$y_1^* \in y^* + \varepsilon D_{Y^*},$$
$$z_2^* \in y^* + \varepsilon D_{Y^*}.$$

Consequently, since $\varepsilon > 0$ is arbitrary, we get that there exist $(x_n, y_n) \overset{\text{Gr} F}{\to} (\bar{x}, \bar{y})$, $(u_n, v_n) \overset{\text{Gr} K}{\to} (\bar{x}, 0)$, $(z_n) \to 0$, $(z_n^*) \subset \partial s_{e,\overline{K}}(z_n)$, (t_n^*), $(v_n^*) \to 0$ and (x_n^*), $(u_n^*) \subset X^*$ such that $x_n^* \in \widehat{D}^* F(x_n, y_n)(z_n^* + t_n^*)$, $u_n^* \in \widehat{D}^* K(u_n, v_n)(z_n^* + v_n^*)$ for every natural number n and $x_n^* + u_n^* \to t^*$. As $z_n^* \in \partial s_{e,\overline{K}}(z_n)$, we obtain from Lemma 8.2.6 that $\|e\|^{-1} \leq \|z_n^*\| \leq$

$(d(e, \text{bd}\overline{K}))^{-1}$ and $(z_n^*) \subset \overline{K}^+$. Since, (z_n^*) is bounded, we can suppose, without losing generality, that it weakly* converges to an element $z^* \in \overline{K}^+$, which satisfies $z^*(e) = 1$, from which we get that it is different from 0_{Y^*}. Further, since $x_n^* + u_n^* \to t^*$, using the same arguments as in Theorem 8.5.5, we get that there exists $x^*, u^* \in X^*$ such that $x_n^* \stackrel{w^*}{\to} x^*$, $u_n^* \stackrel{w^*}{\to} u^*$ and $x^* + u^* = t^*$, therefore $t^* \in D^*F(\bar{x}, \bar{y})(z^*) + D^*K(\bar{x}, 0)(z^*)$ with $z^* \in \overline{K}^+$. Moreover, since $\partial s_{e,\overline{K}}$ is weakly*-closed, the conclusion follows. □

Theorem 8.5.14 *Let X, Y be Asplund spaces, $F, K : X \rightrightarrows Y$ be two multifunctions such that $\text{Gr}\,F$ and $\text{Gr}\,K$ are closed around $(\bar{x}, \bar{y}) \in \text{Gr}\,F$ and $(\bar{x}, 0) \in \text{Gr}\,K$, respectively. Assume that the following assumptions are satisfied:*

(i) $(\bar{x}, \bar{y})$ is a weak ψ-sharp local nondominated point (with the constant $\mu > 0$) for the problem (P);

(ii) the sets C_1 and C_2 are allied at $(\bar{x}, \bar{y}, 0)$, where the sets C_1, C_2 are given in (8.27) for $F_1 = F$, $F_2 = K$ and $k = 2$;

(iii) ψ is differentiable at 0 with $\psi'(0) > 0$;

(iv) K is lower semicontinuous at $\bar{x}$.

Then for every $t^* \in \mu\psi'(0)D_{X^*} \cap \widehat{N}(W, \bar{x})$ there exists $z^* \in D_{Y^*} \cap (K(\bar{x}))^+$ such that

$$t^* \in D^*F(\bar{x}, \bar{y})(z^*) + D^*K(\bar{x}, 0)(z^*). \tag{8.44}$$

Proof. Using Definition 8.3.7, there exist $\mu > 0$ and a neighborhood U of $\bar{x}$ such that for every $x \in U$, $y \in F(x)$, $z \in K(x)$ one has

$$\mu\psi(d(x, W)) \leq \Delta(y - \bar{y}, -K(x)) \leq d(y - \bar{y}, -K(x)) \leq \|y - \bar{y} + z\|,$$

from which we get that $(\bar{x}, \bar{y}, 0)$ is a minimum point for the function

$$X \times Y \times Y \ni (x, y, z) \mapsto \|y - \bar{y} + z\| - \mu\psi(d(x, W)) \in \mathbb{R}$$

on $(U \times Y \times Y) \cap (C_1 \cap C_2)$. In the following, we can use the same technique as in the proof of Theorem 8.5.13, with $\|y - \bar{y} + z\|$ instead $s_{e,\overline{K}}(y + z - \bar{y})$, with the obvious modifications. The conclusion follows. □

Remark 8.5.15 *Note that, in Theorem 8.5.14, we do not assume that $\text{int}\,\overline{K} \neq \emptyset$ as in Theorem 8.5.13. However, in Theorem 8.5.14 we do not know if the multiplier z^* that we get is nonzero. Nevertheless even if $z^* = 0$, Theorem 8.5.14 is not a trivial result, since relation (8.44) holds for every $t^* \in \mu\psi'(0)D_{X^*} \cap \widehat{N}(W, \bar{x})$ and in general, the set $\mu\psi'(0)D_{X^*} \cap \widehat{N}(W, \bar{x})$ does not reduce to $\{0\}$.*

8.6 Bibliographic Notes, Comments, and Conclusions

The penalization technique we use is a hybrid one, in the sense that it is somehow in between two already classical methods to penalize a given (scalar) optimization problem. If one denotes by S the feasible set (the set of constraints), one direct method is the so-called "infinite penalization", where the penalization term is given by the indicator function of S. The other one is the approach of Clarke (see [8, Proposition 2.4.3]), and uses the distance function associated to S.

The latter approach was extended to vector optimization for the fixed order setting in [50] and for the variable order setting in [13] and [12]. Moreover, a new possibility to penalize a scalar and a vector optimization problem was proved in [1], where the penalty term is a function which was studied in [45] under the name of directional minimal time function. Then, in [20], some penalization results for a vector optimization problem in the fixed order case were obtained by the use of the special instance of a minimal time function (studied in [15]) that we use in this work. Observe that, compared to the classic approaches to penalize, the penalization term used in [20] is $T_M(\cdot, S)$, and it could take the value $+\infty$ outside the set S.

Moreover, we mention here some works that, together with the references therein, can offer good knowledge on the subject of minimal time function: [10], [11], [37], [39], [9], [43], [35], [45], [6], [7].

The penalization results obtained in Section 3 generalize some corresponding results in authors' previous works (see [12], [13]), where the penalty term is based on the distance function. However, as mentioned in [20], for instance, one of the main features the minimal time function enjoys is the possibility to capture directional behavior in some situations. Moreover, it is possible to couple this technique with that used in [51], in order to penalize problems with generalized functional constraints.

The Lipschitz type property used in the penalization results generalizes to set-valued maps in the variable order case a notion extensively used in [50]. Moreover, observe that if F is a function $f : X \to \mathbb{R}$ and $K(x) = [0, \infty[$ for every $x \in X$, then the Lipschitz type concept coincides with the classical notion of Lipschitz function. Furthermore, as it is proved in [21], if $K(x) = C$ for every $x \in X$, and there exist $e \in C \setminus \{0\}$ and $L > 0$ such that for all $u, v \in S$,

$$F(u) \subset F(v) - L\|u - v\|e + C,$$

then the epigraphical set-valued map with respect to C associated to F is Lipschitz of rank $L\|e\|$ on S.

Notice as well that the requirement of nontriviality of the intersection of $K(x)$ over a neighborhood of $\bar{x}$ seems to be a natural one in the setting of variable ordering structures, since it was used as well in [22], [31], [4], [13], [12].

Theorem 8.4.1 generalizes several openness theorems from literature as [49, Theorem 1], [22, Theorem 4.4], [19, Theorem 2.6], [13, Theorem 4.8]. Furthermore, Theorem 8.4.2 generalizes several results on openness in literature, as [13, Theorem 4.9], [17, Theorem 3.1], [31, Theorem 5.2], [18, Theorem 3.3], [12, Theorem 1]. In addition to the cited results, in Theorem 8.4.2 we obtain more informations concerning the elements $x_j^* \in \widehat{D}^* F_j(x_j, y_j)(y^* + y_j^*)$ for $j \in \overline{k+1, p}$. More precisely, the assumption (iii) from Theorem 8.4.2 is weakened by the boundedness $x_j^* \in B_{X^*}$ for every $j \in \overline{k+1, p}$, boundedness that plays a major technical role in the developments from Section 5: It is enough to compare the results obtained on dual spaces using the penalization results with the results in [12, Theorem 1] and [13, Theorem 4.9].

Theorem 8.5.5 employs the right mixed qualification condition in order to get a nontrivial positive multiplier for the underlying problem and, in this way, it covers a gap from the proof of [22, Theorem 4.10]. Furthermore, for the objects we use, the mixed qualification condition is a kind of partially sequentially normally compact condition (see [41,

Definition 1.67]) for the sum of the inverses of the involved set-valued maps. Moreover, for more details on Remark 8.5.10, see Lemma 4.9 (i) from [22] and Theorem 5.6 from [31].

Observe that if, in the Definition 8.3.7 of sharp efficiency, one has $F(x) = \{f(x)\}$ for every $x \in X$, where $f : X \to Y$ is a single-valued map, then the concept coincides with the sharp efficiency notion from [31, Definition 2.1]. Since in [31] the characterization of sharp efficiency concept by means of Gerstewitz's scalarizing functional is obtained with the objective map as a single-valued map, Theorem 8.5.13 covers Theorem 4.10 from [31]. Observe as well that Theorems 8.5.14 and 8.5.13 can be formulated for a less restrictive sharp efficiency notion defined in [31] by means of the distance function instead of the oriented distance function (see [31, Theorems 4.1, 4.4, 4.9 and 4.10]). Moreover, if K is constant, then the concept from the previous mentioned definition reduces to the notion of sharp efficiency from the classical case investigated in [16], and if instead of a set-valued map F we take a single-valued map f, then the sharp efficiency introduced in Definition 8.3.7 reduced to the one from [32]. Concerning Theorem 8.5.14, in the particular case where K is constantly equal to a closed convex proper cone C, then from Theorem 8.5.13 we obtain [16, Theorem 3.3], and if we suppose that $F + K$ is closed around $(\bar{x}, \bar{y})$, from Theorem 8.5.14 we get [16, Theorem 4.1].

Furthermore, it is not difficult to check that, once a concept in the variable order setting reduces to a corresponding notion from fixed order case, then the optimality conditions we derive here can be easily rewritten for the latter case and, in this way, generalize known results in literature (see, e.g., [16], [18], [3]).

Acknowledgement. This work was supported by a grant of Romanian Ministry of Research and Innovation, CNCS-UEFISCDI, project number PN-III-P4-ID-PCE-2016-0188, within PNCDI III.

References

[1] M. Apetrii, M. Durea and R. Strugariu. A new penalization tool in scalar and vector optimizations. *Nonlinear Analysis: Theory, Methods and Applications,* 107: 22–33, 2014.

[2] J. P. Aubin and H. Frankowska. *Set-Valued Analysis.* Birkhäuser, Basel, 1990.

[3] T. Q. Bao and B. S. Mordukhovich. Relative Pareto minimizers for multiobjective problems: Existence and optimality conditions. *Mathematical Programming,* 122: 101–138, 2010.

[4] T. Q. Bao and B. S. Mordukhovich. Necessary nondomination conditions in set and vector optimization with variable ordering structures. *Journal of Optimization Theory and Applications,* 162: 350–370, 2014.

[5] T. Q. Bao, B. S. Mordukhovich and A. Soubeyran. Variational analysis in psychological modeling. *Journal of Optimization Theory and Applications,* 164: 164–290, 2015.

[6] M. Bounkhel. Directional Lipschitzness of minimal time functions in Hausdorff topological vector spaces. *Set-Valued and Variational Analysis,* 22: 221–245, 2014.

[7] M. Bounkhel. On subdifferentials of a minimal time function in Hausdorff topological vector spaces. *Applicable Analysis,* 93: 1761–1791, 2014.

[8] F. H. Clarke. *Optimization and Nonsmooth Analysis,* Wiley, New York, 1983.

[9] G. Colombo, V. Goncharov and B. S. Mordukhovich. Well-posedness of minimal time problems with constant dynamics in Banach spaces. *Set-Valued Variational Analysis,* 18: 349–372, 2010.

[10] G. Colombo and P. R. Wolenski. The subgradient formula for the minimal time function in the case of constant dynamics in Hilbert space. *Journal of Global Optimization,* 28: 269–282, 2004.

[11] G. Colombo and P. R. Wolenski. Variational analysis for a class of minimal time functions in Hilbert spaces. *Journal of Convex Analysis*, 11: 335–361, 2004.

[12] M. Durea and E. -A. Florea. A study of generalized vector variational inequalities via vector optimization problems. *Vietnam Journal of Mathematics*, 46: 33–52, 2018.

[13] M. Durea, E. -A. Florea and R. Strugariu. Henig proper efficiency in vector optimization with variable ordering structure. *Journal of Industrial and Management Optimization*, 15: 791–815, 2019.

[14] M. Durea, V. N. Huynh, H. T. Nguyen and R. Strugariu. Metric regularity of composition set-valued mappings: Metric setting and coderivative conditions. *Journal of Mathematical Analysis and Applications*, 412: 41–62, 2014.

[15] M. Durea, M. Panţiruc and R. Strugariu. Minimal time function with respect to a set of directions. Basic properties and applications. *Optimization Methods and Software*, 31: 535–561, 2016.

[16] M. Durea and R. Strugariu. Necessary optimality conditions for weak sharp minima in set-valued optimization. *Nonlinear Analysis: Theory, Methods and Applications*, 73: 2148–2157, 2010.

[17] M. Durea and R. Strugariu. Quantitative results on openness of set-valued mappings and implicit multifunction theorems. *Pacific Journal of Optimization*, 6: 533–549, 2010.

[18] M. Durea and R. Strugariu. On some Fermat rules for set-valued optimization problems. *Optimization*, 60: 575–591, 2011.

[19] M. Durea and R. Strugariu. Optimality conditions in terms of Bouligand derivatives for Pareto efficiency in set-valued optimization. *Optimization Letters*, 5: 141–151, 2011.

[20] M. Durea and R. Strugariu. Generalized penalization and maximization of vectorial nonsmooth functions. *Optimization*, 66: 903–915, 2017.

[21] M. Durea and R. Strugariu. Vectorial penalization for generalized functional constrained problems. *Journal of Global Optimization*, 68: 899–923, 2017.

[22] M. Durea, R. Strugariu and C. Tammer. On set-valued optimization problems with variable ordering structure. *Journal of Global Optimization*, 61: 745–767, 2015.

[23] M. Durea and C. Tammer. Fuzzy necessary optimality conditions for vector optimization problems. *Optimization*, 58: 449–467, 2009.

[24] G. Eichfelder. Optimal elements in vector optimization with a variable ordering structure. *Journal of Optimization Theory and Applications*, 151: 217–240, 2011.

[25] G. Eichfelder. Variable ordering structures in vector optimization. Chapter 4. pp. 95–126. *In*: Q. H. Ansari and J. -C. Yao (eds.). *Recent Developments in Vector Optimization*, Springer, Heidelberg, 2012.

[26] G. Eichfelder. *Variable Ordering Structures in Vector Optimization*, Springer, Heidelberg, 2014.

[27] G. Eichfelder and T. Gerlach. Characterization of properly optimal elements with variable ordering structures. *Optimization*, 65: 571–588, 2016.

[28] G. Eichfelder and T. X. D. Ha. Optimality conditions for vector optimization problems with variable ordering structures. *Optimization*, 62: 597–627, 2013.

[29] G. Eichfelder and R. Kasimbeyli. Properly optimal elements in vector optimization with variable ordering structures. *Journal of Global Optimization*, 60: 689–712, 2014.

[30] G. Eichfelder and M. Pilecka. Set approach for set optimization with variable ordering structures Part I: Set relations and relationship to vector approach, Part II: Scalarization approaches. *Journal of Optimization Theory and Applications*, 171: 931–963, 2016.

[31] E. -A. Florea. Coderivative necessary optimality conditions for sharp and robust efficiencies in vector optimization with variable ordering structure. *Optimization*, 65: 1417–1435, 2016.

[32] F. Flores-Bazán and B. Jiménez. Strict efficiency in set-valued optimization. *SIAM Journal on Control and Optimization*, 48: 881–908, 2009.

[33] C. Gerstewitz (Tammer) and E. Iwanow. Dualität für nichtkonvexe vektoroptimierungsprobleme. *Wissenschaftliche Zeitschrift der Technischen Hochschule Ilmenau*, 31: 61–81, 1985.

[34] H. Gfrerer. On directional metric regularity, subregularity and optimality conditions for nonsmooth mathematical programs. *Set-Valued and Variational Analysis*, 21: 151–176, 2013.

[35] V. Goncharov and F. Pereira. Neighbourhood retractions of nonconvex sets in a Hilbert space via sublinear functionals. *Journal of Convex Analysis*, 18: 1–36, 2011.

[36] A. Göpfert, H. Riahi, C. Tammer and C. Zălinescu. *Variational Methods in Partially Ordered Spaces*. Springer, Berlin, 2003.

[37] Y. He and K. F. Ng. Subdifferentials of a minimum time function in Banach spaces. *Journal of Mathematical Analysis and Applications*, 321: 896–910, 2006.

[38] J. -B. Hiriart-Urruty. New concepts in nondifferentiable programming, analyse non convexe. *Bulletin de la Société Mathématique de France*, 60: 57–85, 1979.

[39] Y. Jiang and Y. He. Subdifferentials of a minimal time function in normed spaces. *Journal of Mathematical Analysis and Applications*, 358: 410–418, 2009.

[40] S. Li, J. -P. Penot and X. Xue. Codifferential calculus. *Set-Valued and Variational Analysis*, 19: 505–536, 2011.

[41] B. S. Mordukhovich. *Variational Analysis and Generalized Differentiation, Vol. I: Basic Theory,* Springer, Berlin, 2006.

[42] B. S. Mordukhovich. *Variational Analysis and Generalized Differentiation, Vol. II: Applications,* Springer, Berlin, 2006.

[43] B. S. Mordukhovich and N. M. Nam. Limiting subgradients of minimal time functions in Banach spaces. *Journal of Global Optimization,* 46: 615–633, 2010.

[44] B. S. Mordukhovich, N .M. Nam and N. D. Yen. Fréchet subdifferential calculus and optimality conditions in nondifferentiable programming. *Optimization,* 55: 685–708, 2006.

[45] N. M. Nam and C. Zălinescu. Variational analysis of directional minimal time functions and applications to location problems. *Set-Valued and Variational Analysis,* 21: 405–430, 2013.

[46] J. -P. Penot. Compactness properties, openness criteria and coderivatives. *Set-Valued Analysis,* 6: 363–380, 1998.

[47] J. -P. Penot. Cooperative behavior of functions, relations and sets. *Mathematical Methods of Operations Research,* 48: 229–246, 1998.

[48] R. T. Rockafellar. Proto-differentiability of set-valued mappings and its applications in optimization. *Annales de l'Institut Henri Poincaré,* 6: 449–482, 1989.

[49] C. Ursescu. Tangency and openness of multifunctions in Banach spaces. *Analele Ştiinţifice ale Universităţii Alexandru Ioan Cuza din Iaşi,* 34: 221–226, 1988.

[50] J. J. Ye. The exact penalty principle. *Nonlinear Analysis: Theory Methods and Applications,* 75: 1642–1654, 2012.

[51] J. J. Ye and X. Y. Ye. Necessary optimality conditions for optimization problems with variational inequality constraints. *Mathematics of Operations Research,* 22: 977–997, 1997.

[52] P. L. Yu. Cone convexity, cone extreme points, and nondominated solutions in decision problems with multiobjectives. *Journal of Optimization Theory and Applications,* 14: 319–377, 1974.

[53] P. L. Yu. *Multiple-Criteria Decision Making. Concepts, Techniques and Extensions,* Plenum Press, New York, 1985.

[54] C. Zălinescu. *Convex Analysis in General Vector Spaces,* World Scientific, Singapore, 2002.

Chapter 9

Vectorial Penalization in Multi-objective Optimization

Christian Günther
Martin Luther University Halle-Wittenberg, Faculty of Natural Sciences II, Institute for Mathematics, 06099 Halle (Saale), Germany.

9.1 Introduction

In multi-objective optimization, one often tries to minimize a vector-valued objective function, say

$$f = (f_1, \ldots, f_m) : E \to \mathbb{R}^m \qquad (m \in \mathbb{N}),$$

over a certain feasible set Ω in a normed space E, i.e., the aim is to study the optimization problem

$$\begin{cases} f(x) = (f_1(x), \cdots, f_m(x)) \to \min \text{ w.r.t. } \mathbb{R}^m_+ \\ x \in \Omega, \end{cases} \qquad (\mathcal{P}_\Omega)$$

where, in this case, the finite-dimensional Euclidean space $\mathbb{R}^m$ is partially ordered by the natural ordering cone

$$\mathbb{R}^m_+ := \{y = (y_1, \cdots, y_m) \in \mathbb{R}^m \mid \forall i \in \{1, \cdots, m\} : y_i \geq 0\}.$$

The componentwise approach is known to be appropriate for studying vector-valued functions with values in $\mathbb{R}^m$. In the first part of the chapter, Ω will be a general set in E with certain useful properties, while, in the second part, Ω will have a more specific structure. More precisely, then the feasible set Ω consists of all points in E that satisfy

$$\begin{cases} g_1(x) \leq 0, \cdots, g_p(x) \leq 0, \\ x \in S, \end{cases} \tag{9.1}$$

where

$$g_1, \cdots, g_p : S \to \mathbb{R} \quad (p \in \mathbb{N})$$

are scalar-valued functions defined on a nonempty, convex set $S \subseteq E$. Usually, one looks for so-called Pareto efficient solutions. A feasible point $x^0 \in \Omega$ is said to be a (global) Pareto efficient solution in Ω if

$$\nexists x^1 \in \Omega \text{ subject to } \begin{cases} \forall i \in I_m : f_i(x^1) \leq f_i(x^0), \\ \exists j \in I_m : f_j(x^1) < f_j(x^0), \end{cases}$$

where the set

$$I_m := \{1, 2, \ldots, m\}$$

consists of all indices of the component functions of f. The concept of Pareto efficiency dates back to the two celebrated works by Edgeworth (1881) and Pareto (1896). For more details, we refer the interested reader to standard books of multi-objective optimization/vector optimization (see, e.g., Ehrgott [13], Göpfert et al. [18], Jahn [24], and Khan, Tammer and Zălinescu [25]).

Multi-objective optimization has become a very active field of research since it has a wide range of applications. Practical problems that are represented by mathematical models often involve certain constraints. In the past years, some penalization techniques in multi-objective optimization have been derived that aim to replace the original constrained optimization problem by some related problems with an easier structured feasible set (or actually by some unconstrained problems, i.e., $\Omega = E$). In particular, the method by vectorial penalization (see the recent works by Durea, Strugariu and Tammer [12], Günther [19, 20], and Günther and Tammer [21, 22]) has turned out to be a useful tool for deriving algorithms for nonconvex multi-objective optimization problems (as illustrated by Günther [19] in the context of location theory, see also Example 9.5.8).

We briefly summarize the structure of the chapter.

In Section 9.2, we start by introducing some appropriate notations that will be used throughout the chapter. Furthermore, we recall important notions from the fields of generalized convexity and multi-objective optimization that are essential for deriving our main results.

Pareto efficiency, with respect to different constraint sets, is studied in Section 9.3. We will extend some results that were derived by Günther and Tammer [21, 22] (see Theorem 9.3.1).

In Section 9.4, we give an overview on penalization approaches that are useful in multi-objective optimization, in particular, we focus on the method by vectorial penalization. Using results derived in Section 9.3, we will generalize some results by Günther and Tammer [21, 22] (see Theorems 9.4.8 and 9.4.13) and also extend some results to the concept of local Pareto efficiency (see Theorems 9.4.10 and 9.4.15).

Penalization in multi-objective optimization with functional inequality constraints (i.e., Ω consists of all points in E that satisfy (9.1)) is studied in Section 9.5 in detail. As pointed out by Günther and Tammer [22], the method by vectorial penalization can also be applied to this class of problems. For the concept of local Pareto efficiency, we are able to formulate corresponding results based on the vectorial penalization approach (see Section 9.5.1). In addition, we study the special case of a convex feasible set given by a system of functional inequality constraints but without convex representation, as considered by Lassere [27, 28, 29]. In Section 9.5.2, by using recent results in Durea and Strugariu [11], we gain new insights in the topic of vectorial penalization by considering appropriate constraint qualifications (such as Slater's condition, Mangasarian-Fromowitz condition, and Lassere's non-degeneracy condition).

We end with some concluding remarks.

9.2 Preliminaries in Generalized Convex Multi-objective Optimization

Throughout this chapter, we will deal with certain standard notions of optimization that we now recall in this section. In what follows, assume that E is a real normed space which is endowed with the norm $||\cdot|| : E \to \mathbb{R}$. For any $\varepsilon > 0$ and $x^0 \in E$, we define the open ball around x^0 with radius ε by

$$B_\varepsilon(x^0) := \{x \in E \mid ||x - x^0|| < \varepsilon\}.$$

For two points $x^0, x^1 \in E$, the closed, open, half-open line segments are given by

$$[x^0, x^1] := \{(1-\lambda)x^0 + \lambda x^1 \mid \lambda \in [0,1]\}, \qquad]x^0, x^1[:= [x^0, x^1] \setminus \{x^0, x^1\},$$
$$[x^0, x^1[:= [x^0, x^1] \setminus \{x^1\}, \qquad\qquad]x^0, x^1] := [x^0, x^1] \setminus \{x^0\}.$$

Considering any nonempty set Ω in E, the interior, the closure and the boundary (in the topological sense) of Ω are denoted by $\text{int}\,\Omega$, $\text{cl}\,\Omega$ and $\text{bd}\,\Omega$, respectively. In addition, the algebraic interior of Ω (or the core of Ω) is given as usual by

$$\text{cor}\,\Omega := \{x \in \Omega \mid \forall v \in E \ \exists \delta > 0 : \ x + [0, \delta] \cdot v \subseteq \Omega\}.$$

It is easily seen that

$$\text{int}\,\Omega \subseteq \text{cor}\,\Omega \subseteq \Omega.$$

Assuming that Ω is a convex set in E, we actually have

$$\text{int}\,\Omega = \text{cor}\,\Omega$$

if one of the following conditions is satisfied (see, e.g., Barbu and Precupanu [6, Sec. 1.1.2], and Jahn [24, Lem. 1.3.2]):

1. $\text{int}\,\Omega \neq \emptyset$;

2. E is a Banach space, and Ω is closed;

3. E has finite dimension.

In this chapter, we will focus on vector-valued functions that are acting between a normed pre-image space E and an image-space given by the finite-dimensional Euclidean space $\mathbb{R}^m$. We assume that $\mathbb{R}^m$ is partially ordered by the natural ordering cone $\mathbb{R}^m_+$. Let the image set of f over any set $\Omega \subseteq E$ be denoted by

$$f[\Omega] := \{f(y) \in \mathbb{R}^m \mid x \in \Omega\}.$$

As already mentioned in the Introduction, a feasible point $x^0 \in \Omega$ of the problem $(\mathcal{P}_\Omega)$ is said to be a global Pareto efficient solution if

$$\nexists x^1 \in \Omega \text{ subject to } \begin{cases} \forall i \in I_m : f_i(x^1) \leq f_i(x^0), \\ \exists j \in I_m : f_j(x^1) < f_j(x^0), \end{cases}$$

which is equivalent to the fact that

$$f[\Omega] \cap (f(x^0) - (\mathbb{R}^m_+ \setminus \{0\})) = \emptyset.$$

In addition, $x^0 \in \Omega$ is called a local Pareto efficient solution for the problem $(\mathcal{P}_\Omega)$ if

$$f[\Omega \cap B_\varepsilon(x^0)] \cap (f(x^0) - (\mathbb{R}^m_+ \setminus \{0\})) = \emptyset \quad \text{for some } \varepsilon > 0.$$

For notational convenience, the set of global Pareto efficient solutions of problem $(\mathcal{P}_\Omega)$ with respect to $\mathbb{R}^m_+$ is defined by

$$\text{Eff}(\Omega \mid f) := \{x^0 \in \Omega \mid f[\Omega] \cap (f(x^0) - (\mathbb{R}^m_+ \setminus \{0\})) = \emptyset\},$$

while the set of local Pareto efficient solutions of problem $(\mathcal{P}_\Omega)$ is given by

$$\text{Eff}_{\text{loc}}(\Omega \mid f) := \{x^0 \in \Omega \mid \exists \varepsilon > 0 : f[\Omega \cap B_\varepsilon(x^0)] \cap (f(x^0) - (\mathbb{R}^m_+ \setminus \{0\})) = \emptyset\}.$$

For deriving our main results, we need the notions of level sets and level lines of scalar functions $h : E \to \mathbb{R}$. Consider any nonempty set Ω in E and let $s \in \mathbb{R}$. Then, we define the following sets:

$L_\leq(\Omega, h, s) := \{x \in \Omega \mid h(x) \leq s\}$ (*lower-level set of h to the level s*);

$L_=(\Omega, h, s) := \{x \in \Omega \mid h(x) = s\}$ (*level line of h to the level s*);

$L_<(\Omega, h, s) := \{x \in \Omega \mid h(x) < s\}$ (*strict lower-level set of h to the level s*);

$L_\geq(\Omega, h, s) := L_\leq(\Omega, -h, -s)$ (*upper-level set of h to the level s*);

$L_>(\Omega, h, s) := L_<(\Omega, -h, -s)$ (*strict upper-level set of h to the level s*).

Notice, for any two sets $\Omega, \Omega' \subseteq E$, we have

$$L_\sim(\Omega \cap \Omega', h, s) = L_\sim(\Omega', h, s) \cap \Omega \quad \text{for all } \sim \in \{\leq, =, <, \geq, >\}.$$

In the next lemma, we recall useful characterizations of (local and global) Pareto efficient solutions by using certain level sets and level lines of the component functions of $f = (f_1, \cdots, f_m) : E \to \mathbb{R}^m$ (cf. Ehrgott [13, Th. 2.30]). Before, for any $x^0 \in \Omega$, we define the intersections of lower-level sets / level lines by

$$S_\leq(\Omega, f, x^0) := \bigcap_{i \in I_m} L_\leq(\Omega, f_i, f_i(x^0));$$

$$S_=(\Omega, f, x^0) := \bigcap_{i \in I_m} L_=(\Omega, f_i, f_i(x^0)).$$

Lemma 9.2.1 *Let $\Omega \subseteq E$ be a nonempty set. For any $x^0 \in \Omega$, the following hold:*

1°. $x^0 \in \text{Eff}(\Omega \mid f)$ *if and only if*

$$S_\leq(\Omega, f, x^0) \subseteq S_=(\Omega, f, x^0).$$

2°. $x^0 \in \text{Eff}_{\text{loc}}(\Omega \mid f)$ *if and only if*

$$S_\leq(\Omega, f, x^0) \cap B_\varepsilon(x^0) \subseteq S_=(\Omega, f, x^0) \quad \text{for some } \varepsilon > 0.$$

Remark 9.2.2 *It is easily seen that*

$$\text{Eff}(\Omega \mid f) \subseteq \text{Eff}_{\text{loc}}(\Omega \mid f) = \bigcup_{x^0 \in \Omega, \, \varepsilon > 0} \text{Eff}(\Omega \cap B_\varepsilon(x^0) \mid f). \tag{9.2}$$

It is known that the inclusion "$\subseteq$" in (9.2) becomes equality under suitable generalized convexity assumptions (for such local-global type properties see, e.g., Bagdasar and Popovici [3, 4, 5] and the references therein).

In order to derive our main results in the next section, we need some well-known generalized convexity notions that are appropriate for studying general vector-valued functions. In particular, the componentwise approach is appropriate for studying vector-valued functions with values in a finite-dimensional Euclidean space.

Assume that $\Omega \subseteq E$ is a nonempty, convex set. Recall that a vector-valued function $f = (f_1, \cdots, f_m) : E \to \mathbb{R}^m$ is said to be

- **componentwise convex** on Ω if, for any $i \in I_m$, f_i is convex on Ω, i.e., for any $x^0, x^1 \in \Omega$ and $\lambda \in [0,1]$, we have

$$f_i((1-\lambda)x^0 + \lambda x^1) \leq (1-\lambda)f_i(x^0) + \lambda f_i(x^1).$$

- **componentwise quasi-convex** on Ω if, for any $i \in I_m$, f_i is quasi-convex on Ω, i.e., for any $x^0, x^1 \in \Omega$ and $\lambda \in [0,1]$, we have

$$f_i((1-\lambda)x^0 + \lambda x^1) \leq \max\{f_i(x^0), f_i(x^1)\}.$$

- **componentwise semi-strictly quasi-convex** on Ω if, for any $i \in I_m$, f_i is semi-strictly quasi-convex on Ω, i.e., for any $x^0, x^1 \in \Omega$ such that $f_i(x^0) \neq f_i(x^1)$, and for any $\lambda \in \,]0,1[$, we have

$$f_i((1-\lambda)x^0 + \lambda x^1) < \max\{f_i(x^0), f_i(x^1)\}.$$

- componentwise explicitly quasi-convex on Ω if it is both componentwise semi-strictly quasi-convex and componentwise quasi-convex on Ω.

The reader should pay attention to the fact that quasi-convexity and semi-strict quasi-convexity are generalizations of convexity, i.e., each convex function $h : \mathbb{R} \to \mathbb{R}$ is explicitly quasi-convex (i.e., both semi-strictly quasi-convex and quasi-convex). The reverse implication is not true. To show this fact, consider the function $h : \mathbb{R} \to \mathbb{R}$ defined by $h(x) := x^3$ for all $x \in \mathbb{R}$. Then, it is easy to check that h is explicitly quasi-convex but not convex. Furthermore, it is known that each lower semi-continuous and semi-strictly quasi-convex function is quasi-convex. For more details in the field of generalized convex multi-objective optimization see, e.g., Bagdasar and Popovici [3, 4, 5], Cambini and Martein [7], Günther [19, 20], Günther and Tammer [21, 22], Luc and Schaible [31], Popovici [35], and Puerto and Rodríguez-Chía [36].

Remark 9.2.3 *For any $i \in I_m$, the following characterizations are useful:*

1°. f_i *is quasi-convex on Ω if and only if $L_\leq(\Omega, f_i, s)$ is convex for all $s \in \mathbb{R}$.*

2°. f_i *is semi-strictly quasi-convex on Ω if and only if for any $s \in \mathbb{R}$, $x^0, x^1 \in \Omega$ such that $x^0 \in L_<(\Omega, f_i, s)$ and $x^1 \in L_=(\Omega, f_i, s)$ we have $]x^0, x^1[\subseteq L_<(\Omega, f_i, s)$.*

According to Popovici [35, Prop. 2], we have the following interesting property for scalar semi-strictly quasi-convex functions.

Lemma 9.2.4 (cf. Popovici [35, Prop. 2]) *Consider any $i \in I_m$. Assume that f_i is semi-strictly quasi-convex on Ω. Then, for any $x^0, x^1 \in \Omega$, the strict upper-level set*

$$L_>(]x^0, x^1[, f_i, \max\{f_i(x^0), f_i(x^1)\})$$

is empty or reduced to a singleton.

9.3 Pareto Efficiency With Respect to Different Constraint Sets

In this section, we analyze the relationships between multi-objective optimization problems of type ($\mathcal{P}_\Omega$) involving different constraint sets (i.e., different choices of $\Omega \subseteq E$). We derive results that will be useful in order to show our main results related to the vectorial penalization approach in the next Section 9.4.

First, we extend the results derived by Günther and Tammer [21, 22] in the next Theorem 9.3.1 (we refer the reader also to the upcoming Proposition 9.3.7).

Theorem 9.3.1 *Let X and Y be sets in E and assume that Y is convex. Then, the following assertions hold:*

1°. *It holds that*

$$X \cap \mathrm{Eff}(Y \mid f) \subseteq \mathrm{Eff}(Y \cap X \mid f).$$

$2°$. If $f : E \to \mathbb{R}^m$ is componentwise semi-strictly quasi-convex on Y, then

$$(Y \cap \operatorname{cor} X) \setminus \operatorname{Eff}(Y \mid f) \subseteq (Y \cap \operatorname{cor} X) \setminus \operatorname{Eff}(Y \cap X \mid f).$$

Proof. The proof of this theorem is similar to the corresponding parts in the proof of Günther and Tammer [22, Lem. 4.4].

$1°$. This assertion follows by the definition of efficient solutions taking into account that $Y \cap X \subseteq Y$.

$2°$. Assume that f is componentwise semi-strictly quasi-convex on Y, and consider any $x^0 \in (Y \cap \operatorname{cor} X) \setminus \operatorname{Eff}(Y \mid f)$. Due to $x^0 \notin \operatorname{Eff}(Y \mid f)$ there exists $x^1 \in L_<(Y, f_j, f_j(x^0)) \cap S_\leq(Y, f, x^0)$ for some $j \in I_m$. Consider the index sets

$$I^< := \{j \in I_m \mid x^1 \in L_<(Y, f_j, f_j(x^0))\},$$
$$I^= := \{i \in I_m \mid x^1 \in L_=(Y, f_i, f_i(x^0))\}.$$

It is easily seen that $I^< \neq \emptyset$ and $I^= \cup I^< = I_m$. In the case $x^1 \in X$, we directly infer $x^0 \in (Y \cap \operatorname{cor} X) \setminus \operatorname{Eff}(Y \cap X \mid f)$. Let us assume $x^1 \in Y \setminus X$. Because of $x^0 \in \operatorname{cor} X$, we get $x^0 + [0, \delta] \cdot v \subseteq X$ for $v := x^1 - x^0 \neq 0$ and some $\delta > 0$. More precisely, $\delta \in]0, 1[$ since $x^0 + v = x^1 \notin X$. So, for $\lambda^* := \delta$, we have $x^\lambda := x^0 + \lambda v \in Y \cap X \cap]x^0, x^1[$ for all $\lambda \in]0, \lambda^*]$. Now, for an arbitrarily $i \in I_m$, we consider two cases:

Case 1: Consider $i \in I^<$. In view of Remark 9.2.3, it is easy to see that the semi-strict quasi-convexity of f_i on Y implies $x^\lambda \in L_<(Y, f_i, f_i(x^0))$ for all $\lambda \in]0, 1]$. Because of $x^\lambda \in Y \cap X$ for all $\lambda \in]0, \lambda^*]$, we get $x^\lambda \in L_<(Y \cap X, f_i, f_i(x^0))$ for all $\lambda \in]0, \lambda^*]$.

Case 2: Consider $i \in I^=$, i.e., $f_i(x^1) = f_i(x^0)$. By the semi-strict quasi-convexity of f_i on Y, we get $\operatorname{card} L_>(]x^0, x^1[, f_i, f_i(x^0)) \leq 1$ by Lemma 9.2.4. In the case $\operatorname{card} L_>(]x^0, x^1[, f_i, f_i(x^0)) = 1$, there exists $\lambda_i \in]0, 1[$ such that $f_i(x^{\lambda_i}) > f_i(x^0)$. Otherwise, we put $\lambda_i := 2\lambda^*$.

Finally, for $\bar{\lambda} := \min\{\lambda^*, 0.5 \cdot \min\{\lambda_i \mid i \in I^=\}\}$, we have $x^{\bar{\lambda}} \in L_\leq(Y \cap X, f_i, f_i(x^0))$ for all $i \in I^=$ as well as $x^{\bar{\lambda}} \in L_<(Y \cap X, f_i, f_i(x^0))$ for all $i \in I^<$. This shows that $x^0 \in (Y \cap \operatorname{cor} X) \setminus \operatorname{Eff}(Y \cap X \mid f)$ by Lemma 9.2.1.

$\square$

As a direct consequence of Theorem 9.3.1 we obtain lower and upper bounds for the set $\operatorname{Eff}(Y \cap X \mid f)$ in the next corollary.

Corollary 9.3.2 *Let X and Y be sets in E and assume that Y is convex. If $f : E \to \mathbb{R}^m$ is componentwise semi-strictly quasi-convex on Y, then*

$$X \cap \operatorname{Eff}(Y \mid f) \subseteq \operatorname{Eff}(Y \cap X \mid f) \subseteq [X \cap \operatorname{Eff}(Y \mid f)] \cup (Y \cap (X \setminus \operatorname{cor} X)).$$

Notice that $X \setminus \operatorname{cor} X \subseteq \operatorname{bd} X$, and if X is convex, $\operatorname{int} X \neq \emptyset$, and X is closed, then $X \setminus \operatorname{cor} X = \operatorname{bd} X$. If, in addition, X is algebraically open (i.e, $\operatorname{cor} X = X$), we get a new representation of the efficient set $\operatorname{Eff}(Y \cap X \mid f)$.

Corollary 9.3.3 *Let X be an algebraically open set and Y be a convex set in E. If $f : E \to \mathbb{R}^m$ is componentwise semi-strictly quasi-convex on Y, then*

$$X \cap \mathrm{Eff}(Y \mid f) = \mathrm{Eff}(Y \cap X \mid f).$$

By considering the convex set $Y \cap Z$ in the role of Y in Theorem 9.3.1, we infer the next result.

Proposition 9.3.4 *Let X, Y and Z be sets in E such that $X \subseteq Y$ and $Y \cap Z$ is convex. Then, the following assertions hold:*

1°. *It holds that*

$$X \cap \mathrm{Eff}(Y \cap Z \mid f) \subseteq \mathrm{Eff}(X \cap Z \mid f).$$

2°. *If $f : E \to \mathbb{R}^m$ is componentwise semi-strictly quasi-convex on $Y \cap Z$, then*

$$(Z \cap \mathrm{cor} X) \setminus \mathrm{Eff}(Y \cap Z \mid f) \subseteq (Z \cap \mathrm{cor} X) \setminus \mathrm{Eff}(X \cap Z \mid f).$$

By Corollary 9.3.4 we get bounds for the set $\mathrm{Eff}(X \cap Z \mid f)$.

Corollary 9.3.5 *Let X, Y and Z be sets in E such that $X \subseteq Y$ and $Y \cap Z$ is convex. If $f : E \to \mathbb{R}^m$ is componentwise semi-strictly quasi-convex on $Y \cap Z$, then*

$$X \cap \mathrm{Eff}(Y \cap Z \mid f) \subseteq \mathrm{Eff}(X \cap Z \mid f) \subseteq [X \cap \mathrm{Eff}(Y \cap Z \mid f)] \cup (Z \cap (X \setminus \mathrm{cor} X)).$$

Assuming, in addition, that X is algebraically open, the following representation of the efficient set $\mathrm{Eff}(X \cap Z \mid f)$ holds:

Corollary 9.3.6 *Let X, Y and Z be sets in E such that $X \subseteq Y$, and $Y \cap Z$ is convex, and X is algebraically open. If $f : E \to \mathbb{R}^m$ is componentwise semi-strictly quasi-convex on $Y \cap Z$, then*

$$X \cap \mathrm{Eff}(Y \cap Z \mid f) = \mathrm{Eff}(X \cap Z \mid f).$$

As a direct consequence of our Theorem 9.3.1 (respectively, Proposition 9.3.4 with $Z := Y$), we get the corresponding result derived by Günther and Tammer in [21, Prop. 1] and [22, Lem. 4.4].

Proposition 9.3.7 (cf. Günther and Tammer [21, 22]) *Let X be a set and Y be a convex set such that $X \subseteq Y \subseteq E$. Then, the following assertions hold:*

1°. *It holds that*

$$X \cap \mathrm{Eff}(Y \mid f) \subseteq \mathrm{Eff}(X \mid f).$$

2°. *If $f : E \to \mathbb{R}^m$ is componentwise semi-strictly quasi-convex on Y, then*

$$(\mathrm{cor} X) \setminus \mathrm{Eff}(Y \mid f) \subseteq (\mathrm{cor} X) \setminus \mathrm{Eff}(X \mid f).$$

In addition, we get the following two straightforward corollaries.

Corollary 9.3.8 *Let X be a nonempty set and Y be a convex set such that $X \subseteq Y \subseteq E$. If $f : E \to \mathbb{R}^m$ is componentwise semi-strictly quasi-convex on Y, then*

$$X \cap \mathrm{Eff}(Y \mid f) \subseteq \mathrm{Eff}(X \mid f) \subseteq [X \cap \mathrm{Eff}(Y \mid f)] \cup (X \setminus \mathrm{cor} X).$$

Corollary 9.3.9 *Let X be a nonempty, algebraically open set and Y be a convex set such that $X \subseteq Y \subseteq E$. If $f : E \to \mathbb{R}^m$ is componentwise semi-strictly quasi-convex on Y, then*

$$X \cap \mathrm{Eff}(Y \mid f) = \mathrm{Eff}(X \mid f).$$

9.4 A Vectorial Penalization Approach in Multi-objective Optimization

In scalar optimization theory, the famous Exact Penalty Principle (see Clarke [8, Prop. 2.4.3], Eremin [14], and Zangwill [39]) is based on the idea of replacing the initial constrained single-objective optimization problem

$$\begin{cases} h(x) \to \min \\ x \in \Omega, \end{cases}$$

where $h : E \to \mathbb{R}$ is the objective function and $\Omega \subseteq E$ is a nonempty feasible set, by a penalized unconstrained single-objective optimization problem

$$\begin{cases} h(x) + \rho \varphi(x) \to \min \\ x \in E, \end{cases}$$

where $\rho > 0$ and $\varphi : E \to \mathbb{R}$ is a scalar-valued function satisfying

$$\varphi(x) = 0 \iff x \in \Omega;$$
$$\varphi(x) > 0 \iff x \in E \setminus \Omega.$$

In some recent works, the Exact Penalty Principle was extended to the multi-objective case. Let us consider the vector-valued objective function $f = (f_1, \cdots, f_m) : E \to \mathbb{R}^m$ that is acting between a normed pre-image space E and a finite-dimensional image space $\mathbb{R}^m$. One approach consists of adding a penalization term in each component function of f (see, e.g., Apetrii, Durea and Strugariu [2], Mäkelä, Eronen and Karmitsa [33, Lem. 4.4], and Ye [37]) and to replace the initial constrained problem ($\mathcal{P}_\Omega$) by an unconstrained problem

$$\begin{cases} (f_1(x) + \rho \varphi(x), \cdots, f_m(x) + \rho \varphi(x)) \to \min \text{ w.r.t. } \mathbb{R}^m_+ \\ x \in E, \end{cases} \quad (9.3)$$

where $\rho > 0$ is a certain positive value. In what follows, we will deal with a different approach, the method by vectorial penalization for solving multi-objective optimization problems involving not necessarily convex constraints. This approach is based on the idea to consider the unconstrained problem

$$\begin{cases} (f_1(x), \cdots, f_m(x), \varphi(x)) \to \min \text{ w.r.t. } \mathbb{R}^{m+1}_+ \\ x \in E, \end{cases} \quad (9.4)$$

i.e., the penalization function $\varphi : E \to \mathbb{R}$ is added to the vector-valued objective function $f : E \to \mathbb{R}^m$ as a new component function. By exploiting the idea of adding new objective functions to multi-objective optimization problems (see, e.g., Klamroth and Tind [26], and Fliege [15]), Günther and Tammer [21] analyzed relationships between constrained and unconstrained multi-objective optimization, and developed a vectorial penalization approach for such problems, where the constraints are assumed to be convex. In addition, Durea, Strugariu and Tammer [12] also used the idea in order to derive a vectorial penalization approach for general vector optimization problems involving not necessarily convex constraints. Also Günther and Tammer [22] extended the results [21] to the more general case with not necessarily convex constraints by using new types of penalization functions. An extended view on this topic can be found in the thesis [20]. The usefulness of the vectorial penalization approach for nonconvex multi-objective optimization problems is pointed out in Günther [19] by providing a complete geometrical description of the set of Pareto efficient solutions to a special nonconvex multi-objective location problem involving some forbidden regions (see Example 9.5.8).

9.4.1 Method by Vectorial Penalization

Throughout this section, we suppose that the following assumptions hold:

$$\begin{cases} \text{let } Y \subseteq E \text{ be a convex set;} \\ \text{let } X \subseteq Y \text{ be a nonempty, closed set with } X \neq Y. \end{cases} \quad (9.5)$$

Remark 9.4.1 *Notice, under the assumptions given in (9.5), we have* $\operatorname{bd} X \neq \emptyset$. *For the case* $\operatorname{bd} X = \emptyset$ *(i.e.,* $\operatorname{int} X = \operatorname{cor} X = X$*), we refer to Corollary 9.3.9.*

In the following, our initial multi-objective optimization problem with not necessarily convex constraints is given by

$$\begin{cases} f(x) = (f_1(x), \cdots, f_m(x)) \to \min \text{ w.r.t. } \mathbb{R}^m_+ \\ x \in X. \end{cases} \quad (\mathcal{P}_X)$$

To the original objective function $f = (f_1, \cdots, f_m) : E \to \mathbb{R}^m$ of the problem $(\mathcal{P}_Y)$ (defined as $(\mathcal{P}_X)$ with Y in the role of X) we now add a real-valued penalization function $\varphi : E \to \mathbb{R}$ as a new component function $f_{m+1} := \varphi$. Then, we are able to consider the penalized multi-objective optimization problem

$$\begin{cases} f^\oplus(x) := (f_1(x), \cdots, f_m(x), \varphi(x)) \to \min \text{ w.r.t. } \mathbb{R}^{m+1}_+ \\ x \in Y. \end{cases} \quad (\mathcal{P}^\oplus_Y)$$

Hence, in $(\mathcal{P}^\oplus_Y)$ the image space dimension is increased exactly by one in comparison to the problems $(\mathcal{P}_X)$ and $(\mathcal{P}_Y)$.

In the sequel, following the approach by Günther and Tammer [22], we will need in certain results some of the following assumptions concerning the lower-level sets / level lines of the penalization function φ:

$$\forall x^0 \in \mathrm{bd}\, X: L_{\leq}(Y, \varphi, \varphi(x^0)) = X, \tag{A1}$$

$$\forall x^0 \in \mathrm{bd}\, X: L_{=}(Y, \varphi, \varphi(x^0)) = \mathrm{bd}\, X, \tag{A2}$$

$$\forall x^0 \in X: L_{=}(Y, \varphi, \varphi(x^0)) = L_{\leq}(Y, \varphi, \varphi(x^0)) = X, \tag{A3}$$

$$L_{\leq}(Y, \varphi, 0) = X, \tag{A4}$$

$$L_{=}(Y, \varphi, 0) = \mathrm{bd}\, X. \tag{A5}$$

Remark 9.4.2 *The following implications are easy to check:*

- (A4) *and* (A5) *implies* (A1) *and* (A2).
- (A3) *implies* (A1).
- (A1) *and* (A2) *imply*

$$\forall x^0 \in \mathrm{bd}\, X: L_{<}(Y, \varphi, \varphi(x^0)) = \mathrm{int}\, X,$$

while under (A3) *we have*

$$\forall x^0 \in X: L_{<}(Y, \varphi, \varphi(x^0)) = \emptyset.$$

- *If* $\mathrm{bd}\, X = \emptyset$, *then* (A1) *and* (A2) *holds if and only if* (A3) *is satisfied.*

Sufficient conditions for the validity of the Assumptions (A1) and (A2) are studied by Günther and Tammer [22, Sec. 6.2] in detail. By taking a closer look on other approaches in the literature, one can see that often a kind of penalization function $\varphi: E \to \mathbb{R}$ (penalty term concerning X) that fulfils Assumption (A3) for $Y = E$ is used (see, e.g., Apetrii *et al.* [2], Durea, Strugariu and Tammer [12], Ye [37], and references therein). More precisely, for any $x^0 \in E$, we have

$$\begin{aligned} x^0 \in X &\iff \varphi(x^0) = 0; \\ x^0 \in E \setminus X &\iff \varphi(x^0) > 0. \end{aligned}$$

In Example 9.4.4, we will consider a penalization function φ which exactly fulfils both conditions formulated above (compare with Clarke's Exact Penalty Principle, as mentioned at the beginning of Section 9.4).

Next, we present some further types of penalization functions that can be used in the vectorial penalization approach under the validity of (9.5).

Example 9.4.3 (Günther and Tammer [21]) *Assume, in addition, that X is convex with $x^1 \in \mathrm{int}\, X$. Let a Minkowski gauge $\mu_B : E \to \mathbb{R}$ with corresponding unit ball $B := -x^1 + X$ be given. Recall that we have*

$$\mu_B(x) := \inf\{\lambda > 0 \mid x \in \lambda \cdot B\} \quad \text{for all } x \in E.$$

Then, the Minkowski gauge penalization function

$$\varphi(\cdot) := \mu_B(\cdot - x^1)$$

fulfils Assumptions (A1) *and* (A2) *for $Y = E$.*

Example 9.4.4 (Günther and Tammer [22, Ex. 4.9]) *Consider the distance function with respect to X, namely $d_X : E \to \mathbb{R}$, that is given by*

$$d_X(x) := \inf\{\|x - x^1\| \mid x^1 \in X\} \quad \text{for all } x \in E.$$

Let us recall some important properties of d_X:

- d_X *is Lipschitz continuous on E of rank 1;*
- d_X *is convex on E if and only if X is convex in E;*
- $L_{\leq}(E, d_X, 0) = L_{=}(E, d_X, 0) = X.$

Therefore, the penalization function

$$\varphi := d_X$$

fulfils Assumptions (A3) *and* (A4) *for $Y = E$.*

Example 9.4.5 (Günther and Tammer [22, Ex. 4.10]) *Using the distance function with respect to X, namely the function $d_X : E \to \mathbb{R}$ in Example 9.4.4, we consider the function $\triangle_X : E \to \mathbb{R}$ that is defined by*

$$\triangle_X(x) := d_X(x) - d_{E \setminus X}(x) = \begin{cases} d_X(x) & \text{for } x \in E \setminus X, \\ -d_{E \setminus X}(x) & \text{for } x \in X. \end{cases}$$

The function $\triangle_X$ is known in the literature as signed distance function or Hiriart-Urruty function $\triangle_X$ (see Hiriart-Urruty [23]). Let us recall some properties of $\triangle_X$ (see Hiriart-Urruty [23], Liu, Ng and Yang [30], and Zaffaroni [38]):

- $\triangle_X$ *is Lipschitz continuous on E of rank 1;*
- $\triangle_X$ *is convex on E if and only if X is convex in E;*
- $L_{\leq}(E, \triangle_X, 0) = X$ *and* $L_{=}(E, \triangle_X, 0) = \operatorname{bd} X.$

We directly get that the Hiriart-Urruty penalization function

$$\varphi := \triangle_X$$

fulfils Assumptions (A1), (A2), (A4) *and* (A5) *for $Y = E$.*

Example 9.4.6 (Günther and Tammer [22, Ex. 4.11]) *In this example, we recall the definition of the so-called Gerstewitz function which can be used as a nonlinear scalarization tool in vector optimization (see Gerstewitz [16], and Gerth and Weidner [17]). Suppose that $C \subseteq E$ is a nontrivial, closed, convex cone, and $k \in \operatorname{int} C$ such that*

$$X - (C \setminus \{0\}) = \operatorname{int} X.$$

Then, the Gerstewitz function $\phi_{X,k} : E \to \mathbb{R}$, defined, for any $x \in E$, by

$$\phi_{X,k}(x) := \inf\{s \in \mathbb{R} \mid x \in sk + X\},$$

fulfils the properties (see Khan, Tammer and Zălinescu [25, Sec. 5.2]):

- $\phi_{X,k}$ is Lipschitz continuous on E;
- $\phi_{X,k}$ is convex on E if and only if X is convex in E;
- $L_\leq(E, \phi_{X,k}, 0) = X$ and $L_=(E, \phi_{X,k}, 0) = \mathrm{bd}\,X$.

We conclude that the Gerstewitz penalization function

$$\varphi := \phi_{X,k}$$

fulfils Assumptions (A1), (A2), (A4) and (A5) for $Y = E$.

Remark 9.4.7 *As can be seen in Examples 9.4.4, 9.4.5 and 9.4.6, the set X considered in the vectorial penalization approach must not be necessarily convex. For any arbitrarily nonempty, closed set $X \subsetneq E$, we know that the Hiriart-Urruty function $\triangle_X$ fulfils Assumptions (A1) and (A2), and moreover, the distance function with respect to X, namely d_X, fulfils Assumption (A3).*

9.4.2 Main Relationships

For any $Z \subseteq E$, we consider the problems $(\mathcal{P}_{X \cap Z})$, $(\mathcal{P}_{Y \cap Z})$, and $(\mathcal{P}_{Y \cap Z}^\oplus)$ (defined as the corresponding problems in Section 9.4.1 with $X \cap Z$ in the role of X and $Y \cap Z$ in the role of Y). Next, we present a second main theorem of the chapter, which generalizes the result in Günther and Tammer [22, Th. 5.1].

Theorem 9.4.8 *Let (9.5) be satisfied. Suppose that $\varphi : E \to \mathbb{R}$ fulfils Assumptions (A1) and (A2). Then, for any $Z \subseteq E$, the following assertions hold:*

1°. *We have*

$$[X \cap \mathrm{Eff}(Y \cap Z \mid f)] \cup \left[(\mathrm{bd}\,X) \cap \mathrm{Eff}(Y \cap Z \mid f^\oplus)\right] \subseteq \mathrm{Eff}(X \cap Z \mid f).$$

2°. *In the case $\mathrm{int}\,X \neq \emptyset$, suppose additionally that Z is convex and that $f : E \to \mathbb{R}^m$ is componentwise semi-strictly quasi-convex on $Y \cap Z$. Then, we have*

$$[X \cap \mathrm{Eff}(Y \cap Z \mid f)] \cup \left[(\mathrm{bd}\,X) \cap \mathrm{Eff}(Y \cap Z \mid f^\oplus)\right] \supseteq \mathrm{Eff}(X \cap Z \mid f).$$

Proof. The proof of this theorem uses ideas that are similar to those given by Günther and Tammer [22, Th. 5.1].

1°. The inclusion $X \cap \mathrm{Eff}(Y \cap Z \mid f) \subseteq \mathrm{Eff}(X \cap Z \mid f)$ follows by Lemma 9.2.1 taking into account that $X \cap Z \subseteq Y \cap Z$ and $X \cap Z \subseteq X$. Consider $x^0 \in (\mathrm{bd}\,X) \cap \mathrm{Eff}(Y \cap Z \mid f^\oplus)$. By Lemma 9.2.1 and due to the validity of the Assumptions (A1) and (A2), we get

$$S_{\leq}(X\cap Z, f, x^0) = S_{\leq}(Y\cap Z, f, x^0)\cap X$$
$$= S_{\leq}(Y\cap Z, f, x^0)\cap L_{\leq}(Y, \varphi, \varphi(x^0))$$
$$= S_{\leq}(Y\cap Z, f, x^0)\cap L_{\leq}(Y\cap Z, \varphi, \varphi(x^0))$$
$$\subseteq S_{=}(Y\cap Z, f, x^0)\cap L_{=}(Y\cap Z, \varphi, \varphi(x^0))$$
$$= S_{=}(Y\cap Z, f, x^0)\cap L_{=}(Y, \varphi, \varphi(x^0))$$
$$= S_{=}(Y\cap Z, f, x^0)\cap \mathrm{bd}\, X$$
$$\subseteq S_{=}(Y\cap Z, f, x^0)\cap X$$
$$= S_{=}(X\cap Z, f, x^0),$$

hence, $x^0 \in \mathrm{Eff}(X\cap Z \mid f)$ by Lemma 9.2.1.

2°. Consider any $x^0 \in \mathrm{Eff}(X\cap Z \mid f) \subseteq X$. In the case $x^0 \in X \cap \mathrm{Eff}(Y\cap Z \mid f)$, the inclusion is true. Now, we consider the case $x^0 \in X \setminus \mathrm{Eff}(Y\cap Z \mid f)$. If $\mathrm{int}\, X = \emptyset$, then clearly we have $x^0 \in \mathrm{bd}\, X$. In the case $\mathrm{int}\, X \neq \emptyset$, we get $x^0 \in X \setminus \mathrm{cor}\, X \subseteq X \setminus \mathrm{int}\, X = \mathrm{bd}\, X$ from Corollary 9.3.5, taking into account the componentwise semi-strict quasi-convexity of f on $Y\cap Z$. So, we have $x^0 \in (\mathrm{bd}\, X) \setminus \mathrm{Eff}(Y\cap Z \mid f)$. By Lemma 9.2.1 and Assumption (A1), we now get

$$S_{\leq}(Y\cap Z, f, x^0)\cap L_{\leq}(Y\cap Z, \varphi, \varphi(x^0)) = S_{\leq}(Y\cap Z, f, x^0)\cap L_{\leq}(Y, \varphi, \varphi(x^0))$$
$$= S_{\leq}(Y\cap Z, f, x^0)\cap X$$
$$= S_{\leq}(X\cap Z, f, x^0)$$
$$\subseteq S_{=}(X\cap Z, f, x^0)$$
$$= S_{=}(Y\cap Z, f, x^0)\cap X.$$

In the next step, we will prove the equation

$$S_{=}(Y\cap Z, f, x^0)\cap X = S_{=}(Y\cap Z, f, x^0)\cap \mathrm{bd}\, X. \tag{9.6}$$

In the case that $\mathrm{int}\, X = \emptyset$, it is easy to see that (9.6) is satisfied. Assuming now that $\mathrm{int}\, X \neq \emptyset$, it is sufficient to prove $S_{=}(Y\cap Z, f, x^0)\cap \mathrm{int}\, X = \emptyset$ in order to derive (9.6). Assume that the contrary assertion holds, i.e., there exists $x^1 \in \mathrm{int}\, X$ with $x^1 \in S_{=}(Y\cap Z, f, x^0)$. We discuss the following two cases:

Case 1: Assume that $x^1 \in (\mathrm{int}\, X) \setminus \mathrm{Eff}(Y\cap Z \mid f)$ and that f is componentwise semi-strictly quasi-convex on $Y\cap Z$. Then, we get $x^1 \in (\mathrm{int}\, X) \setminus \mathrm{Eff}(X\cap Z \mid f)$ by Proposition 9.3.4. We conclude $x^0 \in X \setminus \mathrm{Eff}(X\cap Z \mid f)$ because of $x^1 \in S_{=}(X\cap Z, f, x^0)$, a contradiction to $x^0 \in \mathrm{Eff}(X\cap Z \mid f)$.

Case 2: Let $x^1 \in \mathrm{Eff}(Y\cap Z \mid f)$. So, we infer $x^0 \in \mathrm{Eff}(Y\cap Z \mid f)$ by $x^1 \in S_{=}(Y\cap Z, f, x^0)$ in contradiction to $x^0 \in X \setminus \mathrm{Eff}(Y\cap Z \mid f)$.

This shows that (9.6) holds true.

Due to $x^0 \in \mathrm{bd}\, X$ and condition (A2), it follows

$$S_{=}(Y\cap Z, f, x^0)\cap \mathrm{bd}\, X = S_{=}(Y\cap Z, f, x^0)\cap L_{=}(Y, \varphi, \varphi(x^0))$$
$$= S_{=}(Y\cap Z, f, x^0)\cap L_{=}(Y\cap Z, \varphi, \varphi(x^0)).$$

In view of the above observations and Lemma 9.2.1, we conclude $x^0 \in (\mathrm{bd}\,X) \cap \mathrm{Eff}(Y \cap Z \mid f^\oplus)$ as requested.

$\square$

As a direct consequence of the preceding Theorem 9.4.8 (put $Z := Y$), we get the result by Günther and Tammer [22, Th. 5.1].

Proposition 9.4.9 (cf. Günther and Tammer [22, Th. 5.1]) *Let (9.5) be satisfied. Suppose that $\varphi : E \to \mathbb{R}$ fulfils Assumptions* (A1) *and* (A2). *Then, the following hold:*

1°. *We have*

$$[X \cap \mathrm{Eff}(Y \mid f)] \cup \left[(\mathrm{bd}\,X) \cap \mathrm{Eff}(Y \mid f^\oplus)\right] \subseteq \mathrm{Eff}(X \mid f).$$

2°. *In the case* $\mathrm{int}\,X \neq \emptyset$, *suppose additionally that $f : E \to \mathbb{R}^m$ is componentwise semi-strictly quasi-convex on* Y. *Then, we have*

$$[X \cap \mathrm{Eff}(Y \mid f)] \cup \left[(\mathrm{bd}\,X) \cap \mathrm{Eff}(Y \mid f^\oplus)\right] \supseteq \mathrm{Eff}(X \mid f).$$

Now, we are able to show the results discovered by Günther and Tammer [22] also for local Pareto efficient solutions, as stated in the next theorem.

Theorem 9.4.10 *Let (9.5) be satisfied. Suppose that $\varphi : E \to \mathbb{R}$ fulfils Assumptions* (A1) *and* (A2). *Then, the following assertions hold:*

1°. *We have*

$$[X \cap \mathrm{Eff}_{\mathrm{loc}}(Y \mid f)] \cup \left[(\mathrm{bd}\,X) \cap \mathrm{Eff}_{\mathrm{loc}}(Y \mid f^\oplus)\right] \subseteq \mathrm{Eff}_{\mathrm{loc}}(X \mid f).$$

2°. *In the case* $\mathrm{int}\,X \neq \emptyset$, *suppose additionally that $f : E \to \mathbb{R}^m$ is componentwise semi-strictly quasi-convex on* Y. *Then, we have*

$$[X \cap \mathrm{Eff}_{\mathrm{loc}}(Y \mid f)] \cup \left[(\mathrm{bd}\,X) \cap \mathrm{Eff}_{\mathrm{loc}}(Y \mid f^\oplus)\right] \supseteq \mathrm{Eff}_{\mathrm{loc}}(X \mid f).$$

Proof.

1°. Assume that $x^0 \in X \cap \mathrm{Eff}_{\mathrm{loc}}(Y \mid f)$, i.e., there exists $\varepsilon > 0$ such that $x^0 \in \mathrm{Eff}(Y \cap B_\varepsilon(x^0) \mid f)$. Then, we get $x^0 \in \mathrm{Eff}(X \cap B_\varepsilon(x^0) \mid f)$ by Theorem 9.4.8 (1°) with the convex set $Z := B_\varepsilon(x^0)$, hence, we conclude that $x^0 \in \mathrm{Eff}_{\mathrm{loc}}(X \mid f)$.

Similar arguments show that $(\mathrm{bd}\,X) \cap \mathrm{Eff}_{\mathrm{loc}}(Y \mid f^\oplus) \subseteq \mathrm{Eff}_{\mathrm{loc}}(X \mid f)$.

2°. Consider $x^0 \in \mathrm{Eff}_{\mathrm{loc}}(X \mid f)$, i.e., there exists $\varepsilon > 0$ such that $x^0 \in \mathrm{Eff}(X \cap B_\varepsilon(x^0) \mid f)$. By Theorem 9.4.8 (2°) with $Z := B_\varepsilon(x^0)$, we get

$$\mathrm{Eff}(X \cap B_\varepsilon(x^0) \mid f) \subseteq \left[X \cap \mathrm{Eff}(Y \cap B_\varepsilon(x^0) \mid f)\right] \cup \left[(\mathrm{bd}\,X) \cap \mathrm{Eff}(Y \cap B_\varepsilon(x^0) \mid f^\oplus)\right]$$
$$\subseteq [X \cap \mathrm{Eff}_{\mathrm{loc}}(Y \mid f)] \cup \left[(\mathrm{bd}\,X) \cap \mathrm{Eff}_{\mathrm{loc}}(Y \mid f^\oplus)\right].$$

□

Remark 9.4.11 *Let (9.5) be satisfied and* $\mathrm{int}\,X = \emptyset$. *Suppose that* $\varphi : E \to \mathbb{R}$ *fulfils Assumptions (A1) and (A2), and that* $f : E \to \mathbb{R}^m$ *is componentwise semi-strictly quasi-convex on* $Y \cap Z$, *where* $Z \subseteq E$ *is convex. Then, the inclusion*

$$\mathrm{Eff}(X \cap Z \mid f) \subseteq X \cap \mathrm{Eff}(Y \cap Z \mid f^\oplus) \tag{9.7}$$

does not hold in general (see Günther and Tammer [22] for the case $Z := Y$). *However, if* $\varphi : E \to \mathbb{R}$ *fulfils Assumption (A3), then actually equality holds in (9.7) without a generalized convexity assumption on* f *(see Theorem 9.4.13).*

In the next lemma, we present sufficient conditions for the fact that a solution $x \in \mathrm{Eff}(X \cap Z \mid f)$ is belonging to $\mathrm{Eff}(Y \cap Z \mid f^\oplus)$.

Lemma 9.4.12 *Let (9.5) be satisfied. Suppose that* $\varphi : E \to \mathbb{R}$ *fulfils Assumption (A1). Consider any* $Z \subseteq E$. *If* $x^0 \in \mathrm{Eff}(X \cap Z \mid f)$ *and* $S_=(X \cap Z, f, x^0) \subseteq L_=(Y, \varphi, \varphi(x^0))$, *then* $x^0 \in X \cap \mathrm{Eff}(Y \cap Z \mid f^\oplus)$.

Proof. Consider $x^0 \in \mathrm{Eff}(X \cap Z \mid f)$ and assume that $S_=(X \cap Z, f, x^0) \subseteq L_=(Y, \varphi, \varphi(x^0))$. The following assertions are equivalent:

- $S_=(X \cap Z, f, x^0) \subseteq L_=(Y, \varphi, \varphi(x^0))$.
- $S_=(Y \cap Z, f, x^0) \cap X \subseteq L_=(Y \cap Z, \varphi, \varphi(x^0))$.
- $S_=(Y \cap Z, f, x^0) \cap X \subseteq S_=(Y \cap Z, f, x^0) \cap L_=(Y \cap Z, \varphi, \phi(x^0))$.

Then, by using condition (A1) and Lemma 9.2.1, we have

$$S_\leq(Y \cap Z, f, x^0) \cap L_\leq(Y \cap Z, \varphi, \varphi(x^0)) = S_\leq(Y \cap Z, f, x^0) \cap L_\leq(Y, \varphi, \varphi(x^0))$$
$$= S_\leq(Y \cap Z, f, x^0) \cap X$$
$$= S_\leq(X \cap Z, f, x^0)$$
$$\subseteq S_=(X \cap Z, f, x^0)$$
$$= S_=(Y \cap Z, f, x^0) \cap X$$
$$\subseteq S_=(Y \cap Z, f, x^0) \cap L_=(Y \cap Z, \varphi, \varphi(x^0)),$$

hence, $x^0 \in X \cap \mathrm{Eff}(Y \cap Z \mid f^\oplus)$ again in view of Lemma 9.2.1. □

In the preceding part of the section, for the penalization φ it was always assumed that the Assumptions (A1) and (A2) are fulfilled. The next main result considers the case that the penalization function φ fulfils Assumption (A3).

Theorem 9.4.13 *Let (9.5) be satisfied. Suppose that $\varphi : E \to \mathbb{R}$ fulfils Assumption (A3). Then, for any $Z \subseteq E$, the following assertions are true:*

1°. *It holds that*

$$[X \cap \mathrm{Eff}(Y \cap Z \mid f)] \cup \left[(\mathrm{bd}\, X) \cap \mathrm{Eff}(Y \cap Z \mid f^{\oplus})\right] \subseteq \mathrm{Eff}(X \cap Z \mid f) \quad (9.8)$$

and

$$\mathrm{Eff}(X \cap Z \mid f) = X \cap \mathrm{Eff}(Y \cap Z \mid f^{\oplus}). \quad (9.9)$$

2°. *In the case* $\mathrm{int}\, X \neq \emptyset$, *suppose additionally that Z is convex and that $f : E \to \mathbb{R}^m$ is componentwise semi-strictly quasi-convex on $Y \cap Z$. Then, we have*

$$[X \cap \mathrm{Eff}(Y \cap Z \mid f)] \cup \left[(\mathrm{bd}\, X) \cap \mathrm{Eff}(Y \cap Z \mid f^{\oplus})\right] \supseteq \mathrm{Eff}(X \cap Z \mid f).$$

Proof. The proof of this theorem is similar to Günther and Tammer [22, Th. 5.3].

1°. First, we prove the equality (9.9). For showing "$\supseteq$" in (9.9), consider $x^0 \in X \cap \mathrm{Eff}(Y \cap Z \mid f^{\oplus})$. By Lemma 9.2.1 and Assumption (A3), we have

$$\begin{aligned}
S_{\leq}(X \cap Z, f, x^0) &= S_{\leq}(Y \cap Z, f, x^0) \cap X \\
&= S_{\leq}(Y \cap Z, f, x^0) \cap L_{\leq}(Y, \varphi, \varphi(x^0)) \\
&= S_{\leq}(Y \cap Z, f, x^0) \cap L_{\leq}(Y \cap Z, \varphi, \varphi(x^0)) \\
&\subseteq S_{=}(Y \cap Z, f, x^0) \cap L_{=}(Y \cap Z, \varphi, \varphi(x^0)) \\
&= S_{=}(Y \cap Z, f, x^0) \cap L_{=}(Y, \varphi, \varphi(x^0)) \\
&= S_{=}(Y \cap Z, f, x^0) \cap X \\
&= S_{=}(X \cap Z, f, x^0).
\end{aligned}$$

We conclude $x^0 \in \mathrm{Eff}(X \cap Z \mid f)$ due to Lemma 9.2.1.

To show the reverse inclusion "$\subseteq$" in (9.9), consider $x^0 \in \mathrm{Eff}(X \cap Z \mid f)$. Since

$$\begin{aligned}
S_{=}(X \cap Z, f, x^0) &= S_{=}(Y \cap Z, f, x^0) \cap X \\
&= S_{=}(Y \cap Z, f, x^0) \cap L_{=}(Y \cap Z, \varphi, \varphi(x^0)) \\
&= S_{=}(Y \cap Z, f, x^0) \cap L_{=}(Y, \varphi, \varphi(x^0)) \\
&\subseteq L_{=}(Y, \varphi, \varphi(x^0)),
\end{aligned}$$

we infer $x^0 \in X \cap \mathrm{Eff}(Y \cap Z \mid f^{\oplus})$ by Lemma 9.4.12. Notice that (A3) implies (A1) in view of Remark 9.4.2. The proof of (9.9) is complete

It remains to show (9.8). The inclusion $X \cap \mathrm{Eff}(Y \cap Z \mid f) \subseteq \mathrm{Eff}(X \cap Z \mid f)$ is obvious, while the second inclusion, namely $(\mathrm{bd}\, X) \cap \mathrm{Eff}(Y \cap Z \mid f^{\oplus}) \subseteq \mathrm{Eff}(X \cap Z \mid f)$, is a direct consequence of (9.9) taking into account the closedness of X. So, (9.8) holds true.

We conclude that assertion 1° is valid.

2°. Consider $x^0 \in \mathrm{Eff}(X \cap Z \mid f) \subseteq X$. In the case $x^0 \in X \cap \mathrm{Eff}(Y \cap Z \mid f)$, the inclusion holds true. Assume now that $x^0 \in X \setminus \mathrm{Eff}(Y \cap Z \mid f)$. If $\mathrm{int}\, X = \emptyset$, then we directly get $x^0 \in \mathrm{bd}\, X$. In the case $\mathrm{int}\, X \neq \emptyset$, we have $x^0 \in X \setminus \mathrm{cor}\, X \subseteq X \setminus \mathrm{int}\, X = \mathrm{bd}\, X$ in view of Corollary 9.3.5, taking into account the componentwise semi-strict quasi-convexity of f on $Y \cap Z$. By Lemma 9.2.1 and Assumption (A3), we conclude

$$\begin{aligned}
S_{\leq}(Y \cap Z, f, x^0) \cap L_{\leq}(Y \cap Z, \varphi, \varphi(x^0)) &= S_{\leq}(Y \cap Z, f, x^0) \cap L_{\leq}(Y, \varphi, \varphi(x^0)) \\
&= S_{\leq}(Y \cap Z, f, x^0) \cap X \\
&= S_{\leq}(X \cap Z, f, x^0) \\
&\subseteq S_{=}(X \cap Z, f, x^0) \\
&= S_{=}(Y \cap Z, f, x^0) \cap X \\
&= S_{=}(Y \cap Z, f, x^0) \cap L_{=}(Y, \varphi, \varphi(x^0)) \\
&= S_{=}(Y \cap Z, f, x^0) \cap L_{=}(Y \cap Z, \varphi, \varphi(x^0)).
\end{aligned}$$

Finally, Lemma 9.2.1 yields $x^0 \in (\mathrm{bd}\, X) \cap \mathrm{Eff}(Y \cap Z \mid f^{\oplus})$.

$\square$

By Theorem 9.4.8 (put $Z := Y$), we directly get the result in Günther and Tammer [22, Th. 5.3].

Proposition 9.4.14 (Günther and Tammer [22, Th. 5.3]) *Let* (9.5) *be satisfied. Suppose that* $\varphi : E \to \mathbb{R}$ *fulfils Assumption* (A3). *Then, the following assertions are true:*

1°. *It holds that*

$$[X \cap \mathrm{Eff}(Y \mid f)] \cup \left[(\mathrm{bd}\, X) \cap \mathrm{Eff}(Y \mid f^{\oplus})\right] \subseteq \mathrm{Eff}(X \mid f) = X \cap \mathrm{Eff}(Y \mid f^{\oplus}).$$

2°. *In the case* $\mathrm{int}\, X \neq \emptyset$, *suppose additionally that* $f : E \to \mathbb{R}^m$ *is componentwise semi-strictly quasi-convex on* Y. *Then, we have*

$$[X \cap \mathrm{Eff}(Y \mid f)] \cup \left[(\mathrm{bd}\, X) \cap \mathrm{Eff}(Y \mid f^{\oplus})\right] \supseteq \mathrm{Eff}(X \mid f).$$

In the next theorem, we present corresponding results for local Pareto efficient solutions.

Theorem 9.4.15 *Let* (9.5) *be satisfied. Suppose that* $\varphi : E \to \mathbb{R}$ *fulfils Assumption* (A3). *Then, the following assertions are true:*

1°. *It holds that*

$$[X \cap \mathrm{Eff}_{\mathrm{loc}}(Y \mid f)] \cup \left[(\mathrm{bd}\, X) \cap \mathrm{Eff}_{\mathrm{loc}}(Y \mid f^{\oplus})\right] \subseteq \mathrm{Eff}_{\mathrm{loc}}(X \mid f) = X \cap \mathrm{Eff}_{\mathrm{loc}}(Y \mid f^{\oplus}).$$

2°. *In the case* $\mathrm{int}\, X \neq \emptyset$, *suppose additionally that* $f : E \to \mathbb{R}^m$ *is componentwise semi-strictly quasi-convex on* Y. *Then, we have*

$$[X \cap \mathrm{Eff}_{\mathrm{loc}}(Y \mid f)] \cup \left[(\mathrm{bd}\, X) \cap \mathrm{Eff}_{\mathrm{loc}}(Y \mid f^{\oplus})\right] \supseteq \mathrm{Eff}_{\mathrm{loc}}(X \mid f).$$

Proof. The proof is similar to the proof of Theorem 9.4.10 by using Theorem 9.4.13 (applied for the convex ball $Z := B_\varepsilon(x^0)$, $x^0 \in X$, $\varepsilon > 0$). □

Remark 9.4.16 *Durea, Strugariu and Tammer [12, Prop. 3.1] also proved, for the case X is a nonempty, closed set in E and $Y := E$, the equality*

$$\mathrm{Eff}_{\mathrm{loc}}(X \mid f) = X \cap \mathrm{Eff}_{\mathrm{loc}}(E \mid f^\oplus)$$

for more general vector-valued functions (that are acting between a normed pre-image space and a partially ordered image space) by using a penalization function φ which fulfils (A3) and (A4).

9.5 Penalization in Multi-objective Optimization with Functional Inequality Constraints

In the preceding part of this chapter, we considered the multi-objective optimization problem ($\mathcal{P}_\Omega$) always with a certain feasible set $\Omega \subseteq E$ (sometimes $\Omega := X$ or $\Omega := Y$, where X and Y were given according to (9.5)). For deriving our main results, the set X was always represented by certain level sets of exactly one penalization function $\varphi : E \to \mathbb{R}$ (see the Assumptions (A1), (A3), (A4)). Now, in the second part of the chapter, we consider the problem ($\mathcal{P}_\Omega$) with a more specific feasible set $\Omega \subseteq E$ that consists of all points that satisfy a system of functional inequality constraints as formulated in the Introduction, i.e.,

$$\Omega := \{x \in S \mid g_1(x) \leq 0, \ldots, g_p(x) \leq 0\} = \bigcap_{i \in I_p} L_\leq(S, g_i, 0), \tag{9.10}$$

where $g_1, \ldots, g_p : S \to \mathbb{R}$ and $S \subseteq E$ are given as in (9.1). In the following, we will apply the results from Section 9.4.2 (here for the case $X := \Omega$ and $Y := S$). Following Günther and Tammer [22, Sec. 7], we consider the penalization function

$$\varphi := \max\{g_1, \ldots, g_p\}. \tag{9.11}$$

Then, Assumption (A4) is satisfied, i.e., we have

$$\Omega = \bigcap_{i \in I_p} L_\leq(S, g_i, 0) = L_\leq(S, \varphi, 0). \tag{9.12}$$

According to the vectorial penalization approach presented in Section 9.4.2, we assume that the following assumptions are fulfilled:

$$\begin{cases} \text{let } S \text{ be a convex set;} \\ \text{let } \varphi \text{ be given by (9.11);} \\ \text{let } \Omega = L_\leq(S, \varphi, 0) \text{ be nonempty and closed.} \end{cases} \tag{9.13}$$

Remark 9.5.1 *As mentioned by Günther and Tammer [22, Sec. 7], the following hold:*

- *If S is closed, and $g_1, \cdots, g_p : E \to \mathbb{R}$ are componentwise lower semi-continuous on S, then φ is lower semi-continuous on S.*

- *If S is closed, and φ is lower semi-continuous on S, then the set Ω is closed as well.*

- *If $g_1, \cdots, g_p : E \to \mathbb{R}$ are convex (respectively, quasi-convex) on S, then φ is convex (respectively, quasi-convex) on S.*

- *If Ω is closed, and φ is quasi-convex or semi-strictly quasi-convex on S, then the set Ω is convex.*

When dealing with functional inequality constraints, a well-known constraint qualifications is given by *Slater's condition*, which can be formulated in our case as

$$L_<(S, \varphi, 0) = \bigcap_{i \in I_p} L_<(S, g_i, 0) \neq \emptyset. \tag{9.14}$$

In this section, we are interested in relationships between the initial problem $(\mathcal{P}_\Omega)$ with (not necessarily convex) feasible set Ω and the objective function $f = (f_1, \ldots, f_m)$, and two related problems $(\mathcal{P}_S)$ and $(\mathcal{P}_S^\oplus)$ with convex feasible set S and the objective functions $f = (f_1, \ldots, f_m)$ and

$$f^\oplus = (f_1, \ldots, f_m, \varphi) = (f_1, \ldots, f_m, \max\{g_1, \ldots, g_p\}),$$

respectively.

9.5.1 The Case of a Not Necessarily Convex Feasible Set

We start by presenting results for the general case that $\Omega \subseteq E$ (according to (9.10) and (9.12)) is not necessarily a convex feasible set. The next proposition is a consequence of Proposition 9.4.9.

Proposition 9.5.2 (Günther and Tammer [22, Th. 7.1]) *Let* (9.13) *be satisfied, and assume that φ fulfils Assumption* (A2) *(with $X := \Omega$ and $Y := S$). Then, the following assertions hold:*

1°. *We have*

$$[\Omega \cap \mathrm{Eff}(S \mid f)] \cup \left[(\mathrm{bd}\,\Omega) \cap \mathrm{Eff}(S \mid f^\oplus)\right] \subseteq \mathrm{Eff}(\Omega \mid f).$$

2°. *Let $L_<(S, \varphi, 0)$ be an open set, and let Slater's condition* (9.14) *be satisfied. Suppose that $f : E \to \mathbb{R}^m$ is componentwise semi-strictly quasi-convex on S. Then, we have*

$$[\Omega \cap \mathrm{Eff}(S \mid f)] \cup \left[(\mathrm{bd}\,\Omega) \cap \mathrm{Eff}(S \mid f^\oplus)\right] \supseteq \mathrm{Eff}(\Omega \mid f).$$

Notice that $L_<(S,\varphi,0) = L_<(E,\varphi,0) \cap S$ is an open set if S is open and φ is upper semi-continuous on E.

Under the assumption that the penalization function φ is explicitly quasi-convex on S, we directly get the following result by Proposition 9.5.2.

Corollary 9.5.3 (Günther and Tammer [22, Cor. 7.2]) *Let* (9.13) *be satisfied. Assume that* $L_<(S,\varphi,0)$ *is an open set, and that Slater's condition* (9.14) *holds. Suppose that φ is explicitly quasi-convex on S. If $f: E \to \mathbb{R}^m$ is componentwise semi-strictly quasi-convex on S, then*

$$\mathrm{Eff}(\Omega \mid f) = [\Omega \cap \mathrm{Eff}(S \mid f)] \cup \left[(\mathrm{bd}\,\Omega) \cap \mathrm{Eff}(S \mid f^{\oplus})\right].$$

We conclude, for the special case $S = E$, the following result.

Corollary 9.5.4 (Günther and Tammer [22, Cor. 7.3]) *Let* (9.13) *be satisfied and let $S = E$. Suppose that φ is semi-strictly quasi-convex and continuous on E. Assume that Slater's condition* (9.14) *holds. If $f: E \to \mathbb{R}^m$ is componentwise semi-strictly quasi-convex on E, then*

$$\mathrm{Eff}(\Omega \mid f) = [\Omega \cap \mathrm{Eff}(E \mid f)] \cup \left[(\mathrm{bd}\,\Omega) \cap \mathrm{Eff}(E \mid f^{\oplus})\right].$$

Remark 9.5.5 *The results given in Proposition 9.5.2 and its Corollaries 9.5.3 and 9.5.4 also hold for the concept of local Pareto efficiency taking into account Theorem 9.4.10.*

Next, we study the application of the vectorial penalization approach to certain multi-objective location problems.

Example 9.5.6 *Assume that $E = \mathbb{R}^n$ is endowed with the Euclidean norm $||\cdot||_2$. Let the components of the objective function $f = (f_1, \cdots, f_m): \mathbb{R}^n \to \mathbb{R}^m$ be given by*

$$f_i(x) := ||x - a^i||_2, \ a^i \in \mathbb{R}^n, \quad \text{for all } i \in I_m.$$

In addition, let the feasible set $\Omega \subseteq \mathbb{R}^n$ be nonempty, closed, and convex. For this special class of convex point-objective location problems, it is known that the so-called "projection property" holds (see, e.g., Ndiaye and Michelot [32, Cor. 4.2]), i.e.,

$$\mathrm{Eff}(\Omega \mid f) = \mathrm{Proj}_\Omega(\mathrm{Eff}(\mathbb{R}^n \mid f)) = \mathrm{Proj}_\Omega(\mathrm{conv}\{a^1, \cdots, a^m\}),$$

where $\mathrm{Proj}_\Omega(\mathrm{conv}\{a^1, \cdots, a^m\})$ denotes the projection of the convex hull $\mathrm{conv}\{a^1, \cdots, a^m\}$ onto Ω. According to Ndiaye and Michelot [32, p. 297], the projection property can fail when the Euclidean norm is replaced by a not strictly-convex norm. In particular, in this case, the vectorial penalization approach is useful for deriving geometrical descriptions of solutions sets of convex multi-objective optimization problems. Assume, in addition, that Ω has nonempty interior, i.e., there exists $d \in \mathrm{int}\,\Omega$. Then, we can define the penalization function $\varphi: E \to \mathbb{R}$ by $\varphi(x) := \mu_B(x - d)$ for all $x \in \mathbb{R}^n$, where μ_B is a certain Minkowski gauge function associated to the set $B := -d + \Omega$ (see Example 9.4.3). So, we get

$$\mathrm{Eff}(\Omega \mid f) = [\Omega \cap \mathrm{Eff}(\mathbb{R}^n \mid f)] \cup \left[(\mathrm{bd}\,\Omega) \cap \mathrm{Eff}(\mathbb{R}^n \mid f^{\oplus})\right].$$

Notice that Ω consists of all points $x \in \mathbb{R}^n$ that satisfy the functional inequality constraint $\mu_B(x-d) - 1 \leq 0$. As shown by Günther and Tammer [21] (see also Günther [20] for more details), the unconstrained problems $(\mathcal{P}_{\mathbb{R}^n})$ and $(\mathcal{P}_{\mathbb{R}^n}^\oplus)$ have a similar structure for certain classes of multi-objective location problems involving mixed Minkowski gauges (e.g., point-objective location problems, multi-objective min-sum location problems, or multi-objective min-max location problems). Therefore, $(\mathcal{P}_\Omega)$ can effectively be solved by using known algorithms for the unconstrained case.

Example 9.5.7 (Günther and Tammer [21, Ex. 5]) *Now, let us consider the special case that $E = \mathbb{R}^2$ (i.e., $n = 2$) is endowed with the Manhattan norm $\|\cdot\|_1 : \mathbb{R}^2 \to \mathbb{R}$. Let the components of the objective function $f : \mathbb{R}^2 \to \mathbb{R}^3$ be given by*

$$f_i(x) := \|x - a^i\|_1, \, a^i \in \mathbb{R}^2, \quad \text{for all } i \in I_3.$$

In addition, let Ω consists of all points in $\mathbb{R}^2$ that satisfy

$$g_1(x) := \|x - x'\|_1 - r \leq 0, \, x' \in \mathbb{R}^2, \, r > 0,$$

i.e., Ω is a closed Manhattan ball centered at $x' \in \mathbb{R}^2$ with positive radius $r > 0$, as shown in Figure 9.1. As mentioned in Example 9.5.6, the projection property does not hold here since, in contrast to the Euclidean norm, the Manhattan norm is not strictly convex. However, using the vectorial penalization approach, where the penalization function $\varphi : \mathbb{R}^2 \to \mathbb{R}$ is given by $\varphi(x) := \|x - x'\|_1$ for all $x \in \mathbb{R}^2$, we are able to compute

$$\text{Eff}(\Omega \mid f) = \Big[\Omega \cap \text{Eff}(\mathbb{R}^2 \mid f)\Big] \cup \Big[(\text{bd}\,\Omega) \cap \text{Eff}(\mathbb{R}^2 \mid f^\oplus)\Big]$$

by generating $\text{Eff}(\mathbb{R}^2 \mid f)$ and $\text{Eff}(\mathbb{R}^2 \mid f^\oplus)$ via the Rectangular Decomposition Algorithm developed by Alzorba et al. [1]. Figure 9.1 shows the procedure for computing $\text{Eff}(\Omega \mid f)$ by applying the vectorial penalization approach.

Furthermore, we are going to emphasize the importance of the vectorial penalization approach for deriving solutions of nonconvex multi-objective optimization problems (in particular in multi-objective location theory).

Example 9.5.8 (Günther [19, 20]) *Assume that $E = \mathbb{R}^n$ and consider the objective function $f = (f_1, \cdots, f_m) : \mathbb{R}^n \to \mathbb{R}^m$ as defined in Example 9.5.6. For the functional constraints involved in Ω, we consider*

$$g_i(x) := -\|x - d^i\|_2 + r_i, \, d^i \in \mathbb{R}^n, \, r_i > 0, \text{for all } i \in I_p,$$

i.e., Ω consists of all points in $\mathbb{R}^n$ that satisfy

$$\|x - d^1\|_2 \geq r_1, \cdots, \|x - d^p\|_2 \geq r_p.$$

In view of the literature, such a problem can be seen as a special point-objective location problem involving some forbidden regions that are given by the interiors of the convex sets $D_i := \{x \in \mathbb{R}^n \mid \|x - d^i\|_2 \leq r_i\}$, $i \in I_p$. So, Ω can be written as an intersection of a finite number of reverse convex sets,

$$\Omega = \bigcap_{i \in I_p} \mathbb{R}^n \setminus \text{int}\, D_i.$$

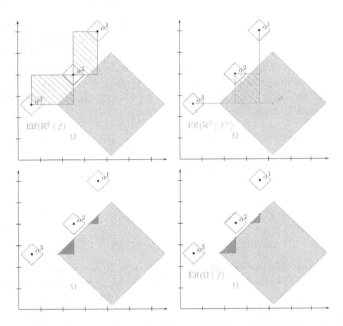

Figure 9.1: Construction of the set of Pareto efficient solutions of a constrained point-objective location problem involving the Manhattan norm.

Adapting the idea by Drezner, Drezner and Schöbel [9] to multi-objective location problems, the special case

$$\begin{cases} m = p, \\ r := r_1 = \cdots = r_m, \\ a^i = d^i \text{ for all } i \in I_m, \end{cases}$$

for our example problem $(\mathcal{P}_\Omega)$ *can be seen as a kind of "Obnoxious Facility Location Model" (i.e., the facility x must be located at least a given distance r from every demand point* a^i*; see also Günther [19, Sec. 5]). Since for p > 1 the location problem involves multiple forbidden regions, it is convenient to introduce a family of penalized multi-objective optimization problems* $\{(\mathcal{P}_{\mathbb{R}^n}^{\oplus_i}) \mid i \in I_p\}$, *where*

$$\begin{cases} f^{\oplus_i}(x) = (f_1(x), \ldots, f_m(x), \varphi_i(x)) \\ \qquad = (||x - a^1||_2, \ldots, ||x - a^m||_2, -||x - d^i||_2) \to \min \text{ w.r.t. } \mathbb{R}_+^{m+1} \\ x \in \mathbb{R}^n, \end{cases} \quad (\mathcal{P}_{\mathbb{R}^n}^{\oplus_i})$$

and $\varphi_i : \mathbb{R}^n \to \mathbb{R}$ *is given by* $\varphi_i(x) := -||x - d^i||_2$ *for all* $x \in \mathbb{R}^n$. *Then, due to Günther [19, Lem. 5.3, Th. 5.1], the following two assertions hold:*

1°. *We have*

$$\text{Eff}(\Omega \mid f) \supseteq [\Omega \cap \text{Eff}(\mathbb{R}^n \mid f)] \cup \left[\bigcup_{i \in I_p} \Omega \cap (\text{bd} D_i) \cap \text{Eff}(\mathbb{R}^n \mid f^{\oplus_i}) \right].$$

2°. *Assume that* $\text{int} D_1, \cdots, \text{int} D_p$ *are pairwise disjoint (i.e.,* $(\text{int} D_i) \cap (\text{int} D_j) = \emptyset$ *for all* $i, j \in I_p$, $i \neq j$). *Consider the index set*

$$J := \{j \in I_p \mid d^j \in \text{rint}(\text{conv}\{a^1, \cdots, a^m\})\},$$

where $\text{rint}(\text{conv}\{a^1, \cdots, a^m\})$ *denotes the relative interior of the convex hull* $\text{conv}\{a^1, \cdots, a^m\}$ *of the given points* $a^1, \cdots, a^m$. *Then, we actually have*

$$\text{Eff}(\Omega \mid f) = [\Omega \cap \text{Eff}(\mathbb{R}^n \mid f)] \cup \left[\bigcup_{i \in I_p} (\text{bd} D_i) \cap \text{Eff}(\mathbb{R}^n \mid f^{\oplus_i}) \right]$$

$$= \Omega \cap \text{conv}\{a^1, \cdots, a^m\}$$

$$\cup \left[\bigcup_{i \in I_p \setminus J} (\text{bd} D_i) \cap \left(\text{conv}\{a^1, \cdots, a^m\} + \text{cone}\left(\text{conv}\{a^1, \cdots, a^m\} - d^i \right) \right) \right]$$

$$\cup \left[\bigcup_{j \in J} \text{bd} D_j \right],$$

where, for any $\Omega' \subseteq \mathbb{R}^n$, *we use the notation* $\text{cone}(\Omega')$ *for the cone generated by the set* Ω'.

Figure 9.2 shows an example (for the case $n = 2$, $m = 3$ *and* $p = 2$) *with*

$$\text{Eff}(\Omega \mid f) = \left[\Omega \cap \text{conv}\{a^1, a^2, a^3\} \right]$$

$$\cup \text{bd} D_1$$

$$\cup \left[(\text{bd} D_2) \cap \left(\text{conv}\{a^1, a^2, a^3\} + \text{cone}\left(\text{conv}\{a^1, a^2, a^3\} - d^2 \right) \right) \right],$$

where $d^1 \in \text{int}(\text{conv}\{a^1, a^2, a^3\})$ *and* $d^2 \notin \text{conv}\{a^1, a^2, a^3\}$ *(i.e.,* $J = \{1\}$).

Remark 9.5.9 *Instead of considering a penalization function given by (9.11) one could also study a vector-valued penalization function* $\varphi_v := (g_1, \cdots, g_p) : E \to \mathbb{R}^p$. *In this way, the penalized problem is*

$$\begin{cases} (f, \varphi_v)(x) = (f_1(x), \cdots, f_m(x), g_1(x), \cdots, g_p(x)) \to \min \text{ w.r.t. } \mathbb{R}_+^{m+p} \\ x \in E. \end{cases} \quad (9.15)$$

Klamroth and Tind [26] discussed some relationships between the problems $(\mathcal{P}_\Omega)$ *and (9.15) for the scalar case* $m = 1$. *Furthermore, Durea, Strugariu and Tammer [12] studied the problem (9.15) (in a more general setting). Using [12, Prop. 3.2], we have the following result:*

Proposition 9.5.10 (cf. Durea, Strugariu and Tammer [12, Prop. 3.2]) *Let (9.13) be satisfied, and put* $S := E$. *Consider* $x^0 \in \Omega$ *such that* $g_1(x^0) = \cdots = g_p(x^0) = 0$. *Then, the following two assertions are equivalent:*

1°. x^0 *is a local Pareto efficient solution of* $(\mathcal{P}_\Omega)$.

234 ■ *Variational Analysis and Set Optimization*

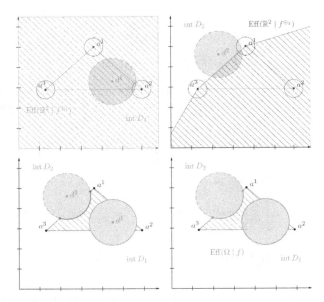

Figure 9.2: Construction of the set of Pareto efficient solutions of a point-objective location problem involving the Euclidean norm, where two forbidden regions are considered in the model.

2°. x^0 *is a local Pareto efficient solution of* (9.15).

By the approach based on the penalization function as in (9.11), *we can give characterizations for local and global Pareto efficient solutions of* ($\mathcal{P}_\Omega$) *(without restriction to the set* $\bigcap_{i \in I_p} L_=(E, g_i, 0)$) *by using problem* ($\mathcal{P}_E^\oplus$) *(see Proposition 9.5.2 and Remark 9.5.5). In contrast to* (9.15), *with the approach based on* ($\mathcal{P}_E^\oplus$), *the image space dimension is always increased only by one.*

9.5.2 The Case of a Convex Feasible Set But Without Convex Representation

In this section, let the following assumptions be fulfilled:

$$\begin{cases} \text{let } E \text{ be a real Banach space;} \\ \text{let } S := E; \\ \text{let } \varphi \text{ be given by (9.11);} \\ \text{let } g_1, \cdots, g_p \text{ be differentiable;} \\ \text{let } \Omega \text{ (as in (9.12)) be nonempty, closed and convex.} \end{cases} \quad (9.16)$$

Lassere studied convex scalar optimization problems in [27, 28, 29] and observed that the geometry of a convex feasible set Ω is more important than its representation. Lassere illustrated this fact by considering

$$\Omega := \{x = (x_1, x_2) \in \mathbb{R}^2 \mid g_1(x) := 1 - x_1 x_2 \leq 0,$$
$$g_2(x) := -x_1 \leq 0,$$
$$g_3(x) := -x_2 \leq 0\},$$

which is a convex set, but g_1 is not convex on $\mathbb{R}^2$. By using a so-called non-degeneracy condition that is, for any $i \in I_p$, we have

$$x \in \Omega, \ g_i(x) = 0 \implies \nabla g_i(x) \neq 0,$$

and Slaters's condition (as given in (9.14)), Lassere [27] proved (for the scalar convex case and $E = \mathbb{R}^n$) that the Karush-Kuhn-Tucker conditions are necessary and sufficient for global optimality. Several interesting recent works (see, e.g., Dutta and Lalitha [10], Durea and Strugariu [11], and Martínez-Legaz [34]) discussed possible extensions of Lassere's results [27, 28, 29]. In order to apply results from Durea and Strugariu [11], we need the celebrated Mangasarian-Fromowitz constraint condition at a point $x \in \Omega$ that is, there exists $u \in \mathbb{R}^p$ such that, for any $i \in I_p$, we have

$$g_i(x) = 0 \implies \langle \nabla g_i(x), u \rangle < 0.$$

Durea and Strugariu [11, Sec. 2] studied the relationships between the Mangasarian-Fromowitz constraint condition (at one point), Slaters's condition, and Lassere's non-degeneracy condition (at one point). The next proposition shows one important result of the work by Durea and Strugariu [11]:

Proposition 9.5.11 (Durea and Strugariu [11, Prop. 2.4]) *Let* (9.16) *be fulfilled. Consider $x^0 \in \Omega$. Then, the following two assertions are equivalent:*

1°. *Lassere's non-degeneracy condition at x^0 and Slater's condition hold.*

2°. *The Mangasarian-Fromowitz constraint qualification condition at x^0 is satisfied.*

According to Durea and Strugariu [11, Prop. 2.6], we get the following interesting characterization of Slater's condition under the validity of Lassere's non-degeneracy condition.

Lemma 9.5.12 (cf. Durea and Strugariu [11, Prop. 2.6]) *Let* (9.16) *be satsified. Assume that Lassere's non-degeneracy condition holds. Then, the following assertions are equivalent:*

1°. *Slater's condition holds.*

2°. $\operatorname{int} \Omega \neq \emptyset$.

3°. $\operatorname{int} \Omega = \{x \in E \mid \forall i \in I_p : g_i(x) < 0\} = L_<(E, \varphi, 0)$.

4°. φ *fulfils Assumptions* (A4) *and* (A5) *(with $Y := E$ and $X := \Omega$).*

Proof. The equivalence between 1°, 2° and 3° is shown in Durea and Strugariu [11, Prop. 2.6, Rem. 2.7]. Moreover, if Ω is closed (i.e., $\mathrm{cl}\,\Omega = \Omega = L_\le(E,\varphi,0)$), then $\mathrm{int}\,\Omega = L_<(E,\varphi,0)$ if and only if $\mathrm{bd}\,\Omega = L_=(E,\varphi,0)$, which shows 3° is equivalent to 4°. □

Now, we are able to state the main theorem of the section.

Theorem 9.5.13 *Let (9.16) be satisfied. Assume that the Mangasarian-Fromowitz constraint qualification condition holds. If $f: E \to \mathbb{R}^m$ is componentwise semi-strictly quasiconvex on E, then*

$$\mathrm{Eff}(\Omega \mid f) = [\Omega \cap \mathrm{Eff}(E \mid f)] \cup \left[(\mathrm{bd}\,\Omega) \cap \mathrm{Eff}(E \mid f^\oplus)\right],$$

$$\mathrm{Eff}_{\mathrm{loc}}(\Omega \mid f) = [\Omega \cap \mathrm{Eff}_{\mathrm{loc}}(E \mid f)] \cup \left[(\mathrm{bd}\,\Omega) \cap \mathrm{Eff}_{\mathrm{loc}}(E \mid f^\oplus)\right].$$

Proof. Follows directly from Proposition 9.5.2 (applied for the case $S := E$), Proposition 9.5.11, Lemma 9.5.12, and Remark 9.5.5 taking into account that (A4) and (A5) imply (A1) and (A2). Notice that $L_<(E,\varphi,0)$ is open since φ is continuous as a maximum of a finite number of continuous functions. □

We end this section by illustrating the previous results for a simple example in $\mathbb{R}^2$.

Example 9.5.14 *Consider two sets in $E = \mathbb{R}^2$, namely a circle $D_1 \subseteq \mathbb{R}^2$ with radius $r_1 > 0$ and center $d^1 := (d_1^1, d_2^1) \in D_1$, and a half-space $D_2 \subseteq \mathbb{R}^2$, according to Figure 9.3. Hence, Ω consists of all points $x = (x_1, x_2) \in \mathbb{R}^2$ that satisfy*

$$g_1(x) := (x_1 - d_1^1)^2 + (x_2 - d_2^1)^2 - r_1^2 \le 0,$$

$$g_2(x) := \alpha x_1 + \beta x_2 - \alpha x_1^0 + \beta x_2^0 \le 0,$$

where $(\alpha, \beta) \in \mathbb{R}^2 \setminus \{(0,0)\}$ and $x^0 = (x_1^0, x_2^0) \in \mathbb{R}^2$. Notice that $\Omega = D_1 \cap D_2$ is nonempty, closed and convex. Assume that $f = f_1 : \mathbb{R}^2 \to \mathbb{R}$ is given by $f(x) := \|x - x'\|_2$ for all $x \in \mathbb{R}^2$. As to see in Figure 9.3,

$$\mathrm{Eff}(\mathbb{R}^2 \mid f) = \{x'\} \not\subseteq \Omega,$$

and, for any $x \in \mathrm{bd}\,\Omega = L_=(\mathbb{R}^2, \varphi, 0)$, we have

$$x \in \mathrm{Eff}(\mathbb{R}^2 \mid (f, \max\{g_1, g_2\}))$$
$$\iff L_\le(\mathbb{R}^2, f, f(x)) \cap L_<(\mathbb{R}^2, \varphi, \varphi(x)) \subseteq L_=(\mathbb{R}^2, f, f(x)) \cap L_=(\mathbb{R}^2, \varphi, \varphi(x));$$
$$\iff L_\le(\mathbb{R}^2, f, f(x)) \cap \Omega \subseteq L_=(\mathbb{R}^2, f, f(x)) \cap \mathrm{bd}\,\Omega;$$
$$\iff x = x^0.$$

So, we conclude

$$\mathrm{argmin}_{x \in \Omega} f(x) = \mathrm{Eff}(\Omega \mid f) = \left[\Omega \cap \mathrm{Eff}(\mathbb{R}^2 \mid f)\right] \cup \left[(\mathrm{bd}\,\Omega) \cap \mathrm{Eff}(\mathbb{R}^2 \mid f^\oplus)\right]$$
$$= (\mathrm{bd}\,\Omega) \cap \mathrm{Eff}(\mathbb{R}^2 \mid f^\oplus)$$
$$= \{x^0\}.$$

Notice, in this example, both Lassere's non-degeneracy condition and Slater's condition hold, or equivalently, the Mangasarian-Fromowitz constraint qualification is satisfied in view of Proposition 9.5.11.

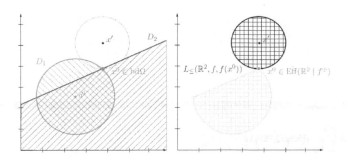

Figure 9.3: The point x^0 is a Pareto efficient solution of the problem $(\mathcal{P}_\Omega)$ given in Example 9.5.14.

9.6 Conclusions

In this chapter, we studied the method of vectorial penalization in multi-objective optimization which is useful for deriving solutions of both convex and nonconvex constrained optimization problems. Important observations for Pareto efficiency, with respect to different constraint sets, allowed us to extend some results derived by Günther and Tammer [21, 22], and Durea, Strugariu and Tammer [12]. For problems involving functional inequality constraints, we extended the results by Günther and Tammer [22] to the concept of local Pareto efficiency. In particular, for the case of a convex feasible set but without convex representation, we derived new insights using recent results obtained by Durea and Strugariu [11]. In view of Example 9.5.14, one can guess that the vectorial penalization approach could also be of significance in the context of constrained single-objective optimization for developing numerical algorithms.

We notice that our derived results, related to global optimality in Sections 9.3, 9.4.2 and 9.5.1, are still valid when the normed space E is replaced by any linear topological space. Concerning local optimality, it is possible to replace E by any locally convex linear space (the corresponding convex neighbourhoods will be used within the definition of $\text{Eff}_{\text{loc}}(\Omega \mid f)$).

Let us end the chapter by pointing out some important research directions:

- To derive a vectorial penalization approach for more general vector optimization problems (that involve not necessarily the natural ordering).

- To identify the relationships between the vectorial penalization approach (see problem (9.4)) and a Clarke-type penalization approach (see problem (9.3)) in multi-objective optimization in detail.

- To consider other types of functional constraints, e.g., functional equality constraints or a mixture of functional inequality and equality constraints.

Acknowledgements. The author thanks Christiane Tammer and Nicolae Popovici for the inspiring joint works in the fields of multi-objective optimization and location theory.

References

[1] S. Alzorba, C. Günther, N. Popovici and C. Tammer. A new algorithm for solving planar multi-objective location problems involving the Manhattan norm. *European J. Oper. Res.*, 258(1): 35–46, 2017.

[2] M. Apetrii, M. Durea and R. Strugariu. A new penalization tool in scalar and vector optimization. *Nonlinear Anal.*, 107: 22–33, 2014.

[3] O. Bagdasar and N. Popovici. Local maximum points of explicitly quasiconvex functions. *Optim. Lett.*, 9(4): 769–777, 2015.

[4] O. Bagdasar and N. Popovici. Local maximizers of generalized convex vector-valued functions. *J. Nonlinear Convex Anal.*, 18(12): 2229–2250, 2017.

[5] O. Bagdasar and N. Popovici. Unifying local-global type properties in vector optimization. *J. Global Optim.*, 72(2): 155–179, 2018.

[6] V. Barbu and T. Precupanu. *Convexity and Optimization in Banach Spaces*. Springer, Netherlands, 2012.

[7] A. Cambini and L. Martein. *Generalized Convexity and Optimization: Theory and Applications*. Springer, Berlin, Heidelberg, 2009.

[8] F. H. Clarke. *Optimization and Nonsmooth Analysis*. Wiley-Interscience, New York, 1983.

[9] T. Drezner, Z. Drezner and A. Schöbel. The Weber obnoxious facility location model: A Big Arc Small Arc approach. *Computers & Operations Research*, 98: 240–250, 2018.

[10] J. Dutta and C. S. Lalitha. Optimality conditions in convex optimization revisited. *Optim. Lett.*, 7: 221–229, 2013.

[11] M. Durea and R. Strugariu. Optimality conditions and a barrier method in optimization with convex geometric constraint. *Optim. Lett.*, 12: 923–931, 2018.

[12] M. Durea, R. Strugariu and C. Tammer. On some methods to derive necessary and sufficient optimality conditions in vector optimization. *J. Optim. Theory Appl.*, 175(3): 738–763, 2017.

[13] M. Ehrgott. *Multicriteria Optimization*. Springer, Berlin, Heidelberg, 2005.

[14] I. I. Eremin. Penalty method in convex programming. *Soviet Math Dolk*, 8: 459–462, 1966.

[15] J. Fliege. The effects of adding objectives to an optimisation problem on the solution set. *Optim. Res. Lett.*, 35(6): 782–790, 2007.

[16] C. Gerstewitz (Tammer). Nichtkonvexe Dualität in der Vektoroptimierung. *Wiss. Zeitschr. TH Leuna-Merseburg*, 25: 357–364, 1983.

[17] C. Gerth (Tammer) and P. Weidner. Nonconvex separation theorems and some applications in vector optimization. *J. Optim. Theory Appl.*, 67(2): 297–320, 1990.

[18] A. Göpfert, H. Riahi, C. Tammer and C. Zălinescu. *Variational Methods in Partially Ordered Spaces*. Springer, New York, 2003.

[19] C. Günther. Pareto efficient solutions in multi-objective optimization involving forbidden regions. *Investigación Oper.*, 39(3): 353–390, 2018.

[20] C. Günther. *On generalized-convex constrained multi-objective optimization and application in location theory*. Dissertation, Martin Luther University Halle-Wittenberg, 2018.

[21] C. Günther and C. Tammer. Relationships between constrained and unconstrained multi-objective optimization and application in location theory. *Math. Meth. Oper. Res.*, 84(2): 359–387, 2016.

[22] C. Günther and C. Tammer. On generalized-convex constrained multi-objective optimization. *Pure Appl. Funct. Anal.*, 3(3): 429–461, 2018.

[23] J. B. Hiriart-Urruty. New concepts in nondifferentiable programming. *Mémoires de la Société Mathématique de France*, 60: 57–85, 1979.

[24] J. Jahn. *Vector Optimization—Theory, Applications, and Extensions*. Springer, Berlin, Heidelberg, 2011.

[25] A. A. Khan, C. Tammer and C. Zălinescu. *Set-valued Optimization: An Introduction with Applications*. Springer, Berlin, Heidelberg, 2015.

[26] K. Klamroth and J. Tind. Constrained optimization using multiple objective programming. *J. Global Optim.*, 37(3): 325–355, 2007.

[27] J. B. Lassere. On representations of the feasible set in convex optimization. *Optim. Lett.*, 4: 1–5, 2010.

[28] J. B. Lassere. On convex optimization without convex representation. *Optim. Lett.*, 5: 549–556, 2011.

[29] J. B. Lassere. Erratum to: On convex optimization without convex representation. *Optim. Lett.*, 8: 1795–1796, 2014.

[30] C. G. Liu, K. F. Ng and W. H. Yang. Merit functions in vector optimization. *Math Program.*, 119(2): 215–237, 2009.

[31] D. T. Luc and S. Schaible. Efficiency and generalized concavity. *J. Optim. Theory Appl.*, 94(1): 147–153, 1997.

[32] M. Ndiaye and C. Michelot. Efficiency in constrained continuous location. *European J. Oper. Res.*, 104(2): 288-298, 1998.

[33] M. M. Mäkelä, V. P. Eronen and N. Karmitsa. On nonsmooth optimality conditions with generalized convexities. *Optimization in Science and Engineering*, 333–357, 2014.

[34] J. E. Martínez-Legaz. Optimality conditions for pseudoconvex minimization over convex sets defined by tangentially convex constraints. *Optim. Lett.*, 9(5): 1017–1023, 2015.

[35] N. Popovici. Pareto reducible multicriteria optimization problems. *Optimization*, 54(3): 253–263, 2005.

[36] J. Puerto and A. M. Rodríguez-Chía. Geometrical description of the weakly efficient solution set for multicriteria location problems. *Ann. Oper. Res.*, 111(1-4): 181–196, 2002.

[37] J. J. Ye. The exact penalty principle. *Nonlinear Anal.*, 75(3): 1642–1654, 2012.

[38] A. Zaffaroni. Degrees of efficiency and degrees of minimality. *SIAM J. Control. Optim.*, 42(3): 1071–1086, 2003.

[39] W. I. Zangwill. Nonlinear programming via penalty functions. *Management Sci.*, 13(5): 344–358, 1967.

Chapter 10

On Classes of Set Optimization Problems which are Reducible to Vector Optimization Problems and its Impact on Numerical Test Instances

Gabriele Eichfelder
Institute for Mathematics, Technische Universität Ilmenau, Po 10 05 65, D-98684 Ilmenau, Germany. `Gabriele.Eichfelder@tu-ilmenau.de`

Tobias Gerlach
Institute for Mathematics, Technische Universität Ilmenau, Po 10 05 65, D-98684 Ilmenau, Germany. `Tobias.Gerlach@tu-ilmenau.de`

10.1 Introduction

Set optimization problems can be considered as a significant generalization and unification of scalar and vector optimization problems and have numerous applications in optimal control, operations research, and economics equilibrium (see for instance [2, 13]). Especially set optimization using the set approach has recently gained increasing interest due to its practical relevance. In this problem class, one studies optimization problems with a set-valued objective map and defines optimality based on a direct comparison of the images of the objective function, which are sets here. Hence one needs relations for comparing sets. A large number of such relations are proposed in the literature [12]. Examples are the possibly less or the certainly less order relation. However, the l-less order relation, the u-less order relation, and most of all the set less order relation are considered to be of highest interest from the practical point of view. For the latter relation one says that a set A is less or equal than a set B if any element of the set A has an element in the set B, which is worse, and if to any element of the set B there exists an element of the set A, which is better. Therefore one also needs a way to compare the elements of the sets. Thus, one assumes in general that the space is equipped with a partial ordering.

It is obvious that it is numerically difficult to compare sets. For convex sets, Jahn has proposed in [10] to use a characterization with supporting hyperplanes. For numerical calculations he proposes to do a single comparison based on the optimal values of 4000 linear optimization problems over these convex sets. This illustrates that the development of numerical algorithms for set optimization problems is very challenging. Nevertheless, some first algorithms have been proposed in the literature.

The first algorithm for unconstrained set optimization problems and the l-less order relation was presented by Löhne and Schrage [17]. This approach is for linear problems only and requires an objective map F with a polyhedral convex graph and a representation by inequalities of this graph. Recently, a derivative free descent method was proposed by Jahn [11], based on the comparison mentioned above. This was extended to nonconvex sets by Köbis and Köbis [14]. Both methods aim for the determination of one single minimal solution of the set optimization problem.

Clearly, set optimization problems, as vector optimization problems, have in general an infinite number of minimal solutions. Hence, one is not only interested in finding one of these solutions but in finding a representation of the set of all optimal solutions. By varying the starting points in the above methods a representation of the set of optimal solutions can be determined. For such approaches it is important to also have test instances in order to verify whether such a representation was successfully obtained.

In the following, we study set optimization problems which are so simple that they have an equivalent reformulation as a vector optimization problem. They can be used to construct new test instances.

In Section 10.2, we state the basic definitions and concepts from set optimization which we need in the following. We also recall a result by Jahn which is helpful for some of the following proofs. Set optimization problems being reducible to vector optimization problems are defined and studied in Section 10.3. Based on this, we make in the final Section 10.4 some suggestions on how the results can be used for the construction of set-valued test instances based on vector-valued or scalar-valued optimization problems.

10.2 Basics of Vector and Set Optimization

Throughout this chapter, we assume that Y is a locally convex real linear space which is partially ordered by a pointed, convex, and closed nontrivial cone C. Recall that a nonempty subset C of Y is called a cone if $y \in C$ and $\lambda \geq 0$ imply $\lambda y \in C$. A cone C is called pointed, if $C \cap (-C) = \{0_Y\}$. We denote by Y^* the topological dual space of Y and by

$$C^* := \{\ell \in Y^* \mid \ell(y) \geq 0 \ \forall y \in C\}$$

the dual cone of C. For the partial ordering introduced by C we write

$$y^1 \leq_C y^2 \Leftrightarrow y^2 - y^1 \in C \text{ for all } y^1, y^2 \in Y.$$

Later, for more specific results, we will often assume that Y is the finite dimensional space $\mathbb{R}^m$ partially ordered by some pointed convex cone or even just by the nonnegative orthant, i.e. $C = \mathbb{R}^m_+$.

We denote an element $\hat{y}$ of a nonempty set $M \subset Y$ nondominated w.r.t. C if

$$(\{\hat{y}\} - C) \cap M = \{\hat{y}\}.$$

This implies the definition of optimal solutions of a vector optimization problem. Let S be a nonempty set and let $f : S \to Y$ be a given vector-valued map. Then $\bar{x} \in S$ is called an efficient solution w.r.t. C of the vector optimization problem

$$\min_{x \in S} f(x) \qquad \text{(VOP}_{f,S}\text{)}$$

if $f(\bar{x})$ is a nondominated element of $f(S) := \{f(x) \mid x \in S\}$ w.r.t. C, i.e., if

$$f(x) \leq_C f(\bar{x}), x \in S \Longrightarrow f(\bar{x}) \leq_C f(x) \qquad (10.1)$$

holds. Clearly, since C is pointed, (10.1) can be formulated as

$$f(x) \leq_C f(\bar{x}), x \in S \Longrightarrow f(\bar{x}) = f(x).$$

The main topic of our chapter is set optimization problems with the set approach. We use the following three set order relations [5, 15, 18]:

(a) the l-less order relation w.r.t. C is defined by: $A \preccurlyeq^l_C B :\Leftrightarrow B \subset A + C$,

(b) the u-less order relation w.r.t. C is defined by: $A \preccurlyeq^u_C B :\Leftrightarrow A \subset B - C$, and

(c) the set less order relation w.r.t. C is defined by: $A \preccurlyeq^s_C B :\Leftrightarrow A \preccurlyeq^l_C B$ and $A \preccurlyeq^u_C B$.

Obviously it holds by definition

$$A \preccurlyeq^l_C B \ \Leftrightarrow \ \forall b \in B \ \exists a \in A : a \leq_C b \quad \text{and} \quad A \preccurlyeq^u_C B \ \Leftrightarrow \ \forall a \in A \ \exists b \in B : a \leq_C b.$$

Based on these set relations, we can easily define minimal elements of a family $\mathcal{A}$ of nonempty subsets of Y. A set $\bar{A}$ is a minimal element of $\mathcal{A}$ w.r.t. the order relation $\preccurlyeq^\star_C$ and $\star \in \{l, u, s\}$ if

$$A \preccurlyeq^\star_C \bar{A}, A \in \mathcal{A} \Longrightarrow \bar{A} \preccurlyeq^\star_C A.$$

Thus, for a set optimization problem

$$\min_{x \in S} F(x) \qquad \text{(SOP}_{F,S}\text{)}$$

with feasible set S and set-valued map $F \colon S \rightrightarrows Y$ with $F(x) \neq \emptyset$ and $F(x) \neq Y$ for all $x \in S$, we denote $\bar{x} \in S$ a minimal solution w.r.t. the order relation $\preccurlyeq_C^\star$ with $\star \in \{l, u, s\}$ if

$$F(x) \preccurlyeq_C^\star F(\bar{x}), \ x \in S \Longrightarrow F(\bar{x}) \preccurlyeq_C^\star F(x) \tag{10.2}$$

holds.

The following theorem gives an important characterization of these set relations by using supporting hyperplanes.

Theorem 10.2.1 *[10, Theorem 2.1] Let A and B be nonempty subsets of Y.*

(i)
$$A \preccurlyeq_C^l B \implies \forall \ell \in C^\star \setminus \{0_{Y^\star}\}: \quad \inf_{y \in A} \ell(y) \leq \inf_{y \in B} \ell(y).$$
$$A \preccurlyeq_C^u B \implies \forall \ell \in C^\star \setminus \{0_{Y^\star}\}: \quad \sup_{y \in A} \ell(y) \leq \sup_{y \in B} \ell(y).$$

(ii) *If the set $A + C$ is closed and convex, then*
$$A \preccurlyeq_C^l B \iff \forall \ell \in C^\star \setminus \{0_{Y^\star}\}: \quad \inf_{y \in A} \ell(y) \leq \inf_{y \in B} \ell(y).$$

If the set $B - C$ is closed and convex, then
$$A \preccurlyeq_C^u B \iff \forall \ell \in C^\star \setminus \{0_{Y^\star}\}: \quad \sup_{y \in A} \ell(y) \leq \sup_{y \in B} \ell(y).$$

(iii) *If the sets $A + C$ and $B - C$ are closed and convex, then*
$$A \preccurlyeq_C^s B \iff \forall \ell \in C^\star \setminus \{0_{Y^\star}\}: \quad \inf_{y \in A} \ell(y) \leq \inf_{y \in B} \ell(y) \quad \text{and} \quad \sup_{y \in A} \ell(y) \leq \sup_{y \in B} \ell(y).$$

If we apply these characterizations to (10.2) then we obtain the following corollary:

Corollary 10.2.2 *[10, Corollary 2.2] Let (SOP$_{F,S}$) be given and let for all $x \in S$ the sets $F(x) + C$ and $F(x) - C$ be closed and convex. Then $\bar{x} \in S$ is a minimal solution of (SOP$_{F,S}$) w.r.t. the order relation $\preccurlyeq_C^s$ if and only if there is no $x \in S$ with*

$$\forall \ell \in C^\star \setminus \{0_{Y^\star}\}: \quad \inf_{y \in F(x)} \ell(y) \leq \inf_{\bar{y} \in F(\bar{x})} \ell(\bar{y}) \quad \text{and} \quad \sup_{y \in F(x)} \ell(y) \leq \sup_{\bar{y} \in F(\bar{x})} \ell(\bar{y})$$

and

$$\exists \hat{\ell} \in C^\star \setminus \{0_{Y^\star}\}: \quad \inf_{y \in F(x)} \hat{\ell}(y) < \inf_{\bar{y} \in F(\bar{x})} \hat{\ell}(\bar{y}) \quad \text{or} \quad \sup_{y \in F(x)} \hat{\ell}(y) < \sup_{\bar{y} \in F(\bar{x})} \hat{\ell}(\bar{y}).$$

Based on this corollary, Jahn proposed in [10] a vector optimization problem which is equivalent to the set optimization problem (SOP$_{F,S}$). This vector optimization problem requires the introduction of a new linear space which is defined by maps. This special space is partially ordered by a pointwise ordering (see the next theorem).

Theorem 10.2.3 *[10, Theorem 3.1] Let* (SOP$_{F,S}$) *be given and let for all* $x \in S$ *the sets* $F(x)+C$ *and* $F(x)-C$ *be closed and convex. Then* $\bar{x} \in S$ *is a minimal solution of* (SOP$_{F,S}$) *w.r.t. the order relation* $\preccurlyeq_C^s$ *if and only if* $\bar{x}$ *is an efficient solution of the vector optimization problem*

$$\min_{x \in S} v(F(x)) \qquad (10.3)$$

where $v(F(x))$ *is for each set* $F(x)$ *a map on* $C^* \setminus \{0_{Y^*}\}$ *which is defined pointwise for each* $\ell \in C^* \setminus \{0_{Y^*}\}$ *by*

$$v(F(x))(\ell) := \begin{pmatrix} \inf_{y \in F(x)} \ell(y) \\ \sup_{y \in F(x)} \ell(y) \end{pmatrix}$$

and the partial ordering in the image space of the vector optimization problem (10.3) is defined by

$$v(F(x^1)) \leq v(F(x^2)) :\Leftrightarrow v(F(x^1))(\ell) \leq_{\mathbb{R}_+^2} v(F(x^2))(\ell) \; \forall \ell \in C^* \setminus \{0_{Y^*}\}.$$

Note that even if the original set optimization problem (SOP$_{F,S}$) is a finite dimensional problem (for instance if $S \subset \mathbb{R}^n$ and $Y = \mathbb{R}^m$) the associated vector optimization problem (10.3) is an infinite dimensional problem. It can then also be interpreted as a multiobjective optimization problem with an infinite number of objectives (two for each $\ell \in C^* \setminus \{0_{Y^*}\}$). Moreover, in order to solve (10.3) an infinite number of scalar-valued optimization problems would have to be solved. This is due to the fact that, according to Theorem 10.2.1 (*iii*), in general already for comparing just two sets $F(x^1)$ and $F(x^2)$ with $x^1, x^2 \in S$ using the set less order relation $\preccurlyeq_C^s$ all (normed) elements ℓ of $C^* \setminus \{0_{Y^*}\}$ are needed. In [11] Jahn showed that (under additional assumptions) this is not the case if $Y = \mathbb{R}^m$ and the sets $F(x)$ are polyhedral for all $x \in S$:

Theorem 10.2.4 *[11, Theorem 2.2] Let S be a nonempty subset of $\mathbb{R}^n$, $Y = \mathbb{R}^m$, C be polyhedral, and let $F: S \rightrightarrows Y$ be a set-valued map defined by*

$$F(x) := \{y \in \mathbb{R}^m \mid A(x) \cdot y \leq b(x)\} \text{ for all } x \in S$$

where $A: S \to \mathbb{R}^{p \times m}$ and $b: S \to \mathbb{R}^p$ are given maps such that $F(x)$ is compact and nonempty for all $x \in S$. Moreover, let for all $x \in S$ and $i \in \{1,\ldots,p\}$ the i-th row of the matrix $A(x)$ be denoted by $a^i(x)$ and let $\mathcal{L}$ be the finite set of all normed extremal directions of C^*. Then for all $x^1, x^2 \in S$ it holds

$$F(x^1) \preccurlyeq_C^s F(x^2)$$
$$\Leftrightarrow \quad \forall i \in \{1,\ldots,p\} \text{ with } \ell^i := -a^i(x^1) \in C^* \setminus \{0_{\mathbb{R}^m}\}: \; b_i(x^1) \geq \max_{y \in F(x^2)} -(\ell^i)^\top y,$$
$$\forall j \in \{1,\ldots,p\} \text{ with } \ell^j := -a^j(x^2) \in C^* \setminus \{0_{\mathbb{R}^m}\}: \; \max_{y \in F(x^1)} (\ell^j)^\top y \leq \max_{y \in F(x^2)} (\ell^j)^\top y, \text{ and}$$
$$\forall \ell \in \mathcal{L}: \; \min_{y \in F(x^1)} \ell^\top y \leq \min_{y \in F(x^2)} \ell^\top y \text{ and } \max_{y \in F(x^1)} \ell^\top y \leq \max_{y \in F(x^2)} \ell^\top y.$$

Note that, in case the cardinality of the set of all rows of the matrices $A(x)$ with $x \in S$ is infinite, an infinite number of functionals $\ell \in C^*$ and their associated scalar-valued optimization problems still have to be considered when all elements $F(x)$ should be comparable by using just the optimal values of these functionals over the sets. For that reason, Jahn proposes in [11] a descent method which uses direct comparisons of two polyhedral sets only.

In contrast to the previous results, we aim in the following for suitable "simple" set optimization problems, such that equivalent reformulations as vector optimization problems in the same or a similar image space or even (for some special cases) as multiobjective optimization problems with a finite number of objectives can be given. In the latter case, this allows for the problems to be made numerically tractable as known techniques from multiobjective optimization can then be applied.

10.3 Set Optimization Problems Being Reducible to Vector Optimization Problems

In this section, we discuss several classes of set optimization problems for which we can show that they are equivalent to vector optimization problems. In all cases we can show that we can reduce the problems to comparatively simple vector optimization problems.

10.3.1 Set-valued Maps Based on a Fixed Set

We start by examining set-valued maps with the most simple structure: Those where the images $F(x)$ are determined by a constant nonempty set $H \subset Y$, which is only moved around in the space by adding $f(x)$ with $f \colon S \to Y$ some vector-valued map. This means we are interested in set-valued maps with the structure

$$F(x) := \{f(x)\} + H = \{f(x) + h \mid h \in H\} \text{ for all } x \in S \qquad (10.4)$$

for some nonempty set S, a nonempty subset H of Y, and $f \colon S \to Y$.

Such a map was for instance studied as a test instance by Köbis and Köbis in [14] for evaluating the properties of their proposed numerical algorithm and was motivated as the basis of the Markowitz stock model.

Example 10.3.1 *[14, Example 4.7] Let $S = Y = \mathbb{R}^2$, the vector-valued map f be defined by*

$$f \colon S \to \mathbb{R}^2 \text{ with } f(x) := \begin{pmatrix} x_1^2 + x_2^2 \\ 2(x_1 + x_2) \end{pmatrix} \text{ for all } x \in S,$$

the set H be given by

$$H := \left\{ \frac{1}{4} \begin{pmatrix} \sin(t) \\ \cos(t) \end{pmatrix} \in \mathbb{R}^2 \,\bigg|\, t \in \left\{ 0, \frac{1}{7}\pi, \ldots, \frac{13}{7}\pi \right\} \right\},$$

and the set-valued map $F \colon S \rightrightarrows \mathbb{R}^2$ be defined according to (10.4). *Using the forthcoming Theorem 10.3.4, the set of all minimal solutions of the corresponding set optimization problem* (SOP$_{F,S}$) *w.r.t. the order relation $\preccurlyeq_{\mathbb{R}_+^2}^\star$ with $\star \in \{l, u, s\}$ is given by* $\{\bar{x} \in \mathbb{R}^2 \mid \bar{x}_1 = \bar{x}_2 \leq 0\}$ *(see the forthcoming Example 10.3.5).*

Also, Hernández and López have studied special set-valued maps of the type (10.4) in [8]. More specifically, they have considered the following special cases:

Set Optimization Problems Reducible to Vector Optimization Problems ■ 247

- $F_1 : S \rightrightarrows \mathbb{R}^m$ with
$$F_1(x) := \{f(x)\} + H$$
for a given nonempty set $H \subset \mathbb{R}^m$ and a linear map $f : \mathbb{R}^n \to \mathbb{R}^m$.

- $F_2 : S \rightrightarrows \mathbb{R}^m$ with
$$\begin{aligned} F_2(x) &:= \{f(x)\} + \{\lambda\, q \in \mathbb{R}^m \mid \lambda \in [0,1]\} \\ &= \{\lambda\, f(x) + (1-\lambda)(f(x)+q) \in \mathbb{R}^m \mid \lambda \in [0,1]\} \end{aligned}$$
for a given map $f : S \to \mathbb{R}^m$ and $q \in \mathbb{R}^m$.

- $F_3 : S \rightrightarrows \mathbb{R}^m$ with
$$F_3(x) := \{f(x)\} + \{y \in \mathbb{R}^m \mid \|y\|_2 \le r\}$$
for a given map $f : S \to \mathbb{R}^m$ and $r \in \mathbb{R}_+$.

For these special classes, they have studied basic properties of the set-valued maps as semicontinuity and convexity in a finite dimensional setting. They also mention that for instance F_1 appears in the literature in the context of Hahn-Banach theorems or in subgradient theory and F_2 is of course related to interval optimization. The map F_3 appears in approximation theory and viability theory (cf. [8]).

We first state a result on the direct comparison of two sets $\{y^1\} + H$ and $\{y^2\} + H$ with $y^1, y^2 \in Y$.

Lemma 10.3.2 *Let H be a bounded nonempty subset of Y. If $y^1, y^2 \in Y$ and the sets A and B are defined by*
$$A := \{y^1\} + H \text{ and } B := \{y^2\} + H.$$
Then it holds
$$A \preccurlyeq_C^s B \iff A \preccurlyeq_C^l B \iff A \preccurlyeq_C^u B \iff y^1 \le_C y^2.$$

Proof. It is easy to see that the assumption $y^1 \le_C y^2$, i.e. $y^2 \in \{y^1\} + C$ and $y^1 \in \{y^2\} - C$, is sufficient for the other statements. Thus we now assume that $A \preccurlyeq_C^l B$. By Theorem 10.2.1.(i) it follows
$$\inf_{y \in A} \ell(y) \le \inf_{y \in B} \ell(y) \text{ for all } \ell \in C^* \setminus \{0_{Y^*}\}$$
which is equivalent to
$$\ell(y^1) + \inf_{h \in H} \ell(h) \le \ell(y^2) + \inf_{h \in H} \ell(h) \text{ for all } \ell \in C^* \setminus \{0_{Y^*}\}.$$
As H is nonempty and bounded it holds
$$-\infty < \inf_{h \in H} \ell(h) \le \sup_{h \in H} \ell(h) < \infty \text{ for all } \ell \in C^* \setminus \{0_{Y^*}\}.$$
Hence, we obtain
$$\ell(y^2 - y^1) \ge 0 \text{ for all } \ell \in C^*$$

and, thus, since C is closed, by [9, Lemma 3.21 (a)] $y^1 \leq_C y^2$.
Using similar arguments, we obtain

$$A \preccurlyeq_C^u B \Rightarrow y^1 \leq_C y^2,$$

and we are done. □

For the proof we need that

$$\inf_{h \in H} \ell(h) \neq -\infty \text{ and } \sup_{h \in H} \ell(h) \neq \infty$$

holds for all $\ell \in C^* \setminus \{0_{Y^*}\}$. Clearly, this is true in case H is bounded. Note that, for an unbounded set, the result of Lemma 10.3.2 might not be true. For instance, for $H = Y$ this can trivially be seen. Additional examples with sets H unequal to the whole space Y are given in the following.

Example 10.3.3 Let $Y = \mathbb{R}^2$, $H^1 := \{y \in \mathbb{R}^2 \mid y_1 \leq 0,\ y_2 = 0\}$, $H^2 := \{y \in \mathbb{R}^2 \mid y_1 \geq 0,\ y_2 = 0\}$, and $H^3 := \{y \in \mathbb{R}^2 \mid y_2 = 0\}$. Then for $y^1 = (2,1)^\top$ and $y^2 = (1,2)^\top$ it holds $y^1 \not\leq_{\mathbb{R}^2_+} y^2$, $\{y^1\} + H^1 \preccurlyeq_{\mathbb{R}^2_+}^l \{y^2\} + H^1$, $\{y^1\} + H^2 \preccurlyeq_{\mathbb{R}^2_+}^u \{y^2\} + H^2$, and $\{y^1\} + H^3 \preccurlyeq_{\mathbb{R}^2_+}^s \{y^2\} + H^3$.

Lemma 10.3.2 directly implies that set optimization problems with set-valued maps defined as in (10.4) are equivalent to vector optimization problems.

Theorem 10.3.4 Let the set optimization problem (SOP$_{F,S}$) be given with an objective map F as defined in (10.4), i.e.

$$F(x) = \{f(x)\} + H \text{ for all } x \in S,$$

and let the set H be a bounded and nonempty subset of Y. Then $\bar{x} \in S$ is a minimal solution of the set optimization problem (SOP$_{F,S}$) w.r.t. the order relation $\preccurlyeq_C^\star$ and $\star \in \{l, u, s\}$ if and only if $\bar{x}$ is an efficient solution of the vector optimization problem (VOP$_{f,S}$) w.r.t. C, i.e. of

$$\min_{x \in S} f(x)$$

w.r.t. the ordering cone C.

Using Theorem 10.3.4, we can verify the set of all minimal solutions of the set optimization problem stated in Example 10.3.1.

Example 10.3.5 We consider again the set optimization problem (SOP$_{F,S}$) defined in Example 10.3.1. Let $\bar{x} \in S = \mathbb{R}^2$ be a minimal solution of (SOP$_{F,S}$) w.r.t. the order relation $\preccurlyeq_{\mathbb{R}^2_+}^\star$ and $\star \in \{l,u,s\}$. By Theorem 10.3.4 this is equivalent to that $\bar{x}$ is an efficient solution of the vector optimization problem

$$\min_{x \in \mathbb{R}^2} \begin{pmatrix} f_1(x) \\ f_2(x) \end{pmatrix} \tag{10.5}$$

w.r.t. $\mathbb{R}^2_+$ and $f_1, f_2 \colon \mathbb{R}^2 \to \mathbb{R}$ with $f_1(x) := x_1^2 + x_2^2$ and $f_2(x) := 2(x_1 + x_2)$ for all $x \in \mathbb{R}^2$. If $\bar{x}_1 > 0$ or $\bar{x}_2 > 0$ then it is easy to see that for $\hat{x} := (-|\bar{x}_1|, -|\bar{x}_2|)^\top \neq \bar{x}$ it holds $f_1(\hat{x}) =$

$f_1(\bar{x})$ and $f_2(\hat{x}) < f_2(\bar{x})$ – contradicting that $\bar{x}$ is an efficient solution of (10.5). Hence, we derive $\bar{x}_1 \leq 0$ and $\bar{x}_2 \leq 0$ for any efficient solution of (10.5). If additionally $\bar{x}_1 \neq \bar{x}_2$ holds, then it follows for

$$\check{x} := \frac{1}{\sqrt{2}} \left(-\sqrt{\bar{x}_1^2 + \bar{x}_2^2}, -\sqrt{\bar{x}_1^2 + \bar{x}_2^2} \right)^\top \neq \bar{x}$$

also $f_1(\check{x}) = f_1(\bar{x})$ and $f_2(\check{x}) < f_2(\bar{x})$. Thus, we obtain $\bar{x}_1 = \bar{x}_2 \leq 0$. Finally, let $\bar{x}' = (x',x')^\top$, $\bar{x}'' = (x'',x'')^\top$, and w.l.o.g. $x' < x'' \leq 0$. Then, it holds $f_1(\bar{x}') > f_1(\bar{x}'')$ and $f_2(\bar{x}') < f_2(\bar{x}'')$. Hence, the set of all minimal solutions of (SOP$_{F,S}$) w.r.t. the order relation $\preccurlyeq_{\mathbb{R}_+^2}^\star$ and $\star \in \{l,u,s\}$ and the set of all efficient solutions of the corresponding vector optimization problem defined by (10.5) w.r.t. $\mathbb{R}_+^2$ is given by $\{\bar{x} \in \mathbb{R}^2 \mid \bar{x}_1 = \bar{x}_2 \leq 0\}$.

Finally, we relate the result of Theorem 10.3.4 to another result from the literature. In [6] the following class of set-valued optimization problems has been studied: Let $F \colon S \rightrightarrows \mathbb{R}^m$ be defined by

$$F(x) := \{f(x+z) \in \mathbb{R}^m \mid z \in Z\}$$

where $S \subset \mathbb{R}^n$ is some nonempty set, $Z \subset \mathbb{R}^n$ is a compact set with $0_{\mathbb{R}^n} \in Z$, and $f \colon \mathbb{R}^n \to \mathbb{R}^m$ is a given vector-valued map. It was assumed that the linear space $\mathbb{R}^m$ is partially ordered by some pointed convex closed cone C with nonempty interior. These problems arise in the study of multiobjective optimization problems, which have some uncertainties in the realization of solutions and when a robust approach is chosen. For the set optimization problem the u-less order relation was used. In [6, Section 5.1], linear objective functions f have been studied. Then, one obtains $F(x) = \{f(x)\} + H$ with $H := \{f(z) \in \mathbb{R}^m \mid z \in Z\}$, which is a bounded nonempty set and Theorem 10.3.4 can be applied.

Other set-valued maps with a simple structure using a constant nonempty set can be defined by multiplication with a scalar-valued function. Thus, we are now interested in set-valued maps with the structure

$$F(x) := \varphi(x)H = \{\varphi(x)h \mid h \in H\} \text{ for all } x \in S \quad (10.6)$$

for some nonempty set S, a nonempty subset H of Y with $H \neq \{0_Y\}$, and $\varphi \colon S \to \mathbb{R}$. We show that such set optimization problems (under strict additional assumptions on the set H) can be formulated equivalently even as scalar-valued optimization problems.

In analogy to Lemma 10.3.2 we want to formulate results on the direct comparison of two sets αH and βH with $\alpha, \beta \in \mathbb{R}$. However, the following example shows that

$$\alpha H \preccurlyeq_C^\star \beta H \Leftrightarrow \alpha \leq \beta \quad (10.7)$$

with $\star \in \{l,u,s\}$ does not hold in general even for a compact set.

Example 10.3.6 *Let $Y = \mathbb{R}$ and $H := [-1,1]$. Then it holds $\gamma H = [-|\gamma|,|\gamma|]$ for all $\gamma \in \mathbb{R}$ and we obtain $\alpha H \preccurlyeq_{\mathbb{R}_+}^l \beta H \Leftrightarrow |\alpha| \geq |\beta|$, $\alpha H \preccurlyeq_{\mathbb{R}_+}^u \beta H \Leftrightarrow |\alpha| \leq |\beta|$, and $\alpha H \preccurlyeq_{\mathbb{R}_+}^s \beta H \Leftrightarrow |\alpha| = |\beta|$ for all $\alpha,\beta \in \mathbb{R}$. See also the forthcoming Lemma 10.3.12 for additional results for this type of set-valued maps.*

The following lemma formulates additional assumptions, under which (10.7) can be guaranteed.

Lemma 10.3.7 *Let H be a bounded nonempty subset of Y with $H \neq \{0_Y\}$, let $\alpha, \beta \in \mathbb{R}$, and let the sets A and B be defined by*

$$A := \alpha H \text{ and } B := \beta H.$$

Then the following holds:

(i) *If $H \subset C$ and $\alpha \leq \beta$, then it holds $A \preccurlyeq_C^\star B$ with $\star \in \{l, u, s\}$.*

(ii) *If $H \subset C$ and there exists an $\hat{\ell} \in C^* \setminus \{0_{Y^*}\}$ such that $\inf_{h \in H} \hat{\ell}(h) > 0$, then it holds*

$$A \preccurlyeq_C^s B \iff A \preccurlyeq_C^l B \iff A \preccurlyeq_C^u B \iff \alpha \leq \beta.$$

(iii) *If $H \subset -C$ and $\alpha \geq \beta$, then it holds $A \preccurlyeq_C^\star B$ with $\star \in \{l, u, s\}$.*

(iv) *If $H \subset -C$ and there exists an $\bar{\ell} \in C^* \setminus \{0_{Y^*}\}$ such that $\sup_{h \in H} \bar{\ell}(h) < 0$, then it holds*

$$A \preccurlyeq_C^s B \iff A \preccurlyeq_C^l B \iff A \preccurlyeq_C^u B \iff \alpha \geq \beta.$$

Proof. We restrict ourselves to the proofs of *(i)* and *(ii)*. Hence $H \subset C$ and note that

$$0 \leq \inf_{h \in H} \ell(h) \leq \sup_{h \in H} \ell(h) < \infty \text{ for all } \ell \in C^* \setminus \{0_{Y^*}\}. \tag{10.8}$$

For the proof of *(i)* let $\alpha \leq \beta$. For any $b \in B$ there exists $h \in H \subset C$ with $b = \beta h$. For $a := \alpha h \in A$ it holds $b - a = (\beta - \alpha)h \in C$. Hence, $a \leq_C b$ and $A \preccurlyeq_C^l B$ are shown. Using similar arguments we obtain $A \preccurlyeq_C^u B$ which proves the assertion.

For *(ii)* it remains to show that $\alpha \leq \beta$ is also necessary for $A \preccurlyeq_C^\star B$ with $\star \in \{l, u, s\}$. If $A \preccurlyeq_C^l B$ then it holds by Theorem 10.2.1.(i)

$$\inf_{h \in H} \alpha \ell(h) \leq \inf_{h \in H} \beta \ell(h) \text{ for all } \ell \in C^* \setminus \{0_{Y^*}\}. \tag{10.9}$$

For $\alpha \geq 0$ and $\beta \geq 0$ it is easy to see that (10.9) is equivalent to

$$\alpha \inf_{h \in H} \ell(h) \leq \beta \inf_{h \in H} \ell(h) \text{ for all } \ell \in C^* \setminus \{0_{Y^*}\}$$

and we obtain $\alpha \inf_{h \in H} \hat{\ell}(h) \leq \beta \inf_{h \in H} \hat{\ell}(h)$. By using $\inf_{h \in H} \hat{\ell}(h) > 0$ and (10.8) it follows $\alpha \leq \beta$. If $\alpha \geq 0$ and $\beta < 0$ then (10.9) is equivalent to

$$\alpha \inf_{h \in H} \ell(h) \leq \beta \sup_{h \in H} \ell(h) \text{ for all } \ell \in C^* \setminus \{0_{Y^*}\}$$

and it follows by (10.8)

$$\alpha \inf_{h \in H} \ell(h) = \beta \sup_{h \in H} \ell(h) = 0 \text{ for all } \ell \in C^* \setminus \{0_{Y^*}\}$$

which contradicts $\beta \sup_{h \in H} \hat{\ell}(h) \leq \beta \inf_{h \in H} \hat{\ell}(h) < 0$.

In case $\alpha < 0$ and $\beta \geq 0$ then $\alpha \leq \beta$ holds trivially.
For $\alpha < 0$ and $\beta < 0$ we obtain as an equivalent formulation of (10.9)

$$\alpha \sup_{h \in H} \ell(h) \leq \beta \sup_{h \in H} \ell(h) \text{ for all } \ell \in C^* \setminus \{0_{Y^*}\}$$

and $\alpha \leq \beta$ follows immediately by using $\sup_{h \in H} \hat{\ell}(h) \geq \inf_{h \in H} \hat{\ell}(h) > 0$ and (10.8) again. Using similar arguments we obtain

$$A \preccurlyeq^u_C B \Rightarrow \alpha \leq \beta,$$

and we are done. □

Remark 10.3.8 *The existence of an $\hat{\ell} \in C^* \setminus \{0_{Y^*}\}$ with $\inf_{h \in H} \hat{\ell}(h) > 0$ can be guaranteed for instance if H is a nonempty, compact, and convex subset of C with $0_Y \notin H$. Since $-C$ is closed and convex there exists in this case by a suitable separation theorem (cf. [9, Theorem 3.20]) an $\ell \in Y^* \setminus \{0_{Y^*}\}$ such that*

$$0 = \ell(0_Y) \leq \sup_{y \in -C} \ell(y) < \inf_{h \in H} \ell(h) < \infty$$

and $\ell \in C^* \setminus \{0_{Y^*}\}$ follows by standard arguments.

We note that, for an unbounded set, the result of Lemma 10.3.7 might not be true.

Example 10.3.9 *Let $Y = \mathbb{R}^2$ and $H := \{y \in \mathbb{R}^2 \mid y = (1+r, 1+r)^\top, r \in \mathbb{R}_+\} \subset \mathbb{R}^2_+$. Then, it holds $\alpha H \preccurlyeq^l_{\mathbb{R}^2_+} \beta H$ for all $\alpha, \beta < 0$ and $\alpha H \preccurlyeq^u_{\mathbb{R}^2_+} \beta H$ for all $\alpha, \beta > 0$.*

Lemma 10.3.7 directly implies that, under the mentioned strict additional assumptions, set optimization problems with set-valued maps defined as in (10.6) can be formulated equivalently as scalar-valued optimization problems. In the following theorem, we restrict ourselves to the case $H \subset C$.

Theorem 10.3.10 *Let the set optimization problem $(\mathrm{SOP}_{F,S})$ be given with an objective map F, as defined in (10.6), i.e.*

$$F(x) = \varphi(x) H \text{ for all } x \in S,$$

and let the set H be a bounded nonempty subset of C with $H \neq \{0_Y\}$. If there exists an $\hat{\ell} \in C^* \setminus \{0_{Y^*}\}$ such that $\inf_{h \in H} \hat{\ell}(h) > 0$, then $\bar{x} \in S$ is a minimal solution of the set optimization problem $(\mathrm{SOP}_{F,S})$ w.r.t. the order relation $\preccurlyeq^\star_C$ and $\star \in \{l, u, s\}$ if and only if $\bar{x}$ is a minimal solution of the scalar-valued optimization problem

$$\min_{x \in S} \varphi(x).$$

10.3.2 Box-valued Maps

The set-valued maps which we study in this section are assumed to have values $F(x)$ which are boxes in Y. For this purpose, let $a, b \in Y$ with $a \leq_C b$ and the corresponding box $[a,b]_C$ be defined by

$$[a,b]_C := (\{a\} + C) \cap (\{b\} - C).$$

Obviously a box is a convex and closed set and we are now interested in set-valued maps with the structure

$$F(x) := [a(x), b(x)]_C \text{ for all } x \in S \qquad (10.10)$$

for some nonempty set S and two vector-valued maps $a, b \colon S \to Y$ with $a(x) \leq_C b(x)$ for all $x \in S$. In a finite dimensional setting, the properties as semicontinuity and convexity of maps as in (10.10) have already been studied in [8]. If for some $x \in S$ it holds $a(x) = b(x)$ then the set $F(x)$ is a singleton. A simple example for such set-valued maps are the so called interval-valued maps in the case $Y = \mathbb{R}$ and $C = \mathbb{R}_+$. For reasons of simplicity, in the case $Y = \mathbb{R}^m$ and $C = \mathbb{R}^m_+$ the usual notation $[a, b]$ instead of $[a, b]_{\mathbb{R}^m_+}$ is used. Another example of such set-valued maps is given in the example below and was provided in [3].

Example 10.3.11 *[3, Example 4.1] Let $S = [0,1]$, $Y = \mathbb{R}^2$, $C = \mathbb{R}^2_+$, and $F \colon S \rightrightarrows \mathbb{R}^2$ be defined by*

$$F(x) := \{y \in \mathbb{R}^2 \mid y_1 = x, \ y_2 \in [x, 2-x]\}.$$

Then the images are boxes with

$$F(x) = (\{(x,x)^\top\} + \mathbb{R}^2_+) \cap (\{(x, 2-x)^\top\} - \mathbb{R}^2_+) = [(x,x)^\top, (x, 2-x)^\top].$$

It is easy to see (cf. for instance the forthcoming Theorem 10.3.13 or Corollary 10.3.15) that $\bar{x} = 0$ is the unique minimal solution of the set optimization problem $(SOP_{F,S})$ w.r.t. the order relation $\preccurlyeq^l_{\mathbb{R}^2_+}$ and the set of all minimal solutions of $(SOP_{F,S})$ w.r.t. the order relation $\preccurlyeq^\star_{\mathbb{R}^2_+}$ with $\star \in \{u, s\}$ is given by the whole set S.

The following lemma formulates a result on the direct comparison of two boxes.

Lemma 10.3.12 *Let two sets A and B in Y be defined by*

$$A := [a^1, b^1]_C \text{ and } B := [a^2, b^2]_C$$

where $a^1, a^2, b^1, b^2 \in Y$ with $a^1 \leq_C b^1$ and $a^2 \leq_C b^2$. Then it holds

$$A \preccurlyeq^l_C B \Leftrightarrow a^1 \leq_C a^2 \text{ and } A \preccurlyeq^u_C B \Leftrightarrow b^1 \leq_C b^2.$$

Proof. If $A \preccurlyeq^l_C B$, then there exists $a \in A \subset \{a^1\} + C$ such that $a \leq_C a^2$ and, thus, $a \in \{a^2\} - C$. Hence, there exist $k^1, k^2 \in C$ with $a = a^1 + k^1 = a^2 - k^2$. It follows $a^2 - a^1 = k^1 + k^2 \in C$ and we obtain $a^1 \leq_C a^2$. Let now $a^1 \leq_C a^2$ and $b \in B \subset \{a^2\} + C$ be arbitrarily chosen. Hence, it holds $a^2 \leq_C b$. By the transitivity of $\leq_C$ for $a := a^1 \in A$ it follows $a \leq_C b$ and, thus, $A \preccurlyeq^l_C B$. Using similar arguments we obtain $A \preccurlyeq^u_C B$ if and only if $b^1 \leq_C b^2$. □

This result is also stated (without proof) in a finite dimensional setting in [8, Remark 1]. Lemma 10.3.12 directly implies that set optimization problems with set-valued maps defined as in (10.10) are equivalent to vector optimization problems.

Theorem 10.3.13 *Let the set optimization problem $(SOP_{F,S})$ be given with an objective map F, as defined in (10.10), i.e.*

$$F(x) = [a(x), b(x)]_C \text{ for all } x \in S.$$

Then the following holds:

(i) $\bar{x} \in S$ is a minimal solution of the set optimization problem (SOP$_{F,S}$) w.r.t. the order relation $\preccurlyeq_C^l$ if and only if $\bar{x}$ is an efficient solution of the vector optimization problem (VOP$_{f,S}$) w.r.t. C for $f \colon S \to Y$ and $f(x) := a(x)$.

(ii) $\bar{x} \in S$ is a minimal solution of the set optimization problem (SOP$_{F,S}$) w.r.t. the order relation $\preccurlyeq_C^u$ if and only if $\bar{x}$ is an efficient solution of the vector optimization problem (VOP$_{f,S}$) w.r.t. C for $f \colon S \to Y$ and $f(x) := b(x)$.

(iii) $\bar{x} \in S$ is a minimal solution of the set optimization problem (SOP$_{F,S}$) w.r.t. the order relation $\preccurlyeq_C^s$ if and only if $\bar{x}$ is an efficient solution of the vector optimization problem (VOP$_{f,S}$) w.r.t. $C \times C := \{(c^1, c^2) \in Y \times Y \mid c^1 \in C,\ c^2 \in C\}$ for $f \colon S \to Y \times Y$ and $f(x) := \begin{pmatrix} a(x) \\ b(x) \end{pmatrix}$, i.e. of

$$\min_{x \in S} \begin{pmatrix} a(x) \\ b(x) \end{pmatrix}$$

w.r.t. the ordering cone $C \times C$.

As a consequence of Theorem 10.3.13, we obtain the following corollary, which confirms the statement of Theorem 10.3.4 for a special case of the set H.

Corollary 10.3.14 *Let $f \colon S \to Y$ be a given map, let $q \in C$, and let the set optimization problem (SOP$_{F,S}$) be given with the objective map F defined by*

$$F \colon S \rightrightarrows Y \text{ and } F(x) := \{f(x)\} + [0_Y, q]_C = [f(x), f(x) + q]_C \text{ for all } x \in S.$$

Then $\bar{x} \in S$ is a minimal solution of the set optimization problem (SOP$_{F,S}$) w.r.t. the order relation $\preccurlyeq_C^\star$ and $\star \in \{l, u, s\}$ if and only if $\bar{x}$ is an efficient solution of the vector optimization problem (VOP$_{f,S}$) w.r.t. C.

Finally, we study boxes which are defined by a multiplication of a scalar-valued function and a fixed box in order to relate our results to those in Theorem 10.3.10. Let $a, b \in Y$ with $a \leq_C b$, $H := [a, b]_C$, $\varphi \colon S \to \mathbb{R}$, and the set-valued map $F \colon S \rightrightarrows Y$ be defined according to (10.6) by $F(x) := \varphi(x)[a, b]_C$ for all $x \in S$. Then it holds for the box-valued map F

$$F(x) = \begin{cases} [\varphi(x)b, \varphi(x)a]_C, & \text{if } \varphi(x) \leq 0 \\ [\varphi(x)a, \varphi(x)b]_C, & \text{if } \varphi(x) \geq 0 \end{cases} \text{ for all } x \in S.$$

An application of Theorem 10.3.13 to the corresponding set optimization problem (SOP$_{F,S}$) is possible only under the additional assumption that either $\varphi(x) \leq 0$ or $\varphi(x) \geq 0$ holds for all $x \in S$. While an application of Theorem 10.3.10 is only possible when we have a compact box $[a, b]_C \subset H$ with $0_Y \notin H$.

We end this subsection by applying Theorem 10.3.13 to the finite dimensional case with the natural ordering:

Corollary 10.3.15 *Let S be a nonempty subset of $\mathbb{R}^n$, let $a, b \colon \mathbb{R}^n \to \mathbb{R}^m$ be given maps with $a_i(x) \leq b_i(x)$ for all $i \in \{1, \ldots, m\}$ and $x \in S$, and let the function $F \colon \mathbb{R}^n \rightrightarrows \mathbb{R}^m$ be defined by*

$$F(x) := [a(x), b(x)] \text{ for all } x \in S.$$

Then $\bar{x} \in S$ is a minimal solution of the set optimization problem (SOP$_{F,S}$) w.r.t. the order relation $\preccurlyeq^l_{\mathbb{R}^m_+} \mid \preccurlyeq^u_{\mathbb{R}^m_+} \mid \preccurlyeq^s_{\mathbb{R}^m_+}$ if and only if $\bar{x}$ is an efficient solution of the vector optimization problem (VOP$_{f,S}$) w.r.t. $\mathbb{R}^m_+ \mid \mathbb{R}^m_+ \mid \mathbb{R}^{2m}_+$ for $f\colon S \to \mathbb{R}^m \mid \mathbb{R}^m \mid \mathbb{R}^{2m}$ and $f := a \mid f := b \mid f := \begin{pmatrix} a \\ b \end{pmatrix}$.

10.3.3 Ball-valued Maps

In this section, let $(Y, \langle \cdot, \cdot \rangle)$ be a Hilbert space, and we write

$$\ell(y) = \langle \ell, y \rangle$$

for the dual pairing. The set-valued maps which we study next are assumed to have values $F(x)$, which are balls with variable midpoints and with variable radii. Therefore, let

$$\mathcal{B}_Y(r) := \{y \in Y \mid \|y\| \leq r\}$$

for $r \in \mathbb{R}_+$ with $\|y\| = \sqrt{\langle y,y \rangle}$. Thus, we are interested in set-valued maps with the structure

$$F(x) := \{c(x)\} + \mathcal{B}_Y(r(x)) \text{ for all } x \in S \tag{10.11}$$

for some nonempty set S, a vector-valued map $c\colon S \to Y$, and a function $r\colon S \to \mathbb{R}_+$. For basic properties of such maps as well as for references to applications where such maps are of interest we refer to [8]. Such a set optimization problem was for instance studied by Jahn:

Example 10.3.16 *[10, Example 3.1] Let $Y = \mathbb{R}^2$, $C = \mathbb{R}^2_+$, $S = [-1, 1]$, and $F\colon S \rightrightarrows \mathbb{R}^2$ be defined by*

$$F(x) := \{y \in \mathbb{R}^2 \mid (y_1 - 2x^2)^2 + (y_2 - 2x^2)^2 \leq (x^2 + 1)^2\} \text{ for all } x \in S.$$

Then the images are balls with

$$F(x) = \{(2x^2, 2x^2)^\top\} + \mathcal{B}_{\mathbb{R}^2}(x^2 + 1)$$

and $\bar{x} := 0$ is the unique minimal solution of the corresponding set optimization problem (SOP$_{F,S}$) *w.r.t. the order relation $\preccurlyeq^s_{\mathbb{R}^2_+}$, cf. Example 10.3.24.*

We need the following results:

Lemma 10.3.17 *Let $r \in \mathbb{R}_+$ and $\ell \in C^*$ with $\|\ell\| = 1$. Then it holds*

$$\min_{y \in \mathcal{B}_Y(r)} \ell(y) = -r \text{ and } \max_{y \in \mathcal{B}_Y(r)} \ell(y) = r.$$

Proof. Let $r \in \mathbb{R}_+$ and $\ell \in C^* \setminus \{0_{Y^*}\}$ with $\|\ell\| = 1$ be arbitrarily chosen. Then it holds by the Cauchy-Schwarz inequality for all y with $\|y\| \leq r$

$$|\ell(y)| = |\langle \ell, y \rangle| \leq \|\ell\| \, \|y\| \leq r$$

and hence

$$-r \leq \langle \ell, y \rangle \leq r.$$

Since for $\underline{y} := -\ell r \in \mathcal{B}_Y(r)$ and $\overline{y} := \ell r \in \mathcal{B}_Y(r)$ it holds
$$\ell(\underline{y}) = \langle \ell, \underline{y} \rangle = -r\langle \ell, \ell \rangle = -r \text{ and } \ell(\overline{y}) = \langle \ell, \overline{y} \rangle = r\langle \ell, \ell \rangle = r$$
we are done. □

Using Theorem 10.2.1(ii) and Lemma 10.3.17 one can easily verify the following lemma:

Lemma 10.3.18 *Let two sets A and B in Y be defined by*
$$A := \{y^1\} + \mathcal{B}_Y(r_1) \text{ and } B := \{y^2\} + \mathcal{B}_Y(r_2) \tag{10.12}$$
where $y^1, y^2 \in Y$ and $r_1, r_2 \in \mathbb{R}_+$. Then it holds

$$A \preccurlyeq_C^l B \iff \forall \ell \in C^* \text{ with } \|\ell\| = 1 : \ell(y^2 - y^1) = \langle \ell, y^2 - y^1 \rangle \geq r_2 - r_1 \text{ and}$$

$$A \preccurlyeq_C^u B \iff \forall \ell \in C^* \text{ with } \|\ell\| = 1 : \ell(y^2 - y^1) = \langle \ell, y^2 - y^1 \rangle \geq -(r_2 - r_1).$$

Using Lemma 10.3.18 we can now prove the following result on the direct comparison of two balls regarding the set less order relation based on only a finite number of inequalities in case the cone C has a finitely generated dual cone.

Lemma 10.3.19 *Let the two sets A and B be defined as in (10.12). If there exist $\ell^1, \ldots, \ell^k \in C^*$ with $\|\ell^i\| = 1$ for all $i \in \{1, \ldots, k\}$ such that $C^* = \text{cone}(\text{conv}(\{\ell^i \mid i \in \{1, \ldots, k\}\}))$, then*
$$A \preccurlyeq_C^s B \iff \langle \ell^i, y^2 - y^1 \rangle \geq |r_2 - r_1| \text{ for all } i \in \{1, \ldots, k\}.$$

Proof. By using Lemma 10.3.18 it holds
$$A \preccurlyeq_C^s B \iff \langle \ell, y^2 - y^1 \rangle \geq |r_2 - r_1| \quad \text{for all } \ell \in C^* \text{ with } \|\ell\| = 1$$
$$\implies \langle \ell^i, y^2 - y^1 \rangle \geq |r_2 - r_1| \quad \text{for all } \ell^i, i \in \{1, \ldots, k\}.$$

Thus, we assume now $\langle \ell^i, y^2 - y^1 \rangle \geq |r_2 - r_1|$ for all $i \in \{1, \ldots, k\}$. Let $\ell \in C^*$ with $\|\ell\| = 1$. Then there exist $\lambda_i \in \mathbb{R}_+, i \in \{1, \ldots, k\}$ such that
$$\sum_{i=1}^k \lambda_i = 1, \ \ell = \frac{1}{\left\|\sum_{i=1}^k \lambda_i \ell^i\right\|} \sum_{i=1}^k \lambda_i \ell^i, \text{ and } \left\|\sum_{i=1}^k \lambda_i \ell^i\right\| \leq \sum_{i=1}^k \lambda_i \|\ell^i\| = 1.$$

Finally, it follows
$$\langle \ell, y^2 - y^1 \rangle = \frac{1}{\left\|\sum_{i=1}^k \lambda_i \ell^i\right\|} \sum_{i=1}^k \lambda_i \langle \ell^i, y^2 - y^1 \rangle \geq \sum_{i=1}^k \lambda_i |r_2 - r_1| = |r_2 - r_1|.$$
□

In the case $Y = \mathbb{R}^m$ and $C = \mathbb{R}_+^m$ it holds
$$C^* = \mathbb{R}_+^m = \text{cone}(\text{conv}(\{e^i \mid i \in \{1, \ldots, m\}\})),$$
where $e^i, i \in \{1, \ldots, m\}$ denotes the i-th unit vector of $\mathbb{R}^m$. Using Lemma 10.3.19 in this special case it follows:

Lemma 10.3.20 *Let $y^1, y^2 \in \mathbb{R}^m$ and $r_1, r_2 \in \mathbb{R}_+$. Then the following holds:*

$$\{y^1\} + \mathcal{B}_{\mathbb{R}^m}(r_1) \preccurlyeq^s_{\mathbb{R}^m_+} \{y^2\} + \mathcal{B}_{\mathbb{R}^m}(r_2) \quad \Leftrightarrow \quad y_i^2 - y_i^1 \geq |r_2 - r_1| \text{ for all } i \in \{1, \ldots, m\}.$$

We use now this result for the formulation of an equivalent vector optimization problem to a set optimization problem with such values of the objective map. For that, we need the ordering cone

$$C^{m+1} := \left\{ y \in \mathbb{R}^{m+1} \,\middle|\, y_i \geq |y_{m+1}| \,\forall\, i \in \{1, \ldots, m\} \right\}. \tag{10.13}$$

It is easy to see that C^{m+1} is a pointed, convex, and closed nontrivial cone and that $\{y^1\} + \mathcal{B}_{\mathbb{R}^m}(r_1) \preccurlyeq^s_{\mathbb{R}^m_+} \{y^2\} + \mathcal{B}_{\mathbb{R}^m}(r_2)$ if and only if

$$\begin{pmatrix} y^1 \\ r_1 \end{pmatrix} \leq_{C^{m+1}} \begin{pmatrix} y^2 \\ r_2 \end{pmatrix}.$$

Hence, we obtain:

Theorem 10.3.21 *Let S be a nonempty subset of $\mathbb{R}^n$, let the maps $c\colon S \to \mathbb{R}^m$ and $r\colon S \to \mathbb{R}_+$ be given, let the set-valued map $F\colon S \rightrightarrows \mathbb{R}^m$ be defined by*

$$F(x) := \{c(x)\} + \mathcal{B}_{\mathbb{R}^m}(r(x)) \text{ for all } x \in S,$$

and let the pointed convex cone C^{m+1} be defined as in (10.13). Then $\bar{x} \in S$ is a minimal solution of the set optimization problem $(\mathrm{SOP}_{F,S})$ w.r.t. the order relation $\preccurlyeq^s_{\mathbb{R}^m_+}$ if and only if $\bar{x}$ is an efficient solution of the vector optimization problem $(\mathrm{VOP}_{f,S})$ w.r.t. C^{m+1} for $f\colon \mathbb{R}^n \to \mathbb{R}^{m+1}$ and $f := \begin{pmatrix} c \\ r \end{pmatrix}$, i.e. of

$$\min_{x \in S} \begin{pmatrix} c(x) \\ r(x) \end{pmatrix}$$

w.r.t. the ordering cone C^{m+1}.

The vector optimization problem in Theorem 10.3.21 is a finite dimensional problem but the ordering cone is not the natural ordering cone. However, it is a finitely generated cone. This can be used to formulate another multiobjective optimization problem to our set optimization problem which is now with respect to the natural (componentwise) ordering. For that, let now

$$\bar{K}^{m+1} := \begin{pmatrix} I_m & 1_m \\ I_m & -1_m \end{pmatrix} \in \mathbb{R}^{2m \times (m+1)} \tag{10.14}$$

where I_m is the m-dimensional identity matrix and 1_m is the m-dimensional all-one vector. It is easy to see that $\mathrm{kernel}(\bar{K}^{m+1}) = \{0_{\mathbb{R}^{m+1}}\}$ and

$$C^{m+1} = \left\{ y \in \mathbb{R}^{m+1} \,\middle|\, \bar{K}^{m+1} y \geq 0_{\mathbb{R}^{2m}} \right\},$$

i.e. C^{m+1} is polyhedral. Using [4, Lemma 1.18] or [16, Lemma 2.3.4] and Theorem 10.3.21 we obtain our main result of this section, which will also be the main result that we use in Section 10.4 for the construction of new test instances for set optimization.

Theorem 10.3.22 *Let S be a nonempty subset of $\mathbb{R}^n$, let the maps $c\colon S \to \mathbb{R}^m$ and $r\colon S \to \mathbb{R}_+$ be given, let the set-valued map $F\colon S \rightrightarrows \mathbb{R}^m$ be defined by*

$$F(x) := \{c(x)\} + \mathcal{B}_{\mathbb{R}^m}(r(x)) \text{ for all } x \in S,$$

and let the matrix $\bar{K}^{m+1}$ be defined as in (10.14). Then $\bar{x} \in S$ is a minimal solution of the set optimization problem (SOP$_{F,S}$) w.r.t. the order relation $\preccurlyeq^s_{\mathbb{R}^m_+}$ if and only if $\bar{x}$ is an efficient solution of the vector optimization problem (VOP$_{f,S}$) w.r.t. $\mathbb{R}^{2m}_+$ for $f\colon \mathbb{R}^n \to \mathbb{R}^{2m}$ and $f := \bar{K}^{m+1} \begin{pmatrix} c \\ r \end{pmatrix}$, i.e. of

$$\min_{x \in S} \bar{K}^{m+1} \begin{pmatrix} c(x) \\ r(x) \end{pmatrix}$$

w.r.t. the ordering cone $\mathbb{R}^{2m}_+$.

As a consequence of Theorem 10.3.22, we obtain the following corollary, which confirms the statement of Theorem 10.3.4 for another special case of the set H.

Corollary 10.3.23 *Let S be a nonempty subset of $\mathbb{R}^n$, let the map $f\colon S \to \mathbb{R}^m$ be given, let $r \in \mathbb{R}_+$, and let the set-valued map $F\colon S \rightrightarrows \mathbb{R}^m$ be defined by*

$$F(x) := \{f(x)\} + \mathcal{B}_{\mathbb{R}^m}(r) \text{ for all } x \in S.$$

Then $\bar{x} \in S$ is a minimal solution of the set optimization problem (SOP$_{F,S}$) w.r.t. the order relation $\preccurlyeq^s_{\mathbb{R}^m_+}$ if and only if $\bar{x}$ is an efficient solution of the vector optimization problem (VOP$_{f,S}$) w.r.t. $\mathbb{R}^m_+$.

Proof. By Theorem 10.3.22 $\bar{x} \in S$ is a minimal solution of the set optimization problem (SOP$_{F,S}$) w.r.t. the order relation $\preccurlyeq^s_{\mathbb{R}^m_+}$ if and only if $\bar{x}$ is an efficient solution of the vector optimization problem

$$\min_{x \in S} \begin{pmatrix} f(x) + r1_m \\ f(x) - r1_m \end{pmatrix}$$

w.r.t. the ordering cone $\mathbb{R}^{2m}_+$. This is equivalent to that $\bar{x}$ is an efficient solution of the vector optimization problem

$$\min_{x \in S} f(x)$$

w.r.t. $\mathbb{R}^m_+$, and we are done. $\square$

Finally, we use our results to verify the unique minimal solution of the set optimization problem stated in Example 10.3.16 by using Theorem 10.3.22.

Example 10.3.24 *We consider again the set optimization problem (SOP$_{F,S}$) defined as in Example 10.3.16. Using Theorem 10.3.22, it holds that $\bar{x} \in S$ is a minimal solution of the set optimization problem (SOP$_{F,S}$) w.r.t. the order relation $\preccurlyeq^s_{\mathbb{R}^2_+}$ if and only if $\bar{x}$ is an efficient solution of the vector optimization problem*

$$\min_{x \in [-1,1]} \begin{pmatrix} 3x^2 + 1 \\ 3x^2 + 1 \\ x^2 - 1 \\ x^2 - 1 \end{pmatrix} \qquad (10.15)$$

w.r.t. the ordering cone $\mathbb{R}_+^4$. Now, it is easy to see that the unique efficient solution of (10.15) w.r.t. $\mathbb{R}_+^4$ and, thus, the unique minimal solution of the corresponding set optimization problem (SOP$_{F,S}$) w.r.t. the order relation $\preccurlyeq_{\mathbb{R}_+^2}^s$ is given by $\bar{x} := 0$.

10.4 Implication on Set-valued Test Instances

In this section, we make some suggestions on how the results of the previous Section 10.3 can be used for the construction of set-valued test instances, based on known vector-valued or scalar-valued optimization problems in the case $X = \mathbb{R}^n$, $Y = \mathbb{R}^m$, and $C = \mathbb{R}_+^m$. In most cases, we will restrict ourselves to $m = 2$.

Such instances for set optimization problems (SOP$_{F,S}$) using a set-valued map $F: S \subset \mathbb{R}^n \to \mathbb{R}^m$ based on a fixed set $H \subset \mathbb{R}^m$ defined as in (10.4) or (10.6), i.e.

$$F(x) = \{f(x)\} + H \text{ or } F(x) = \varphi(x)H \text{ for all } x \in S,$$

can easily be established by directly applying Theorem 10.3.4 or Theorem 10.3.10, respectively. For this, only a suitable set $H \subset \mathbb{R}^m$ in terms of the formulated assumptions in the corresponding theorem and a multi-objective optimization problem $\min_{x \in S} f(x)$ (ideally with known set of all efficient solutions w.r.t. $\mathbb{R}_+^m$) or a scalar-valued optimization problem $\min_{x \in S} \varphi(x)$ (ideally with known set of all minimal solutions) has to be chosen. Using the statements of the mentioned theorems, the set of all minimal solutions of the set optimization problem (SOP$_{F,S}$) w.r.t. the order relation $\preccurlyeq_{\mathbb{R}_+^m}^\star$ and $\star \in \{l,u,s\}$ is given by the set of all efficient solutions of the chosen multi-objective optimization problem w.r.t. $\mathbb{R}_+^m$ or by the set of all a minimal solutions of the chosen scalar-valued optimization problem, respectively.

Moreover, test instances for set optimization problems (SOP$_{F,S}$) using a box-valued map $F: S \subset \mathbb{R}^n \to \mathbb{R}^m$ defined according to (10.10), i.e.

$$F(x) = [a(x), b(x)] \text{ with } a_i(x) \leq b_i(x) \text{ for all } x \in S \text{ and } i \in \{1,\ldots,m\},$$

can be defined by applying Corollary 10.3.15 and by choosing again a suitable multi-objective optimization problem $\min_{x \in S} f(x)$. For instance, if $f: S \subset \mathbb{R}^n \to \mathbb{R}^{2m}$ is a vector-valued function such that $f_i(x) \leq f_{i+m}(x)$ for all $x \in S$ and $i \in \{1,\ldots,m\}$ and we define the vector-valued maps $a, b: S \subset \mathbb{R}^n \to \mathbb{R}^m$ by

$$a(x) := \begin{pmatrix} f_1(x) \\ \vdots \\ f_m(x) \end{pmatrix} \text{ and } b(x) := \begin{pmatrix} f_{m+1}(x) \\ \vdots \\ f_{2m}(x) \end{pmatrix} \text{ for all } x \in S,$$

then the set of all minimal solutions of the set optimization problem (SOP$_{F,S}$) w.r.t. the order relation $\preccurlyeq_{\mathbb{R}_+^m}^l \mid \preccurlyeq_{\mathbb{R}_+^m}^u \mid \preccurlyeq_{\mathbb{R}_+^m}^s$ is given by the set of all efficient solutions of the vector optimization problem $\min_{x \in S} a(x) \mid \min_{x \in S} b(x) \mid \min_{x \in S} f(x)$ w.r.t. $\mathbb{R}_+^m \mid \mathbb{R}_+^m \mid \mathbb{R}_+^{2m}$.

It takes more effort to construct test instances for set optimization problems (SOP$_{F,S}$) with a ball-valued map $F: S \subset \mathbb{R}^n \to \mathbb{R}^m$, i.e., see (10.11), with

$$F(x) := \{c(x)\} + \mathcal{B}_{\mathbb{R}^m}(r(x)) \text{ for all } x \in S$$

with $c\colon S \to \mathbb{R}^m$ and $r\colon S \to \mathbb{R}_+$. We need the following result, which follows with Theorem 10.3.22 and [1, Theorem 3.1(2)]:

Lemma 10.4.1 *Let S be a nonempty subset of $\mathbb{R}^n$ and a map $f\colon S \to \mathbb{R}^m$ be given. Moreover, let the matrix $\bar{K}^{m+1}$ be defined as in (10.14), let $\rho \in \mathbb{R}^m$ be a vector and $H \in \mathbb{R}^{m \times m}$ be a matrix such that*
$$r(x) := \rho^\top f(x) \geq 0 \text{ for all } x \in S, \tag{10.16}$$
and such that the matrix $\bar{H} \in \mathbb{R}^{2m \times m}$ with
$$\bar{H} := \bar{K}^{m+1}\begin{pmatrix} H \\ \rho^\top \end{pmatrix} \text{ has full rank } m, \tag{10.17}$$
and such that
$$\{z \in \mathbb{R}^m \mid \bar{H}z \in \mathbb{R}_+^{2m}\} = \mathbb{R}_+^m. \tag{10.18}$$
Then $\bar{x} \in S$ is a minimal solution of the set optimization problem $(\text{SOP}_{F,S})$ w.r.t. the order relation $\preccurlyeq^s_{\mathbb{R}_+^m}$ and with $F\colon S \rightrightarrows \mathbb{R}^m$ defined by
$$F(x) := \{Hf(x)\} + \mathcal{B}_{\mathbb{R}^m}(r(x)) \text{ for all } x \in S,$$
if and only if $\bar{x} \in S$ is an efficient solution of the vector optimization problem $(\text{VOP}_{f,S})$ w.r.t. $\mathbb{R}_+^m$.

To illustrate how Lemma 10.4.1 can be used for the construction of test instances, we restrict ourselves to the case $m = 2$ and we choose the following matrices $H \in \mathbb{R}^{2 \times 2}$ and the following vectors $\rho \in \mathbb{R}^2$, which guarantee that (10.17) and (10.18) are fulfilled:

(i) $H := \begin{pmatrix} \frac{1}{2} & \frac{1}{2} \\ \frac{1}{2} & \frac{1}{2} \end{pmatrix}$, $\rho := (\frac{1}{2}, -\frac{1}{2})^\top$, and thus $\bar{H} := \bar{K}^3 \begin{pmatrix} H \\ \rho^\top \end{pmatrix} = \begin{pmatrix} 1 & 1 & 0 & 0 \\ 0 & 0 & 1 & 1 \end{pmatrix}^\top$.

(ii) $H := \begin{pmatrix} 0 & \frac{1}{2} \\ 1 & \frac{1}{2} \end{pmatrix}$, $\rho := (0, -\frac{1}{2})^\top$, and thus $\bar{H} := \bar{K}^3 \begin{pmatrix} H \\ \rho^\top \end{pmatrix} = \begin{pmatrix} 0 & 1 & 0 & 1 \\ 0 & 0 & 1 & 1 \end{pmatrix}^\top$.

(iii) $H := \begin{pmatrix} \frac{1}{2} & 1 \\ \frac{1}{2} & 0 \end{pmatrix}$, $\rho := (\frac{1}{2}, 0)^\top$, and thus $\bar{H} := \bar{K}^3 \begin{pmatrix} H \\ \rho^\top \end{pmatrix} = \begin{pmatrix} 1 & 1 & 0 & 0 \\ 1 & 0 & 1 & 0 \end{pmatrix}^\top$.

Furthermore also in all three cases for a map $f\colon S \to \mathbb{R}^2$ it holds for all $x \in S$
$$f_1(x) \geq 0 \text{ and } f_2(x) \leq 0 \Rightarrow r(x) := \rho^\top f(x) \geq 0.$$

To explain our approach, we use in the following two examples which are slight modifications of the test instances from [4, p.145] and [7]. The reason for the slight modifications is to guarantee $r(x) \geq 0$ for all $x \in S$ which is reached by subtracting a suitable constant from the corresponding second objective function. Note that this kind of modification has no influence on the set of all efficient solutions.

We start with an example where the image set of the chosen bicriteria optimization problem is convex.

Example 10.4.2 Let $S := \{x \in \mathbb{R}_+^2 \mid x_1^2 - 4x_1 + x_2 + 1.5 \leq 0\}$, $Y = \mathbb{R}^2$, and the vector-valued map $f : S \to \mathbb{R}^2$ be defined by

$$f(x) = \begin{pmatrix} f_1(x) \\ f_2(x) \end{pmatrix} := \begin{pmatrix} \sqrt{1+x_1^2} \\ x_1^2 - 4x_1 + x_2 \end{pmatrix} \text{ for all } x \in S.$$

For the vector optimization problem $\min_{x \in S} f(x)$ the set of all efficient solutions w.r.t. $\mathbb{R}_+^2$ is given by

$$M := \left\{ x \in \mathbb{R}_+^2 \,\middle|\, x_1 \in \left[2 - \frac{\sqrt{10}}{2}, 2\right], x_2 = 0 \right\}. \tag{10.19}$$

Moreover, $f_1(x) \geq 0$ and $f_2(x) \leq 0$ is satisfied for all $x \in S$.
If now H and ρ are chosen according to **(i)**, **(ii)**, and **(iii)** above, and the three test instances **Test 1**, **Test 2**, and **Test 3** are defined by

$$\min_{x \in S} \{c(x)\} + \mathcal{B}_{\mathbb{R}^2}(r(x))$$

with $\begin{pmatrix} c_1(x) \\ c_2(x) \\ r(x) \end{pmatrix} := \begin{pmatrix} H \\ \rho^\top \end{pmatrix} \begin{pmatrix} f_1(x) \\ f_2(x) \end{pmatrix}$, then we obtain for

(i) Test 1:

$$\begin{pmatrix} c_1(x) \\ c_2(x) \\ r(x) \end{pmatrix} := \begin{pmatrix} \frac{1}{2}\left[\sqrt{1+x_1^2} + x_1^2 - 4x_1 + x_2\right] \\ \frac{1}{2}\left[\sqrt{1+x_1^2} + x_1^2 - 4x_1 + x_2\right] \\ \frac{1}{2}\left[\sqrt{1+x_1^2} - x_1^2 + 4x_1 - x_2\right] \end{pmatrix},$$

(ii) Test 2:

$$\begin{pmatrix} c_1(x) \\ c_2(x) \\ r(x) \end{pmatrix} := \begin{pmatrix} \frac{1}{2}\left[x_1^2 - 4x_1 + x_2\right] \\ \frac{1}{2}\left[2\sqrt{1+x_1^2} + x_1^2 - 4x_1 + x_2\right] \\ -\frac{1}{2}\left[x_1^2 - 4x_1 + x_2\right] \end{pmatrix}, \text{ and}$$

(iii) Test 3:

$$\begin{pmatrix} c_1(x) \\ c_2(x) \\ r(x) \end{pmatrix} := \begin{pmatrix} \frac{1}{2}\left[\sqrt{1+x_1^2} + 2x_1^2 - 8x_1 + 2x_2\right] \\ \frac{1}{2}\sqrt{1+x_1^2} \\ \frac{1}{2}\sqrt{1+x_1^2} \end{pmatrix}.$$

Using Lemma 10.4.1 for all of the three test instances **Test 1**, **Test 2**, and **Test 3** the set of all minimal solution w.r.t. the order relation $\preccurlyeq_{\mathbb{R}_+^2}^s$ is also given by the set M defined in (10.19). In Figure 10.1 we illustrate by the black circles the boundaries of $F(x)$ for some $x \in M$.

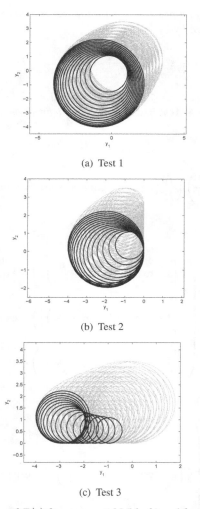

(a) Test 1

(b) Test 2

(c) Test 3

Figure 10.1: Boundaries of $F(x)$ for some $x \in M$ (black) and for some $x \in S \setminus M$ (grey) in Example 10.4.2.

The image set of the chosen bicriteria optimization problem in the following second example is nonconvex. An interesting property of this bicriteria optimization problem is the arbitrary scalability w.r.t. the dimension n of the preimage space $\mathbb{R}^n$.

Example 10.4.3 *Let $n \in \mathbb{N}$, $S := [-4, 4]^n$, $Y = \mathbb{R}^2$, and the vector-valued map $f \colon S \to \mathbb{R}^2$ be defined by*

$$f(x) = \begin{pmatrix} f_1(x) \\ f_2(x) \end{pmatrix} := \begin{pmatrix} 1 - \exp\left(-\sum_{i=1}^{n}\left(x_i - \frac{1}{\sqrt{n}}\right)^2\right) \\ -\exp\left(-\sum_{i=1}^{n}\left(x_i + \frac{1}{\sqrt{n}}\right)^2\right) \end{pmatrix} \quad \text{for all } x \in S.$$

The set of all efficient solutions w.r.t. $\mathbb{R}^2_+$ for the vector optimization problem $\min_{x \in S} f(x)$ is according to [7] given by

$$M := \left\{ x \in \mathbb{R}^n \,\middle|\, x_1 \in \frac{1}{\sqrt{n}}[-1,1],\, x_i = x_1,\, i \in \{2,\ldots,n\} \right\}, \tag{10.20}$$

and $f_1(x) \geq 0$ as well as $f_2(x) \leq 0$ is satisfied for all $x \in S$.
The test instances **Test 4**, **Test 5**, and **Test 6** are defined analogously to Example 10.4.2 by

$$\min_{x \in S=[-4,4]^n} \{c(x)\} + \mathcal{B}_{\mathbb{R}^2}(r(x))$$

with $\begin{pmatrix} c_1(x) \\ c_2(x) \\ r(x) \end{pmatrix} := \begin{pmatrix} H \\ \rho^\top \end{pmatrix} \begin{pmatrix} f_1(x) \\ f_2(x) \end{pmatrix}$ and again in consideration of **(i)**, **(ii)**, and **(iii)**. This leads to

(i) **Test 4:**

$$\begin{pmatrix} c_1(x) \\ c_2(x) \\ r(x) \end{pmatrix} := \begin{pmatrix} \frac{1}{2}\left[1 - \exp\left(-\sum_{i=1}^{n}\left(x_i - \frac{1}{\sqrt{n}}\right)^2\right) - \exp\left(-\sum_{i=1}^{n}\left(x_i + \frac{1}{\sqrt{n}}\right)^2\right)\right] \\ \frac{1}{2}\left[1 - \exp\left(-\sum_{i=1}^{n}\left(x_i - \frac{1}{\sqrt{n}}\right)^2\right) - \exp\left(-\sum_{i=1}^{n}\left(x_i + \frac{1}{\sqrt{n}}\right)^2\right)\right] \\ \frac{1}{2}\left[1 - \exp\left(-\sum_{i=1}^{n}\left(x_i - \frac{1}{\sqrt{n}}\right)^2\right) + \exp\left(-\sum_{i=1}^{n}\left(x_i + \frac{1}{\sqrt{n}}\right)^2\right)\right] \end{pmatrix},$$

(ii) **Test 5:**

$$\begin{pmatrix} c_1(x) \\ c_2(x) \\ r(x) \end{pmatrix} := \begin{pmatrix} -\frac{1}{2}\exp\left(-\sum_{i=1}^{n}\left(x_i + \frac{1}{\sqrt{n}}\right)^2\right) \\ \frac{1}{2}\left[2 - 2\exp\left(-\sum_{i=1}^{n}\left(x_i - \frac{1}{\sqrt{n}}\right)^2\right) - \exp\left(-\sum_{i=1}^{n}\left(x_i + \frac{1}{\sqrt{n}}\right)^2\right)\right] \\ \frac{1}{2}\exp\left(-\sum_{i=1}^{n}\left(x_i + \frac{1}{\sqrt{n}}\right)^2\right) \end{pmatrix}, \text{ and}$$

(iii) **Test 6:**

$$\begin{pmatrix} c_1(x) \\ c_2(x) \\ r(x) \end{pmatrix} := \begin{pmatrix} \frac{1}{2}\left[1 - \exp\left(-\sum_{i=1}^{n}\left(x_i - \frac{1}{\sqrt{n}}\right)^2\right) - 2\exp\left(-\sum_{i=1}^{n}\left(x_i + \frac{1}{\sqrt{n}}\right)^2\right)\right] \\ \frac{1}{2}\left[1 - \exp\left(-\sum_{i=1}^{n}\left(x_i - \frac{1}{\sqrt{n}}\right)^2\right)\right] \\ \frac{1}{2}\left[1 - \exp\left(-\sum_{i=1}^{n}\left(x_i - \frac{1}{\sqrt{n}}\right)^2\right)\right] \end{pmatrix}.$$

Set Optimization Problems Reducible to Vector Optimization Problems ■ **263**

By Lemma 10.4.1 for **Test 4**, **Test 5**, *and* **Test 5** *the set of all minimal solution w.r.t. the order relation* $\preccurlyeq^s_{\mathbb{R}^2_+}$ *is given by the set M defined in* (10.20). *In Figure 10.2, we illustrate for some* $x \in M$ *by the black circles the boundaries of* $F(x)$.

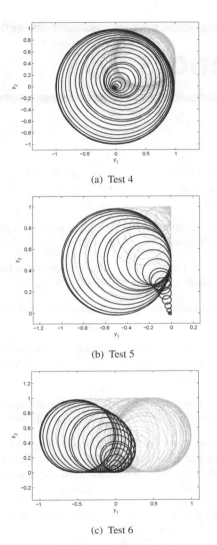

(a) Test 4

(b) Test 5

(c) Test 6

Figure 10.2: Boundaries of $F(x)$ for some $x \in M$ (black) and for some $x \in S \setminus M$ (grey) in Example 10.4.3.

References

[1] S. Dempe, G. Eichfelder and J. Fliege. On the effects of combining objectives in multi-objective optimization. *Mathematical Methods of Operations Research*, 82(1): 1–18, 2015.

[2] S. Dempe and V. Kalashnikov. *Optimization with Multivalued Mappings: Theory, Applications, and Algorithms*. Springer, 2006.

[3] M. Dhingra and C. S. Lalitha. Approximate solutions and scalarization in set-valued optimization. *Optimization*, 66(11): 1793–1805, 2017.

[4] G. Eichfelder. *Adaptive Scalarization Methods in Multiobjective Optimization*. Springer, 2008.

[5] G. Eichfelder and J. Jahn. Vector optimization problems and their solution concepts. pp. 1–27. *In*: Q. H. Ansari and J. -C. Yao (eds.). *Recent Developments in Vector Optimization*, Springer, 2012.

[6] G. Eichfelder, C. Krüger and A. Schöbel. Decision uncertainty in multiobjective optimization. *Journal of Global Optimization*, 69(2): 485-510, 2017.

[7] C. M. Fonseca and P. J. Fleming. Multiobjective genetic algorithms made easy: Selection sharing and mating restriction. *In Proc. First International Conference on Genetic Algorithms in Engineering Systems: Innovations and Applications*, Sheffield, UK, Sep., 45–52, 1995.

[8] E. Hernández and R. López. Some useful set-valued maps in set optimization. *Optimization*, 66(8): 1273–1289, 2017.

[9] J. Jahn. *Vector Optimization—Theory, Applications, and Extensions (2nd ed.)*. Springer, 2011.

[10] J. Jahn. Vectorization in set optimization. *Journal of Optimization Theory and Applications*, 167(3): 783–795, 2015.

[11] J. Jahn. A derivative-free descent method in set optimization. *Computational Optimization and Applications*, 60(2): 393–411, 2015.

[12] J. Jahn and T. X. D. Ha. New order relations in set optimization. *Journal of Optimization Theory and Applications*, 148: 209–236, 2011.

[13] A. A. Khan, C. Tammer and C. Zălinescu. *Set-valued Optimization: An Introduction with Applications*. Springer, 2015.

[14] E. Köbis and M. A. Köbis. Treatment of set order relations by means of a nonlinear scalarization functional: A full characterization. *Optimization*, 65(10): 1805–1827, 2016.

[15] D. Kuroiwa. The natural criteria in set-valued optimization. *RIMS Kokyuroku*, 1031: 85–90, 1998.

[16] Y. Sawaragi, H. Nakayama and T. Tanino. *Theory of Multiobjective Optimization*. Number 176 in Mathematics in Science and Engineering. Academic Press, 1985.

[17] C. Schrage and A. Löhne. An algorithm to solve polyhedral convex set optimization problems. *Optimization*, 62(1): 131–141, 2013.

[18] R. C. Young. The algebra of many-valued quantities. *Mathematische Annalen*, 104: 260–290, 1931.

Chapter 11

Abstract Convexity and Solvability Theorems

Ali Reza Doagooei
Department of Applied Mathematics, Shahid Bahonar University of Kerman, Kerman, Iran. doagooei@uk.ac.ir

11.1 Introduction

One of the main results from convex analysis states that every convex, proper and lower semi-continuous function is the upper envelope of a set of affine functions, that is

$$f(x) = \sup\{h(x) : h \text{ is an affine function and } h \leq f\}. \tag{11.1}$$

This is concluded from the fact that a point out of a closed convex set can be separated from that set by an affine function (geometrically by a hyperplane). Since affine functions are constructed by linear functions, the aforementioned results demonstrate that linear functions are instrumental for studying convexity. This observation stimulates the development of a rich theory of convexity without linearity, known as abstract convexity. Indeed, functions represented as upper envelopes of subsets of a set H, containing sufficiently simple (elementary) functions, are studied in this theory. These functions are called abstract convex, with respect to H or H-convex functions. We refer to three famous books [21, 22, 27] in the field of abstract convexity theory. The set $\text{supp}(f,H) := \{h \in H, h \leq f\}$ of all H-minorants of a function f is called the support set of this function. The support set accumulates global information about the function f in terms of the set of elementary functions H.

In local optimization, the local information of a function is vital in establishing necessary and sufficient conditions for finding a local minimum. Using the classical calculus and modern techniques of nonsmooth analysis, this task, however, is well-done by local approximations of functions. In contrast, finding a global optimum of a function requires the global information of the function. Abstract convexity provides this global information and presents some approaches in order to establish necessary and sufficient conditions for global minimums, see for examples [1, 6, 21, 22, 23, 26]. From theoretical point of view, one of these approaches is solvability theorem.

The minimization of an objective function subject to some inequality constraints can be represented in the following form: A certain inequality is a consequence of the given system of inequalities. The dual representation of this assertion is known as solvability theorem. The solvability approach can be considered as a nonlinear extension of the well-known Farkas lemma in linear programming. Mathematically speaking, consider the maximization problem

$$\text{Maximize } p(x)$$
$$\text{subject to } p_t(x) \leq 0, \qquad (P)$$

where $p, p_t : X \longrightarrow \mathbb{R}$ are arbitrary functions. Then, $x_0 \in X$ is an optimal solution for this problem if and only if

$$x \in X, \ p_t(x) \leq 0 \text{ for all } t \in T \implies p(x) \leq p(x_0). \qquad (P')$$

The dual representation of this assertion in the setting of support sets and subdifferentials for conjugate functions of p and p_t is known as solvability theorem. Using solvability theorems, necessary and sufficient conditions for finding global optimum of the problem (P) are established.

In this chapter, we study solvability theorems for some especial classes of abstract convex functions presented in [19, 20].

The solvability theorems are also well-known as Farkas-type theorems in the literature. This theory has found various versions in both finite dimensional (see for example [9, 12, 13, 15, 24] and references therein) and infinite dimensional (see for example [2, 4, 5, 10, 11, 28] and references therein) spaces. In the following, we briefly state some of them. For a survey of generalizations of Farkass lemma, starting from systems of linear inequalities to a broad variety of non-linear systems, see [3].

A general version of solvability theorems for systems of abstract convex functions is presented in [24], where the set of elementary functions H is a conic set. This means that $h \in H, \lambda > 0$ implies $\lambda h \in H$. In this case, the dual representation of (P') has been presented by the means of support set. Although the results are very general, many sets of elementary functions are not conic sets. One of our goals in this chapter is to present some non-conic class of elementary functions and establish solvability theorems for them.

Farkas-type theorems for systems of inequalities involving positively homogeneous functions in the Euclidean spaces have been investigated in [24] by means of generalized conjugation theory and in [12] by means of support sets. In [4], Farkas lemma and Gale's theorem for infinite systems of inequalities involving positively homogeneous functions defined on a locally bounded topological vector space is presented by means of abstract convexity and the theory of downward sets.

Increasing and Positively Homogeneous (IPH) functions are well-known in mathematical economics. For example, production functions under the assumption of constant returns to scale are IPH. Furthermore, the level sets of IPH functions are required in the study of various models of the economic equilibrium and dynamics, see [14] for more details. Therefore, studying IPH programming problems is important in both optimization theory and applications. In this chapter, we shall establish solvability theorems for infinite systems of inequalities involving IPH functions.

Increasing and CoRadiant (ICR) functions constitute a wide branch of monotonic analysis, see [7, 8, 22]. Every IPH function is ICR and every increasing and convex along rays function is also ICR, provided that it is nonnegative at the origin. More properties and examples of ICR functions may be found in [22, Section 3.2]. Applying corresponding results of IPH functions, we present the solvability theorems for the systems involving ICR functions.

Let $\mathbf{1}$ be a vector in the finite dimensional Euclidean space (regardless of its dimension) whose coordinates are all one. Recall that a function $p : \mathbb{R}^n \longrightarrow \mathbb{R}^m$ is called topical if this function is increasing (i.e. $x \leq y \implies p(x) \leq p(y)$ in which '$\leq$' is the natural order in finite dimensional Euclidean spaces) and plus-homogeneous (i.e. $p(x+\mu\mathbf{1}) = p(x) + \mu\mathbf{1}$ for all $x \in \mathbb{R}^n$ and all $\mu \in \mathbb{R}$). In the next part, we shall develop solvability theorems for infinite systems of inequalities involving topical functions. We shall also define vector-valued topical functions and present similar results for the system involving them.

Now let us demonstrate an admirable application of solvability theorems. As claimed before, solvability theorems study the behavior of a system of functions and present the global information about that. This leads to the presentation of a duality for the problem (P). It is well-known that the unconstrained programming problems are easier to deal with than the constrained ones. There are several methods of transforming a constrained problem to an unconstrained counterpart, which usually need extra variables. However, the solvability theorems provided results to obtain an unconstrained counterpart for the problem (P), namely a dual problem, without adding extra variables. In particular, we apply this technique for standard linear programming problems with nonnegative coefficients and transform these problems to unconstrained concave counterparts.

Throughout this chapter (unless otherwise stated), we assume that X is a topological vector space, ordered by a closed convex cone S. We say that $x \leq_S y$ if and only if $y - x \in S$. For the sake of simplicity we denote $x \leq y$, if there is no confusion. We also assume that S is pointed, i.e., $S \cap (-S) = \{0\}$. $\overline{\mathbb{R}} := [-\infty, +\infty]$ is considered as the set of all extended real numbers and $\mathbb{R}_{+\infty} := \mathbb{R} \cup \{+\infty\}$.

11.2 Abstract Convex Functions

In this section, we study abstract convex functions and present the basic notions of it. Some examples of abstract convex functions are presented. In particular, we study three classes of abstract convex functions, for which we will establish solvability theorems in next sections. The main results of this section are found in [6, 7, 8, 16, 17, 18, 22, 25]. We begin with the definition of abstract convex functions.

Definition 11.2.1 *Let $\mathcal{H}$ be a nonempty set of functions $h : X \longrightarrow \mathbb{R}_{+\infty}$. A function $f : X \longrightarrow \overline{\mathbb{R}}$ is called abstract convex with respect to $\mathcal{H}$ or $\mathcal{H}$-convex if there exists a set $U \subseteq \mathcal{H}$ such that f is the upper envelope of U, that is, for all $x \in X$:*

$$f(x) = \sup_{h \in U} h(x),$$

with the convention $\sup \emptyset = -\infty$. The set $\mathcal{H}$ is called the set of elementary functions.

Before we present an example, let us remind some definitions. Let $f : X \longrightarrow \overline{\mathbb{R}}$. f is called positively homogeneous if $f(\lambda x) = \lambda f(x)$, for all $x \in X$ and $\lambda > 0$. f is called sublinear if f is convex and positively homogeneous.

Example 11.2.2 *Let $\mathcal{H}$ be the set of all continuous linear functionals defined on X. Then $f : X \longrightarrow (-\infty, +\infty]$ is an $\mathcal{H}$-convex function if and only if f is proper, lower semi-continuous and sublinear. Also, f is an $\mathcal{H}_L$-convex function if and only if f is lower semi-continuous and convex, where $\mathcal{H}_L := \{h + c, \ h \in H, \ c \in \mathbb{R}\}$.*

Comparing with (11.1), Example 11.2.2 shows that the concept of abstract convexity is an extension of the concept of convexity. Let us present some more examples of abstract convex functions. Consider the function $l : X \times X \longrightarrow [0, +\infty]$ defined by

$$l(x,y) = \max\{\lambda \geq 0 : \lambda y \leq x\}, \tag{11.2}$$

with the convention $\max \emptyset = 0$. The following properties of this function have been presented in [18]. By a simple calculation, the proof can be concluded from the definition of l. Let $\gamma > 0$, and $x, x', y, y' \in X$. Thus

$$l(\gamma x, y) = \gamma l(x, y), \tag{11.3}$$

$$l(x, \gamma y) = \frac{1}{\gamma} l(x, y), \tag{11.4}$$

$$l(x, y) = +\infty \implies y \in -S, \tag{11.5}$$

$$l(x, x) = 1 \iff x \notin -S, \tag{11.6}$$

$$x \in S, \ y \in -S \implies l(x, y) = +\infty, \tag{11.7}$$

$$x \leq x' \implies l(x, y) \leq l(x', y), \tag{11.8}$$

$$y \leq y' \implies l(x, y) \geq l(x, y'). \tag{11.9}$$

Later, we also need the coupling function $u : X \times X \longrightarrow [0, +\infty]$ defined by

$$u(x, y) := \min\{\mu \geq 0 : \mu y \geq x\}, \tag{11.10}$$

(with the convention $\min \emptyset := +\infty$).

Similar to the function l, we have:

$$u(\gamma x, y) = \gamma u(x, y), \tag{11.11}$$

$$u(x, \gamma y) = \frac{1}{\gamma} u(x, y), \tag{11.12}$$

$$u(x, y) = 0 \iff x \in -S, \tag{11.13}$$

$$u(x, x) = 1 \iff x \notin -S, \tag{11.14}$$

$$x \leq x' \implies u(x, y) \leq u(x', y), \tag{11.15}$$

$$y \leq y' \implies u(x, y) \geq u(x, y'). \tag{11.16}$$

Using the coupling functions l and u, we construct two classes of functions defined on X. Let $y \in X$. Define $l_y(x) := l(x, y)$ and $u_y(x) := u(x, y)$ for all $x \in X$. Set also

$$L := \{l_y : y \in X\} \tag{11.17}$$

and
$$U := \{u_y : y \in X\}. \tag{11.18}$$

The following theorem shows that IPH functions are abstract convex.

Theorem 11.2.3 ([18]) *Let $p : X \longrightarrow [0, +\infty]$ be a function. Then p is IPH if and only if p is L-convex. In this case*

$$p(x) = \sup\{l_y(x) : y \in X, \; p(y) \geq 1\} \; \forall x \in X,$$

with the convention $\sup \emptyset := 0$.

IPH functions constitute a tiny subclass of Increasing and CoRadiant (ICR) functions. Recall that the function $f : X \longrightarrow \overline{\mathbb{R}}$ is called coradiant if $f(\lambda x) \geq \lambda f(x)$, for all $x \in X$ and $\lambda \in [0, 1]$. In the section of Introductions, we presented some example of ICR functions. For another example, let σ be a positive constant. The Cobb-Douglas function $\xi : \mathbb{R}_+^n \longrightarrow \mathbb{R}$ of the form

$$\xi(x) = \sigma x_1^{\gamma_1} x_2^{\gamma_2} \cdots x_n^{\gamma_n}, \text{ with } \gamma \geq 0, \; \sum_i \gamma \leq 1$$

is an ICR function as well. It is shown in [7] (compare with [8]) that there is a class of functions, considered as an extension of the set L, namely Γ, such that every ICR function is Γ-convex. To see this, let the function $l : X \times X \times (0, +\infty) \longrightarrow [0, +\infty]$ defined by

$$l(x, y, \alpha) := \max\{\lambda \in (0, \alpha] : x \geq \lambda y\}, \tag{11.19}$$

with the convention $\max \emptyset = 0$ and $x \geq \lambda y$ if and only if $x - \lambda y \in S$.
For every $x, x', y, y' \in X$, $\mu, \alpha, \alpha' > 0$ and $\gamma \in (0, 1]$ we have

$$l(\mu x, y, \alpha) = \mu l(x, y, \frac{\alpha}{\mu}), \tag{11.20}$$

$$l(x, \mu y, \alpha) = \frac{1}{\mu} l(x, y, \mu \alpha), \tag{11.21}$$

$$x \leq x' \implies l(x, y, \alpha) \leq l(x', y, \alpha), \tag{11.22}$$
$$y \leq y' \implies l(x, y', \alpha) \geq l(x, y, \alpha), \tag{11.23}$$
$$\alpha \leq \alpha' \implies l(x, y, \alpha) \leq l(x, y, \alpha'), \tag{11.24}$$
$$l(\gamma x, y, \alpha) \geq \gamma l(x, y, \alpha), \tag{11.25}$$
$$l(x, \gamma y, \alpha) \leq \frac{1}{\gamma} l(x, y, \alpha), \tag{11.26}$$
$$l(x, y, \alpha) = \alpha \iff \alpha y \leq x. \tag{11.27}$$

Let $y \in X$ and $\alpha \in (0, +\infty)$ be fixed. Define the function $l_{y,\alpha} : X \longrightarrow [0, +\infty]$ by

$$l_{y,\alpha}(x) := l(x, y, \alpha), \quad x \in X. \tag{11.28}$$

It follows from (11.22) and (11.25) that $l_{y,\alpha}$ is an ICR function. Set

$$\Gamma := \{l_{y,\alpha} \: : \: y \in X, \; \alpha > 0\}. \tag{11.29}$$

The following observation demonstrates the Γ-convexity of the ICR functions.

Theorem 11.2.4 *([7, Theorem 3.2]) Let $f : X \longrightarrow [0,+\infty]$ be a function. f is ICR if and only if*

$$f(x) = \sup_{l_{y,\alpha} \in A} l_{y,\alpha}(x),$$

where $A := \{l_{y,\alpha} \in \Gamma : f(\alpha y) \geq \alpha\}$. Hence, f is ICR if and only if f is Γ-convex.

The last example of abstract convex functions presented in the sequel is the abstract convexity of increasing and plus-homogeneous (topical) functions. The corresponding results can be found in [16, 25].

Let **1** be an arbitrary (but fixed) point of int S. $f : X \longrightarrow \bar{\mathbb{R}} := [-\infty, +\infty]$ is called plus-homogeneous if $f(x + \lambda \mathbf{1}) = f(x) + \lambda$ for all $x \in X$ and all $\lambda \in \mathbb{R}$. f is called topical if it is increasing and plus-homogeneous. To construct the set of elementary functions, assume that $\varphi : X \times X \longrightarrow \mathbb{R}$ defined by

$$\varphi(x,y) := \sup\{\lambda \in \mathbb{R} : \lambda \mathbf{1} \leq x + y\}, \quad x, y \in X. \tag{11.30}$$

Let $(X, \|\cdot\|)$ be a Banach space. Recall that cone S is called normal if there exists a constant $m > 0$ such that

$$\|x\| \leq m\|y\|, \tag{11.31}$$

whenever $0 \leq x \leq y$, and $x, y \in X$. During this chapter, when we talk about topical functions, we implicitly assume that X is a Banach space and S is normal.

Notice that the set $B_1 := \{x \in X : -\mathbf{1} \leq x \leq \mathbf{1}\}$ can be considered as the unit ball generated by a certain norm $\|\cdot\|_1$, which is equivalent to the initial norm $\|.\|$ of X.

Assume without loss of generality that $\|\cdot\| = \|\cdot\|_1$. This leads to the fact that the set $\{\lambda \in \mathbb{R} : \lambda \mathbf{1} \leq x + y\}$ is non-empty, closed and bounded above.

For each $y \in X$, define the function $\varphi_y : X \longrightarrow \mathbb{R}$ by

$$\varphi_y(x) := \varphi(x, y), \quad x \in X. \tag{11.32}$$

According to the definition of φ, one has:

$$-\infty < \varphi_y(x) \leq m\|x + y\|, \quad x, y \in X; \tag{11.33}$$

$$\varphi_y(x)\mathbf{1} \leq x + y, \quad x, y \in X; \tag{11.34}$$

$$\varphi_y(x) = \varphi_x(y), \quad x, y \in X; \tag{11.35}$$

$$\varphi_{-x}(x) = \sup\{\lambda \in \mathbb{R} : \lambda \mathbf{1} \leq x - x = 0\} = 0, \quad x \in X \tag{11.36}$$

where m satisfies (11.31). Observe that the function φ_y is topical. In the following we denote by Φ the set of all functions φ_l ($l \in X$) defined by (11.32), that is:

$$\Phi := \{\varphi_y : y \in X\}. \tag{11.37}$$

The following theorem shows the abstract convexity of topical functions with respect to the set Φ.

Theorem 11.2.5 *([16], Theorem 2.1) Let $f : X \longrightarrow \bar{\mathbb{R}} := [-\infty, +\infty]$ be a function and Φ be as (11.37). Then f is topical if and only if f is Φ-convex.*

In the rest of this section, we recall some essential concepts from the theory of abstract convexity. Let $\mathcal{H}$ be a set of elementary functions. For a function $f : X \longrightarrow \mathbb{R}_{+\infty}$, the Fenchel-Moreau $\mathcal{H}$-conjugate function $f^*_{\mathcal{H}}$ of f is defined by

$$f^*_{\mathcal{H}}(h) := \sup_{x \in dom(f)} (h(x) - f(x)), \quad h \in \mathcal{H},$$

where $dom(f) := \{x \in X : f(x) < +\infty\}$. Similarly, the Fenchel-Moreau X-conjugate g^*_X of an extended real valued function g defined on $\mathcal{H}$ is given by

$$g^*_X(x) := \sup_{h \in dom(g)} (h(x) - g(h)), \quad x \in X.$$

The Fenchel-Moreau $\mathcal{H}$-conjugate functions are instrumental to study solvability theorems. In the following lemma, we present a property of Fenchel-Moreau conjugate functions of topical functions which will be used later.

Lemma 11.2.6 *Let $p : X \longrightarrow \mathbb{R}$ be a topical function. Then*

$$(\varphi_l, r) \in \text{epi } p^*_\Phi \iff p(-l + r\mathbf{1}) \geq 0.$$

Proof. Let $(\varphi_l, r) \in \text{epi } p^*_\Phi$. This means that $p^*_\Phi(\varphi_l) \leq r$. Thus, $\varphi_l(x) - p(x) \leq r$ for all $x \in X$. Letting $x := -l + r\mathbf{1}$, one has $r - p(-l + r\mathbf{1}) \leq r$. So $p(-l + r\mathbf{1}) \geq 0$.
Assume now that $p(-l + r\mathbf{1}) \geq 0$ and $x \in X$ is arbitrary. Applying the definition of $\varphi_l(x)$, we have $\varphi_l(x)\mathbf{1} \leq l + x$ which implies that $-l + \varphi_l(x)\mathbf{1} \leq x$. Since p is increasing and plus-homogeneous, $p(-l) + \varphi_l(x) \leq p(x)$. Therefore, $\varphi_l(x) - p(x) \leq -p(-l) \leq r$, for all X. Hence, $p^*_\Phi(\varphi_l) \leq r$. This completes the proof. □

Similar to Fenchel-Moreau $\mathcal{H}$-conjugate functions, we will also define polar functions for IPH and ICR functions. We show that they play an essential role for establishing solvability theorems for the system of inequalities involving IPH function as well as ICR functions.

We finish this section by two definitions. For the function f we denote by $\text{epi } f$, the epigraph of f defined by

$$\text{epi } f = \{(x, r) \in X \times \mathbb{R} : f(x) \leq r\},$$

and by $\text{epi}_s f$, the strict epigraph of f defined by

$$\text{epi}_s f = \{(x, r) \in X \times \mathbb{R} : f(x) < r\}.$$

11.3 Solvability Theorems for Real-valued Systems of Inequalities

In this section, we present solvability theorems for three classes of real-valued abstract convex functions: IPH, ICR and topical functions. Due to the closed relation between IPH and ICR functions, IPH functions are the best tools to study ICR functions. In this section we establish a version of solvability theorem for IPH functions. Then, we apply the obtained results to investigate solvability theorems for ICR functions. To reach these goals, we need to introduce polar functions and study their properties. The main results of Subsection 11.3.1, Subsection 11.3.2 and Subsection 11.3.3 can be found in [7, 18, 19, 22], [19] and [20], respectively.

11.3.1 Polar Functions of IPH and ICR Functions

Whenever an arbitrary function defined on X is at hand, one may develop its domain to $X \times \mathbb{R}$ and obtain a positively homogeneous extension of it. More precisely, let $f : X \longrightarrow [0, +\infty]$ be a function. The positively homogeneous extension of the function f is the function $\hat{f} : X \times \mathbb{R} \longrightarrow [0, +\infty]$ defined by

$$\hat{f}(x, \lambda) = \begin{cases} \lambda f(\frac{x}{\lambda}) & \text{if } (x, \lambda) \in X \times \mathbb{R}_{++} \\ 0 & \text{otherwise,} \end{cases}$$

where $\mathbb{R}_{++} := (0 + \infty)$. Clearly, $\hat{f}(x, 1) = f(x)$ for all $x \in X$. Assume now the natural order relation defined on $X \times \mathbb{R}$ with respect to $S \times \mathbb{R}_+$ by:

$$(x, t) :\leq (y, s) \quad \Leftrightarrow \quad y - x \in S, \, t \leq s,$$

for every $x, y \in X$ and $t, s \in \mathbb{R}$.

It is easy to see that the function f is ICR if and only if its positively homogeneous extension $\hat{f}$ is increasing. Therefore, studying IPH functions helps to find out the behavior of ICR functions. Let $l_{y,\alpha} \in \Gamma$. Define $\tilde{l}_{y,\alpha} : X \times \mathbb{R} \longrightarrow [0, +\infty]$ as follows:

$$\tilde{l}_{y,\alpha}(x, c) = \begin{cases} l_{y, \frac{c}{\alpha}}(x) & (x, c) \in X \times \mathbb{R}_{++}, \\ 0 & o.w. \end{cases}$$

Clearly, $\tilde{l}_{y,\alpha}$ is an IPH function defined on $X \times \mathbb{R}$. Set $\tilde{\Gamma} := \{\tilde{l}_{y, \frac{1}{\alpha}} : l_{y,\alpha} \in \Gamma\}$.

Theorem 11.3.1 *Let $f : X \longrightarrow [0, +\infty]$ be an ICR function. Then, $\hat{f}$ is $\tilde{\Gamma}$-convex. Moreover, for all $x \in X$ and $c > 0$:*

$$\hat{f}(x, c) = cf(\frac{x}{c}) = \sup c \, l_{y,\alpha}(\frac{x}{c}) = \sup l_{y,c\alpha}(x) = \sup \tilde{l}_{y, \frac{1}{\alpha}}(x, c).$$

Proof. The desired result follows from (11.20) and Theorem 11.2.4. □

Similar to the concept of conjugate function, one may define the concept of polar function as follows.

Definition 11.3.2 *Let $f : X \longrightarrow [0, +\infty]$ be a function. The lower polar function of f with respect to the set L is the function $f^\circ : L \longrightarrow [0, +\infty]$ defined by*

$$f^\circ(l_y) = \begin{cases} \sup\{\frac{l_y(x)}{f(x)} : f(x) \neq 0\} & 0 < f(y) < +\infty, \\ +\infty, & f(y) = 0, \\ 0, & f(y) = +\infty. \end{cases}$$

Note that the concept of upper polarity of functions could be similarly defined. We only emphasise that the term "lower" is related to the functions l_y which estimates IPH functions from below. The following result characterizes polar functions for IPH functions.

Proposition 11.3.3 *Let $f : X \longrightarrow [0, +\infty]$ be an IPH function. Then for all $l_y \in L$*

$$f^\circ(l_y) = \frac{1}{f(y)}, \tag{11.38}$$

with the convention $\frac{1}{0} = +\infty$ and $\frac{1}{+\infty} = 0$.

Proof. It follows from the Definition 11.3.2 that

$$f^\circ(l_y) \geq \frac{l_y(y)}{f(y)} = \frac{1}{f(y)}.$$

On the other hand, we have $l_y(x)y \leq x$, whenever $l_y(x) \neq +\infty$. Since f is IPH, a simple calculation (even for the case $l_y(x) = +\infty$) shows that $l_y(x)f(y) \leq f(x)$ for all $x, y \in X$. Therefore, $f^\circ(l_y) \leq \frac{1}{f(y)}$. This completes the proof. □

To avoid confusion, we denote the lower polar of a function f with respect to the set Γ by $f^\diamond$.

Definition 11.3.4 *Let $f : X \longrightarrow [0, +\infty]$ be a function. The lower polar function of f with respect to the set Γ is the function $f^\diamond : \Gamma \longrightarrow [0, +\infty]$ defined by*

$$f^\diamond(l_{y,\alpha}) := \begin{cases} \sup\{\frac{l_{y,\alpha}(x)}{f(x)} : f(x) \neq 0\} & 0 < f(\alpha y) < +\infty, \\ +\infty, & f(\alpha y) = 0, \\ 0, & f(\alpha y) = +\infty. \end{cases}$$

where $l_{y,\alpha} \in \Gamma$ and with the convention $\frac{1}{+\infty} = \frac{0}{0} = 0$.

Notice that the polar functions of ICR functions are defined with respect to the set Γ. Applying a similar argument to the proof of Proposition 11.3.3, if $f : X \longrightarrow [0, +\infty]$ is ICR then

$$f^\diamond(l_{y,\alpha}) = \frac{\alpha}{f(\alpha y)} \quad (\forall l_{y,\alpha} \in \Gamma). \tag{11.39}$$

The following observation shows the relationship between the polarity of an ICR function with respect to the set Γ and the polarity of its positively homogeneous extension with respect to $\widehat{\Gamma}$.

Proposition 11.3.5 *Let $f : X \longrightarrow [0, +\infty]$ be an ICR function. Assume $\hat{f}^\circ$ is the lower polar function of $\hat{f}$, the positively homogeneous extension of f. Then*

$$\hat{f}^\circ(\tilde{l}_{y,\frac{1}{\alpha}}) = \frac{1}{\hat{f}(y, \frac{1}{\alpha})} = f^\diamond(l_{y,\alpha}), \quad (\forall \tilde{l}_{(y,\alpha)} \in \widetilde{\Gamma}).$$

Proof. It is an straightforward consequence from (11.38) and (11.39). □

11.3.2 Solvability Theorems for IPH and ICR Functions

We begin by a definition demonstrating a connection between L and X. Let 2^X denote the set of all subsets of X.

Definition 11.3.6 *The set valued function $\Pi : 2^X \longrightarrow 2^{X \times \mathbb{R}_+}$ is defined by*

$$\Pi(B) := \{(x, t) \in X \times \mathbb{R}_+ : tx \in B\},$$

for all $B \subseteq X$. Also the set valued function $\Theta : 2^{L \times \mathbb{R}} \longrightarrow 2^{X \times \mathbb{R}}$ is defined by

$$\Theta(A) := \{(y, r) \in X \times \mathbb{R} : (l_y, r) \in A\},$$

for all $A \subseteq L \times \mathbb{R}$.

As a simple example, let $X := \mathbb{R}$ and $B := (0,1]$. Thus, $\Pi(B) = \{(x,t) \in \mathbb{R} \times \mathbb{R}_+ : 0 < t \leq \frac{1}{x}, x > 0\} = \text{epi} \frac{1}{(\cdot)} \cap (\mathbb{R}_{++} \times \mathbb{R}_{++})$.

Next result depicts solvability theorem for a system of inequalities consisting of IPH functions defined on a subset of X.

Theorem 11.3.7 *Let $f, f_t : X \longrightarrow [0, +\infty]$ be IPH functions for all $t \in T$ and $C \subseteq X$. Then the following statements are equivalent.*

(i) $x \in C$, $f_t(x) \leq 1$ for all $t \in T \implies f(x) \leq 1$.
(ii) $\Theta(\text{epi}_s f^\circ) \cap \Pi(C) \subseteq \bigcup_{t \in T} \Theta(\text{epi}_s f_t^\circ)$.

Proof. (i) $\implies$ (ii):
Let $(y,r) \in \Theta(\text{epi}_s f^\circ) \cap \Pi(C)$. Thus, $f^\circ(l_y) < r$. Applying Proposition 11.38, we have $\frac{1}{f(y)} < r$, which implies that $f(ry) > 1$. Since $ry \in C$, we get by (i) that there exists a $t_0 \in T$ such that $f_{t_0}(ry) > 1$. Therefore, $f_{t_0}^\circ(l_y) < r$ and so

$$(y,r) \in \Theta(\text{epi}_s f_{t_0}^\circ) \subseteq \bigcup_{t \in T} \Theta(\text{epi}_s f_t^\circ).$$

(ii) $\implies$ (i):
Let $z \in C$ and $f(z) > 1$. Applying Proposition 11.38, one has $f^\circ(l_z) < 1$ and so $(z,1) \in \Theta(\text{epi}_s f^\circ)$. Since $(z,1) \in \Pi(C)$, we get by (ii) that $(z,1) \in \bigcup_{t \in T} \Theta(\text{epi}_s f_t^\circ)$. This implies that there exists a $t_0 \in T$ such that $f_{t_0}^\circ(l_z) < 1$. Thus, $f_{t_0}(z) > 1$. This completes the proof.
□

The following observation is a straightforward consequence of Theorem 11.3.7.

Corollary 11.3.8 *Let $f, f_t : X \longrightarrow [0, +\infty]$ be IPH functions and $a_t, a \in \mathbb{R}_{++}$ for all $t \in T$. Let C be a subset of X. Then the following assertions are equivalent.*
(i) $x \in C$, $f_t(x) \leq a_t$ for all $t \in T \implies f(x) \leq a$.
(ii) $[\Theta(\text{epi}_s a f^\circ) \cap \Pi(C) \subseteq \bigcup_{t \in T} \Theta(\text{epi}_s a_t f_t^\circ)$.

Proof. Notice that $(\mu f)^\circ(l_y) = \frac{1}{\mu f(y)} = \frac{1}{\mu} f^\circ(l_y)$ for all $\mu > 0$. Therefore, the results follow from Theorem 11.3.7.
□

Before we present solvability theorems for IPH functions, we need to define similar concepts to that of Definition 11.3.6.

Definition 11.3.9 *The set valued function $\widetilde{\Pi} : 2^{X \times \mathbb{R}_+} \longrightarrow 2^{X \times \mathbb{R}_+ \times \mathbb{R}_+}$ is defined by*

$$\widetilde{\Pi}(D) := \{(x, \lambda, t) \in X \times \mathbb{R}_+ \times \mathbb{R}_+ : (tx, t\lambda) \in D\},$$

for all $D \subseteq X \times \mathbb{R}_+$. Also the set valued function $\widetilde{\Theta} : 2^{\Gamma \times \mathbb{R}} \longrightarrow 2^{X \times \mathbb{R}_+ \times \mathbb{R}}$ is defined by

$$\widetilde{\Theta}(A) := \{(y, \alpha, r) \in X \times \mathbb{R}_+ \times \mathbb{R} : (l_{y,\alpha}, r) \in A\},$$

for $A \subseteq \Gamma \times \mathbb{R}$.

We finish this subsection by the solvability theorem for the system of inequalities involving ICR functions.

Theorem 11.3.10 Let $f, f_t : X \longrightarrow [0, +\infty]$ be ICR functions for all $t \in T$. Then, the following statements are equivalent.
(i) $x \in X$, $f_t(x) \leq 1$ for all $t \in T \implies f(x) \leq 1$.
(ii) $\widetilde{\Theta}(\text{epi}_s f^\diamond) \cap \{(y, \lambda, \lambda) : y \in X, \lambda > 0\} \subseteq \bigcup_{t \in T} \widetilde{\Theta}(\text{epi}_s f_t^\diamond)$.

Proof. Consider the positively homogeneous extension of the functions f_t and f by $\hat{f}_t$ and $\hat{f}$, respectively. Let $D = \{(x, 1) \in X \times \mathbb{R}_{++}, x \in X\}$. Thus, the following statement and (i) are equivalent:

$$(x, \lambda) \in D, \ \hat{f}_t(x, \lambda) \leq 1, \ \forall t \in T \implies \hat{f}(x, \lambda) \leq 1. \tag{11.40}$$

Applying Theorem 11.3.7, (11.40) is equivalent to the following statement:

$$\widetilde{\Theta}(\text{epi}_s \hat{f}^\diamond) \cap \widetilde{\Pi}(D) \subseteq \bigcup_{t \in T} \widetilde{\Theta}(\text{epi}_s \hat{f}_t^\diamond). \tag{11.41}$$

Furthermore,

$$\begin{aligned}
\widetilde{\Pi}(D) &= \{(x, \lambda, t) \in X \times \mathbb{R}_+ \times \mathbb{R}_+ : (tx, t\lambda) \in D\} \\
&= \{(x, \lambda, \tfrac{1}{\lambda}) : x \in X, \lambda \in \mathbb{R}_{++}\}.
\end{aligned}$$

Using the definition of $\widetilde{\Gamma}$ and Proposition 11.3.5, one has

$$\begin{aligned}
&(y, \tfrac{1}{\alpha}, r) \in \widetilde{\Theta}(\text{epi}_s \hat{f}^\diamond) \cap \widetilde{\Pi}(D) \\
\iff &(y, \tfrac{1}{\alpha}, r) \in \widetilde{\Theta}(\{(\tilde{l}_{y, \tfrac{1}{\alpha}}, r) \in \widetilde{\Gamma} \times \mathbb{R}_{++} : \hat{f}^\diamond(\tilde{l}_{y, \tfrac{1}{\alpha}}) < r, r = \alpha\}) \\
\iff &(y, \alpha, r) \in \widetilde{\Theta}(\{(l_{y, \alpha}, r) \in \Gamma \times \mathbb{R}_{++} : f^\diamond(l_{y, \alpha}) < r, r = \alpha\}) \\
\iff &(y, \alpha, r) \in \widetilde{\Theta}(\text{epi}_s f^\diamond) \cap \{(y, \lambda, \lambda) : y \in X, \lambda > 0\}.
\end{aligned}$$

Similarly, for all $t \in T$ we have

$$(y, \tfrac{1}{\alpha}, r) \in \widetilde{\Theta}(\text{epi}_s \hat{f}_t^\diamond) \iff (y, \alpha, r) \in \widetilde{\Theta}(\text{epi}_s f_t^\diamond).$$

These implications imly that (11.41) is equivalent to

$$\widetilde{\Theta}(\text{epi}_s f^\diamond) \cap \{(y, \lambda, \lambda) : y \in X, \lambda > 0\} \subseteq \bigcup_{t \in T} \widetilde{\Theta}(\text{epi}_s f_t^\diamond).$$

This completes the proof. □

Remark 11.3.11 Theorem 11.3.10 could be proved simply by a routine and direct calculation. However, we prefer to apply the results related to IPH functions in order to show how the statement (ii) of Theorem 11.3.10 is constructed.

11.3.3 Solvability Theorem for Topical Functions

In this section, we study solvability results for inequality systems involving finitely as well as infinitely many topical constraints. In contrast to the previous subsection, the standard conjugate functions are applied. We first recall a well-known definition.

Definition 11.3.12 *Let E be an arbitrary set and $A, B \subseteq E \times \mathbb{R}$. We say that A is vertically closed if each x-section of A, A_x ($A_x := \{\alpha : (x, \alpha) \in A\}$) is a closed subset of $\mathbb{R}$ with respect to the Euclidean norm.*
The vertical closure of B is denoted by $\mathrm{vcl}\, B$ and is the smallest vertically closed subset of $E \times \mathbb{R}$ containing B.

Notice that the x-section of the vertical closure of the set B, $(\mathrm{vcl}\, B)_x$ is equal to $\overline{B_x}$ for all $x \in E$, where $\overline{B_x}$ is the closure of B_x with respect to the Euclidean topology.
Now, we can formulate the main result of this section. For the sake of simplicity, we will drop L and use p^* instead of p_L^*.

Theorem 11.3.13 *Let $p_t, p : X \longrightarrow \mathbb{R}$ be topical functions for all $t \in T$. Then the following statements are equivalent.*
(a) $x \in X$, $p_t(x) \leq 0$ for all $t \in T \implies p(x) \leq 0$.
(b) $\mathrm{epi}\, p^* \subseteq \mathrm{vcl} \bigcup_{t \in T} \mathrm{epi}\, p_t^*$.

Proof. (a) $\implies$ (b):
Assume that $(\varphi_l, r) \in \mathrm{epi}\, p^*$. We have two cases. Either $p(-l + r\mathbf{1}) > 0$ or $p(-l + r\mathbf{1}) = 0$. Let $p(-l + r\mathbf{1}) > 0$. By (a), there is a $t_0 \in T$ such that $p_{t_0}(-l + r\mathbf{1}) > 0$. Applying Lemma 11.2.6, one has $(\varphi_l, r) \in \mathrm{epi}\, p_{t_0}^* \subseteq \bigcup_{t \in T} \mathrm{epi}\, p_t^*$. Now, assume that $p(-l + r\mathbf{1}) = 0$. Thus, for all $\varepsilon > 0$, $p(-l + (r + \varepsilon)\mathbf{1}) > 0$. Using the previous case, we get

$$(\varphi_l, r) + (0, \varepsilon) \in \bigcup_{t \in T} \mathrm{epi}\, p_t^*, \quad (\forall \varepsilon > 0).$$

Thus, $(\varphi_l, r) \in \mathrm{vcl} \bigcup_{t \in T} \mathrm{epi}\, p_t^*$.
(b) $\implies$ (a):
Suppose that there exists an $x_0 \in X$ such that $p(x_0) > 0$, which means that $p(x_0) > \varepsilon$ for some $\varepsilon > 0$. Using Lemma 11.2.6, we obtain $(\varphi_{-x_0}, -\varepsilon) \in \mathrm{epi}\, p^*$. Therefore, for all $\eta > 0$, $(\varphi_{-x_0}, -\varepsilon) + (0, \eta) \in \bigcup_{t \in T} \mathrm{epi}\, p_t^*$. Thus, there exists a $t_0 \in T$ such that $(\varphi_{-x_0}, -\varepsilon + \eta) \in \mathrm{epi}\, p_{t_0}^*$. Choosing $0 < \eta < \varepsilon$, one has

$$p_{t_0}(x_0) \geq \varepsilon - \eta > 0.$$

Hence, the proof is complete. $\square$

Notice that $\bigcup_{t \in T} \mathrm{epi}\, p_t^*$ is not vertically closed in general.

Example 11.3.14 *Let $p_n : \mathbb{R} \longrightarrow \mathbb{R}$ defined by $p_n(x) = x - \frac{1}{n}$ for all $n \geq 1$. Clearly, $\varphi_l(x) = x + l$ and $p_n^*(\varphi_l) = l + \frac{1}{n}$. Assume $l \in \mathbb{R}$ is fixed. Then the φ_l-section of $\bigcup_{n \geq 1} \mathrm{epi}\, p_n^*$ is not closed. Indeed*

$$\left(\bigcup_{n \geq 1} \mathrm{epi}\, p_n^*\right)_{\varphi_l} = \bigcup_{n \geq 1} \{r \in \mathbb{R} : l + \frac{1}{n} \leq r\} = (l, +\infty).$$

Solvability theorems play an important role in studying global optimization problems. However, to reach this goal, one needs to study certain systems of inequalities. In the following we present one of them.

Corollary 11.3.15 *Let $p_t, p : X \longrightarrow \mathbb{R}$ be topical functions and $a_t, a \in \mathbb{R}$ for all $t \in T$. Then the following are equivalent.*
(a) $x \in X$, $p_t(x) \leq a_t$ for all $t \in T \implies p(x) \leq a$.
(b) $\operatorname{epi} p^* + (0, a) \subseteq vcl \bigcup_{t \in T}(\operatorname{epi} p_t^* + (0, a_t))$.

Proof. Consider $q(x) := p(x) - a$ and $q_t(x) := p_t(x) - a_t$ for all $t \in T$. Clearly q and q_t are topical functions and

$$\operatorname{epi} q_t^* = \operatorname{epi} p_t^* + (0, a_t), \; \operatorname{epi} q^* = \operatorname{epi} p^* + (0, a).$$

Hence, the results follow from Theorem 11.3.13. □

The next corollary states the solvability theorem for the case that finitely many topical constraints are at hand. Note that every finite union of vertically closed sets is vertically closed.

Corollary 11.3.16 *Let $p_n, p : X \longrightarrow \mathbb{R}$ be topical functions for $n = 1, 2, ..., k$. Then the following are equivalent.*
(a) $x \in X$, $p_n(x) \leq 0$ for all $n = 1, 2, ..., k \implies p(x) \leq 0$.
(b) $\operatorname{epi} p^* \subseteq \bigcup_{n=1}^{k} \operatorname{epi} p_n^*$.

In order to solve a constrained optimization problem, it is essential to known whether the feasible set is nonempty. The following observation gives us a simple characterization for consistence of a feasible set involving topical functions. This theorem is also known as a generalization of Gale theorem presented for linear systems.

Theorem 11.3.17 *Let $p_t : X \longrightarrow \mathbb{R}$ be a topical function for all $t \in T$. Then, the system $\sigma := \{p_t(x) \leq 0 : t \in T\}$ is inconsistent if and only if $\sup_{t \in T} p_t \equiv +\infty$.*

Proof. Assume that σ is inconsistent, $y \in X$ and $n \in \mathbb{N}$ is arbitrary. If $p_t(y) \leq n$ for all $t \in T$, then $p_t(y - n\mathbf{1}) = p_t(y) - n \leq 0$, which means that σ is consistent, a contradiction. Thus, there is a $t_0 \in T$ such that

$$n < p_{t_0}(y) \leq \sup_{t \in T} p_t(y).$$

As $n \longrightarrow +\infty$, the desired result follows. The converse is trivial. □

11.4 Vector-valued Abstract Convex Functions and Solvability Theorems

Abstract convexity of a real-valued function provides the required global information for finding global minimizers of the function. Therefore, it is natural to develop this theory to find strong, efficient and weakly efficient solutions of vector optimization problems. In the following, we study vector-valued IPH, ICR and topical functions and establish solvability theorems for them. The main results of Subsections 11.4.1 and 11.4.2 are new, while the results from subsection 11.4.3 can be found in [20].

Throughout this section, assume that Y and Z are linear topological vector spaces (unless otherwise stated) partially ordered by the closed convex pointed cone Q and K, respectively. We denote the partial order relation defined on Y (on Z) by $\leq_Q$ (by $\leq_K$). That is:

$$\forall y_1, y_2 \in Y: \quad y_1 \leq_Q y_2 \text{ (or, } y_2 \geq_Q y_1\text{) if and only if } y_2 - y_1 \in Q. \tag{11.42}$$

11.4.1 Vector-valued IPH Functions and Solvability Theorems

In this subsection, we introduce vector-valued IPH functions and establish a solvability theorem for the systems involving them.

Definition 11.4.1 *A function $F: X \longrightarrow Y$ is called a vector-valued positively homogeneous function if $F(\lambda x) = \lambda F(x)$ for all $x \in X$ and $\lambda > 0$. A function $F: X \longrightarrow Y$ is called a vector valued IPH function, if F is vector-valued positively homogeneous and increasing. The latter means, if $x_1 \leq_S x_2$, then $F(x_1) \leq_Q F(x_2)$.*

In the following, two examples of vector-valued IPH functions are given.

Example 11.4.2 *Let $Y := \mathbb{R}^m$ and $Q := \mathbb{R}_+^m$. Assume that $f_1, \ldots, f_m : X \longrightarrow \mathbb{R}_+$ are IPH functions. Then the function $F: X \longrightarrow \mathbb{R}^m$ defined by $F(x) := (f_1(x), \ldots, f_m(x))$ is a vector-valued IPH function.*

Example 11.4.3 *Let $X = Y := \ell^2 = \{\{a_n\}_{n \geq 1} \subset \mathbb{R} : \sum_{n=1}^{+\infty} a_n^2 < +\infty\}$, where $\{a_n\}_{n \geq 1}$ is a sequence of real numbers, and let*

$$S = Q := \{\{a_n\}_{n \geq 1} \in \ell^2 \ : \ a_n \geq 0, \ \forall n \in \mathbb{N}\}.$$

Define the function $F: \ell^2 \longrightarrow \ell^2$ by $F(\{a_n\}_{n \geq 1}) := \{\frac{a_n}{n}\}_{n \geq 1}$. Then F is a vector-valued IPH function.

Consider the function l defined by (11.2). The domain of the function l_y, $y \in X$, defined by $l_y(\cdot) := l(\cdot, y)$ is X. To avoid confusion, we denote this function by l_y^X, for all $y \in X$. We define similar functions on the spaces Y and Z as follows:

$$\forall w \in Y: \ l_w^Y(y) := \max\{\lambda \geq 0 : \lambda w \leq_Q y\}, \quad (y \in Y) \tag{11.43}$$

$$\forall w \in Z: \ l_w^Z(z) := \max\{\lambda \geq 0 : \lambda w \leq_K z\}, \quad (z \in Z). \tag{11.44}$$

Let us denote $L_X := \{l_w^X : w \in X\}$, $L_Y := \{l_w^Y : w \in Y\}$, $L_Z := \{l_w^Z : w \in Z\}$. We apply the same discussion for the function u defined by (11.10). Therefore, we have functions u_w^X, u_w^Y and u_w^Z defined on the spaces X, Y and Z, respectively. Similarly, denote $U_X := \{u_w^X : w \in X\}$, $U_Y := \{u_w^Y : w \in Y\}$, $U_Z := \{u_w^Z : w \in Z\}$.
Next, we study the relationships between real valued and vector-valued IPH functions.

Theorem 11.4.4 *Let $F : X \longrightarrow Y$. Then the following are equivalent.*
(a) F is a vector-valued IPH function.
(b) For each $l_w^Y \in L_Y$, $l_w^Y \circ F : X \longrightarrow [0, +\infty]$ is an IPH function.
(c) For each $u_w^Y \in U_Y$, $u_w^Y \circ F : X \longrightarrow [0, +\infty]$ is an IPH function.

Proof. (a) $\Rightarrow$ (b) and (a) $\Rightarrow$ (c) are trivial.
(b) $\Rightarrow$ (a): By hypothesis, for all $x \in X$ and $\lambda > 0$:

$$\begin{aligned} l_{F(x)}^Y(F(\lambda x)) &= (l_{F(x)}^Y \circ F)(\lambda x) \\ &= \lambda (l_{F(x)}^Y \circ F)(x) \\ &= \lambda l_{F(x)}^Y(F(x)) \\ &= \lambda. \end{aligned}$$

Using the definition of l_w^Y, $F(\lambda x) - \lambda F(x) \in Q$.
Replace λ and x by $\frac{1}{\lambda}$ and λx, respectively. Thus, $F(x) - \frac{1}{\lambda} F(\lambda x) \in Q$. Since Q is a pointed cone, $F(\lambda x) = \lambda F(x)$. Thus, F is a vector-valued positively homogeneous function.
Assume now that $x_1 \leq x_2$. Then $(l_w^Y \circ F)(x_1) \leq (l_w^Y \circ F)(x_2)$ for all $w \in Y$. Choosing $w = F(x_1)$, one has

$$1 = l_{F(x_1)}^Y(F(x_1)) \leq l_{F(x_1)}^Y(F(x_2)).$$

By definition of $l_{F(x_1)}^Y(F(x_2))$, we get

$$F(x_1) \leq F(x_2).$$

Hence, F is a vector-valued IPH function. The implication (c) $\Rightarrow$ (a) is done similarly. $\square$

Now, we are ready to establish a new version of solvability results for inequality systems involving vector-valued IPH constraints. In the following, the set-valued functions Π and Θ are those defined by Definition 11.3.6. As stated above, we also have

$$u_w^Z(x) = \min\{\lambda \geq 0 \ : \ \lambda w \geq_S x\},$$

for all $x, w \in Z$.

Theorem 11.4.5 *Let $\mathcal{T}$ be an arbitrary index set. Let $F : X \longrightarrow Z$ and $F_t : X \longrightarrow Y$ be vector-valued IPH functions for all $t \in \mathcal{T}$. Assume that $a_t \in Y \setminus (-Q)$ for all $t \in \mathcal{T}$, $a \in Z \setminus (-K)$, $C \subseteq X$, then the following are equivalent.*
 (a) $x \in C$, $F_t(x) \leq_Q a_t$, for all $t \in \mathcal{T} \implies F(x) \leq_K a$.
(b) $\Theta(\mathrm{epi}_s(u_a^Z \circ F)^\circ) \cap \Pi(C) \subseteq \bigcup_{t \in \mathcal{T}} \Theta(\mathrm{epi}_s(u_{a_t}^Y \circ F_t)^\circ).$

Proof. It is easy to see that $u_y^Y(y) = 1$ if and only if $y \notin -Q$, and $u_z^Z(z) = 1$ if and only if $z \notin -K$. Using the definition of $u_{a_t}^Y$, one has

$$\{x \in X : F_t(x) \leq_Q a_t\} = \{x \in X : (u_{a_t}^Y \circ F_t)(x) \leq 1\}. \tag{11.45}$$

Similarly,

$$\{x \in X : F(x) \leq_K a\} = \{x \in X : (u_a^Z \circ F)(x) \leq 1\}.$$

Applying Theorem 11.4.4 and Theorem 11.3.7, the results follow. $\square$

11.4.2 Vector-valued ICR Functions and Solvability Theorems

Similar to vector-valued IPH functions, vector-valued ICR functions are defined as follows.

Definition 11.4.6 *A vector-valued function $F : X \longrightarrow Y$ is called coradiant if $\lambda F(x) \leq_Q F(\lambda x)$ for all $x \in X$ and $\lambda \in [0,1]$. $F : X \longrightarrow Y$ is called a vector-valued ICR function, if F is a vector-valued coradiant and increasing function.*

In the following, some examples of vector-valued ICR functions are given.

Example 11.4.7 *Let Y and Q be as in Example 11.4.2. Let $f_1, \ldots, f_m : X \longrightarrow \mathbb{R}_+$ be ICR functions. The function $F : X \longrightarrow \mathbb{R}^m$ defined by $F(x) := (f_1(x), \ldots, f_m(x))$ is a vector-valued ICR function.*

Example 11.4.8 *Let $F : X \longrightarrow Y$ be a Q-concave function; that is:*

$$F(tx + (1-t)y) - tF(x) - (1-t)F(y) \in Q,$$

for all $t \in [0,1]$ and $x, y \in X$. If F is increasing and $F(0) \in Q$, then F is a vector-valued ICR function. Indeed, for $x \in X$ and $\lambda \in [0,1]$:

$$F(\lambda x) = F(\lambda x + (1-\lambda)0) \geq_Q \lambda F(x) + (1-\lambda)F(0) \geq_Q \lambda F(x).$$

To avoid confusion, let us consider $\Gamma_X := \{l_{y,\alpha}^X : y \in X, \alpha > 0\}$ instead of Γ defined by (11.29), where the function $l_{y,\alpha}^X := l_{y,\alpha}$ is defined on the space X. Similarly, we define the same functions on the spaces Y and Z as follows:

$$\forall w \in Y, t \in \mathbb{R}_{++} : l_{w,t}^Y(y) := \max\{0 < \lambda \leq t : \lambda w \leq_Q y\}, \quad (y \in Y)$$

$$\forall w \in Z, t \in \mathbb{R}_{++} : l_{w,t}^Z(z) := \max\{0 < \lambda \leq t : \lambda w \leq_K z\}, \quad (z \in Z).$$

We set $\Gamma_Y := \{l_{w,t}^Y : w \in Y, t > 0\}$ and $\Gamma_Z := \{l_{w,t}^Z : w \in Z, t > 0\}$. Similar to the function l defined by (11.19), the function u could be defined (on the three spaces X, Y and Z) as follows:

$$\forall w \in X, \beta \in \mathbb{R}_{++} : u_{w,\beta}^X(x) := \min\{\lambda \geq \beta : \lambda w \geq_S x\}, \quad (x \in X)$$

$$\forall w \in Y, \beta \in \mathbb{R}_{++} : u_{w,\beta}^Y(y) := \min\{\lambda \geq \beta : \lambda w \geq_Q y\}, \quad (y \in Y)$$

$$\forall w \in Z, \beta \in \mathbb{R}_{++} : u_{w,\beta}^Z(z) := \min\{\lambda \geq \beta : \lambda w \geq_K z\}, \quad (z \in Z)$$

The aforementioned scalarizing functions play an important role to study solvability theorems for vector-valued ICR functions. Observe that they are ICR functions, however, they have similar properties comparing the function l defined by (11.19). Let us set

$$\Lambda_X := \{u_{w,\beta}^X : w \in X, \beta > 0\},$$
$$\Lambda_Y := \{u_{w,\beta}^Y : w \in Y, \beta > 0\},$$
$$\Lambda_Z := \{u_{w,\beta}^Z : w \in Z, \beta > 0\}.$$

The following results indicate the relations between real valued and vector-valued ICR functions.

Theorem 11.4.9 *Let $F : X \longrightarrow Y$. Then the following are equivalent.*
(a) F is a vector-valued ICR function.
(b) $l^Y_{w,t} \circ f : X \longrightarrow [0,+\infty]$ is an ICR function for every $l^Y_{w,t} \in \Gamma_Y$.
(c) $u^Y_{w,\beta} \circ f : X \longrightarrow [0,+\infty]$ is an ICR function for every $u^Y_{w,\beta} \in \Lambda_Y$.

Proof. Clearly, (a) $\Longrightarrow$ (b) and (a) $\Longrightarrow$ (c).
(b) $\Longrightarrow$ (a): By hypothesis, for all $x \in X$, $\lambda \in [0,1]$ and $t=1$:

$$\begin{aligned} l^Y_{F(x),1}(F(\lambda x)) &= (l^Y_{F(x),1} \circ F)(\lambda x) \\ &\geq \lambda (l^Y_{F(x),1} \circ F)(x) \\ &= \lambda l^Y_{F(x),1}(F(x)) \\ &= \lambda. \end{aligned}$$

Using the definition of $l^Y_{F(x),1}$, since $\lambda \leq 1$, it concludes that $F(\lambda x) - \lambda F(x) \in Q$. Thus, F is a vector-valued coradiant function.
Assume now that $x_1 \leq x_2$. Then $(l^Y_{w,t} \circ F)(x_1) \leq (l^Y_{w,t} \circ F)(x_2)$ for all $w \in Y$ and $t > 0$. Choosing $w = F(x_1)$ and $t=1$, one has

$$1 = l^Y_{F(x_1),1}(F(x_1)) \leq l^Y_{F(x_1),1}(F(x_2)).$$

This together with the fact that $l^Y_{F(x_1),1}(F(x_2)) \leq 1$ implies that $l^Y_{F(x_1),1}(F(x_2)) = 1$. By definition of $l^Y_{F(x_1),1}(F(x_2))$, we get

$$F(x_1) \leq F(x_2).$$

Therefore, F is a vector-valued ICR function.

The assertion (c) $\Longrightarrow$ (a) is similar. Hence, the proof is complete. $\square$

In the rest of this section, we study solvability theorems for inequality systems involving vector-valued ICR constraints. Notice that the set-valued mapping $\widetilde{\Theta}$ used in the following theorem is defined by definition 11.3.9.

Theorem 11.4.10 *Let $\mathcal{T}$ be an arbitrary index set. Let $F : X \longrightarrow Z$ and $F_t : X \longrightarrow Y$ be vector-valued ICR functions for all $t \in \mathcal{T}$. Assume that $a_t \in \operatorname{int} Q$ for all $t \in T$ and $a \in \operatorname{int} K$. Then the following statements are equivalent.*
(a) $x \in X$, $F_t(x) \leq_Q a_t$, for all $t \in \mathcal{T}$ $\Longrightarrow$ $F(x) \leq_K a$.
(b) $\widetilde{\Theta}(\operatorname{epi}_s[u^Z_{a,1} \circ F]^\circ) \cap \{(y, \lambda, \lambda) : y \in X, \lambda > 0\} \subseteq \bigcup_{t \in T} \widetilde{\Theta}(\operatorname{epi}_s[u^Y_{a_t,1} \circ F_t]^\circ)$.

Proof. Using the definition of $u^Y_{a_t,1}$, one has

$$\{x \in X : F_t(x) \leq_Q a_t\} = \{x \in X : (u^Y_{a_t,1} \circ F_t)(x) \leq 1\}. \tag{11.46}$$

Similarly,

$$\{x \in X : F(x) \leq_K a\} = \{x \in X : (u^Z_{a,1} \circ F)(x) \leq 1\}.$$

Applying Theorem 11.4.9 and Theorem 11.3.10, the results follow. $\square$

11.4.3 Vector-valued Topical Functions and Solvability Theorems

Throughout this subsection, we assume that X, Y and Z are real Banach spaces ordered by pointed closed convex cones S, Q, K, respectively. We also assume that S, Q and K are normal (see Section 2). Let $\mathbf{1}_S \in \operatorname{int} S$, $\mathbf{1}_Q \in \operatorname{int} Q$ and $\mathbf{1}_K \in \operatorname{int} K$ be fixed points. Similar to Definitions of (11.30) and (11.32), we define

$$\forall w \in X: \ \varphi_w^X(x) := \max\{\lambda \in \mathbb{R} : \lambda \mathbf{1} \leq_S w + x\}, \quad (x \in X)$$

$$\forall w \in Y: \ \varphi_w^Y(y) := \max\{\lambda \in \mathbb{R} : \lambda \mathbf{1} \leq_Q w + y\}, \quad (y \in Y)$$

$$\forall w \in Z: \ \varphi_w^Z(z) := \max\{\lambda \in \mathbb{R} : \lambda \mathbf{1} \leq_K w + z\}, \quad (z \in Z).$$

We set $\Phi_X := \{\varphi_w^X : w \in X\}$, $\Phi_Y := \{\varphi_w^Y : w \in Y\}$ and $\Phi_Z := \{\varphi_w^Z : w \in Z\}$. We also need the following scalarizing functions:

$$\forall w \in X \ : \ \psi_w^X(x) := \min\{\lambda \in \mathbb{R} : \lambda \mathbf{1} \geq_S x + w\}, \quad (x \in X)$$

$$\forall w \in Y \ : \ \psi_w^Y(y) := \min\{\lambda \in \mathbb{R} : \lambda \mathbf{1} \geq_Q y + w\}, \quad (y \in Y)$$

$$\forall w \in Z \ : \ \psi_w^Z(z) := \min\{\lambda \in \mathbb{R} : \lambda \mathbf{1} \geq_K z + w\}, \quad (z \in Z).$$

Observe that these six aforementioned scalaraizing functions are topical and inherit similar properties stated in (11.33)-(11.36) for the function φ. Set $\Psi_X := \{\psi_w^X : w \in X\}$, $\Psi_Y := \{\psi_w^Y : w \in Y\}$ and $\Psi_Z := \{\psi_w^Z : w \in Z\}$.

Definition 11.4.11 *A function $P : X \longrightarrow Y$ is called a vector-valued plus-homogeneous function if $P(x + \lambda \mathbf{1}_X) = P(x) + \lambda \mathbf{1}_Y$ for all $x \in X$ and $\lambda \in \mathbb{R}$. A function $P : X \longrightarrow Y$ is called a vector valued topical function if P is a vector-valued plus-homogeneous function and P is increasing, i.e. $x_1 \leq_S x_2$ implies $P(x_1) \leq_Q P(x_2)$.*

Example 11.4.12 *Let $Y = \mathbb{R}^m$ and $Q = \mathbb{R}^m_+$ with $\mathbf{1}_{\mathbb{R}^n} := (1,1,...,1) \in \operatorname{int} Q$. Let $p_i : X \longrightarrow \mathbb{R}$ be topical functions for $i = 1,2,...,n$. Define $P : X \longrightarrow \mathbb{R}^n$ by $P(x) := (p_1(x),...,p_n(x))$. Then P is a vector-valued topical function.*

The following example shows that we can deal with many topical functions using different points as $\mathbf{1}$. We will use this fact in the next section.

Example 11.4.13 *Let $X = Y = \mathbb{R}^3$ and $S = Q = \mathbb{R}^3_+$. Assume that $\mathbf{1}_X = (1,2,3)$ and $\mathbf{1}_Y = (\frac{1}{2}, \frac{1}{3}, \frac{1}{4})$. Define $P : X \longrightarrow Y$ by $P(x,y,z) = (\frac{1}{2}x - 1, \frac{1}{6}y - 2, \frac{1}{12}z - 3)$. Then P is a vector-valued topical function.*

Let $f, g : X \longrightarrow (-\infty, +\infty]$. Recall that inf-convolution of f and g, $f \Box g$, is defined by

$$(f \Box g)(x) := \inf_{x_1 + x_2 = x} f(x_1) + g(x_2).$$

Example 11.4.14 *Consider $X = Y = B(\mathbb{R}^n)$, the set of all bounded real-valued functions defined on $\mathbb{R}^n$, equipped with the norm $\|.\|$ defined by*

$$\|f\| = \sup_{x \in \mathbb{R}^n} |f(x)|, \quad f \in B(\mathbb{R}^n).$$

Assume that $S = Q = \{f : f \in X, f(x) \geq 0, \forall x \in \mathbb{R}^n\}$, and $\mathbf{1}_X = \mathbf{1}_Y = 1$ where 1 is the constant function with the value 1 defined on $\mathbb{R}^n$. Clearly $\mathbf{1}_S \in \text{int}\, S$ and $\mathbf{1}_Q \in \text{int}\, Q$. Fix $f_0 \in X$ and define $P: X \longrightarrow Y$ by $P(f) := f \square f_0$. It is easy to see that P is a vector-valued topical function.

The following results show the relationship between real-valued and vector-valued topical functions. The proof is similar to that of Theorem 11.4.9.

Theorem 11.4.15 *Let $P: X \longrightarrow Y$. Then the following statements are equivalent.*
(a) P is a vector-valued topical function.
(b) $\varphi_w^Y \circ P : X \longrightarrow \mathbb{R}$ is topical for all $\varphi_w^Y \in \Phi_Y$.
(c) $\psi_w^Y \circ P : X \longrightarrow \mathbb{R}$ is topical for all $\psi_w^Y \in \Psi_Y$.

To reach the main results of this section, let us first bring the following lemma. Similar results are also satisfied for the cones S and K with the corresponding scalarizing functions.

Lemma 11.4.16 $Q = \{y \in Y : \varphi_0^Y(y) \geq 0\}$ and $-Q = \{y \in Y : \psi_0^Y(y) \leq 0\}$.

Proof. We only prove $Q = \{y \in Y : \varphi_0^Y(y) \geq 0\}$. The other one is similar. Since φ_0^Y is increasing and $\varphi_0^Y(0) = 0$, $Q \subseteq \{y \in Y : \varphi_0^Y(y) \geq 0\}$. Assume that $\varphi_0^Y(y) \geq 0$ for some $y \in Y$. Since φ_0^Y is finite and $\mathbf{1}_Y \in \text{int}\, Q$, one has

$$0 \leq_Q \varphi_0^Y(y) \mathbf{1}_Y \leq_Q y,$$

which means that $y \in Q$. $\square$

Now, we establish a solvability result for inequality systems involving vector-valued topical constraints. This result will be applied in vector topical optimization in Subsection 11.5.2 in order to derive a new method of solving linear programming problems with positive multipliers.

Theorem 11.4.17 *Let $P: X \longrightarrow Z$ and $P_t : X \longrightarrow Y$ be vector-valued topical functions for all $t \in \mathcal{T}$. Then the following statements are equivalent.*
(a) $x \in X$, $P_t(x) \leq_Q 0$, for all $t \in \mathcal{T} \implies P(x) \leq_K 0$.
(b) $\text{epi}(\psi_0^Z \circ P)^* \subseteq \text{vcl} \bigcup_{t \in \mathcal{T}} \text{epi}(\psi_0^Y \circ P_t)^*$.

Proof. Using the fact that $-Q = \{y \in Y : \psi_0^Y(y) \leq 0\}$ from Lemma 11.4.16, one has

$$\{x \in X;\ P_t(x) \leq_Q 0\} = \{x \in X;\ (\psi_0^Y \circ P_t)(x) \leq 0\}. \tag{11.47}$$

Moreover, $-K = \{z \in Z : \psi_0^Z(z) \leq 0\}$ implies that

$$\{x \in X;\ P(x) \leq_K 0\} = \{x \in X;\ (\psi_0^Z \circ P)(x) \leq 0\}.$$

Now, the results follow from Theorem 11.4.15 and Theorem 11.3.13. $\square$

In the following, we characterize the consistence of systems involving vector-valued topical functions.

Theorem 11.4.18 *Let $P_t : X \longrightarrow Y$ be vector topical functions for all $t \in \mathcal{T}$. Then the system $\{P_t(x) \leq_Q 0;\ t \in \mathcal{T}\}$ is inconsistent if and only if $\sup_{t \in \mathcal{T}}(\psi_0^Y \circ P_t) \equiv +\infty$.*

Proof. Using (11.47), the results follow from Theorems 11.4.15 and 11.3.17. $\square$

11.5 Applications in Optimization

In this section X is a topological vector space partially ordered by a closed pointed convex cone. We apply the results obtained in previous sections in studying maximization programming problems. In particular, we transform certain constrained maximization problems to an unconstrained counterparts. The main results of Subsection 11.5.1 could be found in [19], while the main results of Subsection 11.5.2 could be found in [20].

11.5.1 IPH and ICR Maximization Problems

Consider the following maximization problem:

$$\text{maximize } f(x) \quad \text{subject to } x \in X, \; f_t(x) \leq 1, \; \forall t \in T. \quad (IPHP)$$

Here T is an arbitrary index set and $f, f_t : X \longrightarrow [0, +\infty]$ are IPH functions, for all $t \in T$. $(IPHP)$ is considered as the primal problem. Hence, we shall introduce and study a dual problem for it. Let $(IPHP)$ be the optimal value of problem $(IPHP)$, that is:

$$v(IPHP) := \sup\{f(x) \; : \; x \in X, \; f_t(x) \leq 1, \; \forall t \in T\},$$

with the convention $\sup \emptyset = 0$. Define the function $H : X \longrightarrow [0, +\infty]$ by:

$$H(x) := \begin{cases} \frac{f(x)}{\sup_{t \in T} f_t(x)}, & f_t(x) \neq 0, \; \exists t \in T, \\ 0, & f(x) = 0 = \sup_{t \in T} f_t(x) \text{ or } \sup_{t \in T} f_t(x) = +\infty, \\ +\infty, & f(x) \neq 0 = \sup_{t \in T} f_t(x). \end{cases} \quad (11.48)$$

We consider the following unconstrained problem as the dual problem for $(IPHP)$:

$$\sup\{H(x) : x \in X\}. \quad (D)$$

Define the optimal value of the dual problem (D) by $v(D)$, i.e.

$$v(\text{D}) := \sup_{x \in X} H(x).$$

First we show that the weak duality condition holds. To see this, let $F := \{x \in X : f_t(x) \leq 1, t \in T\}$. Then, $x \in F$ and $\sup_{t \in T} f_t(x) \neq 0$ imply that $\frac{f(x)}{\sup_{t \in T} f_t(x)} \geq f(x)$. This means that $H(x) \geq f(x)$ for all $x \in F$. Therefore,

$$v(D) \geq \sup_{x \in F} H(x) \geq v(IPHP). \quad (11.49)$$

The following theorem shows the duality zero gap for problems $(IPHP)$ and (D).

Theorem 11.5.1 *Let T be an arbitrary nonempty index set, $f, f_t : X \longrightarrow [0, +\infty]$ be IPH functions for all $t \in T$ and H be defined by (11.48). Then $v(IPHP) = v(D)$.*

Proof. Due to (11.49), it suffices to show that $v(IPHP) \geq v(D)$.
First we investigate the case that $v(D) = +\infty$. If there is a vector $x \in X$ such that $f_t(x) = 0 \neq f(x)$ for all $t \in T$, then $H(x) = +\infty$ and consequently $v(D) = +\infty$. But in this

case, $x \in F$ and $v(IPHP)$ becomes $+\infty$ as well. Let $\sup_{t \in T} f_t(x) \neq 0$ for all $x \in X \setminus \{0\}$ and $v(D) = +\infty$. Then, there exists a sequence $\{x_n\} \subset X$ such that $H(x_n) \to +\infty$. This implies that there is a subsequence of $\{x_n\}$ (say without lose of generality $\{x_n\}$) such that $\sup_{t \in T} f_t(x_n) \neq +\infty$ for all $n \geq 1$. Put $y_n := \frac{x_n}{\sup_{t \in T} f_t(x_n)}$. Clearly $y_n \in F$ for all $n \geq 1$. According to the definition of H, $H(x_n) = f(y_n)$. Thus, $\lim_{n \to +\infty} f(y_n) = \lim_{n \to +\infty} H(x_n) = +\infty$. Hence, for the case that $v(D) = +\infty$, we get that $v(D) = v(IPHP)$. Assume now that $v(D) < +\infty$. Then $v(IPHP) \leq \alpha$ for some $\alpha \in \mathbb{R}_+$. This implies that $f(x) < +\infty$ for all $x \in X$. Therefore, either $H(x) = \frac{f(x)}{\sup_{t \in T} f_t(x)}$ or $H(x) = 0$, for every $x \in X$. In addition, $v(IPHP) \leq \alpha$ leads to the following implication:

$$x \in X, \; f_t(x) \leq 1, \; t \in T \implies f(x) \leq \alpha.$$

Applying Corollary 11.3.8 and the fact that $\Pi(X) = X \times \mathbb{R}_+$, one has

$$\Theta(\mathrm{epi}_s(\alpha f^\circ)) \subseteq \bigcup_{t \in T} \Theta(\mathrm{epi}_s f_t^\circ). \tag{11.50}$$

Let $z \in X$ such that $f(z) \neq 0$ and $\varepsilon > 0$ be arbitrary. Using (11.50), there exists a $t_0 \in T$ such that

$$(z, \alpha f^\circ(l_z) + \varepsilon) \in \Theta(\mathrm{epi}_s f_{t_0}^\circ),$$

which implies that $f_{t_0}^\circ(l_z) < \alpha f^\circ(l_z) + \varepsilon$. This together with the Proposition 11.3.3 imply that

$$\frac{f(z)}{\sup_{t \in T} f_t(z)} \leq \frac{f(z)}{f_{t_0}(z)} < \alpha + \varepsilon f(z).$$

Since ε is arbitrary, we get that

$$\frac{f(z)}{\sup_{t \in T} f_t(z)} \leq \alpha. \tag{11.51}$$

Clearly (11.51) implies that $H(z) \leq \alpha$ for all $z \in X$. Hence, $v(D) \leq \alpha$. Since this inequality holds for all α with $v(IPHP) \leq \alpha$, we get that $v(D) \leq v(IPHP)$. This completes the proof. $\square$

Remark 11.5.2 *Notice that in order to prove the zero duality gap in Theorem 11.5.1, we do not assume any constraint qualification or a slater-type condition. It is also worth mentioning that the Corollary 11.3.8 is used not only to prove the zero duality gap, but also helps to construct the dual problem. Indeed, it leads to the inequality (11.51) and by using that we could discover the form of dual problem (D).*

Next, we present some examples to illustrate Theorem 11.5.1.

Example 11.5.3 *Consider the following IPH problem:*

$$\begin{aligned} \text{maximize} \quad & \frac{xy}{x+y} \\ \text{subject to} \quad & x+y \leq 3, \\ & x \geq 0, \; y \geq 0, \; (x,y) \neq (0,0). \end{aligned}$$

Define $f, f_1 : \mathbb{R}^2 \longrightarrow \mathbb{R}_+$ by

$$f(x,y) := \begin{cases} \frac{xy}{x+y} & (x,y) \in \mathbb{R}_+^2 \setminus \{(0,0)\}, \\ 0 & \text{Otherwise}, \end{cases}$$

and

$$f_1(x,y) := \begin{cases} \frac{1}{3}x + \frac{1}{3}y & (x,y) \in \mathbb{R}_+^2 \setminus \{(0,0)\}, \\ 0 & \text{Otherwise}. \end{cases}$$

Clearly, f and f_1 are both IPH functions. Thus, $H(x,y) := \frac{3xy}{(x+y)^2}$ for all $(x,y) \in \mathbb{R}_+^2 \setminus \{(0,0)\}$ and $H(x,y) = 0$ for other cases. A simple calculation shows that (s,s) is a critical point of function H for every $s > 0$. Therefore, the maximal value of H is $\frac{3s^2}{(s+s)^2} = \frac{3}{4}$. So, in this case, $v(\text{IPHP}) = \frac{3}{4}$.

Remark 11.5.4 Let $v(D) = H(x_0)$ for some $x_0 \in X$. Note that μx_0 is a maximizer for the function H for all $\mu > 0$. Because $H(\mu x) = H(x)$ for all $x \in X$ and $\mu > 0$. To find a maximizer for (IPHP), it is enough to find a $\mu > 0$ such that $\sup_{t \in T} f_t(\mu x) = 1$.

Using Remark 11.5.4, in the Example 11.5.3, $(\frac{3}{2}, \frac{3}{2})$ is the maximizer for (IPHP).

Example 11.5.5 Let $c, d \in \mathbb{R}_+^n$ and $b > 0$. Consider the following linear programming problem:

$$\begin{aligned} \text{maximize} \quad & c^t x \\ \text{subject to} \quad & d^t x \leq b, \\ & x \geq 0. \end{aligned}$$

Let

$$f(x) := \begin{cases} c^t x & x \in \mathbb{R}_+^n, \\ 0 & \text{Otherwise}, \end{cases}$$

and

$$f_1(x) := \begin{cases} \frac{1}{b} d^t x & x \in \mathbb{R}_+^n, \\ 0 & \text{Otherwise}. \end{cases}$$

Clearly f and f_1 are IPH functions. If $c = 0$, then $H \equiv 0$ and $v(D) = 0 = v(\text{IPHP})$. If $d_j = 0 \neq c_j$ for some $1 \leq j \leq n$, then $H(x) = +\infty$. The same statement is also true if $d = 0$ and $c \neq 0$. Applying Theorem 11.48, $v(\text{IPHP}) = +\infty$.
Assume that $d_i \neq 0$ for $i = 1, 2, \ldots, n$. We have

$$H(x) = b \frac{c^t x}{d^t x}$$

for all $x \in \mathbb{R}_+^n \setminus \{0_n\}$. By a simple calculation,

$$\sup_{x \in \mathbb{R}^n} H(x) = b \max_{1 \leq i \leq n} \frac{c_i}{d_i}.$$

Let e_i be the unit vector whose components are zero, except the ith component which is one. Let $\frac{c_j}{d_j} = \max_{1 \leq i \leq n} \frac{c_i}{d_i}$, then λe_j is the maximizer of H for every $\lambda > 0$. Applying Remark 11.5.4, one has $f_1(\frac{b}{d_j} e_j) = 1$. Hence, $\frac{b}{d_j} e_j$ is a maximizer of (IPHP).

In the rest of this section, we exploit Theorem 11.5.1 in order to study a constrained ICR programming problem via an unconstrained problem. Let T be an arbitrary index set. Consider the following ICR programming problem:

$$\max \{f(x): x \in X, \ f_t(x) \leq 1 \ \forall t \in T\}, \quad (ICRP)$$

where $f, f_t : X \longrightarrow \mathbb{R}_+$ are nonnegative and real valued ICR functions, for all $t \in T$. Similar to (11.48), the function $F : X \longrightarrow [0, +\infty]$ is defined by

$$F(x) := \begin{cases} \frac{f(x)}{\sup_{t \in T} f_t(x)}, & f_t(x) \neq 0, \ \exists t \in T, \\ 0, & f(x) = 0 = \sup_{t \in T} f_t(x) \text{ or } \sup_{t \in T} f_t(x) = +\infty, \\ +\infty, & f(x) \neq 0 = \sup_{t \in T} f_t(x). \end{cases} \quad (11.52)$$

From now on, we assume that X is a normed linear space. Let

$$C := \{x \in X \ : \ f_t(x) \leq 1, \ t \in T\},$$

and

$$G := \{x \in X \ : \ \sup_{t \in T} f_t(x) = 1\}.$$

A dual problem for $(ICRP)$ is defined as follows:

$$\sup\{F(x) : x \in X\}. \quad (D')$$

Let $w(D') := \sup_{x \in X} F(x)$ and $v(ICRP) := \sup_{x \in C} f(x)$. Now, we are going to investigate the zero duality gap for problems $(ICRP)$ and (D'). It is easy to see that $f(x) \leq F(x)$ for all $x \in X$. Therefore

$$v(ICRP) = \sup_{x \in C} f(x) \leq \sup_{x \in C} F(x) \leq \sup_{x \in X} F(x) = w(D'). \quad (11.53)$$

For the converse inequality we need first some definitions. Recall that the algebraic interior of the set A, A^i, is defined by:

$$A^i := \{a \in X \ : \ \forall x \in X, \ \exists \delta > 0, \ \forall \lambda \in [0, \delta] : \ a + \lambda x \in A\}.$$

The function F is called increasing along rays on C, if the function $F_z : [0, +\infty) \longrightarrow [0, +\infty)$ defined by $F_z(t) := F(tz)$ is increasing for all $z \in C$.

Constraint qualification. We say that the constraint qualification holds for the system $\{f_t : t \in T\}$, if C is bounded, $0 \in C^i$ and $\sup_{t \in T} f_t$ is continuous on C.

In the following, we study $(ICRP)$ problem in which $\sup_{t \in T} f_t$ is an IPH function.

Theorem 11.5.6 *Let $f, f_t : X \longrightarrow \mathbb{R}_+$ be nonnegative and real valued ICR functions. Assume that the function F defined in (11.52) is real-valued and increasing along rays on C. Let constraint qualification hold for the system $\{f_t : t \in T\}$ and $\sup_{t \in T} f_t$ be IPH. Then $v(ICRP) = w(D')$.*

Proof. We are going to show that the inverse inequality of (11.53) is fulfilled. Since $0 \in C^i$, one has $X = \{\mu x : \mu \geq 1, x \in X\}$. Therefore,

$$\sup_{x \in X} \frac{f(x)}{\sup_{t \in T} f_t(x)} = \sup_{x \in C,\, \mu \geq 1} \frac{f(\mu x)}{\sup_{t \in T} f_t(\mu x)} = \sup_{x \in C,\, \mu \geq 1} \frac{\frac{f(\mu x)}{\mu}}{\sup_{t \in T} f_t(x)} \leq \sup_{x \in C} \frac{f(x)}{\sup_{t \in T} f_t(x)},$$

because $\sup_{t \in T} f_t$ is IPH.
On the other hand, F is increasing along rays and f is increasing. These facts imply that

$$\sup_{x \in C} \frac{f(x)}{\sup_{t \in T} f_t(x)} = \sup_{x \in G} \frac{f(x)}{\sup_{t \in T} f_t(x)} = \sup_{x \in G} f(x) = \sup_{x \in C} f(x),$$

which the first and the last equalities follow from the fact that $bd(C) \subseteq G$ and $bd(C) \cap \{\lambda x : \lambda \geq 1\} \neq \emptyset$ for all $x \in C$. Hence, the proof is complete. $\square$

In the following theorem, we study problem (*ICRP*) in which the objective function f is IPH. Recall that F is decreasing along rays on $X \setminus C$ if the function $F_z : [0, +\infty) \longrightarrow [0, +\infty)$ defined by $F_z(t) := F(tz)$ is decreasing for all $z \in X \setminus C$.

Theorem 11.5.7 *Let $f, f_t : X \longrightarrow \mathbb{R}_+$ be nonnegative and real valued ICR functions. Assume that the function F defined in (11.52) is real-valued and decreasing along rays on $X \setminus C$. Let constraint qualification hold for the system $\{f_t : t \in T\}$ and f also be IPH on $X \setminus C$. Then $w(D') = v(ICRP)$.*

Proof. Since $0 \in C^i$ and C is bounded, we have $X = \{\lambda x : x \in X \setminus C,\, 0 \leq \lambda \leq 1\}$. Therefore,

$$\sup_{x \in X} \frac{f(x)}{\sup_{t \in T} f_t(x)} = \sup_{x \in X \setminus C,\, \mu \leq 1} \frac{f(\mu x)}{\sup_{t \in T} f_t(\mu x)}$$

$$= \sup_{x \in X \setminus C,\, \mu \leq 1} \frac{f(x)}{\sup_{t \in T} \frac{f_t(\mu x)}{\mu}}$$

$$\leq \sup_{x \in X \setminus C} \frac{f(x)}{\sup_{t \in T} f_t(x)}.$$

The rest of the proof is similar to that of Theorem 11.5.6. $\square$

11.5.2 A New Approach to Solve Linear Programming Problems with Nonnegative Multipliers

In this section, X, Y, Z, S, Q and K are those defined in Subsection 11.4.3. The main idea of this subsection is to transform especial constraint maximization problems into unconstrained ones, without adding extra variables such as Lagrange multipliers. This will be done by using appropriate versions of solvability theorems for systems involving topical functions. Especially, these results in some cases of a linear optimization problem, lead to a concave unconstrained problem which can be solved by many standard algorithms. Consider the Topical maximization Problem (*TP*) as follows:

$$\text{maximize } \{p(x) : x \in X,\ p_t(x) \leq 0,\ \forall t \in T\}, \qquad (TP)$$

where T is an arbitrary index set and $p, p_t : X \longrightarrow \mathbb{R}$ are topical functions, for all $t \in T$. Also, consider the Vector Topical maximization Problem (*VTP*) as

$$\text{maximize}\{P(x) : x \in X, \ P_t(x) \leq_Q 0, \ \forall t \in \mathcal{T}\}, \qquad (VTP)$$

where $\mathcal{T}$ is an arbitrary index set, $P : X \longrightarrow Z$ and $P_t : X \longrightarrow Y$ are vector topical functions for all $t \in \mathcal{T}$.

Lemma 11.5.8 *Let $p : X \longrightarrow \mathbb{R}$ be a topical function. Then*

$$p^*(\varphi_l) = -p(-l), \quad \forall l \in X. \qquad (11.54)$$

Proof.

$$p^*(\varphi_l) = \sup_{x \in X} \varphi_l(x) - p(x) \geq \varphi_l(-l) - p(-l) = -p(-l).$$

On the other hand, due to the definition of φ_l, we have $\varphi_l(x)\mathbf{1} - l \leq x$, for all $l, x \in X$. Since p is increasing and plus-homogeneous, one has

$$\varphi_l(x) + p(-l) \leq p(x),$$

which implies that $\varphi_l(x) - p(x) \leq -p(-l)$. This completes the proof. □

Theorem 11.5.9 *Consider the problem (TP) and let $v(TP) := \sup\{p(x) : x \in X, \ p_t(x) \leq 0, \ \forall t \in T\}$. Then*

$$v(TP) = \sup_{x \in X}[p(x) - \sup_{t \in T} p_t(x)].$$

Proof. By Theorem 11.3.17, the system $\{p_t(x) \leq 0, \ t \in T\}$ is inconsistent if and only if $\sup_{t \in T} p_t \equiv +\infty$. In this case

$$\sup_{x \in X}[p(x) - \sup_{t \in T} p_t(x)] = -\infty = v(TP).$$

Now let the system $\{p_t(x) \leq 0, \ t \in T\}$ be consistent. Since the inequality

$$\sup_{x \in X}[p(x) - \sup_{t \in T} p_t(x)] \geq \sup_{p_t(x) \leq 0, \ t \in T} p(x) = v(TP) \qquad (11.55)$$

always holds, it suffices to prove the reverse inequality. To do this, if $v(TP) = +\infty$, there is nothing to prove. Consider $m \in \mathbb{R}$ and $m \geq v(TP)$, which is equivalent to the following implication

$$x \in X, \ p_t(x) \leq 0 \text{ for all } t \in T \implies p(x) \leq m.$$

Applying Corollary 11.3.15, for arbitrary $l \in X$ and $\varepsilon > 0$, we conclude that $(\varphi_l, p^*(\varphi_l)) + (0, m) + (0, \varepsilon) \in \bigcup_{t \in T} \text{epi } p_t^*$, which means there exists $t_0 \in T$ such that

$$p^*(\varphi_l) + m + \varepsilon \geq p_{t_0}^*(\varphi_l). \qquad (11.56)$$

According to (11.54) and (11.56),

$$m + \varepsilon \geq p(-l) - p_{t_0}(-l) \geq p(-l) - \sup_{t \in T} p_t(-l).$$

Since both l and ε are arbitrary, we get
$$m \geq p(x) - \sup_{t \in T} p_t(x), \quad x \in X.$$
Hence, the proof is complete. □

To illustrate Theorem 11.5.9, the following simple example is presented.

Example 11.5.10 *Consider the following maximization problem:*
$$\begin{aligned} \text{maximize } \quad p(x,y) &:= \frac{3}{2}x + \frac{1}{2}y \\ \text{subject to } \quad p_1(x,y) &:= 2x \leq 0 \\ p_2(x,y) &:= x + y - 1 \leq 0. \end{aligned}$$

Let $\mathbf{1} := (\frac{1}{2}, \frac{1}{2})$ and $S := \mathbb{R}_+^2$. Then all functions p, p_1 and p_2 are topical. Applying Theorem 11.5.9, one has:
$$\begin{aligned} p(x_0, y_0) &= \sup_{x,y \in \mathbb{R}} [p(x,y) - \sup_{i \in \{1,2\}} p_i(x,y)] \\ &= \sup_{x,y \in \mathbb{R}} [\frac{3}{2}x + \frac{1}{2}y - \max\{2x, x+y-1\}] \\ &= \sup_{x,y \in \mathbb{R}} \frac{1}{2} - \frac{1}{2}|x - y + 1| \\ &= \frac{1}{2}. \end{aligned}$$

Hence $(0,1)$ is an optimal solution.

Applying Theorem 11.4.17 together with a similar argument to the proof of Theorem 11.5.9, we conclude the following result for the problem (VTP). Notice that ψ_0^Y and ψ_0^Z are defined in subsection 11.4.3.

Theorem 11.5.11 *Let $P : X \longrightarrow Z$ and $P_t : X \longrightarrow Y$ be vector topical functions for all $t \in \mathcal{T}$. Then*
$$\sup\{(\psi_0^Z \circ P)(x) : P_t(x) \leq_Q 0, x \in X, t \in \mathcal{T}\} = \sup_{x \in X}[(\psi_0^Z \circ P)(x) - \sup_{t \in \mathcal{T}}(\psi_0^Y \circ P_t)(x)].$$

In example 11.5.10, all functions are plus-homogeneous with respect to the unique vector $\mathbf{1} = (\frac{1}{2}, \frac{1}{2})$. This, however, is a restricting condition, because there are several topical functions which are plus-homogeneous with their own $\mathbf{1}$.

Let $c \in \mathbb{R}^n$, A be an $m \times n$ matrix and $b \in \mathbb{R}^m$. Assume that all components of c and all entries of A are nonnegative. Consider the following linear optimization problem
$$\text{maximize } \{c^t x : Ax \leq b, x \in \mathbb{R}^n\}. \tag{LP}$$

In the sequel, we try to transform problem (LP) to an unconstrained concave problem without adding extra variables. Let
$$X := \mathbb{R}^n, \ Y := \mathbb{R}^m, \ Z := \mathbb{R}, \ S := \mathbb{R}_+^n, \ Q := \mathbb{R}_+^m, \ K := \mathbb{R}_+$$

and $T := \{1\}$. Let the function $P_1 : \mathbb{R}^n \longrightarrow \mathbb{R}^m$ be defined by $P_1(x) := Ax - b$ and the function $P : \mathbb{R}^n \longrightarrow \mathbb{R}$ be defined by $P(x) := c^t x$. Assume that $a_1, ..., a_m$ are the m rows of matrix A.

To construct $\mathbf{1}_{\mathbb{R}^n}$ and $\mathbf{1}_{\mathbb{R}^m}$ in such a way that all functions are topical, we use the following process:

1. Select an arbitrary vector from the interior points of $\mathbb{R}^n_+$. Call it $\mathbf{1}_{\mathbb{R}^n}$.
2. Put $u_i := a_i^t \mathbf{1}_{\mathbb{R}^n}$, $i = 1, 2, \ldots, m$.
3. Put $\mathbf{1}_{\mathbb{R}^m} := (u_1, \ldots, u_m)$.
4. Put $\mathbf{1}_{\mathbb{R}} := c^t \mathbf{1}_{\mathbb{R}^n}$.

Since $a_i \geq 0$, $a_i \neq 0$ and $\mathbf{1}_{\mathbb{R}^n} \in \text{int } \mathbb{R}^n_+$, we conclude that u_i defined in Step 2 is positive for all $i = 1, 2, \ldots, m$. Therefore, $\mathbf{1}_{\mathbb{R}^m} \in \text{int } \mathbb{R}^m_+$.

P_1 is a vector topical function. Indeed, $P_1(x + \mu \mathbf{1}) = P_1(x) + \mu \mathbf{1}_{\mathbb{R}^m}$. This means that P_1 is vector plus-homogeneous. On the other hand, since a_i, $i = 1, 2, \ldots, m$ are nonnegative, P_1 is increasing.

Now observe that $P(x + \mu \mathbf{1}_{\mathbb{R}^n}) = P(x) + \mu c^t \mathbf{1}_{\mathbb{R}^n}$. Thus, it follows that P is vector plus-homogeneous.

Theorem 11.5.12 *Consider the linear programming problem (LP). Let P, P_1, u_i ($i = 1, \ldots, m$), $\mathbf{1}_{\mathbb{R}^n}$, $\mathbf{1}_{\mathbb{R}^m}$ and $\mathbf{1}_{\mathbb{R}}$ be defined as above. Consider the following unconstrained concave maximization problem:*

$$\text{maximize } \{c^t x - \max_{1 \leq i \leq n} \frac{\mathbf{1}_{\mathbb{R}}}{u_i}(a_i^t x - b_i) : x \in \mathbb{R}^n\}. \quad (CP)$$

Let α and β be the optimal values of problems (LP) and (CP), respectively. Then $\alpha = \beta$.

Proof. It follows from the definition of ψ_0^Y and ψ_0^Z that for all $x \in \mathbb{R}^n$

$$\begin{aligned}
(\psi_0^Y \circ P_1)(x) &= \min\{\lambda \in \mathbb{R} : \lambda \mathbf{1}_{\mathbb{R}^m} \geq_Q Ax - b\}, \\
&= \min\{\lambda \in \mathbb{R} : \lambda(u_1 \ldots, u_m) \geq_{\mathbb{R}^m_+} a_i^t x - b_i\}, \\
&= \max_{1 \leq i \leq m} \frac{1}{u_i}(a_i^t x - b_i),
\end{aligned}$$

and

$$(\psi_0^Z \circ P)(x) = \frac{1}{\mathbf{1}_{\mathbb{R}}} c^t x.$$

Therefore,

$$\sup_{x \in X}[(\psi_0^Z \circ P)(x) - \sup_{t \in \mathcal{T}}(\psi_0^Y \circ P_t)(x)] = \max_{x \in \mathbb{R}^n} \frac{1}{\mathbf{1}_{\mathbb{R}}} c^t x - \max_{1 \leq i \leq n} \frac{1}{u_i}(a_i^t x - b_i)$$

and

$$\sup\{(\psi_0^Z \circ P)(x) : P_t(x) \leq_Q 0, x \in X, t \in \mathcal{T}\} = \max\{\frac{1}{\mathbf{1}_{\mathbb{R}}} c^t x : Ax \leq b, x \in \mathbb{R}^n\}.$$

According to Theorem 11.5.11, the result follows. $\square$

In Example 11.5.10, we tried to find a vector **1** for which all functions p, p_1, p_2 are topical. However, the above argument shows that it is redundant. In fact, one can choose any vector whose components are all positive as the vector **1**. Let us reformulate Example 11.5.10 as follows:

$$\text{maximize} \quad \frac{3}{2}x + \frac{1}{2}y$$
$$\text{subject to} \quad x \leq 0, \ x + y - 1 \leq 0.$$

Let $a_1 := (1,0)$, $a_2 = (1,1)$ and $b = (0,1)$. Consider $\mathbf{1}_{\mathbb{R}^2} := (1,2)$. So $u_1 := a_1^t(1,2) = 1$ and $u_2 := a_2^t(1,2) = 3$. Moreover, $\mathbf{1}_\mathbb{R} := c^t(1,2) = \frac{5}{2}$. By problem (CP)

$$\begin{aligned}
\beta &= \max_{\mathbf{x} \in \mathbb{R}^2} c^t\mathbf{x} - \max_{1 \leq i \leq 2} \frac{\mathbf{1}_\mathbb{R}}{u_i}(a_i^t\mathbf{x} - b_i) \\
&= \max_{x,y \in \mathbb{R}} (\frac{3}{2}x + \frac{1}{2}y) - \frac{5}{2}\max\{x, \frac{1}{3}(x+y-1)\} \\
&= \frac{1}{2}.
\end{aligned}$$

Thus, $\alpha = \beta = \frac{1}{2}$. Now, similar to Example 11.5.10, let us choose $\mathbf{1}_{\mathbb{R}^2} := (\frac{1}{2}, \frac{1}{2})$. Then $u_1 := a_1^t(\frac{1}{2}, \frac{1}{2}) = \frac{1}{2}$, $u_2 := a_2^t(\frac{1}{2}, \frac{1}{2}) = 1$ and $\mathbf{1}_\mathbb{R} := c^t(\frac{1}{2}, \frac{1}{2}) = 1$. In this case, we get

$$\begin{aligned}
\beta &= \max_{\mathbf{x} \in \mathbb{R}^2} c^t\mathbf{x} - \max_{1 \leq i \leq 2} \frac{\mathbf{1}_\mathbb{R}}{u_i}(a_i^t\mathbf{x} - b_i) \\
&= \max_{x,y \in \mathbb{R}}[\frac{3}{2}x + \frac{1}{2}y - \max\{2x, x+y-1\}] \\
&= \frac{1}{2}.
\end{aligned}$$

References

[1] S. Bartz and S. Reich. Abstract convex optimal antiderivatives. *Ann. Inst. H. Poincaré Anal. Non Linéaire,* 29: 435–454, 2012.

[2] R. I. Boţ and G. Wanka. Farkas-type results with conjugate functions. *SIAM Journal of Optimization,* 15: 540–554, 2005.

[3] N. Dinh and V. Jeyakumar. Farkas lemma: Three decades of generalizations for mathematical optimization. *Top.,* 22: 1–22, 2014.

[4] A. R. Doagooei. Farkas-type theorems for positively homogeneous systems in ordered topological vector spaces. *Nonlinear Analysis: Theory, Methods and Applications,* 17: 5541–5548, 2012.

[5] A. R. Doagooei and H. Mohebi. Dual characterization of the set containment with strict cone-convex inequalities in Banach spaces. *Journal of Global Optimization,* 43: 577–591, 2008.

[6] A. R. Doagooei and H. Mohebi. Optimization of the difference of topical functions. *Journal of Global Optimization,* 57: 1349–1358, 2013.

[7] A. R. Doagooei and H. Mohebi. Monotonic analysis over ordered topological vector spaces: IV. *Journal of Global Optimization,* 45: 355–369, 2009.

[8] J. Dutta, J. E. Martinez-Legaz and A. M. Rubinov. Monotonic analysis over cones: III. *Journal of Convex Analysis,* 15: 561–579, 2008.

[9] V. Jeyakumar. Characterizing set containments involving infinite convex constraints and reverse-convex constraints. *SIAM Journal of Optimization,* 13: 947–959, 2003.

[10] V. Jeyakumar, G. M. Lee and N. Dinh. Chararcterizations of solution sets of convex vector minimization problems. *European Journal of Operation Research,* 174: 1380–1395, 2006.

[11] V. Jeyakumar, S. Kum and G. M. Lee. Necessary and sufficient conditions for Farkas lemma for cone systems and second-order cone programming duality. *Journal of Convex Analysis,* 15: 63–71, 2008.

[12] M. A. Lopez and J. E. Martinez-Legaz. Farkas-type theorems for positively homogeneous semi-infinite systems. *Optimization*, 54: 421–431, 2005.

[13] M. A. Lopez, A. M. Rubinov and V. N. Vera de Serio. Stability of semi-infinite inequality systems involving min-type functions. *Numerical Functional Analysis and Optimization*, 26: 81–112, 2005.

[14] V. L. Makarov, M. J. Levin and A. M. Rubinov. *Mathematical Economic Theory: Pure and Mixed Types of Economic Mechanisms*, Elsevier, Amsterdam, 1995.

[15] O. L. Managasarian. Set containment characterization. *Journal of Global Optimization*, 24: 473–480, 2002.

[16] H. Mohebi. Topical functions and their properties in a class of ordered Banach spaces. *Applied Optimization*, 99: 343–361, 2005.

[17] H. Mohebi and A. R. Doagooei. Abstract convexity of extended real valued increasing and positively homogeneous functions. *Journal of Dcdis B*, 17: 659–674, 2010.

[18] H. Mohebi and H. Sadeghi. Monotonic analysis over ordered topological vector spaces I. *Optimization*, 56: 305–321, 2007.

[19] V. Momenaee and A. R. Doagooei. Farkas type theorems for increasing and coradiant functions. *Numerical Functional Analysis and Optimization*, 1–15, 2018.

[20] V. Momenaei Kermani and A. R. Doagooei. Vector topical functions and Farkas type theorems with applications. *Optimizataion Letters*, 9: 359–374, 2015.

[21] D. Pallaschke and S. Rolewicz. *Foundations of Mathematical Optimization (Convex Analysis without Linearity)*, Kluwer Academic Publishers, Dordrecht, 1997.

[22] A. M. Rubinov. *Abstract Convex Analysis and Global Optimization*. Kluwer Academic Publishers, Boston, Dordrecht, London, 2000.

[23] A. M. Rubinov and M. Y. Andramonov. Minimizing increasing star-shaped functions based on abstract convexity. *Journal of Global Optimization*, 15: 19–39, 1999.

[24] A. M. Rubimov, B. M. Glover and V. Jeyakumar. A general approach to dual characterizations of solvability of inequality systems with applications. *Journal of Convex Analysis*, 2: 309–344, 1995.

[25] A. M. Rubinov and I. Singer. Topical and sub-topical functions, downward sets and abstract convexity. *Optimization*, 50: 307–351, 2001.

[26] A. M. Rubinov and A. Uderzo. On global optimality conditions via separation functions. *Journal of Optimization Theory and Applications*, 109: 345–370, 2001.

[27] I. Singer. *Abstract Convex Analysis*. John Wiley and Sons, New York, 1997.

[28] X. K. Sun. Regularity conditions characterizing Fenchel-Lagrange duality and Farkas-type results in DC infinite programming. *Journal of Mathematical Analysis and Applications*, 414: 590–611, 2014.

Chapter 12

Regularization Methods for Scalar and Vector Control Problems

Baasansuren Jadamba
School of Mathematical Sciences, Rochester Institute of Technology, Rochester, New York, USA. bxjsma@rit.edu

Akhtar A. Khan
School of Mathematical Sciences, Rochester Institute of Technology, Rochester, New York, USA. aaksma@rit.edu

Miguel Sama
Departamento de Matemática Aplicada, E.T.S.I. Industriales, Universidad Nacional de Educación a Distancia. Madrid, Spain. msama@ind.uned.es

Christiane Tammer
Institute of Mathematics, Martin-Luther-University of Halle-Wittenberg, D-06120 Halle-Saale, Germany. christiane.tammer@mathematik.uni-halle.de

12.1 Introduction

This work aims to give an overview of some of the recent developments in regularization methods for optimal control of partial differential equations and variational inequalities with pointwise state constraints. We will also provide a few research directions that we believe to be of interest. We will address optimal control problems with the scalar-valued as well as the vector-valued objectives. Although we will discuss a variety of regularization methods, our main focus will be on the developments related to the conical regularization which was introduced in [43] and provides a unified framework for studying optimization problems for which a Slater-type constraint qualification fails to hold due to the empty interior of the ordering cone associated with the inequality constraints. The failure of a Slater-type constraint qualification is a common hurdle in numerous branches of applied mathematics, including optimal control, inverse problems, nonsmooth optimization, and variational inequalities, among others. Recent results on the conical regularization and its variants gave convincing evidence that the conical regularization has a potential to be a valuable theoretical and computational tool in different disciplines suffering due to the lack of a Slater-type constraint qualification. We note that there has been extensive research on control problems, most of which focused on control-constrained problems, see [16, 23]. In recent years, however, the focus has shifted to optimal control problems with pointwise state constraints that frequently appear in applications (see [10, 27]). For instance, while optimizing the process of hot steel profiles by spraying water on their surfaces, it is necessary to control the temperature differences in order to avoid cracks, which leads to pointwise state constraints, see [25]. A similar situation arises in the production of bulk single crystals, where the temperature in the growth apparatus must be kept between prescribed bounds, see [59]. Moreover, during local hyperthermia in cancer treatment, pointwise state constraints are imposed in order to ensure that only the tumor cells get heated and not the nearby healthy ones see [22].

To illustrate the obstacles in a satisfactory treatment of pointwise state constraints, we consider the following control problem

$$(Q) \begin{cases} \text{Minimize } \frac{1}{2}\int_\Omega (y_u - z_d)^2 dx + \frac{\kappa}{2}\int_\Omega (u - u_d)^2 dx, \\ \text{subject to} \\ -\Delta y_u(x) + y_u(x) = u(x) \text{ in } \Omega, \ \partial_n y_u = 0 \text{ on } \partial\Omega, \\ y_u(x) \leq w(x), \text{ a.e. in } \Omega, \ u \in L^2(\Omega), \end{cases}$$

where $\Omega \subset \mathbb{R}^d$ ($d = 2, 3$) is an open, bounded, and convex domain with boundary $\partial\Omega$, $\kappa > 0$ is a fixed parameter, z_d, u_d, and w are given elements, and ∂_n is the outward normal derivative to $\partial\Omega$. If the control u is taken in $L^2(\Omega)$, then the state y_u is in $H^2(\Omega)$. Since $H^2(\Omega) \subset C(\bar{\Omega})$, the state can be viewed as a continuous function. This fact has an impact on deriving optimality conditions for (Q). In particular, a Lagrange multiplier exists due to the following well-known Slater constraint qualification:

$$\text{there exists } \bar{u} \in L^2(\Omega) \text{ such that } y_{\bar{u}} - w \in -\operatorname{int} C_+(\bar{\Omega}), \tag{12.1}$$

The difficulty, however, is that the multiplier now are in the dual of $C_+(\bar{\Omega})$, where $C_+(\bar{\Omega})$ denotes the natural ordering cone in $C(\bar{\Omega})$, which is in the space of Radon measures (see [1, 11]). This causes low-regularity in the control and has an adverse effect on both the analytical level when deriving optimality conditions for the control problem and on the numerical level when performing discretization. In fact, the appearance of measures is inevitable in such studies. Explicit examples of (Q) where the multipliers are Radon measures can be found in [41].

To provide another viewpoint of the difficulties in control problems with pointwise state constraints, we consider an abstract problem. Let U, Y, and H be Hilbert spaces equipped with the norms $\|\cdot\|_U$, $\|\cdot\|_Y$ and $\|\cdot\|_H$. Let $C \subset Y$ be a closed, convex, and pointed cone which is not necessarily solid, that is, it may have an empty interior. The cone C induces a partial ordering on Y which is denoted by $\leq_C$. In order words, $y_1 \leq_C y_2$ is equivalent to $y_2 - y_1 \in C$. Let $C^+ = \{\lambda^* \in Y^* : \forall c \in C, \lambda^*(c) \geq 0\}$ be the positive dual of C. Given linear and bounded maps $S : U \to H$ and $G : U \to Y$, and a parameter $\kappa > 0$, we consider:

$$(P) \quad \begin{cases} \text{Minimize } J(u) := \frac{1}{2}\|Su - z_d\|_H^2 + \frac{\kappa}{2}\|u - u_d\|_U^2, \\ \text{subject to} \\ G(u) \leq_C w, \; u \in U. \end{cases}$$

Problem (P) represents a wide variety of PDE constrained optimization and optimal control problems. Notice that for (P), the existence of Lagrange multipliers and reliable and efficient computational schemes can only be developed through a careful selection of the state space U and the control space Y. We note that (Q) is a special case of (P) with $U = H = L^2(\Omega)$, $S : L^2(\Omega) \to L^2(\Omega)$ and $G : L^2(\Omega) \to Y$. For the control-to-state map $(u \to y_u)$, there are two natural choices of the range space, namely, $Y = L^2(\Omega)$ and $Y = C(\bar{\Omega})$. However, as already noted above, to ensure the existence of Lagrange multipliers for the L_2-controls, the choice $Y = C(\bar{\Omega})$ seems natural, as the Slater constraint qualification (12.1) holds. The main drawback is that the multipliers are Radon measures, making the numerical treatment of the optimality conditions quite challenging. On the other hand, for the choice $Y = L^2(\Omega)$, the main technical hindrance is that the cone of positive functions $L^2_+(\Omega)$ has an empty interior. Consequently, no general Karush-Kuhn-Tucker theory is available. From the mathematical programming point of view, the lack of regularity can be attributed to the fact that both the ordering cone C in (P), and correspondingly $L^2_+(\Omega)$ in (Q), are not solid. It has been long recognized that the inequality constraints defined by an ordering cone that has an empty interior are a major obstacle in a satisfactory treatment of optimization and optimal control problems.

In recent years, a variety of regularization methods have been used extensively for control problems of type (Q) in order to handle the low regularity of the multipliers. Regularization methods, in the context of control problems, consist of relating to the original problem, a family of perturbed problems with more regular Lagrange multipliers. We emphasize that in the regularization methods, the constraints (that is, either the underlying equations or the explicit constraints on the state) are being regularized. The methods that fall into this category are the penalty method (see [32]), interior point methods (see [72]), Moreau-Yosida regularization (see [31]), virtual control regularization (see [51, 49]), the Lavrentiev regularization (see [60]), the conical regularization (see [43], and the Half-space regularization (see [41]). We remark that, whereas the Lavrentiev regularization is designed for specific PDEs, the conical regularization provides a general framework that is by no means restricted to the control problems with pointwise state constraints. This particular case, however, is of importance as pointwise state constraints commonly appear in numerous applied models. In the subsequent sections, we give a detailed discussion of some of the regularization methods. In passing, we would like to mention some other works where a regularization played a central role. Recently, Pörner and Wachsmuth [65] employed generalized Bregman distances in order to propose an iterative regularization method for an optimal control problem with box constraints and gave convergence results, and a priori error estimates. In another article [66], the same authors investigated

Tikhonov regularization of an optimal control problem of semilinear partial differential equations with box constraints on the control. See also [73, 78].

12.2 Lavrentiev Regularization

The Lavrentiev regularization has its origin in the study of ill-posed operator equations (see [53]). The well-known Tikhonov regularization investigates an ill-posed operator equation as a minimization problem. On the other hand, the Lavrentiev regularization method directly regularizes the ill-posed equation through a family of singularly perturbed operator equations (see [53, 75]). The Lavrentiev regularization holds a place of pride in the subject of ill-posed problems, and numerous researchers have contributed to various aspects of this approach, ranging from theoretical results and algorithmic schemes to a wide range of applications.

To ensure the existence of regular multipliers, in 2006, a novel Lavrentiev regularization approach was proposed and analyzed by Meyer, Rösch, and Tröltzsch [60]. Their method, in the context of pointwise state constraints, consists of replacing the inequality constraint by its regularized analog. Two critical observations motivated their approach. Firstly, as we have already noted, the Lagrange multipliers associated with pointwise state constraints need not be measurable functions. However, the situation becomes different for the mixed constraints where, under natural assumptions, the Lagrange multipliers can be assumed to be functions from $L^2(\Omega)$. Therefore, the pointwise state constraints could be regularized, in the Lavrentiev sense, to obtain a family of mixed constraints, which would give a well-behaved family of the associated Lagrange multipliers. The second motivation comes from the observation that, due to the compact imbeddings of the involved spaces, the control-to-target map can be viewed as an ill-posed operator equation and, hence, the Lavrentiev regularization should be useful.

We shall now summarize the main results from Meyer, Rösch, and Tröltzsch [60] where the focus was on the following optimal control problem:

$$(R) \quad \begin{cases} \text{Minimize } \widetilde{J}(u) := \frac{1}{2}\|\widetilde{S}u\|_H^2 + \int_\Omega (a(x)u(x) + \frac{\kappa}{2}u(x)^2)dx, \\ \text{subject to} \\ \widetilde{G}u \leq_C w, \\ 0 \leq_C u, \ u \in U, \end{cases}$$

with Ω a suitable domain, and $\widetilde{S}: U \to H$, $\widetilde{G}: U \to Y$ linear bounded maps.

For $\varepsilon > 0$, define the regularized map $\widetilde{G}_\varepsilon : U \to Y$ by $\widetilde{G}_\varepsilon := \widetilde{G} - \varepsilon I$, where I is the identity map. In [60], the following family of regularized problems were studied:

$$(R_\varepsilon) \quad \begin{cases} \text{Minimize } \widetilde{J}(u) := \frac{1}{2}\|\widetilde{S}u\|_H^2 + \int_D a(x)u(x) + \frac{\kappa}{2}u(x)^2)dx, \\ \text{subject to} \\ \widetilde{G}_\varepsilon(u) \leq_C w, \\ 0 \leq_C u, \ u \in U. \end{cases}$$

For $\varepsilon = 0$, set $Q_0 \equiv Q$, and by u_0 denote the unique solution of problem (Q). The main results of [60] can be summarized in the following:

Theorem 12.2.1 *Assume that $U = Y = L^2(D)$ and $C = L^2_+(D)$. Let $\widetilde{S}$, $\widetilde{G}$ be compact. Then:*

1. *Problem (R) has the unique solution u_0.*

2. *For every $\varepsilon > 0$, there exists the unique solution u_ε of (R_ε). For each sequence $\{\varepsilon_n\} \to 0^+$, the corresponding sequence of regularized solutions $\{u_{\varepsilon_n}\}$ converges strongly to u_0.*

3. *If $-\widetilde{G}$ is non-negative, that is, if $0 \leq_C u$ implies that $\widetilde{G} u \leq_C 0$, then (R_ε) is regular for every $\varepsilon > 0$. That is, there exist regular multipliers μ^1_ε, $\mu^2_\varepsilon \in L^2_+(D)$. Moreover, the multipliers $(\mu^1_\varepsilon, \mu^2_\varepsilon) \in (L^2_+(D))^2$ are unique and solve the dual problem*

$$\varphi(\mu^1_\varepsilon, \mu^2_\varepsilon) = \inf_{(\mu^1,\mu^2) \in (L^2_+(D))^2} \varphi(\mu^1, \mu^2)$$

$$:= -\langle w, \mu^1 \rangle_Y + \frac{1}{2} \left\| a - \varepsilon \mu^1 + \widetilde{G}^* \mu^1 - \mu^2 \right\|^2_\Lambda,$$

where $\|\cdot\|_\Lambda$ is the norm in Y given by $\|d\|^2_\Lambda = \langle d, \Lambda^{-1} d \rangle_Y$, and $\Lambda = \widetilde{S}^* \widetilde{S} + \kappa I$.

The existence of multipliers for (R_ε) was proved by means of the dual problem where the non-negativity of $-\widetilde{G}$ played a fundamental role.

The above approach has been pursued rigorously and has attracted tremendous interest in recent years. To mention some of the related research, we begin by noting that Prüfert and Tröltzsch [67] used Lavrentiev regularization for state-constrained problems and studied a primal-dual interior point method. In [76], Tröltzsch and Yousept presented the Lavrentiev regularization for elliptic boundary control problems with pointwise state constraints. The idea is to seek controls in the range of the adjoint control-to-state mapping. The authors introduced a semi-smooth Newton method in order to solve the regularized problems. In [77], Tröltzsch and Yousept investigated an optimal boundary control problem of linear elliptic equations with state constraints. The authors approximated this problem by a sequence of auxiliary problems, where the boundary control is defined as the trace of the solution of an auxiliary elliptic equation. The source term in this auxiliary equation serves as a distributed control and is employed in the Lavrentiev regularization.

Neitzel, Prüfert, and Slawig [62] showed how time-dependent optimal control for partial differential equations could be realized in a modern high-level modeling and simulation package. Neitzel and Tröltzsch [63] focused on control problems for parabolic equations and pointwise state constraints. The authors studied the distributed and boundary control case separately. A state-constrained problem is regularized utilizing a Lavrentiev regularization. In the case of boundary control, the authors regularized the state constraint with the help of an additional adjoint problem. Neitzel and Tröltzsch [64] studied the behavior of Moreau-Yosida and Lavrentiev regularization for control problems governed by semilinear parabolic equations with control and state constraints. Strong convergence of global and local solutions is addressed. Based on a second-order sufficient optimality condition, the authors showed strong regularity of the optimality system for the Lavrentiev regularization and the local uniqueness of local minima.

Hömberg, Meyer, Rehberg, and Ring [35] focused on the state-constrained control for a thermistor problem, a quasi-linear coupled system of a parabolic and elliptic PDE with mixed boundary conditions. This system models the heating of conducting material employing a direct current. Existence, uniqueness, and continuity for the state system are derived by employing maximal elliptic and parabolic regularity. These results allowed the authors to derive first-order necessary conditions for the optimal control problem. Meyer and Yousept [61] considered an optimal control problem for an elliptic equation with non-local interface conditions. The optimal control problem involved pointwise control and state constraints. The authors considered a Moreau-Yosida regularization of the state constraints. Krumbiegel and Rösch [50] studied constrained Neumann optimal control problems governed by second-order elliptic partial differential equations. Furthermore, the control is constrained on the domain boundary as well. The authors gave estimates of the error between the optimal solutions of the original and the regularized problem.

Deckelnick and Hinze [20] gave error estimates for a finite-element discretization of a pure state-control problem with Neumann boundary conditions by considering a variational discretization approach. Several authors have considered finitely many state constraints, instead of general pointwise state constraints. This line of investigation has been initiated by Casas [12] for the case of finitely many integral state constraints and by Casas and Mateos [13] for finitely many pointwise state constraints, where, in both contributions, a semilinear state equation was considered. Recently, Du, Ge, and Liu [24] studied an optimal control problem where the control constraints are given in both the pointwise and the integral sense. Using adaptive finite element methods, the authors derived both a priori and a posteriori error estimates. In a related paper [4], the authors studied semilinear elliptic control problems with pointwise state constraints and gave a posteriori error estimates of a dual weighted residual type. Recently, Leykekhman and Vexler [56] gave general error estimates for optimal control of parabolic problems. In an exciting paper, Casas, Mateos, and Vexler [14], for optimal control problem governed by an elliptic partial differential equation and box constrained state constraints, showed high regularity of multiplier under general conditions. They also supplied error estimates.

Returning to the contribution of Meyer, Rösch, and Tröltzsch [60], we note that the imposed compactness and nonnegativity are natural hypotheses in the context of optimal control of linear PDEs. Indeed, by using the usual compact embeddings and maximum principles associated with the solution operators of the PDEs, these properties can easily be verified. The existence of multipliers of problems (Q_ε) is established by proving the solvability of the dual problem, where the non-negativity of the operator $-\widetilde{G}$ played a fundamental role. Nonetheless, for the general case when $U \neq Y$, a direct extension of the above regularization approach does not appear self-evident. In fact, even when U is compactly embedded into Y and we replace in a natural way the identity with the injection operator, the multipliers for the regularized problems may fail to exist. The following example justifies this claim.

Example 12.2.2 [43] Assume that $U = H = \mathbb{R}$, $Y = L^2(0,1)$, and $C = L^2_+(0,1)$. Assume that $G: \mathbb{R} \to L^2[0,1]$ is such that for every $x \in \mathbb{R}$, the element $G(x)$ is the constant map given by $G(x) = -x_F$, where $x_F(x) = x$ for every $x \in (0,1)$. We also assume that the upper bound $w \in L^2(0,1)$ is defined by $w(t) = -t$ for every $t \in [0,1]$. Furthermore, for every $\varepsilon \geq 0$, we consider a regularization map by

$$G_\varepsilon(u) := G(u) - \varepsilon u_F = -(1+\varepsilon)u_F.$$

Clearly, G is linear and compact. Moreover, $-G$ is nonnegative, in the sense $-G(\mathbb{R}_+) \subset C$. In a trivial way, $\mathbb{R} \hookrightarrow L^2(0,1)$ is compactly embedded by the injection operator $i(x) = x_F$.

Let us now consider the family of the following optimization problems:

$$(Q_\varepsilon) \quad \begin{cases} \text{Minimize } J(u) := \tfrac{1}{2}u^2 \\ \text{subject to} \\ G_\varepsilon(u) \leq_C w, \\ 0 \leq_C u_F,\ u \in U. \end{cases}$$

Notice that the feasible set is $[1/(1+\varepsilon), \infty)$, and consequently $u_\varepsilon = 1/(1+\varepsilon)$ is the solution of (Q_ε). We observe that $u_\varepsilon \to u_0$. On the other hand, there are no multipliers for these points. For the sake of argument, assume that there are positive functions μ_ε^1, $\mu_\varepsilon^2 \in L_+^2[0,1]$ such that

$$u_\varepsilon - (1+\varepsilon) \int_0^1 \mu_\varepsilon^1(s)ds - \int_0^1 \mu_\varepsilon^2(s)ds = 0, \tag{12.2a}$$

$$\left\langle \mu_\varepsilon^1, -(1+\varepsilon)(u_\varepsilon)_F - w \right\rangle_Y = 0, \tag{12.2b}$$

$$\left\langle \mu_\varepsilon^2, (u_\varepsilon)_F \right\rangle_Y = 0. \tag{12.2c}$$

Since $u_\varepsilon \neq 0$, by (12.2c), we have $\mu_\varepsilon^2 = 0$, and, therefore, (12.2) is equivalent to

$$\int_0^1 \mu_\varepsilon^1(s)ds = \frac{1}{(1+\varepsilon)^2}, \tag{12.3a}$$

$$\int_0^1 \mu_\varepsilon^1(s)(s-1)ds = 0. \tag{12.3b}$$

However, this equation is unsolvable. In fact, since the multiplier from the first equation is positive, we deduce that

$$\mu_\varepsilon^1(x) \leq 1/(1+\varepsilon)^2, \text{ a.e.,}$$

and hence,

$$\int_0^1 \mu_\varepsilon^1(s)s\,ds \leq 1/[2(1+\varepsilon)^2] < 1/(1+\varepsilon)^2 = \int \mu_\varepsilon^1(s)ds,$$

which implies that $\int_0^1 \mu_\varepsilon^1(s)(s-1)ds < 0$, contradicting (12.3b). $\square$

12.3 Conical Regularization

We shall now discuss the conical regularization. This approach was proposed in [43] in order to circumvent the difficulties caused by the empty interior of the ordering cone in PDE constraint optimization problems. In this approach, the regularization is conducted by replacing the ordering cone with a family of dilating cones. This process leads to a family of optimization problems with regular multipliers. In contrast with the Lavrentiev regularization for control of PDEs with pointwise state constraints, the conical regularization remains valid in general Hilbert spaces, and it does not require any compactness or positivity condition on the involved maps. One of the significant advantages of this approach is that it is readily amenable to numerical computations.

Conical regularization is based on the following notion of the dilating cones:

Definition 12.3.1 *Let $\delta > 0$. Given a closed, and convex cone $C \subset Y$, a family of solid, closed, and convex cones $\{C_\varepsilon\}_{0 < \varepsilon \leq \delta}$ is said to be a family of dilating cones of C if it satisfies the following three conditions:*

1. $C \backslash \{0\} \subset \text{int}(C_\varepsilon)$, *for every* $0 < \varepsilon \leq \delta$.
2. $C_{\varepsilon_1} \subseteq C_{\varepsilon_2}$ *for* $\varepsilon_1 \leq \varepsilon_2$.
3. $C = \bigcap\limits_{0 < \varepsilon} C_\varepsilon$.

In Section 12.6, we will present two specific reconstructions of the dilating cones.

For a sequence $\varepsilon_n \to 0$, we denote the family of dilating cones $\{C_{\varepsilon_n}\}$ by $\{C_n\}$. We consider the following family of conically regularized problems:

$$(P_n) \begin{cases} \text{Minimize } J(u) := \frac{1}{2}\|Su - z_d\|_H^2 + \frac{\kappa}{2}\|u - u_d\|_U^2, \\ \text{subject to} \\ Gu \leq_{C_n} w, \ u \in U. \end{cases}$$

For the subsequent discussion, we formulate the following condition:

$$\text{There exists } \tilde{u} \in U \text{ such that } G(\tilde{u}) \leq_C w \text{ and } G(\tilde{u}) \neq w. \tag{12.4}$$

Following [43], the following results hold for the regularized family (P_n):

Theorem 12.3.2 *For every $n \in \mathbb{N}$, the problem (P_n) is uniquely solvable. Moreover, the following statements hold:*

1. *If the assumption (12.4) holds then for each $n \in \mathbb{N}$, the regularized optimization problem (P_n) is regular. That is, u_n is a minimizer of (P_n), if and only if, there exists Lagrange multiplier $\mu_n \in C_n^+$ such that*

$$S^*(Su_n - z_d) + \kappa(u_n - u_d) + G^*\mu_n = 0,$$
$$\langle \mu_n, Gu_n - w \rangle_Y = 0,$$
$$Su_n \leq_{C_n} w.$$

2. *If assumption (12.4) holds, then the sequence $\{G^*\mu_n\}$ converges strongly to $-S^*(Su_0 - z_d) - \kappa(u_0 - u_d)$.*

3. *If the sequence $\{\|\mu_n\|_Y\}$ is bounded, then*

   (a) *(P) is regular.*

   (b) *If the map G^* is injective, then $\{\mu_n\}$ converges weakly to a regular multiplier μ_0.*

4. *If the map G is one-to-one, then (P) is regular and the regular multiplier μ_0 is unique. Furthermore, the sequence of unique multipliers $\{\mu_n\}$ converges strongly to μ_0.*

The above results show that the conical regularization enjoys nice features. Furthermore, it turns out that Example 12.2.2, for which the regularization approach of [60] failed, can be solved by the conical regularization.

Example 12.3.3 We apply conical regularization to Example 12.2.2. For this we choose $U = \mathbb{R}$, $H = L^2(0,1)$, the product space $Y = L^2(0,1) \times L^2(0,1)$, $C = L^2_+(0,1) \times L^2_+(0,1)$. Let $G : L^2(0,1) \to L^2(0,1) \times L^2(0,1)$ be the operator defined by $G(u) = (\widetilde{G}(u), -u)$, and let the upper bound be given by $w \equiv (w, 0)$. Notice that the feasibility condition (12.4) is verified. Indeed by taking $\widetilde{u} = 1$, we have

$$G(\widetilde{u}) = (-1_F, -1_F) \leq_C w, \ G(\widetilde{u}) \neq w. \qquad (12.5)$$

Given dilating cones $\{C_n\}$ of C, the conical regularized problems read:

$$(P_n) \begin{cases} \text{Minimize } J(u) := \frac{1}{2} u^2 \\ \text{subject to} \\ \widetilde{G}(u) \leq_{C_n} 1_F, \\ 0 \leq_{C_n} u_F, \ u \in U. \end{cases}$$

By Theorem 12.3.2, the solution u_n of (P_n) exists for every $n \in \mathbb{N}$, and the corresponding solutions $\{u_n\} \subset \mathbb{R}$ converge in norm to the solution $u_0 = 1$ of (P). Furthermore, since (12.5) is verified by Theorem 12.3.2, each solution u_n is one solution of the following optimality system: find $\mu_n^1, \mu_n^2 \in C_n^+$ such that

$$u_n - \int_0^1 \mu_n^1 ds - \int_0^1 \mu_n^2 ds = 0,$$

$$\int_0^1 \mu_n^1 (u_n - s) ds = 0,$$

$$u_n \left(\int_0^1 \mu_n^2 ds \right) = 0,$$

as claimed. □

12.4 Half-space Regularization

The Half-space regularization, an analog of the conical regularization, was proposed by Jadamba, Khan, and Sama [41]. This approach is based on performing the regularization by employing certain half-spaces and it makes use of an important idea that every closed and convex cone can be represented as an intersection of half-spaces. In the following, we first discuss the half-space regularization in the context of a general optimization problem and will later apply it to optimal control problems.

Recall that we are working with real separable Hilbert spaces U and Y with $\langle \cdot, \cdot \rangle_U$ and $\langle \cdot, \cdot \rangle_Y$ as their inner products, and $\|\cdot\|_U$ and $\|\cdot\|_Y$ as the associated norms. We also recall that $C \subset Y$ is a pointed, closed, and convex cone. We will identify the dual Y^* of Y with Y. Given linear and bounded maps $S : U \to H$ and $G : U \to Y$, and a parameter $\kappa > 0$, we continue to focus on the following minimization problem:

$$(P) \begin{cases} \text{Minimize } J(u) := \frac{1}{2} \|Su - z_d\|_H^2 + \frac{\kappa}{2} \|u - u_d\|_U^2, \\ \text{subject to} \\ G(u) \leq_C w, \ u \in U. \end{cases}$$

The half-spaced regularization approach is based on the fundamental result that for every closed and convex cone $C \subset Y$, it holds that

$$C = \bigcap_{\mu^* \in C^+} H_{\mu^*},$$

where C^+ is the positive dual cone of C and H_{y^*}, where $y^* \in Y^*$, denotes the associated (positive) half-space given by:

$$H_{y^*} = \{y \in Y : y^*(y) \geq 0\}.$$

Therefore, we give the following concept:

Definition 12.4.1 *Let I be an arbitrary nonempty index-set and let $\{y_i^*\}_{i \in I} \subset Y^*$. The collection of half-spaces $\{H_{y_i^*}\}_{i \in I}$ is said to be a representation of C if*

$$C = \bigcap_{i \in I} H_{y_i^*}.$$

Since the space Y is separable, the index set I can be considered numerable (see [8]). As a consequence, in the following, $\{H_{\mu_i^*}\}_{i \in I}$ will represent a numerable representation of C.

For each $n \in \mathbb{N}$, by C_n we denote the closed and convex cone

$$C_n = \bigcap_{i=1}^{n} H_{\mu_i^*}.$$

The cone C_n is an outer approximation of C.

By replacing C by C_n, we consider the regularized optimization problems:

$$(P_n) \quad \begin{cases} \text{Minimize } J(u) := \dfrac{1}{2}\|Su - z_d\|_H^2 + \dfrac{\kappa}{2}\|u - u_d\|_U^2, \\ \text{subject to} \\ G(u) \leq_{C_n} w, \ u \in U. \end{cases}$$

We summarize the main results from [41] on the Half-space regularization.

Theorem 12.4.2 *Problem (P) has the unique solution u_0. Moreover, for each $n \in \mathbb{N}$, optimization problem (P_n) has the unique solution u_n. Moreover, $\{u_n\}$ converges strongly to u_0.*

It has been shown that under a mild constraint qualification, optimization problem (P_n) admits Lagrange multipliers, independently of the existence of multipliers for the original problem (P). In the following, we assume that f and g are continuously differentiable. We first formulate the following:

Definition 12.4.3 *A point $c \in C$ is called a quasi-interior point if $\mu^*(c) > 0$, for any $\mu^* \in C^+ \setminus \{0\}$. The set of all quasi-interior points of C is denoted by $\mathrm{qint} C$.*

The quasi-interior of a solid cone coincides with the interior, but it may be nonempty in non-solid cone. In fact, the existence of quasi-interior points can be assured for every closed, convex, and pointed cone of a separable Hilbert space (see [44]). We formulate the following constraint qualification:

$$\exists \, \tilde{x} \in X \text{ such that } G(\tilde{x}) \in -\operatorname{qint} C. \tag{12.6}$$

Condition (12.6) is an analog of the Slater constraint qualification where the quasi-interior has replaced the interior. Variants of this condition have been used in the context of applied optimization problems posed in infinite-dimensional spaces where the ordering cone has a non-empty interior. However, we emphasize that condition (12.6), even for linear problems, does not by itself guarantee the existence of multipliers, see [41, Example 2.1]. In the context of the half-space regularization approach, however, (12.6) ensures the existence of Lagrange multipliers for regularized problems (P_n). In other words, if condition (12.6) holds, then (P_n) is regular for each $n \in \mathbb{N}$, as shown below.

Theorem 12.4.4 *Assume that* (12.6) *holds. If u_n is a minimizer of (P_n), then there exists $(\mu_1^n, \ldots, \mu_n^n) \in \mathbb{R}_+^n$, such that for $i = 1, \ldots, n$, we have*

$$DJ(u_n) + \sum_{i=1}^{n} \mu_i^n G^* \mu_i^* = 0, \tag{12.7a}$$

$$\sum_{i=1}^{n} \mu_i^n \left(\mu_i^* \circ G \right)(u_n) = 0, \tag{12.7b}$$

$$\mu_i^* \circ G(u_n) \leq 0. \tag{12.7c}$$

In the following, we set $\mu_{(n)}^* := \sum_{i=1}^{n} \mu_i^n \mu_i^*$, where $(\mu_1^n, \ldots, \mu_n^n) \in \mathbb{R}_+^n$ is a sequence that verifies Theorem 12.4.4 and corresponds to a multiplier in C_+^n for problem (P_n). The following result gives a useful consequence of the boundedness of regularized multipliers.

Theorem 12.4.5 *Assume that* (12.6) *holds. A necessary condition for the existence of multipliers for problem (P) is the boundedness of the sequence $\|\mu_{(n)}^*\|_Y$.*

Remark 12.4.6 A comparison of the half-space regularization with the Lavrentiev regularization, examples, and numerical results showing the usefulness of the approach are available in [41]. In Lavrentiev regularization, each regularized problem is an infinite dimensional control problem akin to the original problem. However, in the half-spaces based decomposition approach, regularized problems are simpler. In particular, for an application to the control problems we obtain finitely many state constraints.

12.5 Integral Constraint Regularization

Integral constraint regularization refers to the specific case of the half-space regularization when the half-space representation of the cone of positive functions in the space of the square integrable functions is used. The particular form leads to control problems with a unique structure that deserves special attention.

Our focus remains on the pointwise state-constrained control problem

$$(Q) \begin{cases} \text{Minimize } \frac{1}{2}\int_\Omega (y_u - z_d)^2 dx + \frac{\kappa}{2}\int_\Omega (u - u_d)^2 dx, \\ \text{subject to} \\ -\Delta y_u(x) + y_u(x) = u(x) \text{ in } \Omega, \ \partial_n y_u = 0 \text{ on } \partial\Omega, \\ y_u(x) \le w(x), \ \text{a.e. in } \Omega, \ u \in L^2(\Omega), \end{cases}$$

where $\Omega \subset \mathbb{R}^d$ ($d = 2, 3$) is an open, bounded, and convex domain with boundary $\partial\Omega$, $\kappa > 0$ is a fixed parameter, z_d, $u_d \in H^2(\Omega)$, and $w \in W^{2,\infty}(\Omega)$ are given elements, and ∂_n is the outward normal derivative to $\partial\Omega$. We suppose that the control variable u is taken in $L^2(\Omega)$ and as a consequence, we have $y_u \in H^2(\Omega)$.

The study of errors estimates is one of the most prominent research issues for the regularization methods. Derivation of numerical error is commonly carried out in two steps. The first step estimates the error between the solution of the original problem and the regularized solution in terms of the regularization parameter. The second step requires determining the error between the regularized solution and a discrete regularized solution. To derive conclusive error estimates from the two estimates, it is necessary that the discretization and the regularization parameters be coupled optimally. For example, Hinze and Hintermuller [31] followed this idea to give error estimates for the Moreau-Yosida regularization. In the same fashion, Cherednichenko, Krumbiegel, and Rösch, [15] and Hinze and Meyer [33] obtained error estimates for the Lavrentiev regularization, whereas Hinze and Schiela [34] gave error estimates for the barrier methods.

We would also like to review contributions where error estimates for a direct finite element discretization of a state-constrained the problem is given. We first note the interesting contribution by Deckelnick and Hinze [20] that gave error estimates for a finite-element discretization of a state-constrained control problem with Neumann boundary conditions by devising a variational discretization framework, whereas, in [21], the same authors gave similar results for control and state constrained problems. They obtained an order of $O(h^{1-\varepsilon})$ in 2d and $O(h^{\frac{1}{2}-\varepsilon})$ in 3d for the control discretization in the L^2-norm by assuming smooth domains and considering curved finite elements. In related research, Meyer [58] provided comparable results for a control problem with mixed pointwise state and control constraints. Analogously, Rösch and Steinig [69] studied a problem with control and state constraints and Dirichlet conditions and obtained $O(h^{\frac{3}{4}})$ in the 3-d case and $O(h|\log h|)$ under additional regularity assumption on the control. Recently Casas, Mateos, and Vexler [14], by assuming an additional hypothesis, obtained $O(h|\log h|)$ in both dimensions for a two-sided state-constrained problem with Dirichlet boundary conditions. To describe error estimates for problems with finitely many state constraints, we note that for a linear elliptic problem with Dirichlet boundary conditions, Liu, Yang, Yuan [57] obtained $O(h^2)$ for the control discretization of a problem with one integral constraint. Recently, Leykekhman, Meidner, and Vexler [55] obtained $O(h)$ for finitely many pointwise state constraints. We also point out the contributions by Casas [12] for the case of finitely many integral state constraints and by Casas and Mateos [13] for finitely many pointwise state constraints, where the focus was on a semilinear state equation.

In the following, we summarize error estimates proved recently by Jadamba, Khan, and Sama [39] for the integral constraint regularization method. Let $\{\Delta^\delta\}_{\delta>0}$ be a family of convex partitions of Ω, where δ is a real parameter denoting the diameter of each partition Δ^δ. That is, $\{\Delta^\delta\}$ consists of a finite number of closed and convex sets $\{\Delta_i^\delta\} \subset \Omega$ ($i = 1, ..., T(\delta)$) satisfying

$$\sum_{\Delta_i^\delta \in \Delta^\delta} |\Delta_i^\delta| = |\Omega|,$$

where $T(\delta) = \#\Delta^\delta$ is the cardinality of $\{\Delta^\delta\}$ and $|\Omega|$ is the Lebesgue measure of Ω. We assume that the diameter of the family tends to zero, that is,

$$\mathrm{diam}(\Delta^\delta) = \max_{i=1,...,T(\delta)} \mathrm{diam}(\Delta_i^\delta) \xrightarrow[\delta \to 0^+]{} 0.$$

An example of such a partition is a family of triangulations of the domain Ω.

The integral constraint regularization approach relates to $\delta > 0$, the following control problem with finitely many integral state constraints.

$$(Q_\delta) \begin{cases} \text{Minimize } \frac{1}{2} \int_\Omega (y_u - z_d)^2 dx + \frac{\kappa}{2} \int_\Omega (u - u_d)^2 dx, \\ \text{subject to} \\ \int_{\Delta_i^\delta} y_u(s) ds \leq \int_{\Delta_i^\delta} w(s) ds \text{ for every } i = 1, ..., T(\delta), u \in L^2(\Omega). \end{cases}$$

It was shown in [39] that (Q_δ) constitutes a family of regularized problems of (Q). That is, for every δ, viewed as the regularization parameter, there exists a unique regularized solution u_δ of (Q_δ), the regularized solutions $\{u_\delta\}$ converge strongly to the solution u_0 of (Q), the regularized solutions can be obtained by the KKT optimality system, and the multipliers are L^2-functions.

In [39], the following error estimates are available

$$\frac{\kappa}{2} \|u_\delta - u_0\|_{2,\Omega}^2 + \frac{1}{2} \|y_\delta - y_0\|_{\infty,\Omega}^2 \leq C \int_\Omega (y_\delta - w) d\mu_0, \tag{12.8}$$

where $y_\delta \equiv y_{u_\delta}$ is the optimal state and C is a constant. The left-hand side term in (12.8) specifies a measure of the infeasibility of the regularized state with respect to the measure $d\mu_0$ associated to the multiplier μ_0^* of the original problem. As a result of (12.8), under an additional mild assumption, it holds that $\|u_\delta - u_0\|_{2,\Omega} = O(\delta^{\frac{1}{2}})$. This estimate can be improved to $\|u_\delta - u_0\|_{2,\Omega} = O(\delta)$, if the multiplier or the solution is more regular.

Given a finite element discretization of the control-to-state map $S_h : L^2(\Omega) \to V_h$, we consider the following discretization of (Q_δ):

$$(Q_{\delta,h}) \begin{cases} \text{Minimize } \frac{1}{2} \int_\Omega (S_h u_h - I_h z_d)^2 dx + \frac{\kappa}{2} \int_\Omega (u_h - I_h u_d)^2 dx, \\ \text{subject to} \\ \int_{\Delta_i^\delta} S_h u_h ds \leq \int_{\Delta_i^\delta} I_h w ds \text{ for every } i = 1, ..., T(\delta), u_h \in L^2(\Omega). \end{cases}$$

Problem $(Q_{\delta,h})$ is uniquely solvable and the unique solution $u_{\delta,h}$ can be easily computed. By the variational discretization approach, $(Q_{\delta,h})$ is a fully finite-dimensional problem. In [39], assuming that both of the parameters be coupled to satisfy $h^2 \delta^{\frac{-d}{2}} \leq C$, the authors proved

$$\|u_\delta - u_{\delta,h}\|_{2,\Omega} \leq C\sqrt{\varphi(h)}, \tag{12.9}$$

where $\varphi : \mathbb{R}_+ \to \mathbb{R}_+$ is an abstract formulation of the convergence rate verifying

$$\max\left\{\|(S-S_h)(u_\delta)\|_{\infty,\Omega}, \|(S-S_h)(u_{\delta,h})\|_{\infty,\Omega}\right\} \leq C\varphi(h). \tag{12.10}$$

In (12.9) and (12.10), constraints must be independent of both parameters. This is crucial, because by using regularization error estimates, it follows that $(Q_{\delta,h})$ is a full discretization of problem (Q). Furthermore,

$$\|u_0 - u_{\delta,h}\|_{2,\Omega} = O\left(\sqrt{\delta + \varphi(h)}\right),$$
$$\|u_0 - u_{\delta,h}\|_{2,\Omega} = O\left(\delta + \sqrt{\varphi(h)}\right),$$

where the second estimate was proved under additional regularity either on the solution u_0 or on the associated multiplier.

12.6 A Constructible Dilating Regularization

The conical regularization is defined using an abstract notion of a family of dilating cones, whereas the half-space regularization is devised using a half-space representation of a convex cone. In this section, we discuss the construction of a family of dilating cones, which is based on a halfspace decoupling of the given cone and the notion of the Henig dilating cone which we discuss first.

Let Y be a Hilbert space and $C \subset Y$ is a closed, convex, and pointed cone. We will construct a family of dilating cones $\{C_\varepsilon\}$ of C, where $\varepsilon \in (0,1)$. We assume that C has a closed and convex base $\Theta \subset C$ such that $0 \notin \Theta$ and

$$K = \bigcup_{\lambda \geq 0}\{\lambda\theta : \theta \in \Theta\}.$$

Furthermore, without loss of generality (see for [44, Theorem 2.2.12]), we assume that the base Θ is given by a strictly positive functional $\beta^* \in K^\sharp := \{y^* \in Y^* : y^*(k) > 0, \forall k \in K\setminus\{0\}\}$, that is, $\Theta = \{y \in K : \beta^*(y) = 1\}$, where we normalize to $\|\beta^*\|_{Y^*} = 1$.

Given $\varepsilon > 0$, the Henig dilating cone is then given by

$$K_\varepsilon := \mathrm{cl}\left[\mathrm{cone}\left(\Theta + \varepsilon B_Y\right)\right]$$

where $B_Y = \{v \in Y : \|v\|_Y \leq 1\}$ is the closed unit ball.

The definition given above is closely related to the notion introduced by M. Henig in the finite-dimensional setting with the objective of proposing a new notion of proper efficiency (see [30]). Later, Borwein and Zhuang [9] extended this notion to normed spaces and presented a systematic study of its useful properties. The family of dilating cones exists for every closed, and convex cone in a separable normed space.

Given $y^* \in Y^*$, by $y_\varepsilon^* \in Y^*$ we denote the functional

$$y_\varepsilon^* := \varepsilon\beta^* + (1-\varepsilon)y^*, \tag{12.11}$$

which, for sufficiently small ε, belongs to K_ε^+. In fact, it holds that if $y^* \in C^+$, $\varepsilon \in (0,1)$, and $\|y^*\|_{Y^*} \leq 1$, then $y_\varepsilon^* \in K_\varepsilon^*$ with $\|y_\varepsilon^*\|_{Y^*} \leq 1$.

In the following, $\{H_{\lambda_i^*}\}_{i\in I}$ is a fixed representation of C, where without loss of generality, $\{\lambda_i^*\}_{i\in I}$ are normalized so that $\|\lambda_i^*\|_{Y^*} = 1$, for every $i \in I$. Following (12.11) and the given representation $\{H_{\lambda_i^*}\}_{i\in I}$ of C, for each $\lambda_i^* \in C^+$, we define, for any $i \in I$:

$$\lambda_{i,\varepsilon}^* = \varepsilon \beta^* + (1-\varepsilon)\lambda_i^*. \tag{12.12}$$

For any $i \in I$ and $0 < \varepsilon < 1$, and, hence, $\|\lambda_{i,\varepsilon}^*\|_{Y^*} \leq 1$. For $0 < \varepsilon < 1$, we define

$$C_\varepsilon := \bigcap_{i\in I} H_{\lambda_{i,\varepsilon}^*}. \tag{12.13}$$

The following result shows that $\{C_\varepsilon\}$ is a family of dilating cones:

Theorem 12.6.1 *[40] For $\varepsilon \in (0,1)$, C_ε is a solid, closed, convex, and pointed cone such that*

1. $K_\varepsilon \subset C_\varepsilon$.

2. $C\setminus\{0\} \subset \mathrm{int}(C_\varepsilon)$.

3. $C = \bigcap_{\varepsilon>0} C_\varepsilon$.

The following example shows that $K_\varepsilon \subset C_\varepsilon$ is in general strict:

Example 12.6.2 *[40]*. We set $Y = \mathbb{R}^2$ equipped with its Euclidean norm and the ordering $C = \mathbb{R}_+ \times \{0\}$. By identifying elements of the dual space with vectors, we consider the strictly positive functional $\beta^* \equiv (1,0)$ and the representation $\{H_{\lambda_i^*}\}_{i\in\{1,2,3\}}$ given by vectors $\lambda_1^* \equiv (0,1)$, $\lambda_2^* \equiv (0,-1)$, $\lambda_3^* \equiv (1,0)$. Then,

$$K_\varepsilon = \left\{(x,y) \in \mathbb{R}^2 : \varepsilon\left(1-\varepsilon^2\right)^{\frac{-1}{2}} x \geq y \geq -\varepsilon\left(1-\varepsilon^2\right)^{\frac{-1}{2}} x\right\},$$

$$C_\varepsilon = \left\{(x,y) \in \mathbb{R}^2 : \varepsilon(1-\varepsilon)^{-1} x \geq y \geq -\varepsilon(1-\varepsilon)^{-1} x\right\},$$

which confirms that $K_\varepsilon \subsetneq C_\varepsilon$ for $\varepsilon \in (0,1)$.

Our focus continues to be on an abstract optimization problem introduced earlier. For convenience, we recall that U, H and Y are real Hilbert spaces, with $\|\cdot\|_U$, $\|\cdot\|_H$, and $\|\cdot\|_Y$ be their norms. Let $C \subset Y$ be a closed, convex, and pointed cone, let Y^* be the dual space of Y, and let C^+, be the positive dual of C.

We focus on the following convex minimization problem (P):

$$(P) \begin{cases} \text{Minimize } J(u) := \dfrac{1}{2}\|Su - z_d\|_H^2 + \dfrac{\kappa}{2}\|u - u_d\|_U^2 \\ \text{subject to} \\ Gu \leq_C w,\ u \in U. \end{cases}$$

Here $S : U \to H$, $G : U \to Y$ are linear bounded operators, $\kappa > 0$ is a given parameter, and $u_d \in U$, $z_d \in H$, and $w \in Y$ are given elements. Clearly, problem (P) has a unique solution $\bar{u}$.

The above preparation permits to define two families of regularized problems. The first one is the Henig conical regularization

$$(P_\varepsilon^H) \begin{cases} \text{Minimize } J(u) := \frac{1}{2}\|Su - z_d\|_H^2 + \frac{\kappa}{2}\|u - u_d\|_U^2 \\ \text{subject to} \\ Gu \leq_{K_\varepsilon} w, \ u \in U, \end{cases}$$

and the second one is the constructible regularized problems:

$$(P_\varepsilon^C) \begin{cases} \text{Minimize } J(u) := \frac{1}{2}\|Su - z_d\|_H^2 + \frac{\kappa}{2}\|u - u_d\|_U^2 \\ \text{subject to} \\ Gu \leq_{C_\varepsilon} w, \ u \in U. \end{cases}$$

Based on the recent work by Jadamba, Khan, and Sama [40], we give a summary of some stability estimates for the regularization error in both of the cases. We will denote by u_ε, u_ε^H, and u_ε^C, the solutions to (P_ε), (P_ε^H), and (P_ε^C), respectively. We will use the notation $q_\varepsilon = Gu_\varepsilon - w$, $q_\varepsilon^H = Gu_\varepsilon^H - w$, and $q_\varepsilon^C = Gu_\varepsilon^C - w$, for $\varepsilon \geq 0$ sufficiently small. In the rest of the section, we assume that

$$q_0 \neq 0, \tag{12.14}$$

which prevents problem (P) from being a pure inequality constraint problem. Furthermore, (12.14) yields a nontrivial feasible point so that the regularized optimality system is solvable.

To give a stability estimate for the Henig conical regularization, let $\mu_{H,\varepsilon}^* \in Y^*$ be a multiplier associated with Henig regularized problem (P_ε^H) for ε small enough. Without loss of generality, we assume $\mu_{H,\varepsilon}^* \neq 0$. Indeed it follows from the KKT conditions that otherwise we would have the trivial case $u_\varepsilon^H = \bar{u}$.

Using the notation $\delta_\varepsilon^H := \varepsilon \|\mu_{H,\varepsilon}^*\|_{Y^*}$, we have the following estimate:

Theorem 12.6.3 *The sequence* $\{\delta_\varepsilon^H\} \to 0$ *as* $\varepsilon \to 0$, *and the following holds:*

$$\frac{\kappa}{2}\|\bar{u} - u_\varepsilon^H\|_U^2 + \frac{1}{2}\|S\bar{u} - Su_\varepsilon^H\|_V^2 \leq J(\bar{u}) - J(u_\varepsilon^H) - \delta_\varepsilon^H|\beta^*(q_0)|. \tag{12.15}$$

Moreover, if (P) is regular, then there is constant $c > 0$ such that

$$\frac{\kappa}{2}\|\bar{u} - u_\varepsilon^H\|_U^2 + \frac{1}{2}\|S\bar{u} - Su_\varepsilon^H\|_V^2 \leq c\varepsilon. \tag{12.16}$$

For the constructible case, the following result, which is an analogue Theorem 12.6.3, holds:

Theorem 12.6.4 *For $\varepsilon \to 0$, we have $\{\delta_\varepsilon^C\} \to 0$. Furthermore,*

$$\frac{\kappa}{2}\|\bar{u} - u_\varepsilon^C\|_U^2 + \frac{1}{2}\|S\bar{u} - Su_\varepsilon^C\|_V^2 \leq J(\bar{u}) - J(u_\varepsilon^C) - \delta_\varepsilon^C|\beta^*(q_0)|.$$

Furthermore, if C has non-empty interior and a Slater constraint qualification holds, then there exists a constant $c > 0$ such that

$$\|u - u_\varepsilon^C\|_U^2 \leq c\varepsilon.$$

12.7 Regularization of Vector Optimization Problems

In this section, we discuss a prototype of PDE-constrained vector optimal control problems. In recent years, such problems have attracted a lot of attention because they provide a natural setting for various applied models where there is a need to incorporate multiple and often conflicting goals.

Let U be a real Hilbert space equipped with the inner product $\langle \cdot, \cdot \rangle_U$ and let $\|\cdot\|_U$ be the associated norm. Let Y be a real Banach space, let Y^* be the dual of Y, and let $C \subset Y$ be a pointed, closed, and convex cone, inducing a partial ordering $\leq_C$ on Y. Let C^+ be the positive dual of C. In the Euclidean space $\mathbb{R}^p$, the set of all vectors with nonnegative coordinates is denoted by $\mathbb{R}^p_+$ and the set of all vectors with positive coordinates is denoted int $\mathbb{R}^p_+$.

We consider the following convex vector optimization problem, which is a generalization of the scalar problem considered earlier:

$$(P) \begin{cases} \text{Minimize} J(u) = (J_1(u), J_2(u), \ldots, J_p(u))^\top \\ \text{subject to} \\ u \in D := \{u \in U : Gu \leq_C w\}, \end{cases}$$

where $G : U \to Y$ is a linear and bounded map, and the component objective maps $J_i : U \to \mathbb{R}$, for $i \in \{1, \ldots, p\}$, are continuous, coercive, and strictly convex.

An element $\bar{u} \in D$ is said to be a minimal element of (P), if $J(u) \leq_{\mathbb{R}^p_+} J(\bar{u})$, for some $u \in D$ implies that $J(u) = J(\bar{u})$. Analogously, an element $\bar{u} \in D$ is called a weakly minimal element of (P), if there is no $u \in D$ such that $J(u) <_{\mathbb{R}^p_+} J(\bar{u})$. In the following, the set of all minimal elements of (P) will be denoted by Min(P) and the set of all weakly minimal elements of (P) is denoted by WMin(P).

We note that the vector optimization problems posed in general Banach spaces, besides retaining the main challenges of the classical PDE constrained optimization problems with pointwise state constraints, present new technical difficulties. One of the main impediments stems from the presence of the infinite-dimensional linear constraint, for which the ordering cone C often lacks interior points. To be specific, a commonly used technique for solving problems like (P) is the coupling of Lagrange multiplier rules, aka KKT conditions, with scalarization techniques. Unfortunately, when the constraint cone has an empty interior, the well-known constraint qualifications either fail to hold or are challenging to verify. We note that such problems where the solutions cannot be solved via the KKT conditions are commonly termed as non-regular. This situation is more persistent in infinite dimensional problems.

The extension of the regularization methods from the scalar optimal control problems to vector optimal control problem is by no means trivial and is an active area of research, see [7, 28, 36, 48, 52, 54, 68, 70]. To the best of our knowledge, there are only a handful of works devoted to the linearly point-wise state constrained multiobjective problems, and even fewer are focused on extending the well-known regularization methods from the scalar to the vector scenario. The notable contributions for regularization methods are [46, 47, 71].

Developing a regularization framework for vector optimization problems with infinite-dimensional constraints poses some non-trivial challenges. Firstly, even for a coercive objective, there is no unique solution but a set of solutions. Secondly, in vector optimization problems, there are different solution concepts, such as minimal elements, weakly minimal elements, properly minimal elements and strongly minimal elements, among others, and it is entirely unclear which solution concept suits best the applied models. Thirdly, the KKT conditions in vector optimization problems involve a scalarization of the problem and, depending on the scalarization parameters, we may find different solutions. In fact, for the vector case, it is unclear how to define the notion of a non-regular problem. To clarify, we first recall that, given a minimal element $\bar{u} \in \text{Min}(P)$, a Lagrange multiplier rule holds if there exists a Lagrange multiplier vector $(\lambda, \bar{\mu}_\lambda) = ((\lambda_1, ..., \lambda_p), \bar{\mu}_\lambda) \in \mathbb{R}^p_+ \times C^*$, with $\lambda \neq 0$, such that

$$\sum_{i=1}^{p} \lambda_i D J_i(\bar{u}) + \bar{\mu}_\lambda \circ G = 0, \qquad (12.17a)$$

$$\bar{\mu}_\lambda (G\bar{u} - w) = 0, \qquad (12.17b)$$

$$G\bar{u} \leq_C w. \qquad (12.17c)$$

Multiplier λ is associated with the scalarization vector and $\bar{\mu}_\lambda$ with the Lagrange dual element is related to the constraint. On the other hand, depending on the scalarization vector λ, the solution of optimality conditions (12.17) corresponds to different kind of solutions, see for example [42]. Consequently, the regularity might hold for one solution concept but not for another.

Recently, Jadamba, Khan, Lopez, and Sama [38] developed a conical regularization approach for problem (P). To summarize the results given in [38], the following family of regularized problems is defined by replacing C in problem (P) by a family of dilating cones C_ε:

$$(P_\varepsilon) \begin{cases} \text{Minimize} J(u) = (J_1(u), J_2(u), \ldots, J_p(u))^\top \\ \text{subject to} \\ u \in D_\varepsilon := \{u \in U : Gu \leq_{C_\varepsilon} w\}. \end{cases}$$

The regularized family (P_ε) is a real one-parametric family of vector problems where the ordering cone C_ε has a nonempty interior. Due to the well-known Slater constraint qualification, the minimal elements of the regularized problems can be computed via the KKT conditions. In other words, the perturbed problems (P_ε) are regular for every $\varepsilon > 0$.

In [38], the authors proved the equivalence of the minimal elements and the weakly minimal elements for the regularized problems and established that the weak limit of the regularized solutions is the set of minimal elements of the problem (P). They scalarized the original problem and the regularized analog and proved that the scalarized solutions parameterize the regularized minimal elements. Furthermore, they established the convergence in Kuratowski-Mosco sense of the conically regularized solutions to the minimal elements of the original problem. In the particular case of the Henig regularization, they gave an a priori Hölder continuity estimate for the regularized efficient set-valued map when the original problem is regular, and in addition, a norm boundedness property of the Lagrange multipliers set holds. They presented numerical examples in finite and infinite dimensions, illustrating the efficacy of the developed framework.

12.8 Concluding Remarks and Future Research

We presented an overview of some of the recent developments in the regularization methods for pointwise state-constrained optimal control problems. In the future, we would like to focus on some of the following issues:

12.8.1 Conical Regularization for Variational Inequalities

An extension of the conical regularization to optimal control problem of variational inequalities with pointwise state constraints seems promising. Optimal control problems become significantly challenging when a variational inequality replaces the constraint state equation. An example is the following:

$$(S) \begin{cases} \text{Minimize } J(u,y) := \frac{1}{2}\int_\Omega (y-z_d)^2 dx + \frac{\kappa}{2}\int_\Omega (u-u_d)^2 dx, \\ \text{subject to} \\ y \in K, \ \langle -\triangle y(x), z-y \rangle_{L^2(\Omega)} \geq \langle u, z-y \rangle_{L^2(\Omega)}, \ \forall z \in K, \ y = 0 \text{ on } \partial\Omega, \end{cases}$$

where $K := \{y \in L^2(\Omega) : y(x) \leq w(x), \text{ a.e. in } \Omega\}$.

The primary obstacle in deriving optimality conditions is the nonsmoothness of the control-to-state map $u \to y_u$. Several approaches have been used in order to tackle this difficulty. For instance, penalization methods have been investigated by Barbu [3] and recently by Ito and Kunisch [37], whereas the regularization of the feasible domain has been devised by Bergounioux [5]. A general treatment of control problems for variational inequalities with state constraint is available in Serovajsky [74] (see also [79]).

The optimal control problem (S) can be written in the equivalent form (Q):

$$(Q) \begin{cases} \text{Minimize } J(u,y) := \frac{1}{2}\text{int}_\Omega (y-z_d)^2 dx + \frac{\kappa}{2}\text{int}_\Omega (u-u_d)^2 dx, \\ \text{subject to} \\ -\triangle y(x) = y + \xi, \text{ in } \Omega, \\ y \leq_{L^2_+(\Omega)} w, \ 0 \leq_{L^2_+(\Omega)} \xi, \ \langle \xi, y-w \rangle_{L^2(\Omega)} = 0, y = 0 \text{ on } \partial\Omega. \end{cases}$$

The above control problem can easily be written in the general optimization framework. The main challenge is the fact that $L^2_+(\Omega)$ has an empty interior, and consequently the existence of regular Lagrange multipliers cannot be proved. In fact, Bermudez and Saguez [6] proved optimality conditions by using measures, making the numerical treatment complicated. Furthermore, due to the nonconvex nature of the term $\langle \xi, y-w \rangle_{L^2(\Omega)} = 0$, problem (Q) presents additional difficulties. Bergounioux [5] regularized $\langle \xi, y-w \rangle_{L^2(\Omega)} = 0$ by $\langle \xi_\varepsilon, y_\varepsilon - w \rangle_{L^2(\Omega)} \leq \varepsilon$, where ε is a real parameter tending to 0. Although, one can assure the convergence of the solutions of the regularized problem, due to the unboundedness of ξ_ε, it is only possible after a penalization of the regularized problem. This coupling of regularization and penalization also appears in [37].

Given a family of dilation cones $\{C_n\}$ associated with $L^2_+(\Omega)$, the conical regularization approach by replacing the variational inequality

$$y \leq_{L^2_+(\Omega)} w, \quad 0 \leq_{L^2_+(\Omega)} \xi, \quad \langle \xi, y-w \rangle_{L^2(\Omega)} = 0,$$

in (Q) by its conically regularized analogue

$$y \leq_{C_n} w, \quad 0 \leq_{C_n} \xi, \quad \langle \xi, y-w \rangle_{L^2(\Omega)} = 0.$$

12.8.2 Applications to Supply Chain Networks

The conical regularization has a natural application to an optimal control problem for a mathematical model describing supply chains on a network. This mathematical model involves conditions on each vertex of the network, which results in a system of nonlinear conservation laws coupled with ordinary differential equations. Herty Göttlich, and Klar [29] recently presented this model. However, a more suitable version of this problem is the optimal control problem recently proposed by D'Apice, Kogut, and Manzo [18, 19], where the considered formulation necessitates pointwise state constraints. Interestingly, the authors in [18] suggest that the dilating cones can remedy the problem caused by the lack of any Slater-type constraint qualification for L^2-controls.

12.8.3 Nonlinear Scalarization for Vector Optimal Control Problems

We note that dilating cones and half-spaces have been used successfully in nonsmooth and set-valued optimization in order to derive abstract theoretical results. The described regularization methods supplement those studies by providing a computational framework, making them readily applicable to applied models. It is natural to consider optimal control problems in general spaces and make use of the recent developments that have taken place in the dynamic field of nonlinear scalarization (see [45]). Another interesting direction is to consider an optimal control problem with non-coercive variational problems as the constraints. The use of elliptic regularization then results in a well-posed optimal control problem. A coupling of the conical regularization and the elliptic regularization ought to raise a wealth of interesting ideas that should be explored.

12.8.4 Nash Equilibrium Leading to Variational Inequalities

More recently, numerous works have been devoted to the study of economic models in an infinite dimensional setting. The obstacle that has prevented the study of economic models in general spaces is precisely that the ordering cone in these models has an empty interior. Recently, a new separation theorem that uses the notion of quasi-interior (see [17]) to allow for new results ensuring the existence of multipliers (see [2]) has been formulated. However, the results obtained so far are purely of a theoretical nature. This project gives an alternate path to revisit those models and give new theoretical and computational tools. Recently, Faraci and Raciti [26] studied a generalized Nash equilibrium problem in an infinite-dimensional setting. To prove the existence of Lagrange multipliers, they relied on recent results in the duality theory. Since the main challenge in [26] is the fact that the ordering cone has a non-empty interior, the use of conical regularization seems a natural candidate to explore. The advantage is that the conical regularization is readily amenable to numerical computations, whereas the duality arguments are primarily of a theoretical nature.

Acknowledgements. Baasansuren Jadamba and Akhtar Khan are supported by National Science Foundation Grant Number 1720067. Miguel Sama's work is partially supported by Ministerio de Economía y Competitividad (Spain), project MTM2015-68103-P and 2017-MAT12 (ETSI Industriales, UNED).

References

[1] J. J. Alibert and J. P. Raymond. Boundary control of semilinear elliptic equations with discontinuous leading coefficients and unbounded controls. *Numer. Funct. Anal. Optim.*, 18(3-4): 235–250, 1997.

[2] A. Barbagallo and A. Maugeri. Duality theory for the dynamic oligopolistic market equilibrium problem. *Optimization*, 60(1-2): 29–52, 2011.

[3] V. Barbu. *Optimal control of variational inequalities*, volume 100 of *Research Notes in Mathematics*. Pitman, Boston, MA, 1984.

[4] O. Benedix and B. Vexler. A posteriori error estimation and adaptivity for elliptic optimal control problems with state constraints. *Comput. Optim. Appl.*, 44(1): 3–25, 2009.

[5] M. Bergounioux. Optimal control of an obstacle problem. *Appl. Math. Optim.*, 36(2): 147–172, 1997.

[6] A. Bermudez and C. Saguez. Optimal control of variational inequalities. *Control Cybernet.*, 14(1-3): 9–30 (1986), 1985.

[7] H. Bonnel and C. Y. Kaya. Optimization over the efficient set of multi-objective convex optimal control problems. *J. Optim. Theory Appl.*, 147(1): 93–112, 2010.

[8] J. M. Borwein and J. D. Vanderwerff. Constructible convex sets. *Set-Valued Anal.*, 12(1-2): 61–77, 2004.

[9] J. M. Borwein and D. Zhuang. Super efficiency in vector optimization. *Trans. Amer. Math. Soc.*, 338(1): 105–122, 1993.

[10] S. C. Brenner, L.-Y. Sung, and Y. Zhang. Post-processing procedures for an elliptic distributed optimal control problem with pointwise state constraints. *Appl. Numer. Math.*, 95: 99–117, 2015.

[11] E. Casas. Control of an elliptic problem with pointwise state constraints. *SIAM J. Control Optim.*, 24(6): 1309–1318, 1986.

[12] E. Casas. Error estimates for the numerical approximation of semilinear elliptic control problems with finitely many state constraints. *ESAIM Control Optim. Calc. Var.*, 8: 345–374, 2002.

[13] E. Casas and M. Mateos. Numerical approximation of elliptic control problems with finitely many pointwise constraints. *Comput. Optim. Appl.*, 51(3): 1319–1343, 2012.

[14] E. Casas, M. Mateos, and B. Vexler. New regularity results and improved error estimates for optimal control problems with state constraints. *ESAIM Control Optim. Calc. Var.*, 20(3): 803–822, 2014.

[15] S. Cherednichenko, K. Krumbiegel, and A. Rösch. Error estimates for the Lavrentiev regularization of elliptic optimal control problems. *Inverse Problems*, 24(5): 055003, 21, 2008.

[16] F. Clarke, Y. Ledyaev, and M. R. de Pinho. An extension of the Schwarzkopf multiplier rule in optimal control. *SIAM J. Control Optim.*, 49(2): 599–610, 2011.

[17] P. Daniele, S. Giuffrè, G. Idone, and A. Maugeri. Infinite dimensional duality and applications. *Math. Ann.*, 339(1): 221–239, 2007.

[18] C. D'Apice, P. I. Kogut, and R. Manzo. On relaxation of state constrained optimal control problem for a PDE-ODE model of supply chains. *Netw. Heterog. Media*, 9(3): 501–518, 2014.

[19] Ciro D'Apice, Peter I. Kogut, and Rosanna Manzo. On optimization of a highly re-entrant production system. *Netw. Heterog. Media*, 11(3): 415–445, 2016.

[20] K. Deckelnick and M. Hinze. Convergence of a finite element approximation to a state-constrained elliptic control problem. *SIAM J. Numer. Anal.*, 45(5): 1937–1953, 2007.

[21] K. Deckelnick and M. Hinze. *A finite element approximation to elliptic control problems in the presence of control and state constraints*. Fachbereich Mathematik, Univ., 2007.

[22] P. Deuflhard, A. Schiela, and M. Weiser. Mathematical cancer therapy planning in deep regional hyperthermia. *Acta Numer.*, 21: 307–378, 2012.

[23] A. L. Dontchev and William W. Hager. The Euler approximation in state constrained optimal control. *Math. Comp.*, 70(233): 173–203, 2001.

[24] N. Du, L. Ge, and W. Liu. Adaptive finite element approximation for an elliptic optimal control problem with both pointwise and integral control constraints. *J. Sci. Comput.*, 60(1): 160–183, 2014.

[25] K. Eppler and F. Tröltzsch. Fast optimization methods in the selective cooling of steel. In *Online optimization of large scale systems*, pages 185–204. Springer, Berlin, 2001.

[26] F. Faraci and F. Raciti. On generalized Nash equilibrium in infinite dimension: The Lagrange multipliers approach. *Optimization*, 64(2): 321–338, 2015.

[27] H. Frankowska. Optimal control under state constraints. In *Proceedings of the International Congress of Mathematicians. Volume IV*, pages 2915–2942. Hindustan Book Agency, New Delhi, 2010.

[28] V. V. Gorokhovik, S. Ya. Gorokhovik, and B. Marinković. First and second order necessary optimality conditions for a discrete-time optimal control problem with a vector-valued objective function. *Positivity*, 17(3): 483–500, 2013.

[29] S. Göttlich, M. Herty, and A. Klar. Modelling and optimization of supply chains on complex networks. *Commun. Math. Sci.*, 4(2): 315–330, 2006.

[30] M. I. Henig. Proper efficiency with respect to cones. *J. Optim. Theory Appl.*, 36(3): 387–407, 1982.

[31] M. Hintermüller and M. Hinze. Moreau-Yosida regularization in state constrained elliptic control problems: Error estimates and parameter adjustment. *SIAM J. Numer. Anal.*, 47(3): 1666–1683, 2009.

[32] M. Hintermüller and K. Kunisch. Path-following methods for a class of constrained minimization problems in function space. *SIAM J. Optim.*, 17(1): 159–187, 2006.

[33] M. Hinze and C. Meyer. Variational discretization of Lavrentiev-regularized state constrained elliptic optimal control problems. *Comput. Optim. Appl.*, 46(3): 487–510, 2010.

[34] M. Hinze and A. Schiela. Discretization of interior point methods for state constrained elliptic optimal control problems: Optimal error estimates and parameter adjustment. *Comput. Optim. Appl.*, 48(3): 581–600, 2011.

[35] D. Hömberg, C. Meyer, J. Rehberg, and W. Ring. Optimal control for the thermistor problem. *SIAM J. Control Optim.*, 48(5): 3449–3481, 2009/10.

[36] L. Iapichino, S. Ulbrich, and S. Volkwein. Multiobjective pde-constrained optimization using the reduced-basis method. 2013.

[37] K. Ito and K. Kunisch. Optimal control of elliptic variational inequalities. *Appl. Math. Optim.*, 41(3): 343–364, 2000.

[38] B. Jadamba, A. Khan, R. López, and M. Sama. Conical regularization for multi-objective optimization problems. *Submitted*, pages 1–16, 2018.

[39] B. Jadamba, A. Khan, and M. Sama. Error estimates for integral constraint regularization of state-constrained elliptic control problems. *Comput. Optim. Appl.*, 67(1): 39–71, 2017.

[40] B. Jadamba, A. Khan, and M. Sama. Stable conical regularization by constructible dilating cones with an application to lp-constrained optimization problems. *Taiwanese Journal of Mathematics*, pages 1–23, 2018.

[41] B. Jadamba, A. A. Khan, and M. Sama. Regularization for state constrained optimal control problems by half spaces based decoupling. *Systems Control Lett.*, 61(6): 707–713, 2012.

[42] J. Jahn. *Vector optimization*. Springer, 2009.

[43] A. A. Khan and M. Sama. A new conical regularization for some optimization and optimal control problems: Convergence analysis and finite element discretization. *Numerical Functional Analysis and Optimization*, 34(8): 861–895, 2013.

[44] A. A. Khan, C. Tammer, and C. Zălinescu. *Set-valued optimization*. Springer, Heidelberg, 2015.

[45] E. Köbis, T. T. Le, C. Tammer, and J.-C. Yao. A new scalarizing functional in set optimization with respect to variable domination structures. *Appl. Anal. Optim.*, 1(2): 301–326, 2017.

[46] P. Kogut, G. Leugering, and R. Schiel. On Henig regularization of material design problems for quasi-linear p-biharmonic equation. *Applied Mathematics*, 7(14): 1547, 2016.

[47] P. I. Kogut, G. Leugering, and R. Schiel. On the relaxation of state-constrained linear control problems via henig dilating cones. *Control & Cybernetics*, 45(2), 2016.

[48] P. I. Kogut and R. Manzo. On vector-valued approximation of state constrained optimal control problems for nonlinear hyperbolic conservation laws. *J. Dyn. Control Syst.*, 19(3): 381–404, 2013.

[49] K. Krumbiegel, I. Neitzel, and A. Rösch. Regularization for semilinear elliptic optimal control problems with pointwise state and control constraints. *Comput. Optim. Appl.*, 52(1): 181–207, 2012.

[50] K. Krumbiegel and A. Rösch. On the regularization error of state constrained Neumann control problems. *Control Cybernet.*, 37(2): 369–392, 2008.

[51] K. Krumbiegel and A. Rösch. A virtual control concept for state constrained optimal control problems. *Comput. Optim. Appl.*, 43(2): 213–233, 2009.

[52] A. Kumar and A. Vladimirsky. An efficient method for multiobjective optimal control and optimal control subject to integral constraints. *J. Comput. Math.*, 28(4): 517–551, 2010.

[53] M. M. Lavrentiev, A. V. Avdeev, M. M. Lavrentiev, Jr., and V. I. Priimenko. *Inverse problems of mathematical physics*. Inverse and Ill-posed Problems Series. VSP, Utrecht, 2003.

[54] G. Leugering and R. Schiel. Regularized nonlinear scalarization for vector optimization problems with PDE-constraints. *GAMM-Mitt.*, 35(2): 209–225, 2012.

[55] D. Leykekhman, D. Meidner, and B. Vexler. Optimal error estimates for finite element discretization of elliptic optimal control problems with finitely many pointwise state constraints. *Comput. Optim. Appl.*, 55(3): 769–802, 2013.

[56] D. Leykekhman and B. Vexler. Optimal a priori error estimates of parabolic optimal control problems with pointwise control. *SIAM J. Numer. Anal.*, 51(5): 2797–2821, 2013.

[57] W. Liu, D. Yang, L. Yuan, and C. Ma. Finite element approximations of an optimal control problem with integral state constraint. *SIAM J. Numer. Anal.*, 48(3): 1163–1185, 2010.

[58] C. Meyer. Error estimates for the finite-element approximation of an elliptic control problem with pointwise state and control constraints. *Control Cybernet.*, 37(1): 51–83, 2008.

[59] C. Meyer, P. Philip, and F. Tröltzsch. Optimal control of a semilinear PDE with nonlocal radiation interface conditions. *SIAM J. Control Optim.*, 45(2): 699–721 (electronic), 2006.

[60] C. Meyer, A. Rösch, and F. Tröltzsch. Optimal control of PDEs with regularized pointwise state constraints. *Comput. Optim. Appl.*, 33(2-3): 209–228, 2006.

[61] C. Meyer and I. Yousept. Regularization of state-constrained elliptic optimal control problems with nonlocal radiation interface conditions. *Comput. Optim. Appl.*, 44(2): 183–212, 2009.

[62] I. Neitzel, U. Prüfert, and T. Slawig. Strategies for time-dependent PDE control with inequality constraints using an integrated modeling and simulation environment. *Numer. Algorithms*, 50(3): 241–269, 2009.

[63] I. Neitzel and F. Tröltzsch. On convergence of regularization methods for nonlinear parabolic optimal control problems with control and state constraints. *Control Cybernet.*, 37(4): 1013–1043, 2008.

[64] I. Neitzel and F. Tröltzsch. On regularization methods for the numerical solution of parabolic control problems with pointwise state constraints. *ESAIM Control Optim. Calc. Var.*, 15(2): 426–453, 2009.

[65] F. Pörner and D. Wachsmuth. An iterative Bregman regularization method for optimal control problems with inequality constraints. *Optimization*, 65(12): 2195–2215, 2016.

[66] F. Pörner and D. Wachsmuth. Tikhonov regularization of optimal control problems governed by semi-linear partial differential equations. *Math. Control Relat. Fields*, 8(1): 315–335, 2018.

[67] U. Prüfert and F. Tröltzsch. An interior point method for a parabolic optimal control problem with regularized pointwise state constraints. *ZAMM Z. Angew. Math. Mech.*, 87(8-9): 564–589, 2007.

[68] A. M. Ramos, R. Glowinski, and J. Periaux. Nash equilibria for the multiobjective control of linear partial differential equations. *Journal of optimization theory and applications*, 112(3): 457–498, 2002.

[69] A. Rösch and S. Steinig. A priori error estimates for a state-constrained elliptic optimal control problem. *ESAIM Math. Model. Numer. Anal.*, 46(5): 1107–1120, 2012.

[70] T. Roubíček. On nash equilibria for noncooperative games governed by the burgers equation. *Journal of Optimization Theory and Applications*, 132(1): 41–50, 2007.

[71] R. Schiel. *Vector Optimization and Control with Partial Differential Equations and Pointwise State Constraints*. PhD thesis, FAU Erlangen-Nrnberg, 2014.

[72] A. Schiela. Barrier methods for optimal control problems with state constraints. *SIAM J. Optim.*, 20(2): 1002–1031, 2009.

[73] C. Schneider and G. Wachsmuth. Regularization and discretization error estimates for optimal control of ODEs with group sparsity. *ESAIM Control Optim. Calc. Var.*, 24(2): 811–834, 2018.

[74] S. Serovajsky. State-constrained optimal control of nonlinear elliptic variational inequalities. *Optim. Lett.*, 8(7): 2041–2051, 2014.

[75] U. Tautenhahn. On the method of Lavrentiev regularization for nonlinear ill-posed problems. *Inverse Problems*, 18(1): 191–207, 2002.

[76] F. Tröltzsch and I. Yousept. A regularization method for the numerical solution of elliptic boundary control problems with pointwise state constraints. *Comput. Optim. Appl.*, 42(1): 43–66, 2009.

[77] F. Tröltzsch and I. Yousept. Source representation strategy for optimal boundary control problems with state constraints. *Z. Anal. Anwend.*, 28(2): 189–203, 2009.

[78] N. von Daniels. Tikhonov regularization of control-constrained optimal control problems. *Comput. Optim. Appl.*, 70(1): 295–320, 2018.

[79] Y. Xu and S. Zhou. Optimal control of pseudoparabolic variational inequalities involving state constraint. *Abstr. Appl. Anal.*, pages Art. ID 641736, 9, 2014.

Index

A

Abstract convexity 266, 267, 269, 271, 278
applications to behavioral sciences 20

C

conical regularization 297, 298, 302–304, 309, 311, 313–315
Constrained optimization 211, 237, 278
convex cones 73–78, 80–83, 86, 88
convex functions 74, 75, 78, 80, 81, 83–85, 87, 88
Cournot duopoly 46

D

directional closedness 93, 94, 99
dynamic potential games 49

E

efficiency 159, 160, 167–169, 172–174, 179, 180, 192, 193, 195, 197, 200, 205

F

Financial problems 25, 27, 28, 36, 41
fuzzy theory 85, 88

G

Generalized convexity 211, 214, 225

H

half-space regularization 304–306, 309
Henig dilating cones 309

I

ICR functions 268, 270, 272–275, 281, 282, 288, 289
integral constraint regularization 306–308
IPH functions 267, 268, 270, 272, 273, 275, 276, 279–281, 285, 287, 288

L

Lavrentiev regularization 298–300, 302, 306, 307
Lipschitz set-valued maps 143

M

mixed openness 180
multi-criteria decision-making 28
Multi-objective optimization 210–212, 215, 218, 219, 228, 230–232, 237
multi-period portfolio selection problems 41

N

Nash equilibrium 46, 47, 49, 50, 59
nontranslative functions 93, 103, 104
normals 2, 3, 18, 19, 21

O

Optimal control 112, 113, 117, 125, 296–301, 304, 312, 314, 315
optimality conditions 158–160, 163, 165, 179, 180, 192–195, 199, 200, 205
optimization 1, 2, 5, 7, 8, 19–21
Order relations 242–246, 248, 249, 251–260, 263

P

Pareto efficiency 211, 212, 215, 230, 237
penalization 160, 161, 172, 174, 197, 200, 201, 203, 204
pointwise state constraints 296–302, 307, 312, 314, 315
Potential games 45, 48–50, 52, 53, 59
proximal alternating linearized method 44, 45, 60

S

Scalarization functions 73, 74, 78–80, 85, 87, 88
Set approach 242, 243
Set optimization 241–246, 248, 249, 251–254, 256–258, 259
set relations 130, 134
set-valued inequality 85
Set-valued maps 143
Solvability theorems 266–268, 272, 274, 275, 277–279, 281–283, 289
state constraints 112, 113, 119, 123, 125
subgradients 2, 6, 8–10, 12–14, 16, 17, 20, 21, 23, 24
System of inequalities 267, 272, 275

T

Test instances 241, 242, 246, 256, 258, 259, 260, 262

Topical functions 268, 271, 272, 277–279, 283, 284, 289–292
translative functions 92, 93, 98–101, 103, 106

U

Uniform sublevel sets 92–94, 97, 100, 103–105

V

variable ordering structure 158, 160, 172, 193, 204
Variational analysis 1–3, 5, 8–11, 13, 14, 16–21, 44, 45, 50, 53
variational and extremal principles 2
variational inequalities 28, 35–38, 40, 41, 296, 297, 314, 315
variational rationality 1–5, 9, 11, 20, 21, 44, 45, 52–55
Vector optimization 241–246, 248, 249, 252–254, 256, 257–260, 262
Vectorial penalization approach 212, 215, 218–220, 222, 228, 230, 231, 237

W

worthwhile moves 21